PRIMITIVE ORIBATIDS OF THE PALAEARCTIC REGION

THE SOIL MITES OF THE WORLD

Series Editors:

J. BALOGH and S. MAHUNKA

Volume 1 Primitive Oribatids of the Palaearctic Region
by J. Balogh and S. Mahunka

PRIMITIVE ORIBATIDS OF THE PALAEARCTIC REGION

by

J. BALOGH

Member of the Hungarian Academy of Sciences

and

S. MAHUNKA D. Sc. (Biol.)

ELSEVIER

AMSTERDAM – OXFORD – NEW YORK 1983

Joint edition published by

Elsevier Science Publishers B. V., Amsterdam, The Netherlands and Akadémiai Kiadó, The Publishing House of the Hungarian Academy of Sciences, Budapest, Hungary

Translated by
L. GOZMÁNY

Illustrated by
L. MAHUNKA—PAPP

The distribution of this book is being handled by the following publishers for the U.S.A. and Canada

Elsevier Science Publishing Company, Inc.
52 Vanderbilt Avenue
New York, NY 10017, U.S.A.

for the East European Countries, Korean People's Republic, Cuba, The People's Republic of Vietnam and Mongolia

Kultura, Hungarian Foreign Trading Company
P.O. Box 149
H-1389 Budapest 62, Hungary

for all remaining areas

Elsevier Science Publishers B. V.
Molenwerf 1
P.O. Box 211, 1000 AE Amsterdam, The Netherlands

Library of Congress Cataloging in Publication Data
Main entry under title:

Primitive Oribatids of the Palaearctic Region.

(The Soil mites of the world; v. 1)
Bibliography: p.
Includes index.
1. Oribatidae—Identification. 2. Arachnida—Identification. I. Balogh, János. II. Mahunka, S. III. Series.
QL458.2.Ó74P73 1983 595.4'2 83-8884
ISBN 0-444-99655-9 (U.S.)

ISBN 0-444-99655-9 (Volume 1)
ISBN 0-444-99654-0 (Series)

Printed in Hungary

CONTENTS

INTRODUCTION

This volume is the first part of the brief characterizations of and identification keys to the Oribatid taxa of the world, dealing with the primitive Oribatids of the Palaearctic Region. The second and third parts, comprising the Picnonotic and Poronotic Oribatids, are in preparation. We are convinced that the time is ripe for a concise survey and a summary of the accumulated knowledge, in order to facilitate further investigations and syntheses.

Like every identificatory work of a compilative nature, ours too is confined within certain limits. Since we became acquainted during our studies with only a part of the Palaearctic Oribatida, a number of identification keys had to be based merely on descriptions available in literature. Accordingly, the reliability of this part of the book depends primarily on that of the literature source. It is but natural that descriptions originating from diverse periods, and that the works of scholars occupying various levels of erudition and capability, differ in value. Identifications of older species made by recourse to this book should, therefore—in case of any doubt—be regarded as preliminary information only, to be checked by a study of the original diagnoses.

Our basic endeavour was the interpretation, as far as possible, of the contributions published by contemporary or previous Oribatidologists. We do not invariably agree with systematists who elaborate an animal group exclusively by the total re-evaluation of type-materials preserved in museums. Any such attempt fails inevitably because, in the case of Oribatid mites, older type-materials are either no more extant or in a very bad state of preservation; nor is the number of active Oribatidologists sufficient to undertake a work of redescription of this immense scope. We believe, however, that a healthy subdivision of labour has already evolved among the Oribatidologists of the world: some authors, following F. Grandjean's concepts, dedicate their time to very detailed studies of one (or more) species, including also their developmental stages, while others describe the faunas of unexplored regions, and others again investigate the Oribatid faunas of their mother countries. We made every attempt to include in our work the results of all of these projects and thus let the identification keys to the Palaearctic Oribatids reflect the present state of Oribatidology, with all its concurrent advantages and disadvantages.

The identification keys contain every hitherto described species (the manuscript was closed in June, 1980) whose diagnosis and/or figure permits separation from every congener. If the available information was here considered as inadequate, we placed the species among the "species inquirendae". Part of these species will probably prove to be synonyms, senior or junior, another part distinct species. The respective decision will probably rest with Oribatidologists who reside in the country of origin of the taxa in question. We are also aware of the fact that despite every precaution in certain cases, for instance in the Phthiracarids, our keys and diagnoses represent the figures and descriptions of the species rather than actual representatives of the species themselves. For groups like these, revisions are highly desirable, but not in the currently prevailing form of a painstaking scrutiny, using some recently discovered features, to differentiate some few species of a very large genus, but without extending the application of these characteristics to evaluate the entire generic group in question.

Our work was already in the last stages of preparation when the volume treating the Oribatids of the Soviet Union (Fauna SSSR) was published. Although this work differs in many respects (e.g. concept and structure) from our own, it was of immense help concerning the fauna of the eastern and southeastern territories. The intense oribatidological work done by the specialists of the Nearctic Region is of a similarly invaluable nature, rendering available a number of new data for species of a Holarctic distribution.

It is highly probable that, despite all our efforts, this work will be found to harbour a number of errors and lacunae. We can only ask the Oribatidologists of the world to communicate to us—in view also of an eventual second edition—their corrections and remarks. The book is principally intended to help young Oribatidologists by the establishment of a basis for taxonomic studies and to promote investigations more extensive than the present ones. If this state of affairs is attained, the identification keys can in one or two decades be superseded by an essentially enlarged and more complete opus on the Palaearctic Oribatids, written collectively.

However, even the present work is not merely a string of identification keys—yet neither is it a monograph. The basic concept was to submit, in the frames at our

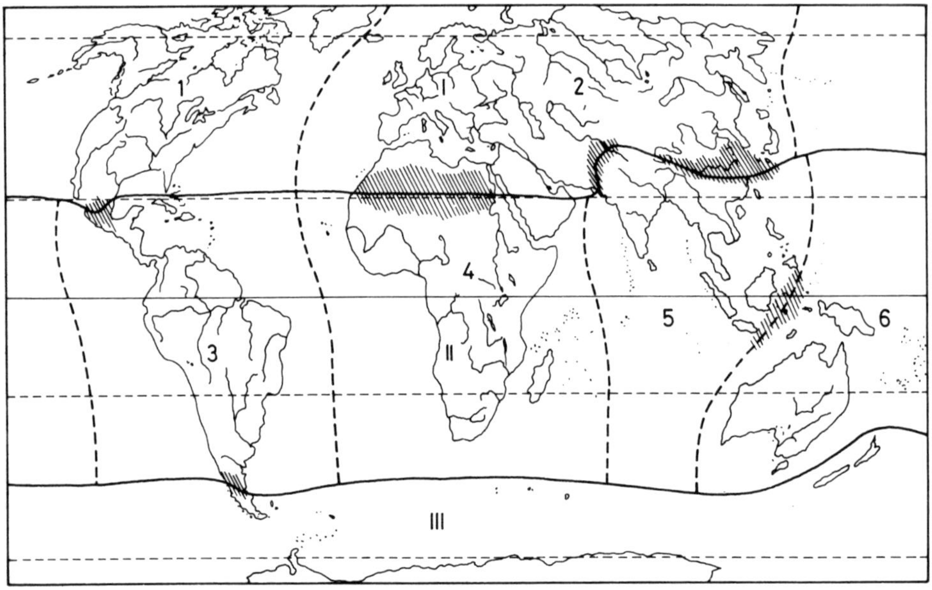

Map 1. The zoogeographical division of the world. I: Holarctic Region, 1: Nearctic Region, 2: Palaeartic Region; II: Holotropical Region, 3: Neotropical Region, 4: Ethiopian Region, 5: Oriental Region, 6: Pacific Region; III: Holantarctic Region

disposal, as much informative material—beyond the identification keys and the concurrent illustrations—as possible, in order to supply the essentials for any further work, e.g. distribution data, lists of species, and the main literature references.

For the species presented in the identification keys, we cite the locus typicus as the basic datum of distribution, gleaned from the original description; if this was unavailable, we appended special remarks. Attempts have been made to confine distribution data to correct ones only, but we are aware of the sources of errors (or omissions) thereby invoked.

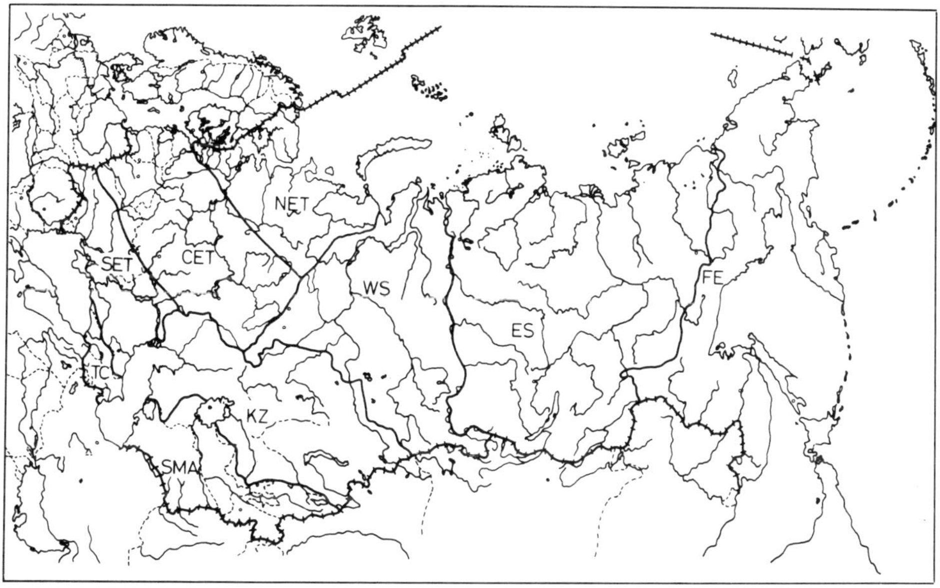

Map 2. The zoogeographical divisions of the Soviet Union. NET: North European territory, CET: Central European territory, SET: South European territory, TC: Transcaucasus, KZ: Kazakh SSR, SMA: Soviet Middle Asia, WS: West Siberia, ES: East Siberia, FE: Far East

In the catalogue sections—and this holds especially for the "species interpretabiles", i.e. the species or subspecies which can be safely interpreted on the basis of literature, or by the knowledge of actual specimens, as representing the animal diagnosed by the author—at most four main literature references (exclusively those of synonyms) are given, namely that of the original description, of the currently applied combination, and eventually one or two of special interest: of redescriptions, of additional data, etc.

References to basic works of a comprehensive character are, however, omitted, otherwise they ought to have been constantly cited for most of the taxa involved. The five works are those of Balogh (1972), Ghilarov and Krivolutsky (1975), Sellnick (1928, 1960) and Willmann (1931).

In the delimitation of the zoogeographical regions, an inherently basic feature of the entire work, we have applied our own concept, with considerations given mainly to P. Müller and H. Rapoport's treatises; the main divisions appear in Map 1. Since the greatest part of the Palaearctic Region lies in the Soviet Union, we deemed it useful to give the respective subdivisions in Map 2.

We are heavily indebted to Dr. L. Gozmány, translator and Dr. Á. Soós and L. Zombori readers as well as technical advisers, furthermore to Mrs. L. Mahunka–Papp, illustrator. It is our pleasant duty to express our sincere thanks for their generous help also in this place.

TERMINOLOGICAL SURVEY

The morphologically highly diverse moss mites or Oribatids constitute the order richest in species in the subclass Acari; by their very numbers, both individual and specific, they are among the most important terricolous arthropodan groups: their activity might be of decisive importance in the life of the soil and the processes of energy turnover.

Grandjean and his followers have worked out in detail the morphology of these mites; in the following an attempt is made to present the most important terms, widely used in the relevant literature, indispensable for identification.

GENERAL

Mites of minute, medium or large body-size alike may belong here: their length may vary between 110—300 μm, but the great majority of the species fall into the size interval of 300—900 μm and only very rarely exceed 1000 μm. Their shape is highly variable, but characteristic and constant within a taxon. In general, they are dorsoventrally flattened, rarely cylindrical (Holonota) or laterally compressed (Ptyctima). According to the degree of sclerotization and pigmentation, their colour varies between opaline white and black: in the lower groups (Palaeosomata, Arthronota) it is light, at most yellow, while in the higher ones at least yellow, but in the majority of the species the brown hues predominate and a fair number of dark brown or even black species are also known.

The body of the larvae and nymphs is covered by an elastic, thin and mostly rugulose cuticle. In a few imagoes (some Palaeosomata) it persists, but will partly transform into sclerites. Those of the dorsal side enlarge gradually into transverse shieldlets and then a uniform dorsal shield; while on the ventral side the so-called anogenital region displays various grades of shield differentiation in the primitive Oribatids. The ventral plates of the higher Oribatids also coalesce and only the genital pair of plates (covering the genital opening) and the anal pair of plates (covering the anal opening) remain differentiated from the homogenous ventral shield. The surface of the cuticle may be shiny or mat, and of various ornamentation: dotted, pitted, striolate, tuberculate, rugose, with a polygonal or reticulate, regularly or irregularly arranged, chitinous structure. The body surface may also be covered by a layer of secretion or cerotegument. The imagoes of some groups carry the larval or nymphal exuviae on their backs, fixed in concentric or excentric arrangements.

Acarologists use several kinds of terms to denote the somatic regions of the Oribatids. The borders and names of the regions can be read easily from the diagram (fig. 1). The sutures delimiting the body regions and the regions themselves are usually denoted in the literature as follows:

— *sejugal suture (sej.):* the furrow separating the proterosoma from the hysterosoma; dorsally decurrent between the prodorsum and the notogaster, ventrally between the propodosoma and the metapodosoma;
— *dorsosejugal suture (dsej.):* the dorsal section of the sejugal furrow, between the prodorsum and the notogaster;
— *disjugal suture (disj.):* the furrow between the posterior part of the pleural region and the notogaster;
— *abjugal suture (abj.):* the furrow between the anterior part of the pleural region and the prodorsum;
— *prosoma:* the anterior part of the body including the four pairs of legs, delimited dorsally by the dorsosejugal suture, laterally and ventrally by the disjugal furrow as well as the anterior margin of the genital plate, composed of the gnathosoma, the prodorsum and the podosoma;
— *proterosoma:* the anterior part of the body including the first two pairs of legs, composed of the gnathosoma, the prodorsum and the propodosoma;
— *hysterosoma:* the posterior part of the body including the last two pairs of legs, delimited dorsally by the dorsosejugal suture, laterally and ventrally by the sejugal furrow;
— *opisthosoma:* the posterior part of the body without legs, delimited dorsally by the dorsosejugal suture, laterally and ventrally by the disjugal furrow and the anterior margin of the genital plate;

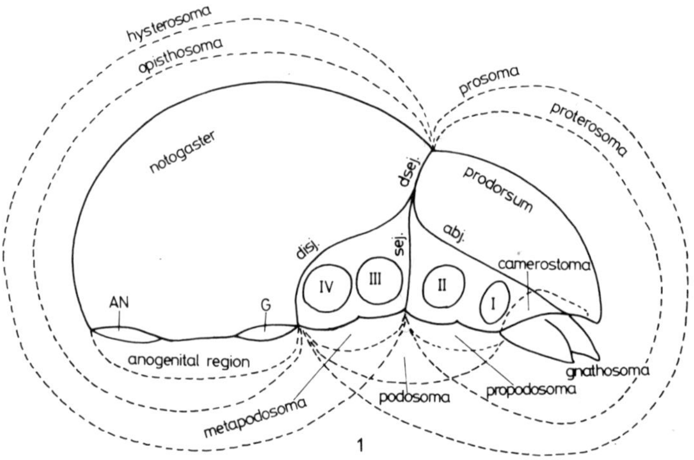

1

Fig. 1. Body regions of the Oribatids (schematically): *AN* = anal plate, *G* = genital plate, *abj.* = abjugal suture, *disj.* = disjugal suture, *dsej.* = dorsosejugal suture, *sej.* = sejugal suture, I—IV = acetabuli of legs I—IV (modified after Grandjean)

— *gnathosoma:* comprising the oral organs;
— *camerostoma:* the anterior cavity of the body containing the gnathosoma;
— *podosoma:* the ventral anterior part of the body containing the coxae of legs I—IV;
— *propodosoma:* the anterior part of the podosoma containing coxae I—II;
— *metapodosoma:* the posterior part of the podosoma containing coxae III—IV;
— *prodorsum:* the shield covering the first two segments (A and B) of the body;
— *notogaster:* the shield covering (or containing) the median and posterior segments *(C—PS)* of the body;
— *anogenital region:* the ventral posterior part of the body comprising the genital, aggenital, anal and adanal plates.

ARTICULATION TYPES

The body parts enumerated above are generally immovably fused, but some sutures represent a movable articulation. In taxonomical practice, three main articulatory types are used for the differentiation of larger groups. Accordingly, the articulation may be:

— *dichoid* if the body articulates between legs II and III, that is, between the proterosoma and the hysterosoma, in this case the comparatively wide, soft, thin, rugose chitinous membrane between the two parts afford them some restricted mobility (e.g. Palaeosomata);

— *ptychoid* if the body articulates behind legs IV, that is, between the prosoma and the opisthosoma: movability can go so far that the animal can be "closed" like a penknife, allowing the inclusion of the legs, too as in Ptyctima: Phthiracaridae ("box" mites) and Protoplophoroidea;

— *holoid* if every somatic region rigidly coalesced to form a uniform chitinous sheath: the borders of the constituting parts are indicated merely by sutures (dorsosejugal, sejugal and disjugal).

THE HYPOTHETICAL BODY OF THE PRIMORDIAL FORM

Segmentation. Morphologically, the Oribatids can be reduced to a single hypothetical primordial form (fig. 2), allowing the differentiation of 11 segments. The first eight segments can be found already in the larval form, while the last three appear (or may appear) only in the subsequent stages of development. The eight larval segments are as follows: *A, B, C, D, E, F, H and PS.* It was Grandjean who applied upper case letters of the Latin alphabet to designate the segments; *G* was used for the genital segment, while *PS* derives from the pseudoanal plate surrounding the anal plate in the larvae and, therefore, applied to the respective segment. The so-called postlarval segments, appearing subsequent to the larval stage, are as follows: the adanal *(AD)* segment appearing first on the protonymph, the anal *(AN)* segment developed first on the deutonymph, and the peranal *(PA)* segment appearing first on the tritonymph in certain cases (*Parhypochthonius, Pterochthonius, Phyllochthonius,*

Brachychthonoid genera). The segments *AD, AN* and occasionally segment *PA* shift, together with the anal opening, to the ventral side in most Oribatids, where they more or less join the genital region and constitute the so-called anogenital region.

Setation. Each segment bears 1—4 pairs of setae. They are designated by lower case letters corresponding to the "inital(s)" of the segment and provided also by a lower index number (from 1 to 4), the sequence based on the location outwards from the midline of the body. The setae *ad* and *an*, shifted to the ventral side, are provided with index numbers in accordance with their original position, from the ventral side towards the dorsal side.

The homologies of the six segments and the pairs of setae of the prodorsum are uncertain, therefore, their designation with the letters *a* and *b* is not widely distributed. The generally accepted designation and the assumed homologies of the setae are as follows:

$ro = a_1$ (rostral seta),
$le = a_2$ (formerly: *la*, lamellar seta),
$exa = a_3$ (anterior exostigmatal seta),
$in = b_1$ (interlamellar seta),
$ss = b_2$ (sensillus, sensillar seta),
$exp = b_3$ (posterior exostigmatal seta).

The hypothetically primordial setae, appearing on segments 3—11, are as follows:

C: c_1, c_2, c_3, *cp* (of uncertain position; some authors had formerly placed it as d_3 on the next segment!),
D: d_1, d_2
E: e_1, e_2
F: f_1, f_2
H: h_1, h_2, h_3
PS: ps_1, ps_2, ps_3
AD: $ad_1, ad_2, ad_3, (ad_4)$

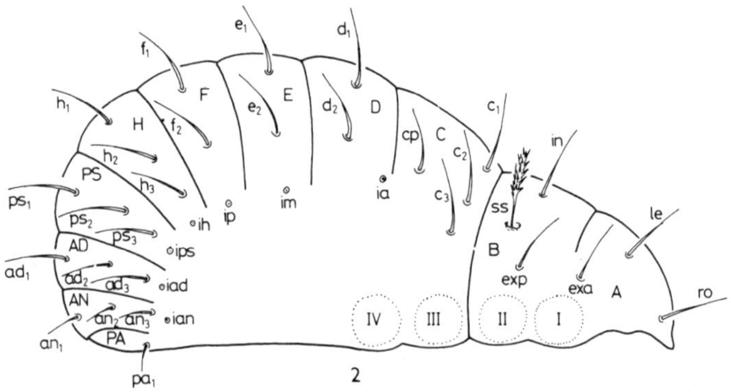

Fig. 2. Hypothetical segmentation of ancient Oribatids. Segments: *A, B, C, D, E, F, H, PS, AD, AN* and *PA*. Setae: $ro (= a_1)$, $le (= a_2)$, $exa (= a_3)$ of segment *A*; $in (= b_1)$, $ss (= b_2)$, $exp (= b_3)$ of segment *B*; c_1, c_2, c_3, *cp* of segment *C*; d_1, d_2 of segment *D*; e_1, e_2 of segment *E*; f_1, f_2 of segment F; h_1, h_2, h_3 of segment *H*; ps_1, ps_2, ps_3 of segment *PS*; ad_1, ad_2, ad_3 of segment *AD*, an_1, an_2, an_3 of segment *AN*; pa_1 of segment *PA*. Lyrifissures: *ia* of *D*, *im* of *E*, *ip* of *F*, *ih* of *H*, *ips* of *PS*, *iad* of *AD*, *ian* of *AN* (modified after Grandjean)

AN: an_1, an_2, an_3, (an_4)
PA: pa_1

Lyrifissures. On segments 4—10 and in largely corresponding positions, organs of still unclarified functions appear: they are termed fissure-organs. Their position and designation are as follows: *D*: *ia*, *E*: *im*, *F*: *ip*, *H*: *ih*, *PS*: *ips*, *AD*: *iad*, *AN*: *ian*. Their maximum number is 7, frequently less.

Sclerotization. The morphology of the hypothetical primordial form discussed here is regarded as a model and the detailed morphology following below as its complementation.

MORPHOLOGICAL CONSIDERATIONS

The general morphology of the Oribatids will be shown in a primitive [*Perlohmannia* sp. (figs. 3—4) and *Euphthiracarus* sp. (figs. 5—8)] and a higher [*Ceratozetes* sp. (figs. 9—11)] Oribatid. Certain repetitions cannot be obviated in the discussion of details.

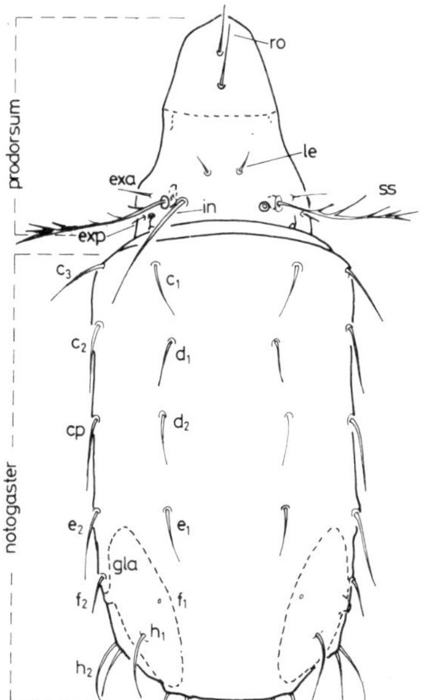

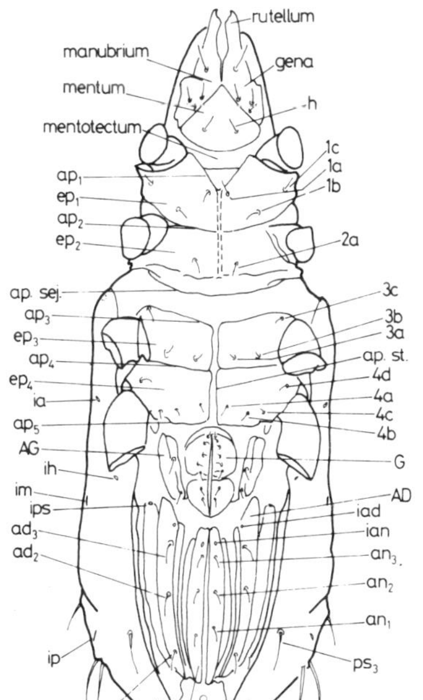

Figs. 3—4. Body of primitive Oribatids of the dichoid type. 3 = dorsal view: *ro* = rostral seta, *le* = lamellar seta, *in* = interlamellar seta, *exa* = anterior exobothridial seta, *exp* = posterior exobothridial seta, *ss* = sensillus; c_1—c_3, *cp*, d_1—d_2, e_1—e_2, f_1—f_2, h_1—h_3, ps_1—ps_2 = notogastral setae; *gla* = oil gland; — 4 = ventral view: *AD* = adanal plate, *AG* = aggenital plate, *AN* = anal plate, *G* = genital plate, *PA* = postanal plate, ap_1—ap_5, *ap. sej.*, *ap. st.* = apodemes, ep_1—ep_4 = epimeres, *ia*, *ih*, *im*, *ips*, *iad*, *ian* = lyrifissures, ad_1—ad_3 = adanal setae, an_1—an_3 = anal setae, ps_3 = notogastral seta, *h* = hypostomal seta, *1a*—*1c* = setae of epimere 1, *2a* = seta of epimere 2, *3a*—*3c* = setae of epimere 3, *4a*—*4d* = setae of epimere 4 (after Grandjean)

Body regions. In the overwhelming majority of the Oribatids the shieldlets of the primitive forms evolved into larger shields protecting the various body regions. The proterosoma is covered dorsally by the prodorsum, the hysterosoma by the notogaster; the former projects anteriorad and usually covers also the gnathosoma. The proterosoma and the prosoma, respectively, is anteroventrally delimited by the cavity of the camerostoma largely enveloping the gnathosoma. Behind this, a uniform shield is formed in the majority of the species: the ventral plate whose anterior part, covering the podosoma and joining legs I—IV, is defined as the coxosternal or epimeral region (plate), while the posterior part, comprising also the genital and anal openings, as the anogenital region (plate). In the taxic diagnoses and identification keys these are the usually recurrent main units (prodorsum, notogaster, gnathosoma, epimeral and anogenital regions). In the Oribatids, there are also ambulatory legs consisting of 5—6 joints, three pairs in the larvae and four pairs in the nymphs and mature animals.

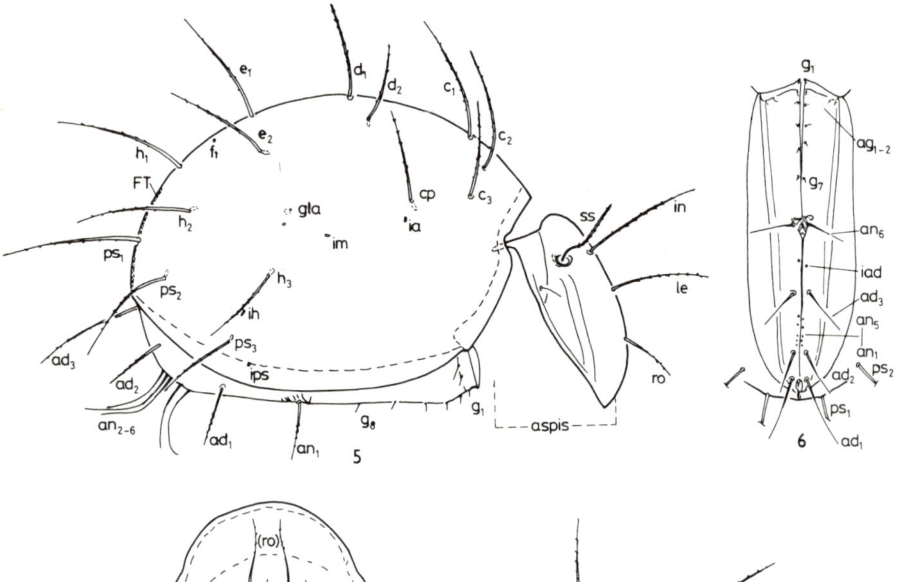

Fig. 5—8. Body of primitive Oribatids of the ptychoid type. 5 = lateral view: *FT* = terminal fissure, *ex* = exobothridial seta, g_1—g_8 = genital setae; — 6 = anogenital region: ag_1—ag_2 = aggenital setae; — 7 = aspis in dorsal view; — 8 = aspis in lateral view. (For rest of symbols see figs. 3—4.) (After Märkel)

PRODORSUM. An anteriorly tapering and roughly triangular shield, with a frontally pointed portion (rostrum) extending also over the mouthparts; it often bears teeth or tubercles and displays one or more incisions. In some groups (Ptyctima), the prodorsum is called aspis (figs. 5—8). The prodorsal surface may be ornamented with chitinous excrescences and setae or organs of setose origin, the so-called trichobothria. The chitinous excrescences (lamella, costula, tutorium, genal tooth, pedotecta, tubercles, bothridial and other protuberances, condyli, etc.) characterize mainly the higher taxa (fig. 9). The centrally decurrent longitudinal pairs of laths, originating from the basal part of the prodorsum and from the bothridia, respectively, are called

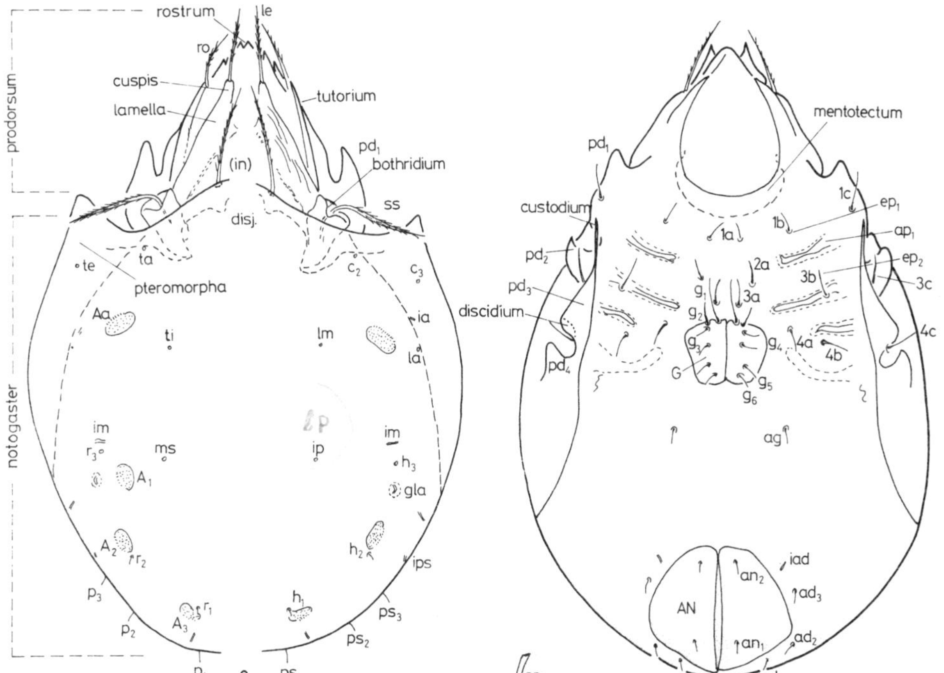

Figs. 9—11. Body of higher Oribatids. 9 = dorsal view: *Aa*, A_1—A_3 = areae porosae, *ta, te, ti, ms*, r_1—r_3, p_1—p_3, *la, lm* = setae of notogaster, pd_1 = pedotectum 1; — 10 = ventral view; — 11 = prosoma in lateral view: *Ah, Al* = areae porosae. (For rest of symbols see figs. 1—8.) (After Menke)

the prodorsal lamellae. They have two types: if they are flat and ribbon-like or lamelliform, or at least partly free of the prodorsal surface, they are termed true lamellae, but if they show no lamelliform character, but appear merely as longitudinal laths or ridges and have no free section (disattached) from the prodorsal surface, they are distinguished as costulae. The distal ends of the true lamellae are often elongated and free, these apices are called cuspides and they bear (or are adjacent to) the lamellar setae. The lamellae are often connected by transverse bands, the translamellae; it (or they) may appear as a thin line or as a wide band. Prolamellae may extend from the distal ends of the lamellae towards the rostrum.

On the basal part of the prodorsum, and often adnate to the bothridial margin, there is a robust condyle, termed the bothridial condyle or condylus prodorsalis lateralis *(co. pl.)*; there may often appear a similar pair of condyli medially, designated as inner condyle or condylus prodorsalis medialis *(co. pm.)*. In some higher groups (e.g. *Ceratozetoides*), the prodorsal margin may bear a robust longitudinal chitinous lamella or ridge, the tutorium; it may also have a free apex. The higher Oribatids are further characterized by a pointed spine on the side of the propodosoma, the genal tooth and/or the discreting incision (fig. 11) on the border of the rostrum and the camerostoma.

Between the prodorsum and the epimeral region, on the side of the podosoma, there appear the pedotecta behind the legs and partially covering or protecting their bases. The pedotecta are squamose or lamelliform, largely tri- or quadrangular, usually ornamented with cuspides or tubercles; there are generally two, but also one may occur. Pedotecta 1 are usually much larger than pedotecta 2.

Prodorsal setae. Of the setae, or similar formations of a setose origin, the rostral pair of setae *(ro)* is situated near the rostrum on the prodorsal surface or margins; more internally the lamellar *(le)* and the interlamellar *(in)*, more marginally the exobothridial *(ex)*, setae can be distinguished. In the primitive types, there are two pairs of the latter one, definable as anterior *(exa)* and posterior *(exp)* exobothridial setae, while the higher Oribatids possess only one pair of exobothridial hairs. For the perception of vibrations, the sixth (in the higher Oribatids the fifth) pair of setae is modified into trichobothridia (pseudostigmata); this seta (sensillus) often appears in a peculiar shape (figs. 12—20): it can be clavate, pectinate, setiform, lamelliform, spatulate, caliciform, fusiform, etc., while its base emerges from a sacculiform or cup-shaped hollow, the bothridium, with a collar-like rim. The other setae may also be smooth or rough, pubescent, squamose, ciliate, pectinate, or shaped like a seta, rod, sword, leaf, spatula, spoon, lance, scalpel, brush, etc.

The prodorsum is separated from the hysterosoma by the dorsosejugal and sejugal sutures, generally forming a continuous line, rarely interrupted in the middle, or only exceptionally absent.

NOTOGASTER. Hysterosoma: its structure and the arrangement of the setae differs considerably between the primitive and the higher Oribatids. In the most primitive forms, either no closed shield evolves and the notogaster or even the original segments remain discernible, or though the notogaster develops (the uniform shield dorsally covering the hysterosoma) there occurs also a secondary differentiation (but still following the traces of the original segments) so that the shield is subdivided by furrows or sutures into 2—4 parts.

The hysterosoma—as also the notogaster—coalesced from a total of 10 segments; of these, evidence of the traces of merely six segments (*C, D. E. F. H* and PS) are present in the imagoes.

The anterior part of the notogaster may bear various chitinous excrescences: spines (spinae adnatae), condyles (singles or paired), lateral and medial condyles, *co. nl.* and *co. nm.;* as in the case of the prodorsum), teeth, posteriorly decurrent lamellae and ridges (cristae).

Pteromorphae. The higher Oribatids may frequently display in the "shoulder" area of the notogaster some lamelliform excrescences under which the animal may in some cases be able to draw in its legs. This formation is designated as the humeral lobe (pteromorpha) which is either immovable or joined to the hysterosoma by a hinge and thus movable. The movable pteromorphae possess a distinct musculature in the front of the notogaster, with a chitinous myoinsertion surface. This is well discernible as a dark, usually obscure and triangular, characteristic pair of spots behind the dorsosejugal suture.

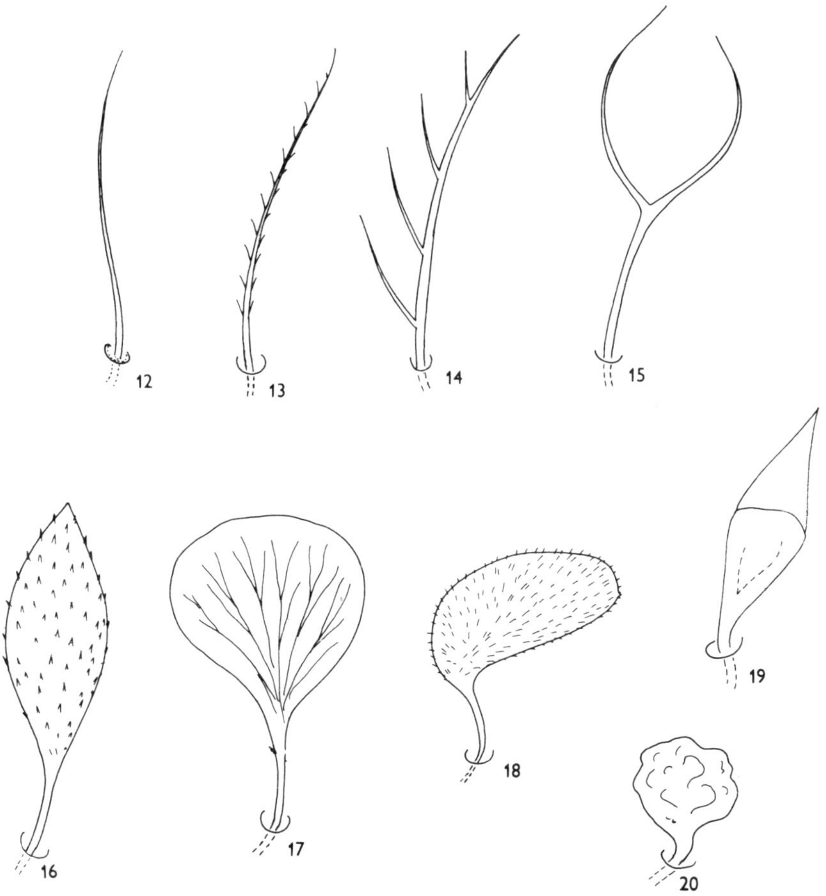

Figs. 12—20. Various types of sensillus

On the basis of the shape, size and movability of the pteromorphae several types can be distinguished, but also some transitions may occur:

— *aptera:* the pteromorpha is wholly absent, the "shoulder" is rounded;
— *immoptera:* not movable, shape and size various, but never auriculate and without a long, spiniform anterior appendage;
— *oxyptera:* the pteromorphae are immovable, large, anteriorly with a long, acute, spiniform appendage;
— *euptera:* the pteromorphae are movable, the hinge and the robust musculature are well discernible; their shape is not auriculate and bears no acute, spiniform appendage;
— *umbellata:* both the anterior and the posterior margins of the pteromorphae are arcuate, they are large, auriculate or umbelliform.

Setation of notogaster. Among the notogastral formations it is the setae (chaetom) which have best preserved the original segmentation. The number of setae of the Oribatids is generally constant fluctuating between 1 and 16 pairs (orthotrichia), but in some groups secondary setae (neotrichia) may also exceptionally occur.

In the primitive forms (fig. 3) the setae appear in six rows according to the six original segments, as follows: c_1, c_2, c_3, c_p—d_1, d_2—e_1, e_2—f_1, f_2—h_1, h_2, h_3—ps_1, ps_2, ps_3,* then, by the elimination of the f setae, the types with 15 and 14 setae evolve.

In the higher Oribatids, the basic type is considered the one with 10 pairs of setae (fig. 9). The homology of the setae with the former ones cannot be proved, therefore, a different notation is used: $ta+te+ti$—ms—$r_1+r_2+r_3$—$p_1+p_2+p_3$. Any other number of setae, e.g. 11—14, is in our opinion secondary, hence a homologization with the original chaetotaxy is also incorrect. In these types longitudinal and transverse rows of setae are respectively distinguished or denoted: $c_1+c_2+c_3$—$da+dm+dp$—la_1+lm_1+lp—$h_1+h_2+h_3$—$p_1+p_2+p_3$.

Octotaxic organ. Besides setae and other chitinous excrescences, the notogastral surface displays the openings of partly stigmata, partly glands, and partly other sense organs (lyrifissures).

Three types of the respiratory organs, called also octotaxic organs on the basis of their number, are known. In the higher Oribatids, the notogaster may show pore fields (areae porosae), minute pockets (sacculi), or simple pores (pori). The areae porosae are, with some exceptions, perforated spots (with also a thinner cuticle) of definite shape on the notogastral shield. There are usually four (more rarely three or more) pairs present. Those beside the pteromorphae are the areae porosae adalares (Aa), the rest are indicated by numerals: A_1, A_2, A_3. In some groups also the dorsosejugal region may contain a pair of areae porosae, or certain pore fields may exceptionally be divided into two or more parts and thus their numbers increase. The sacculi have similar locations; they are minute pockets sunk into the body and opening by pores or small clefts. The pori are simple respiratory apertures; their recognition is difficult.

The notogaster may also possess a pair of oil glands, with the opening of their ducts (gla) in the posterior part of the shield; in the primitive forms even the entire gland may also be visible. In some groups, the gland opens by a tubiform excrescence.

* In the superfamily Nothroidea further denotations also occur for some of the setae.

The lyrifissures are the openings of organs of still unknown function; there are a maximum of 7 pairs present: *ia*, *im*, *ip*, *ih*, *ips*, *iad* and *ian*.

GNATHOSOMA. A well differentiable somatic part containing the oral organs and occupying the camerostoma. Towards the epimeral region, the articulation is covered or protected by the mentotectum (fig. 4). The basal part of the gnathosoma is the infracapitulum, basally consisting of the unpaired mentum (or hysterostoma), dorsally of the "neck region" (cervix), bilaterally of the paired genae and their continuation in the rutelli of setal origin. It contains the mouth and the immediate mouthparts: the chelicerae, the labium, the oral opening, and the pharynx as its continuation. These parts may fuse or be articulated (figs. 21—25).

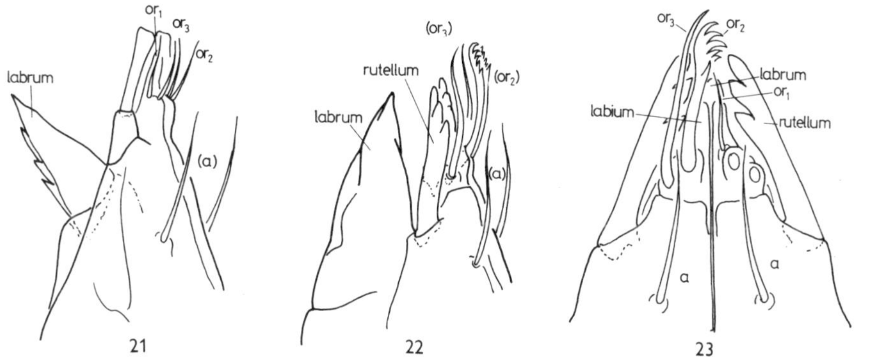

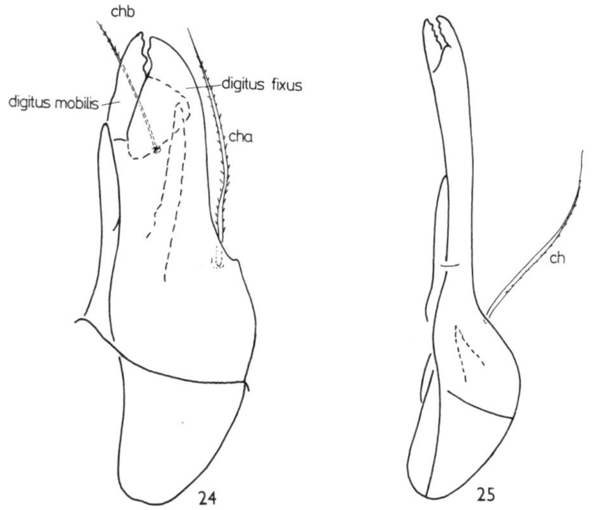

Figs. 21—25. Parts of the gnathosoma. 21—22 = apical part of gnathosoma in lateral view: *a* = seta of maxilla, or_1—or_3 = oral setae; — 23 = apical part of gnathosoma in ventral view. — 24—25 = two main types of chelicera. — 24 = normal: *cha*, *chb* = cheliceral setae; — 25 = peloptoid: *ch* = cheliceral seta (after Grandjean)

Labiogenal articulation. From the points of view of classification or the separation of taxa, the presence or absence of especially the labiogenal articulation (between the mentum and the genae) is especially important; on this basis, as well as the form of the mentum and rutelli, four types are distinguished, namely:

— *anarthric* (fig. 26): not specialized, anteriorly usually differentiated;
— *stenarthric* (fig. 27); the labiogenal articulation is present, it is directed posteriorad hence the mentum is triangular, the genae small, their rutelli similar to those in the preceding group;
— *diarthric* (fig. 28): the labiogenal articulation is present, the articulatory suture is transverse, the mentum large, quadrangular, the genae are large, the rutelli too, wide and meeting;
— *suctorial* (fig. 29): no labiogenal articulation is present, the mentum and the genae are fused, also the rutelli are modified into a tube, a part of the suctorial mouth.

The shape of the infracapitulum is also influenced by the chelicerae. They have two types: 1. a squat and wide chewing (Fig. 24) and 2. an elongated pricking (peloptoid) form (Fig. 25). The shape of the chelicerae reflects the method of feeding or the kind of food, respectively.

Among the several plates, the mentum bears one pair *(h)*, the genae one pair of robust *(a)* and two pairs of minute (m_{1-2}) setae, the rutellum is a modified seta itself. The chelicerae have two pairs *(cha, chb)* of setae, but only one on the peloptoid type. The palps consist of 2—5 joints, their structure and chaetotaxy are related to those of the ambulatory legs.

Epimeral region. The epimeral or coxosternal region (figs. 4, 10) is that part of the ventral plate which is connected with legs I—III and extends to the genital opening. This surface derives from the coxal shields and the sternal shield — the distinct sclerites can still be recognized in some primitive forms and in the nymphal stage. The fields (=epimeres: $ep_1 - ep_4$) connected with the legs are generally easily distinguishable; they are bordered (or) separated from each other) by the epimeral furrows and their continuations: the elevated ridges (=muscular insertions) or apodemes. These latter are directed towards the legs, and bear the corresponding notations: apodeme 1 *(ap_1)* belongs to leg I, apodeme 2 *(ap_2)* to leg II, etc. The sejugal furrow is backed by the sejugal apodeme *(ap.sej.)*, often connected longitudinally by the sternal apodeme *(ap.st.)*. The dark bands following the apodemes on the surface are called also bordures.

The ventral plate forms laterally the discidia or projections of various shapes but usually triangular which protect, similarly to the function of the pedotecta, the bases of legs IV. The custodia are also appendages of the ventral plate, forming a wedge-shaped crista from legs IV and decurrent attenuating anteriorad and there terminating in a free ridge or cuspis, respectively. The tuberculi are surface excrescences: they usually appear in opposing positions on the epimeres.

The epimeres bear setae, of a number characteristic to the taxon. Their totality is given by the epimeral setal formula, e.g. 3 — 1 — 2 — 4, which means that epimere 1 bears 3 setae, epimere 2 bears one seta, epimere 3 bears two setae, and epimere 4 bears four setae.

Anogenital region. It is the posterior part of the ventral plate bordered anteriorly by the epimeral region and the bases of legs IV; it includes also the genital and anal openings. Concerning its structure, it can be macropylic (fig. 4) if the genital and anal openings or the lamellae covering them are long, and usually meeting and filling the entire length of the region in a U- or Y-shape. They often have adjacent aggenital and adanal lamellae of the same length, indeed, some other adventitious plates (peranal, preanal and postanal) may also occur. These plates bear also a comparatively great number of setae, thus the genital plate may possess a maximum of 10 setae. The number of aggenital *(ag)*, anal *(an)* and adanal *(ad)* setae is also characteristic. A part

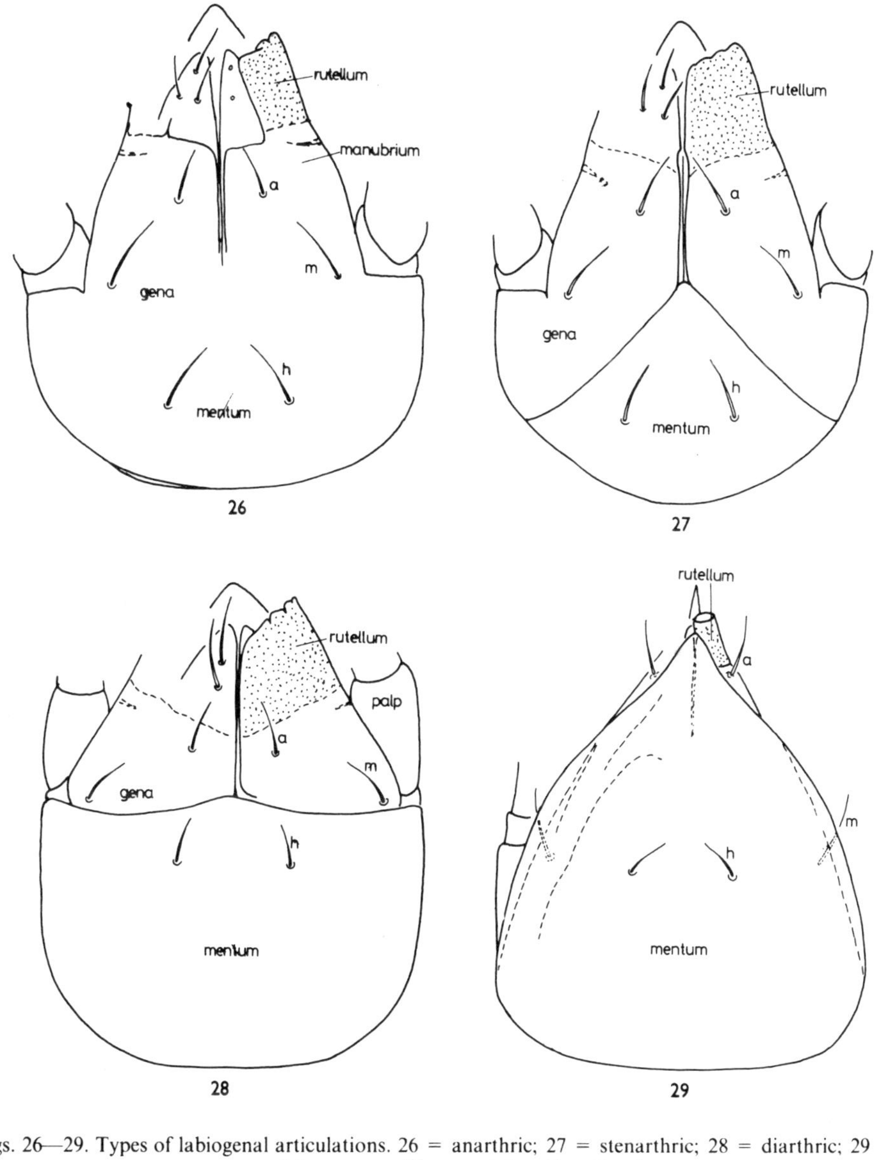

Figs. 26—29. Types of labiogenal articulations. 26 = anarthric; 27 = stenarthric; 28 = diarthric; 29 = suctorial: *m* = maxillar seta (after Grandjean)

of the plates had coalesced, but further differentiations can also be traced, e.g. the genital plate is subdivided by a transverse suture. In the brachypilic type (fig. 10) the genital and anal openings are visibly separated, because they are surrounded by a closed, uniform plate. Separately they are round, trapezoid, oblong or pyriform. The aggenital, adanal and other adventitious plates are invariably absent, that is, they are fused with the ventral plate and their former existence and position are indicated only by their setae. The genital and anal plates bear comparatively few setae. The brachypylic aggenital region displays four subtypes:

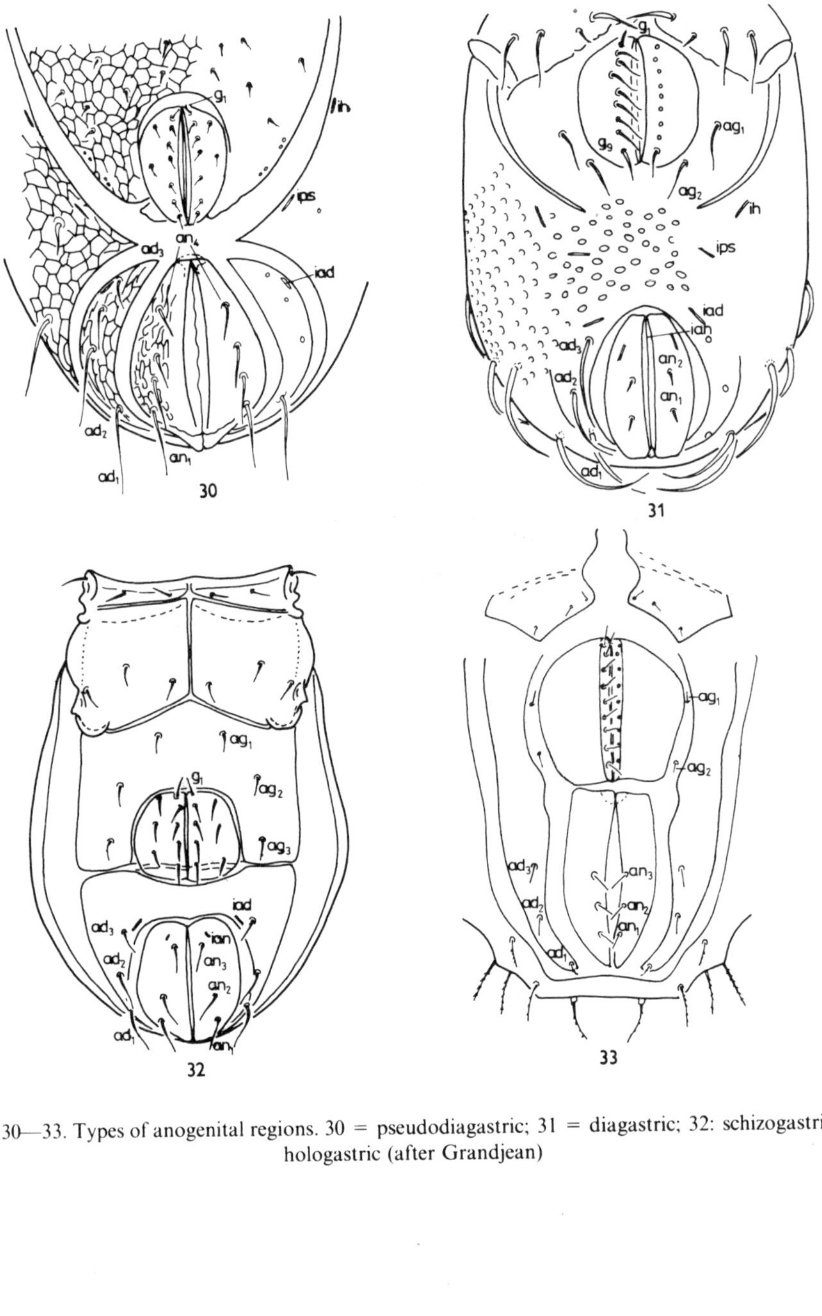

Figs. 30—33. Types of anogenital regions. 30 = pseudodiagastric; 31 = diagastric; 32: schizogastric; 33: hologastric (after Grandjean)

— *pseudodiagastric,* where the genital and anal openings are separated by a parabolic line (fig. 30);
— *diagastric,* where the seperating line between the genital and anal plates disappears medially, (fig. 31);
— *schizogastric,* where the separating line between the genital and anal openings is straight (fig. 32);
— *hologastric,* where the ventral plate is undivided.

The first two subtypes are known only among the primitive forms, the hologastric one (fig. 33) is characteristic of the higher Oribatids and occurs only exceptionally

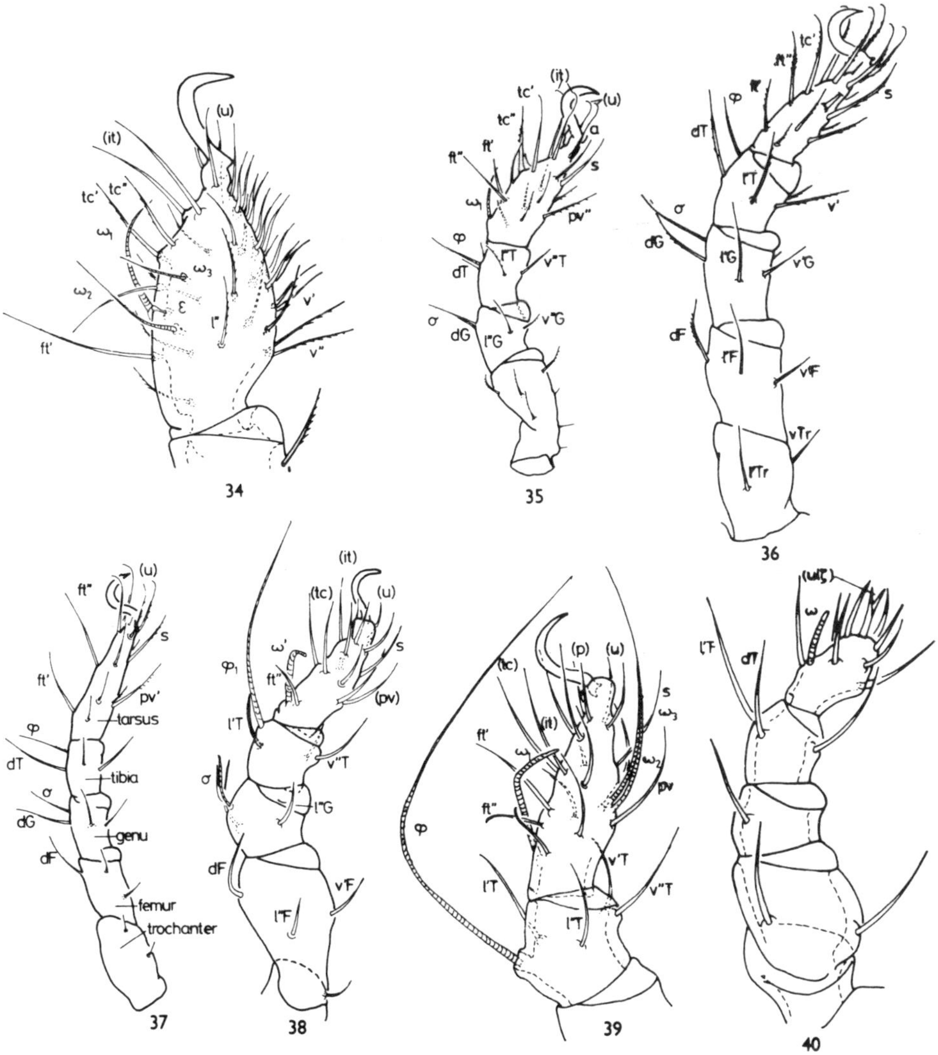

Figs. 34—40. Legs and palp of primitive Oribatids. 34—37 = Tarsus and legs of *Perlohmannia* sp.; 38—40 = Legs and palp of *Brachychthonius* sp. (Latin letters refer to normal setae, Greek letters to solenidia) (After Grandjean)

among the primitive forms (Epilohmannioides; Hermanniidae), but the genital plate still shows two rows, either 7(4+3) or 9(6+3), of genital setae.

The genital and anal plates, or in the higher Oribatids also the ventral plate, may also possess porous fields (e.g. areae porosae postanales) or slit-organs, as the nearly always present pore *iad* in the higher Oribatids.

LEGS. The legs (figs. 34—40) articulate into the acetabuli and usually consist of five free joints (trochanter, femur, genu, tibia and tarsus); terminally the tarsus has 1—3 claws on a short peduncle. In some groups, the femur is a composition and appears as "two femora", because the coxae had not fused to the body surface; in such cases there are six free joints.

The articulation of the joints is usually simple (fig. 41), but in some groups (Gymnodamaeidae) there is a cuff-like articulation (crispin) between the genu, tibia and tarsus (fig. 42). The length of the legs, the shape of the joints and their chitinous excrescences, as well as their cuticle, are extremely variable, adapted to the life of the animal and characteristic to the taxon.

Chaetom of legs. The legs bear setae (chaetom) which can be subdivided into four subgroups according to their structure:

- *simple setae:* homogeneous, with at most a very small median cavity, the robust chitinous axis is surrounded by an external surface showing usually granules, squamae, cilia; they have a basal bulb;
- *acanthoids:* special setae on the tarsi of legs I and II, they are canaliculate inside and completely smooth outside;
- *solenidia:* tubiform setae, with annuliform incrassations in the wall of the tube; smooth outside, withont a basal bulb;
- *famulus:* a modified hair on the tarsus of leg I, of special form (figs. 43—49); the inner narrow channel cuts through even the root of the hair.

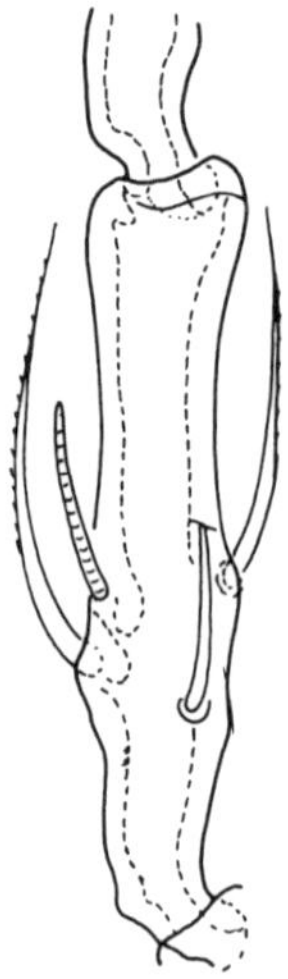
41

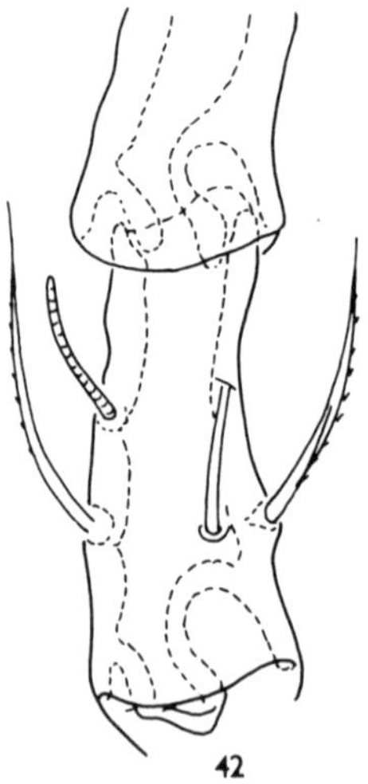
42

Figs. 41—42. Types of articulation between joints of legs. 41 = normal; 42: crispin (after Grandjean)

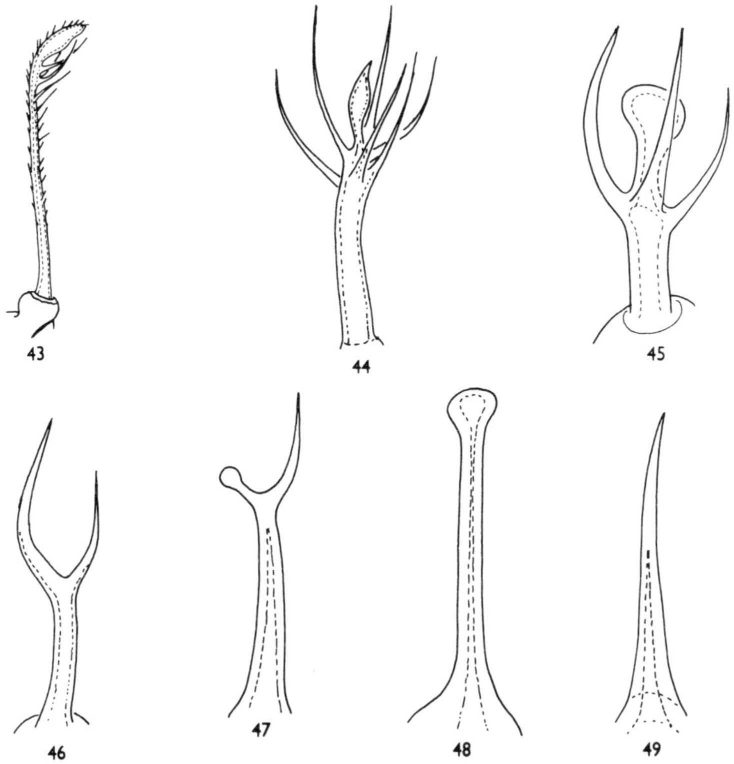

Figs. 43—49. Various types of famulus (ε) (after Grandjean)

The number and position of the hairs on the leg joints are constant. For their designation, the letters of the Greek alphabet are used for the solenidia, and those of the Latin for the rest.

Excepting the tarsus, the setae are in a verticillate position on the joints; accordingly, there can be dorsal *(d)*, ventral *(v)* and lateral *(l)* hairs. On the tarsi, this arrangement cannot be followed, therefore, they are indicated from an initial basic dorsal position distally. These pairs of setae are termed as follows: fastigial *(ft)*, tectal *(tc)*, proral *(p)*, unguinal *(u)*, primiventral *(pv)*. There are also some complementary paired *(it*, pl_1 and *ad)* and unpaired *(s,v)* setae. Legs II — IV have considerably fewer setae than leg I described above. Solenidia may occur on the genu (indicated as δ), the the tibia (φ) and the tarsus (ω), the accompanying lawer index numeral (e.g. ω_1) refers to their appearance during ontogenesis.

The chaetom, especially the number of setae per joint, is an important classificatory character, so it is also given as a setal formula; e.g. I: 0 — 5 — 3 + 1 — 4 + 2 — 19 + 2 — 1, which means, that the trochanter has no setae, the femur has five setae, the genu possesses three setae and a solenidium, etc; the last member indicates the number of claws.

Genital organs, reproduction. Genital organs: the Oribatids are bisexual. The genital organs have no special role in taxonomy. The females possess a long ovipositor

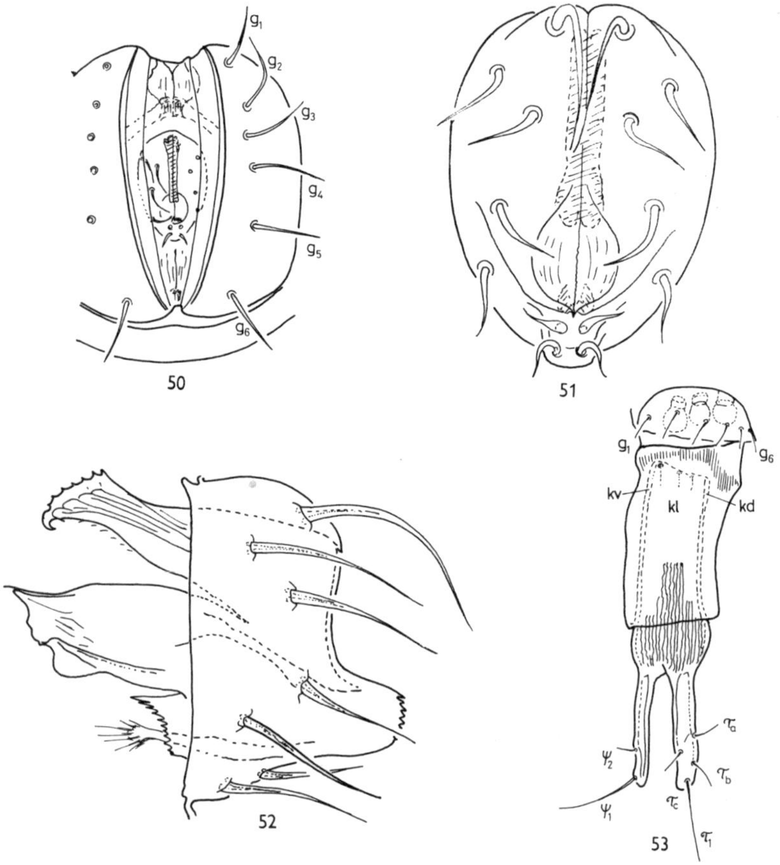

Figs. 50—53. Parts of genital organs. 50 = male genital plate; 51 = penis in ventral view; 52 = penis in lateral view; 53 = ovipositor (after Grandjean)

(fig. 53), well discernible also within the body in satisfactorily bleached specimens. The male penis is shorter but more complex in construction (figs. 50—52).

A secondary sexual dimorphism is very rare (Amerobelbidae, Galumnidae) and will be mentioned only for some species.

Reproduction and ontogenesis are also complex; owing to their obscure way of life, these phenomena are known only in the case of a few species, and these too mostly from laboratory breedings. The developmental stages do not resemble the imagoes, their identification is difficult and in many cases still impossible. They will not be discussed in detail, only some general outlines will be given.*

* The developmental stages, especially the nymphal stage, can, however, be advantageously used in the separation of higher systematical categories. Since, however, the developmental stages and imagoes live often in different habitats and at different times, usually only the imagoes are described. There is, therefore, still a certain duplicity in the Oribatid system: some groups, especially the primitive ones, are relegated on the basis of the developmental stages to the various families, while the great majority of the higher Oribatids solely by their imagoes. One of the most important tasks of future classification will be the detection and description of the unknown developmental stages.

ONTOGENESIS

Egg — larve — protonymph — deutonymph — tritonymph — imago; between these stages there is always a so-called pronymph stage, when the animals are immobile and prepare themselves for moulting.

The eggs are elliptical, often well discernible in the body of the female. Their chorion is usually thin and transparent, but sometimes with a rough sculpture *(Ptyctima)*.

The larvae are characterized by three pairs of legs and the absence of the genital opening. The hysterosoma consists of eight segments. The Claparède's organ is a special characteristic of the larval stage.

The protonymph has four pairs of legs, and the primordium of the genital opening also appears (but there is only one pair of genital papillae). The Claparède's organ disappears. The hysterosoma is composed of mine segments.

In the deutonymph stage the hysterosoma is completed into 10 segments. The number of central setae significantly increases and two pairs of genital papillae appear.

The tritonymph has three pairs of genital papillae. The chaetom characteristic of the imago develops in the coxosternal region.

KEY TO SYMBOLS

A — 1st somatic segment
Aa — area porosa adalaris
a — 1. genal seta, 2. subunguinal seta
A_{1-3} — notogastral areae porosae
abj. — abjugal suture
AD — adanal plate (or 9th somatic segment)
ad_{1-3} — adanal setae
AG — aggenital plate
ag_{1-3} — aggenital setae
Ah — area porosa hypoalaris
Al — area porosa lamellaris
AN — anal plate (or 10th somatic segment)
an_{1-4} — anal setae
ap.sej. — sejugal apodeme
ap.st. — sternal apodeme
ap_{1-5} — apodemes
B — 2nd somatic segment
C — 3rd somatic segment
ch — seta of chelicera
cha — posterior seta of chelicera
chb — anterior seta of chelicera
cp — notogastral seta
c_{1-3} — notogastral setae
D — 4th somatic segment
dF — dorsal seta of femur
dG — dorsal seta of genu
disj. — disjugal suture
dsej. — dorsosejugal suture
dT — dorsal seta of tibia
d_{1-2} — notogastral setae
E — 5th somatic segment
ex — exobothridial seta
exa, *exp* — anterior and posterior exobothridial setae
e_{1-2} — notogastral setae
ep_{1-4} — epimeres
F — 6th somatic segment
ft'–ft'' — fastigial setae
f_{1-2} — notogastral setae
G — genital plate
gla — oil gland
g_{1-9} — genital setae
H — 7th somatic segment
h — hypostomal or mental seta

h_{1-3} — notogastral setae
ia, iad, ian, ih, im, ip, ips, — lyrifissures or pori
in — interlamellar seta
it — seta of tarsus
kd, kl, kv — dorsal, lateral and ventral setae in the median part of ovipositor
l'-l'' — lateral setae of tarsus
l'T — lateral seta of tibia
la — notogastral seta
le — lamellar seta
l''G — lateral seta of genu
lm — notogastral seta
m — median infracapitular seta
ms — notogastral seta
or_{1-3} — adoral setae
PA — postanal plate (or 11th somatic segment)
PeA — peranal plate
PS — 8th somatic segment
p, p_{1-3}, ps_{1-3} — notogastral setae
pe — peranal seta
pd_{1-4} — pedotecta
pv'' — seta of tarsus
ro — rostral seta
r_{1-3} — notogastral setae
s — seta of tarsus
ss — sensillus
sej. — sejugal suture
tc'-tc'' — setae of tarsus
ta, te, ti — notogastral setae
u — unguinal seta of tarsus
v'-v'' — setae of tarsus
ul — seta of palp
v''G — seta of genu
v''T — seta of tibia

1a-c, 2a, 3a-c, 4a-d — epimeral setae
ε — famulus of tarsus
φ — solenidium of tibia
δ — solenidium of genu
ω_{1-3} — solenidia of tarsus
ψ_1-ψ_2, τ_1, τ_{a-c} — setae at apical part of ovipositor
( ) — setae in pairs
' — designating inner seta of leg
'' — designating outer seta of leg

COLLECTING AND STUDY TECHNIQUE

The Oribatids occur in the whole world; the highest individual and specific numbers are attained in the forest litter and the immediate soil surface rich in organic substances. Masses can be collected in moss cushions and lichen crusts. Their distribution follows the plant cover towards the poles and the mountain tops: great numbers can thus be found in the moss and lichen incrustations of arctic and antarctic regions as well as at the highest elevations still permissive of this vegetation (2000—5000 m). Most species inhabit the moss, lichen and soil layer, fewer live on tree trunks, branches and foliage, in ants' nests, caves, and exceptionally in the intertidal zone of seashore, on maritime algae, freshwater plants, and in other special habitats. They occur in great masses or very large numbers mostly in moist, humid, dusky or dark sites rich in organic detritus.

To collect mites in these habitats is extremely simple; samples of litter, moss, lichens, humus are placed in extractors of the Berlese or the Tullgren type and desiccated either at room temperature or by the application of light bulbs. When the humidity content of the sample diminishes below a critical point, the majority of the Oribatids present will in a short time leave it and fall into the glass vial, containing the preservation fluid, affixed under the exhaustor. By this simple method an enormous amount of Oribatids can be had, so that most Oribatidologists possess a multiple amount of Oribatid mites of what he can study or identify.

This extremely quick and easy method is the cause why most Oribatidologists use it exclusively and neglect very special, or extremely dry, habitats, or those poor in organic substances and humidity. Investigations made by Grandjean, Travé, and others show, however, that although the Oribatid faunas of these very dry or special habitats are poor in individual numbers they usually harbour highly interesting species.

Since the animals, dropping out from the samples, concurrently remove also bits of detritus, grains of the soil substrate, etc., the material arriving in the alcohol container is still rather contaminated. Earthen grains and other detritus can be easily removed by the careful siphoning off of the alcohol and a subsequent transfer of the entire material into a concentrated NaCl solution. After a few seconds or one or two minutes of mixing the animals and a part of the plant detritus will float to the surface of the salt solution and they can then be spilled into a funnel lined with filter paper. After rinsing in water (if rinsing is omitted, salt crystals will form in the alcohol!), the cleared material will be preserved in 75% alcohol, until later study.

Most Oribatidologists preserved, until the beginnings of the 1960s, all elaborated material in permanent slides. The so-called Berlese—Hoyer fluid, polyvinyl alcohol, glycerine, or various modifications of these substances have been used as imbedding media. Since all of these substances suffer losses in volume with time, the slides must be framed. Experience showed, however, that the Oribatids of even carefully framed slides disintegrate and become unsuitable for study. (The single known safe medium is Canada balsam, but the animals must first be dehydrated either in absolute alcohol or in xylol. During this process, however, the hairs fall out easily, or the specimens suffer deformations. In addition, the light refraction index of Canada balsam is very high, so that the finer parts can be studied only by recourse to special optical processes.)

Mainly by Grandjean's initiative, nearly every research worker — excepting a few conservative Oribatidologists — applies Grandjean's open-slide method. In essence, this is as follows.

Both the unidentified Oribatid material and the collection of species are preserved in 75% alcohol. Alcohol should also contain some drops of glycerine so that in case of an eventual evaporation (container breakage or cracks in the glass!) the animals will be saved from desiccation in the glycerine.

Microscopic examination is made with the so-called half-open slide. Specimens to be examined are placed in alcohol in a porcelain bowl and some drops of lactic acid added depending on the size or the amount of specimens. Concurrently with the evaporation of the alcohol, the lactic acid penetrates the body of the animal and renders it transparent. This process takes about 1—3 days, depending on the size of the specimen, the grade of chitinization and the temperature of the room. They are then suitable for microscopic study.

The shallow central hollow of the slide must now be half covered by the cover glass, so that its perpendicular margin is aligned with the median line (diameter) of the circular excation. A drop of lactic acid will then be placed carefully in the middle of the hollow, so that capillary action will draw it in under the cover glass. Lactic acid will be added until the covered half of the hollow is filled with it and none remains in the open part. Any lactic acid residue remaining there will be carefully removed by filter paper or wiped away. In this state, the half-open slide is ready to receive the specimen to be studied.

The specimen to be studied is taken from the lactic acid by the tip of a fine brush and then deposited in the centre of the excavation of the slide and carefully moved under the edge of the cover glass over the lactic acid..The animal must then be pushed "upward" until it becomes stuck between the cover glass and the slide. Minute mites become rather fixed there, while the bigger ones get stuck nearer to the centre of the hollow. Subsequently, the animal will be rotated by the tip of the brush or a careful horizontal sliding of the covering glass into positions suitable for study.

After examination, the specimens will be taken out by the tip of the brush again, placed in their original vial of alcohol and thereafter into the collection.

In the case of very small, weakly chintinized and pigmented specimens, e.g. the Brachychthoniid species, it is also advisable to make some permanent slides for rapid comparison purposes. The specimens must be placed in a dorsoventral situation. For imbedding, we use Berlesse-Hoyer fluid and frame every slide with some good quality substance. However, Brachychthoniid type-material is preserved in alcohol. Specimens demanding examination from every body aspect (e.g. *Suctobelbidae*) should not be embedded in permanent slides.

IDENTIFICATION KEYS

Key to suborders

1 (2) At least one of the following characteristics unequivocally and well visible: ptychoidy, specimens laterally compressed (when dead lying invariably on their sides); macropily, combined with a dorsoventrally flattened or cylindrical body; brachypily, combined with pseudodiagarstry or schizogastry; quite exceptionally hologastry with 14 or 16 pairs of notogastral setae and 7(4 + 3) *(Epilohmannoides)* or 8(5 + 3), 9(6 + 3) *(Hermanniidae)* genital setae arranged in two rows; number of genital setae only exceptionally fewer than seven pairs; number of notogastral setae usually 14—16 pairs

suborder 1: **Archoribatida**

2 (1) None of the above characteristics present; therefore; body mostly oval or rotund combined with a vaulted shape; no ptychoidy (dead specimens lying always on their backs or venters); brachypily, always combined with hologastry; quite exceptionally a partial diagastry, but then combined with 9—12 pairs of genital setae arranged in a single row *(Liodidae)*; number of genital setae only exceptionally more than six pairs (usually 6, 4 or 5 pairs); number of notogastral setae only exceptionally more than 10 pairs (usually 10, more rarely 14, exceptionally 13, 12 or 11, very rarely even less, or more than 16 pairs of setae)

suborder 2: **Euoribatida**

Suborder 1: ARCHORIBATIDA

Archoribatida Balogh et Mahunka, 1979 : 279.

In one or more features retaining primordial characteristics of organisation. Hysterosoma may thus be archeonotic, taht is, without evolvement of notogaster; sclerites discrete, occasionally partly or entirely absent, or separated from each other by wide intervals; in such cases body hardly pigmented, whitish, occasionally with long, movable, densely ciliate black setae. Legs 6-jointed, two femora present. In other cases notogaster arthronotic, with 1—3 transverse furrows, 2—4 tergites, or holonotic, that is, completely uniform, whole Ptychoidy, dichoidy and holoidy may equally occur. Anogenital region usually macropilic, in the oldest forms considerably complex: genital, aggenital, anal, peranal, adanal, preanal and postanal plates may be present, and during development also variously fused. Brachypilic anogenital region occuring only rarely, combined with pseudodiagastry or schizogastry; in two cases *(Epilohmannoides, Hermanniidae)* also hologastry known, with the arrangement of the setae indicating the primitive relationship.

This suborder contains a number of ancient, very isolated forms, therefore, monotypical genera and families, represented by a single species, are rather frequent. They are relegated to five supercohorts.

Key to supercohorts

1 (4) Body ptychoid (fig. 5), that is, prosoma and opisthosoma can be folded like a penknife; body laterally more or less compressed (dead specimens lying on their sides).

2 (3) Either notogaster with well discernible furrows on posterior half of body (pl. 6 : A), or ventral side with distinct ventral plate showing genital and anal openings well separated by brachypily (pl. 6: B)
supercohort 2: **Arthronota** Balogh, 1972 (partim)

3 (2) Neither posterior notogastral furrows nor a brachypilic ventral plate displaying separated genital and anal openings present. Genital and anal openings large; either wide and jointly resembling a "window" (pl. 71 : B), or narrow, similar to a V-shaped cleavage (pl. 106 : I)
supercohort 4: **Mixonomata** Grandjean, 1969 (partim)

4 (1) Body not ptychoid, that is, prosoma and opisthosoma never foldable like a penknife, both parts at most slightly movable. Body dorsoventrally compressed (dead specimens lying in this position).

5 (6) Hysterosoma weakly sclerotized and hardly pigmented, whitish. No uniform (whole) hologaster evolved. Smaller sclerites either absent or widely separated from each other (pl. 1 : A). These latter and hysterosoma — with some exceptions — carrying one or several pairs of conspicuous, black, densely ciliate setae. Legs composed of six free joints
supercohort 1: **Palaeosomata** Grandjean, 1969

6 (5) Hysterosoma without the above features; usually covered with a heavily sclerotized shield, either completely fused or subdivided by 2—4 furrows, the notogaster. This latter never bearing densely ciliate black setae. Legs composed of merely five free joints.

7 (10) Notogaster arthronotic, that is, a shield subdivided by 2—4 furrows (pl. 6 : A).

8 (9) A lateroabdominally situated oil gland present, well discernible also in superior view (pl. 54 : A). Notogaster with one furrow, but latter never with insertion points origin of reduced setae
supercohort 3: **Parhyposomata** Balogh et Mahunka, 1979

9 (8) No oil gland present. Notogaster subdivided by 1—3 furrows; if merely one furrow present, then always with points of insertion of two pairs of setae
supercohort 2: **Arthronota** Balogh, 1972 (partim)

10 (7) Notogaster holonotic, that is, whole, without transverse furrows.

11 (12) Body dichoid, that is, with a less sclerotized region between legs II and III on ventral size (fig. 3), therefore both parts slightly movable. Anogenital region macropilic, or if rarely brachypilic then this appears as either pseudodiagastry (fig. 30) or schizogastry (fig. 32). In one case *(Epilohmannoides)* also hologastry occurs.
supercohort 4: **Mixonomata** Grandjean, 1969 (partim)

12 (11) Body holoid, that is, rigidly fused also between legs II and III on ventral side. Anogenital region usually macropilic; if exceptionally brachypilic then combined with diagastry (fig. 31). In one case also hologastry occurs *(Hermanniidae)*, but then 9 (6 + 3) genital setae arranged in two rows on genital plate (pl. 131 : C)

supercohort 5: **Holosomata** Grandjean, 1969

Supercohort 1: PALAEOSOMATA Grandjean, 1969

Palaeacariformes Trägårdh, 1932 : 1.
Palaeosomata Grandjean, 1969 : 141.
Bifemorata Balogh, 1972 : 32.
Palaeosomata Balogh et Mahunka: 1979 : 279.

No continuous shield (notogaster) evolved on hysterosoma, all sclerites either absent, or if present discernible only by recourse to special examination techniques. Contrarily to most Oribatids, they are not pigmented, whitish or yellowish animals, often with striking, densely ciliated, large black setae. Apices of chelicerae similarly black, or even entire chelicerae, tarsal claws and genital papillae. Even basal parts of chelicerae and palps visible from above, left free by prodorsum. Prodorsum with two pairs of exobothridial setae. Hysterosoma with at least 17 pairs of setae, often more; setae f_1 and f_2 invariably present. Oil gland and lyrifisurres absent. Legs composed of six joints. Tarsi with two or three claws, in case of three claws, median one appearing as a claw-shaped empodium, though often reduced. Famulus *(ε)* on tarsus of leg I often modified, apically arcuate, clavate or ciliate, occassionally setiform.

The mites live in the lower layers of a thick litter cover, in a constantly moist environment. The three superfamilies contain 14 genera and about 20 species, with 11 genera and 19 species occurring in the Palaearctic Region.

Key to superfamilies

1 (6) Hysterosomatic setae divers in shape: the majority simple, but 1—4 pairs conspicuously long, thick, densely ciliate, black, usually the setae e_1 and f_2.
2 (5) Legs II—IV with three claws, hence empodium developed (pl. 2 : D).
3 (4) Hysterosoma with several sclerites (pl. 2 : A); seta c_1 multiply longer and robuster than seta c_2

superfamily 1: **Archeonothroidea** Grandjean, 1954

4 (3) Hysterosoma with three large but transparent and hardly recognizable sclerites (pl. 4 : A); setae c_1 and c_2 equal or nearly equal in length

superfamily 3: **Ctenacaroidea** Grandjean, 1932 (partim)

5 (2) Tarsi of legs II—IV with two claws, empodium reduced (pl. 4 : E)

superfamily 2: **Palaeacaroidea** Grandjean, 1932

6 (1) Hysterosomatic setae uniform, simple, thin, only slightly ciliate; no conspicuously long, black and densely ciliate setae present (pl. 5 : A)

superfamily 3: **Ctenacaroidea** Grandjean, 1932 (partim)

Superfamily 1: ARCHEONOTHROIDEA Grandjean, 1954

Archeonothridae Grandjean, 1954: 428.
Archeonothroidea Grandjean, 1969: 141.

Hysterosoma with several small notogastral sclerites. 1—4 pairs among the setae conspicuously long, incrassate, often densely ciliate and blackish, mostly setae e_1 and f_2. Some pairs arising on a common sclerite or their bases connected by chitinous bridges. Tarsi with small, uncinate empodium, thus legs apparently 3-clawed.

Six genera are relegated to the two families, but of the species only three have been found in the Palaearctic Region.

Key to families

1 (2) Setae c_1 and c_2 on a common transverse sclerite, setae e_1 arising near each other, their bases connected by a chitinous lath
family 1: **Archeonothridae** Grandjean, 1954

2 (1) Setae c_1 and c_2 on a separate sclerite each, setae e_1 arising far from each other
family 2: **Acaronychidae** Grandjean, 1932

Family 1: ARCHEONOTHRIDAE Grandjean, 1954

Archeonothridae Grandjean, 1954: 428.
Archeonothrinae Grandjean, 1954: 196.

Setae c_1 and c_2 arising on a common biscuit-shaped transverse sclerite; setae e_1 originating also on a transverse sclerite, with their insertion points nearer to each other than in the case of setae d_2.

Two genera are known from the Palaearctic Region.

Key to genera

1 (2) Setae ps_2 simple, setiform, ciliate. Digitus fixus of chelicera with 10 minute teeth, digitus mobilis with 6—7 very small teeth
genus 2: **Amuracarus** Lange, 1975

2 (1) Setae ps_2 modified: widened lanceolate or phylliform (pl. 1:D), smooth. Digitus fixus of chelicera with 3—4 teeth, digitus mobilis with two minute teeth
genus 1: **Zachvatkinella** Lange, 1954

Genus 1: ZACHVATKINELLA Lange, 1954

Zachvatkinella Lange, 1954: 1042.
Himalacarus Sheals, 1965: 3 (Lange: 1972: 1889).

Some of the hysterosomatic sclerites coalesced, thus setae c_1 and c_2 as well as setae e_1 situated on a transverse sclerite each. Seta ps_2 modified, usually lanceolate, smooth.

Seta e_1 essentially longer than the other setae. Digitus fixus of chelicera with 3—4, digitus mobilis with two minute teeth. Tarsal empodium small, but always well discernible. Famulus (ε) on tarsus of leg I ciliate.

Type-species: *Zachvatkinella belbiformis* Lange, 1954.

Four closely related species are known from the Palaearctic Region, the fifth is recorded from the Himalaya.

Key to species

1 (2) Seta ps_2 knife-shaped, comparatively wide (pl. 1: D), margins throughout slightly convex. — Seta f_2 shorter even than half length of seta e_1 (pl. 1: A, C) and than seta e_2. Solenidium ω_p on tarsus of leg I directed laterally, transversely to tarsus.

Locus typicus: Sikhote Alin range, Soviet Union. Distribution: Soviet Union (ES, FE)

belbiformis Lange, 1954

2 (1) Seta ps_2 relatively narrow, dagger-shaped, gradually attenuating apicad, margins straight or finely concave.

3 (4) Seta ps_2 (pl. 1 : G) gradually narrowing then slightly broadening again, hence margins medially finely concave. — Solenidium ω_p on tarsus of leg I (pl. 1:I) transverse, directed outwards (as in preceding species). Seta f_2 slightly longer than half length of e_1.

Locus typicus: the Khumni lake along the Amur, Soviet Union. Only the type-locality known

doryura Lange, 1972

4 (3) Seta ps_2 with straight margins, these latter never concave. Solenidium ω_p on tarsus of leg I always directed apicad (pl. 1 : F).

5 (6) Seta ps_2 narrow, dagger-shaped, with nearly parallel margins, terminally not elongated into a lengthy apex (pl. 1:J). — Seta f_2 one-fourth shorter than e_1, but seta e_2 essentially shorter than f_2.

Locus typicus: near the Baikal lake, Soviet Union. Only the type-locality known

baicalica Lange, 1972

6 (5) Seta ps_2 terminally very long, last third drawn out into an attenuating apex (pl. 1:K). — Setae f_2 and e_2 as well as seta *cp* equal in length, and shorter than e_1. Dentition of cheliceral digitus fixus as in (pl. 1 : L).

Locus typicus: Gagra, Caucasus, Soviet Union. Distribution: Soviet Union (TC)

caucasica Lange, 1972

Species interpretabiles

1. baicalica Lange, 1972: 1891, fig. 9.
2. belbiformis Lange, 1954: 1042, figs. 1—14.
   belbiformis Lange: 1972: 1889, fig. 14.
3. caucasica Lange, 1972: 1891, figs. 10—13.
4. doryura Lange, 1972: 1889, figs. 1—8.

Genus 2: AMURACARUS Lange, 1975*

Amuracarus Lange, 1975: 50.

Seta ps_2 simple, setiform, ciliate. All hysterosomatic setae comparatively long, setae e_1 not shorter than body, their insertion points connected by a chitinous lath. Setae e_2, f_2 and h_1 only slightly shorter than e_1. Digitus fixus of chelicera serrulate, with 10 minute teeth (pl. 2: I). Median claws (empodia) of tarsi vestigial, hardly discernible on tarsi II and III. Famulus (ε) on tarsus of leg I ciliate.

Type-species: *Amuracarus voskresenskii* Lange, 1975.

A single species.

— Characterized as above. No measurement data are available.

Locus typicus: the Lower Amur, Soviet Union. Distribution: only the type-locality known.

voskresenskii Lange, 1975

Species interpretabilis

1. voskresenskii Lange, 1975: 50, fig. 39.

Family 2: ACARONYCHIDAE Grandjean, 1932

Acaronychidae Grandjean, 1932: 426.
Acaronychinae Grandjean, 1954: 196.

Setae c_1 and c_2 arising on discrete sclerites; setae e_1 also carried on distinct sclerites and their points of isertion further removed from each other than those of setae d_2.

A single genus.

Genus 1: ACARONYCHUS Grandjean, 1932

Acaronychus Grandjean, 1932: 421.

Most hysterosomatic setae, thus c_1, c_2, e_1, originating on distinct sclerites, but insertions of c_1 on a small transverse shieldlet. Setae e_1 and f_2 thick, obtuse, throught minutely ciliate, all other hairs acute, smooth or finely pubescent. Seta ps_2 lanceolate or dagger-shaped, incrassate. Tarsi with small incinate empodia. Apex of famulus (ε) on leg I slightly flattened.

Type-species: *Acaronychus tragardhi* Grandjean, 1932.

Three species are known, of which two present in the Palaearctic Region.

Key to species

1 (2) Sensillus filiform, exclinate then at half its length rapidly angulate and directed posteriorad. Body setae short, thick (pl. 2: A, C); lamellar setae about half as

* We are acquainted with this taxon only from Lange's text in the identification book edited by Ghilarov and Krivolutsky: 1975.

long as prodorsum. Ciliate end of famulus *(ε)* on tarsus of leg I hardly flattened. Chaetom of anogenital region as shown in pl. 2: B.

Locus typicus: Tangier, Cape Spartel, Morocco. Distribution: Algeria, Morocco, France, Soviet Union (SMA)

tragardhi Grandjean, 1932

2 (1) Sensillus fusiform (pl. 2: E). Body setae slightly longer but thinner; lamellar setae hardly shorter than prodorsum. Ciliate apex of famulus on tarsus of leg I spatulate.

Locus typicus: Caucasus, Soviet Union. Distribution: only the type-locality known

grandjeani Lange, 1975

Species interpretabiles

1. grandjeani Lange, 1975: 46, fig. 31.
2. Trägårdhi Grandjean, 1932: 421, figs. 8—10.
   Trägardhi Grandjean: 1954: 214, figs. 6—10.

Superfamily 2: Palaeacaroidea Grandjean, 1932

Palaeacaridae Grandjean, 1932: 426.
Palaeacaroidea Grandjean, 1969: 139.

Hysterosoma with three notogastral sclerites; two of which dorsal and one pygidial. Four pairs (d_2, e_1, h_2 and ps_2) or three pairs (d_2, e_1 and ps_2) of hysterosomatic setae long, thick and densely ciliate, setae d_2 and e_1 especially long and movable. Sensillus filiform. Seta c_2 longer than c_1; solenidium ω_p on tarsus of leg I not accumbent to surface of leg joint. Famulus *(ε)* setiform, apically eventually bent. Tarsi 2-clawed.

Composed of a single family.

Family 1: Palaeacaridae Grandjean, 1932

Characterized by the superfamilial features. Containing 2 genera.

Key to genera

1 (2) Four pairs of longer, thick, ciliate setae present: d_2, e_1, h_2 and ps_2. End of body with setae h_1 and ps_1 lanceolately incrassate

genus 1: **Palaeacarus** Trägårdh, 1932

2 (1) Only three pairs of longer, thick and ciliate setae present: d_2, e_1 and ps_2. Setae h_1 and ps_1 on end of body short, simple, thin

genus 2: **Palaeacaroides** Lange, 1972

Genus 1: PALAEACARUS Trägårdh, 1932

Palaeacarus Trägårdh, 1932 : 2, figs. 2, 4, 6.
Tragardhacarus Zachvatkin, 1945: 674 (Bulanova-Zachvatkina: 1968: 99).
Palaeacarus Grandjean: 1954: 199, figs. 1—5.

Four pairs of hysterosomatic setae (d_2, e_1, h_2 and ps_2) resilient, comparatively long, apically considerably elongated, attenuating; setae h_1 and ps_1 widely lanceolate. All adanal setae uniform. Only a slight difference in size between proximal and distal setae of chelicera. Famulus *(ε)* on tarsus of leg I setiform, straight or arcuate. Tarsus of palp with cleft acanthoids.

Type-species: *Palaeacarus hystricinus* Trägårdh, 1932.

Of the seven known species or subspecies, the latest revision (Lange, 1975) considers four as synonyms or species inquirendae although the known localities are removed rather far from each other and also some smaller morphological differences can be shown. We follow this concept only partly, but a final decision can only be made after an examination of the type-specimens.

Key to species

1 (2) Seta c_3 absent. Solenidia on tarsus of leg I long (pl. 3 : G), generally longer than half length of tarsus.
Locus typicus: Pitsunda, Soviet Union. Distribution: Soviet Union (TC, SET)
caucasicus Lange, 1972

2 (1) Seta c_3 present. Solenidia on tarsus of leg I shorter than half length of tarsus.

3 (6) Seta 1*a* only half as long as the other three pairs arising ventrally on trochanter of leg I (pl. 3 : K). Tarsus of palp with two claft acanthoids (pl. 3: E).

4 (5) Famulus *(ε)* on tarsus of leg I bent posteriorad at nearly a right angle, its distal end widened. Setae h_1 and ps_1 on end of body swollen, hence their margins convex (pl. 3: C). — Sensillus (pl. 3 : B) smooth, at most basally slightly ciliate. Body and seta proportions shown in pl. 3 : A, B. Length: 330—440 µm.
Locus typicus: Dalby, Scania, Sweden*. Distribution: England, France, Sweden, Japan
hystricinus Trägårdh, 1932

5 (4) Famulus *(ε)* on tarsus of leg I at most finely arcuate, not widening distally. Setae h_1 and ps_1 on end of body spiniform. Sensillus ciliate.
Locus typicus: Paramusir, Kurili Islands, Soviet Union. Distribution: only the type-locality known
orientalis Bulanova-Zachvatkina, 1967

6 (3) Setae arising ventrally on trochanter of leg I equal in length (pl. 3 : L). Tarsus of palp without cleft acanthoid, latter either simple or trifurcate.

7 (8) Merely simple, terminally undivided acanthoids present on tarsus of palp. Seta c_1 of notogaster four times longer than c_3. — Famulus *(ε)* on tarsus of leg I obtuse, situated perpendicularly to surface of leg joint. Length: 444—475 µm.

*See Grandjean, 1954: 200.

Locus typicus: Kharkov, Soviet Union. Distribution: only the type-locality known.

lapshovi Zachvatkin, 1945

8 (7) A trifurcate acanthoid present on tarsus of palp, all other setae simple and setiform. Seta c_1 of notogaster merely two and a half times longer than c_3. — Famulus *(ε)* on tarsus of leg I apically acute, strongly curved (pl. 3: I). Setae h_1 and h_2 on end of body spiniform (pl. 3: J). Length: 447 µm.

Locus typicus: Orichevsk, Kirov district, Soviet Union. Distribution: Soviet Union (CET, SET, WS, FE)

kamenskii Zachvatkin, 1945

Species interpretabiles

1. caucasicus Lange, 1972: 103, figs. A—D.
2. hystricinus Trägårdh, 1932: 2, figs. 2, 4, 6.
   hystricinus Grandjean: 1954: 199, figs. 1—5.
   gallicus Grandjean, 1939: 64 (nom. nudum) (Grandjean: 1954: 199).
3. kamenskii Zachvatkin, 1945: 674, *(Tragardhacarus)*.
4. lapshovi Zachvatkin, 1945: 674 *(Tragardhacarus)*.
5. orientalis Bulanova-Zachvatkina, 1967: 101, fig. I: 4.

Species inquirendae

— appalachicus Jacot, 1938: 127, figs. 1—2.
  hystricinus appalachicus Grandjean: 1954: 213.
— hystricinus västerbottensis Grandjean, 1964: 212.

Genus 2: Palaeacaroides Lange, 1972

Palaeacaroides Lange, 1972: 103.

The three pairs of thick and densely ciliate hysterosomatic setae *(d_2, e_2, ps_2)* not flexible, comparatively short, their apices smooth. Adanal setae ad_2 considerably longer than the other adanal setae. Distal seta on chelicera twice longer than proximal one. Apical part of famulus *(ε)* on tarsus of leg I lanceolately widened, nearly straight. Acanthoids on tarsus of palp not cleft.

Type-species: *Palaeacaroides pacificus* Lange, 1972.

Only a single species from the Far East is known.

— With the preceding characterization (pl. 4: D—F).

Locus typicus: Kunasir, Kurili Islands, Soviet Union. Distribution: only the type-locality known

pacificus Lange, 1972

Species interpretabilis

1. pacificus Lange, 1972: 103, fig. E—1.

Superfamily 3: Ctenacaroidea Grandjean, 1954

Ctenacaridae Grandjean, 1954: 197.
Ctenacaroidea Grandjean: 1969: 141.

Hysterosoma without small notogastral sclerites: nearly completely covered by three large plates extending downwards to ventral side, although plates transparent and difficult to discern. All intervening narrow cuticle rugose. Hysterosomatic chaetom of two types: either all setae uniform in appearance: thin, setiform, or two pairs (d_2, e_1) conspicuously dark, thick, ciliate, all other pairs simple, excepting guttiform setae ps_1.

The superfamily consists of three families and six genera, all monotypical. All of them are known from the Palaearctic Region, but some of the species have been found also in other faunal regions.

Key to families

1 (2) Hysterosomatic chaetom heterogeneous: two pairs of thick, black and densely ciliate setae (d_2, e_2) and one pair (ps_1) guttiform to phylliform, shortened. Empodium of tarsi small, uncinate, eventually absent on tarsus I
family 1: **Ctenacaridae** Grandjean, 1954

2 (1) Hysterosomatic chaetom homogeneous, consisting of fine, simple and thin setae, without modified setae. Empodium large, hardly smaller than exterior claws and similar in shape.

3 (4) Two pairs of minute genital papillae present. Sensillus heavily incrassate, usually arcuate, blackish
family 3: **Aphelacaridae** Grandjean, 1954

4 (3) Three pairs of genital papillae present. Sensillus not or only slightly thickened, straight
family 2: **Adelphacaridae** Grandjean, 1954

Family 1: CTENACARIDAE Grandjean, 1954

Ctenacaridae Grandjean, 1954: 197.

Hysterosomatic chaetom heterogeneous: two pairs of thick, blackish, densely ciliate setae (d_2, e_1) and usually phylliform, shortened seta *ps*. Empodium of tarsi small, uncinate, eventually absent from tarsus I.

Four monotypical genera known.

Key to genera

1 (2) Tarsus of leg I with three claws. Setae d_2 and e_1 thick, with a wide margin as that of a hunting knife
genus 1: **Ctenacarus** Grandjean, 1939

2 (1) Tarsus of leg I with merely two claws. Setae d_2 and e_1 densely ciliate, without a wide margin.

3 (4) Seta ps_1 setiform, with large cilia (fig. 4: G)
genus 2: **Beklemishevia** Zachvatkin, 1945

4 (3) Seta ps_1 lanceolate or phylliform, smooth (fig. B: H)

5 (6) Anal region with four pairs of anal and five pairs of adanal setae. Tarsus of palp with one bifurcate seta

genus 3: **Neoctenacarus** Moritz, 1974

6 (5) Anal region with five pairs of anal and six pairs of adanal setae. Tarsus of palp with two bifurcate seta

genus 4: **Gilarovella** Lange, 1974*

Genus 1: CTENACARUS Grandjean, 1939

Ctenacarus Grandjean, 1939 : 543.
Grandjeanacarus Zachvatkin, 1949 : 67.
Ctenacarus Grandjean: 1954 : 248.

Sensillus slightly incrassate, falcate anteriorad. Setae d_2 and e_1 long, with wide margins as in a hunting knife; ps_1 widened, lanceolate. Empodium on tarsus of leg I large, only slightly differring from exterior claws.

Type-species: *Palaeacarus araneola* Grandjean, 1932.

Two species known, one of them occurring also in the Palaearctic Region.

— Median hysterosomatic sclerites fused with pygidial one, their border indicated by merely a thin line (pl. 4: A). No essential difference in length between setae c (pl. 4: C) d_2 originating nearly aligned with them, far anteriorly. Chaetom of anogenital region as in pl. 4: B.

Locus typicus: Tangier, Morocco. Distribution: North Africa, South Europe,Soviet Union (SET), South America

araneola (Grandjean, 1932)

Species interpretabilis

1. araneola Grandjean, 1932 : 417, figs. 5—7 (*Palaeacarus*).
araneola Grandjean; 1954 : 248, figs. 22—25.
araneola Schubart: 1968 : 254, figs. 25—26.

Genus 2: BEKLEMISHEVIA Zachvatkin, 1945

Beklemishevia Zachvatkin, 1945 : 60.
Beklemishevia Zachvatkin, 1945 : 675.
Beklemishevia Lange: 1975 : 43.

Sensillus slightly incrassate basally, otherwise thin, filiform, arcuate, very finely ciliate. Setae d_2 and e_1 thick, long, densely ciliate; seta ps_1 also densely ciliate, but essentially shorter. Empodium reduced on tarsus of leg I, therefore tarsus 2-clawed.

Type-species: *Beklemishevia galeodula* Zachvatkin, 1945.

Only the type-species known, described from the Soviet Union.

* Its relationship to the preceding genus needs further clarification; though separable on the basis of the description, it should be proven by comparisons.

— Hysterosomatic sclerites well differentiated, included areas narrow with a lineolate pattern (pl. 4: G). Position and proportions of seta *c* approximately as in preceding species.

Locus typicus: Kharkov, Soviet Union. Distribution: only the type-locality known.

galeodula Zachvatkin, 1945

Species interpretabilis

1. galeodula Zachvatkin, 1945: 60, figs. 1—10.
galeodula Zachvatkin, 1945: 675.
galeodula Lange: 1975: 43, figs. 26b—b.

Genus 3: Neoctenacarus Moritz, 1974

Neoctenacarus Moritz, 1974: 1.

Sensillus thin, but medially slightly widened, weakly fusiform, straight, directed slightly latero-posteriorad. Seta *ps* widely phylliform; setae d_2 and c_1 long, densely ciliate up to apex; four pairs of anal and five pairs of adanal setae present. Tarsus of palp with one bifurcate acanthoid.

Type-species: *Neoctenacarus hastilis* Moritz, 1974.

Only a single species known.

— Anterior exobothridial seta *(exa)* essentially longer than posterior one and prodorsal setae. No essential difference in length between the hysterosomatic setae not mentioned above (pl. 5: H, L). Sensillus visibly widening, especially in lateral view (pl. 5: J). All setae originating on ventral side (pl. 5: I) thin, simple, excepting genital seta of inner row. Legs 2-clawed, but exceptionally a vestigial empodium may occur (pl. 5: K). Length: 212—245 µm, width: 90—95 µm.

Locus typicus: Etzdorf, Saalkreis, GDR. Distribution: only the type-locality known

hastilis Moritz, 1974

Species interpretabilis

1. hastilis Moritz, 1974: 3, figs. 1—9.

Genus 4: Gilarovella Lange, 1974

Gilarovella Lange, 1974: 18.

Sensillus weakly widening, fusiform, latero-posteriorad. Seta ps_1 short, widely phylliform; setae d_2 and e_1 incrassate, long, densely ciliate. Tarsus of palp with two bifurcate acanthoids (fig. 4: H). Five pairs of anal and six pairs of adanal setae.

Type-species: *Gilarovella demetrii* Lange, 1974.

Only a single species known from the Soviet Union.

— Very similar to the type-species of the preceding genus. However, sensillus (fig. 4 : I) less elongate, paraaxial genital seta acutely pointed on both sides. Chaetom of tarsus (pl. 5 : K) of leg 1 also rather agrees with that of the preceding species. Length: 250—280 µm.

Locus typicus: Badhüz, Turkmenia, Soviet Union. Distribution: only the type-locality known

demetrii Lange, 1974

Species interpretabilis

1. demetrii Lange, 1974 : 8, figs. A-K.

Family 2: ADELPHACARIDAE Grandjean, 1954

Adelphacaridae Grandjean, 1954 : 198.
Adelphacaridae Grandjean: 1969 : 141.

Only a single genus known.

Genus 1: ADELPHACARUS Grandjean, 1952

Adelphacarus Grandjean, 1952 : 460.

Sensillus long, slightly incrassate, bacilliform. All hysterosomatic setae thin, only e_2 longer than the others, but on dissimilar, blackish, densely ciliate seta present. Empodia claw-shaped, hence tarsi apparently 3-clawed.

Type-species: *Adelphacarus sellnicki* Grandjean, 1952.

Only a single species known.

— Body short, not much longer than wide. Anterior and median shields of hysterosoma separated by a narrow zone bearing a lineate pattern. Chaetom of body and legs shown in pl. 2 : F—H.

Locus typicus: Abisko Östra, Sweden. Distribution: Sweden, GFR*

sellnicki Grandjean, 1952

Species interpretabilis

1. sellnicki Grandjean, 1952 : 460, figs. 1—3.

Family 3: APHELACARIDAE Grandjean, 1954

Aphelacaridae Grandjean, 1954 : 198.
Aphelacaridae Grandjean: 1969 : 141.

The family contains a single genus.

* A so far unpublished new datum: Ville, Nordrhein-Westfalen (GFR), leg. Dr. V. Hermosilla.

Genus 1: Aphelacarus Grandjean, 1932

Aphelacarus Grandjean, 1932: 412.
Aphelacarus Grandjean: 1954: 226.

Sensillus heavily incrassate, arcuate, usually falcate, dark. Hysterosomatic setae divers in length, none of a different colour or ciliation. Interlamellar seta conspicuously long. Empodia large, thick, the true claws thin, nearly setiform.

Type-species: *Parhypochthonius acarinus* Berlese, 1910.

Two species of the genus have been described; one is widely distributed (some authors regard it as cosmopolitan), while the second needs corroboration (Lange: 1975).

— Body very long, more than thrice longer than wide. Hysterosomatic sclerites transparent, smooth, their borders hardly discernible. Shape and chaetom of body and leg I shown in pl. 5: A—G. Length: 440 μm.

Locus typicus: Palermo, Sicily, Italy. Distribution: South Europe, North Africa, Soviet Union (TC, KZ, SMA), Japan. Allegedly cosmopolitan
acarinus (Berlese, 1910)

Species interpretabilis

1. acarinus Berlese, 1910: 219, fig. 42 *(Parhypochthonius)*
acarinus Grandjean: 1932: 412, figs. 1—4.
acarinus Grandjean: 1954: 226, figs. 11—21.
acarinus Van der Hammen: 1959: 9.

Species inquirenda

— rossicus Zachvatkin, 1945: 675.

Supercohort 2: Arthronota Balogh, 1972

Arthronota Balogh, 1972: 33.

Usually well or only exceptionally weakly chitinized species. Body well discernibly subdivided, notogaster separated by 1—3 furrows into 2—4 shields (*Na, Nm, Py:* pl. 6: A); if furrow absent, then body ptychoid, but ventral plate brachypilic (Mesoplophoroidea). Notogaster with 16 pairs (or with 1—2 pairs less) setae. Setae *d* or *e* arising occasionally in furrows, eventually only their insertion points recognizable, f_1 and f_2 invariably present as hairs. Notogastral setae often modified, in some groups phylliform, ensiform, umbellate, with long cilia, or resembling a palm-leaf or a feather. Occasionally two or more pairs of setae longer than the rest, rigid, more robust and movable, usually setae *e* and *f*. One or two pairs of exobothridial setae present. Oil gland absent, also slit-organs *iad* and *ian* reduced. Ventral plate of divers evolvement, but usually macropilic. A peranal plate present or absent. Legs with 1 femur.

On the basis of the presence or absence of ptychoidy, the taxon can be subdivided into two subgroups (cohorts).

Key to cohorts

1 (2) Body not ptychoid
cohort 1: **Euarthronota** Balogh et Mahunka, 1979
2 (1) Body ptychoid
cohort 2: **Arthroptyctima** Grandjean, 1967

Cohort 1: Euarthronota Balogh et Mahunka, 1979

Euarthronota Balogh et Mahunka, 1979 : 281.

Body never ptychoid, otherwise characterized as in the diagnosis of the supercohort.

Nine superfamilies are relegated to this cohort, most of them monotypical at the family level.

Key to superfamilies

1 (14) Notogastral and leg setae either simple, or if modified always palmate, plumose or phylliform, never umbelliform and never entirely covering body surface. If exceptionally some very large, palmate setae present *(Ovochthonius)*, then setae *c* and *d* always simple.
2 (7) Notogaster divided by a transverse furrow into two shields (pl. 40 : A).
3 (4) 16 pairs of notogastral setae present, all arising on surface of shields, none in furrow itself. Setae *c* and *d* arising on shield *Na*, setae *e*, *f*, *h* and *p* on shield *Py* (pl. 41 : C)
superfamily 4: **Eniochthonioidea** Grandjean, 1947
4 (3) Only 14 pairs of true notogastral setae present, 2 further pairs recognizable only by their insertion points within furrow.
5 (6) Shield *Na* with two rows of setae *(c* and *d)*, furrow with insertion points of setae *e*. Notogastral setae simple or at most widening like a blade, never T-shaped. Notogaster flattened, elongate or pentagonal, smooth or punctate
superfamily 3: **Hypochthonioidea** Berlese, 1910
6 (5) Shield *Na* with merely one row of seta *(c)*. Furrow with insertion points of setae *d*.* Notogastral setae T-shaped, branching, pectinately ciliate, covered with cerotegument grains (pl. 47 : A). Notogaster rounded, convex, with a polygonal ornamentation composed of chitinous laths
superfamily 6: **Sphaerochthonioidea** Grandjean, 1947
7 (2) Notogaster subdivided by 2 or 3 furrows into 3 or 4 shields.
8 (11) Notogaster at least with four pairs of setae essentially longer than the rest, rigid and movable, arising in the transverse furrows (mostly setae *e* and *f*) (pl. 48: A—J).

* The furrow is usually covered by the shield *Na*, therefore, the alveoli appear conspicuously in swollen specimens!

9 (10) Six pairs of long, rigid setae present: d_1,d_2,e_1,e_2, f_1 and f_2. Setae d_1 and d_2 accumbent to notogastral surface, setae *e* and *f* erect. Prodorsum with 2 or 3 black "eyes" (pigment spots) (pl. 48 : A)
superfamily 7: **Heterochthonioidea** Grandjean, 1954

10 (9) Only four pairs of long and rigid setae present: e_1,e_2,f_1 and f_2 (fig. 42:A). Prodorsum without "eye"-spote, setae different
superfamily 5: **Cosmochthonioidea** Grandjean, 1947

11 (8) Notogaster without rigid setae longer than the rest. All setae arising on the shields.

12 (13) Notogaster with 3 furrows and therefore with 4 shields. Shield *Na* bearing only setae *c*. No peranal plate present (pl. 6 : B)
superfamily 1: **Haplochthonioidea** Van der Hammen, 1959

13 (12) Notogaster with merely two furrows and the shields. Shield *Na* bearing two rows of setae *(c* and *d)*. A peranal plate present (pl. 11 : D)
superfamily 2: **Brachychthonioidea** Thor, 1934

14 (1) Notogaster and legs with large, partly phylliform partly umbelliform, expanded setae, covering major part of body (pl. 49 : A—C).

15 (16) Notogaster with 2 furrows and, therefore, with 3 shields. Seta *d* absent, hence notogaster with merely 14 pairs of setae, among them also long, blade-like setae also present. Six pairs of genital hairs
superfamily 8: **Atopochthonioidea** Grandjean, 1948

16 (15) Notogaster with 3 furrows and therefore with 4 shields. All notogastral setae evolved, hence 16 pairs present, all very large, umbelliform or fungiform. 8 pairs of genital setae
superfamily 9: **Pterochthonioidea** Grandjean, 1950

Superfamily 1: Haplochthonioidea Van der Hammen, 1959

Haplochthoniidae Van der Hammen, 1959 : 13.
Haplochthonioidea Balogh et Mahunka, 1979 : 282.

Body whitish, weakly chitinized. Especially notogaster elongated, narrow. Setae simple, thin, or slightly fusiform. Seta *c* arising on shield *Na*, seta *d* on Nm_1, seta *e* on Nm_2, and setae *f* and *h* on shield *Py*. Peranal plate absent.

Only a single family known.

Family 1: Haplochthoniidae Van der Hammen, 1959

Haplochthoniidae Van der Hammen, 1959 : 13.

Characterizable as the superfamily.

Two genera known from the Holarctic Region. The occurrences were reported from Europe, North Africa, Asia and North America.

Key to genera

1 (2) Rostral seta present. Ventral region with 7 pairs of genital, 4 pairs of anal and 4 pairs of adanal, setae (pl. 6 : B). Epimeral setal formula: 3—2—2—4

genus 1: **Haplochthonius** Willmann, 1930

2 (1) Rostral seta absent. Ventral region only with 4 pairs of genital, 3 pairs of anal and 3 pairs of adanal setae present (pl. 6 : K). Epimeral setal formula: 3—2—2—3

genus 2: **Amnemochthonius** Grandjean, 1948

Genus 1: Haplochthonius Willmann, 1930

Haplochthonius Willmann, 1930: 2 (subgenus).
Haplochthonius Grandjean, 1947: 224.

Narrow, elongate, minute, colourless animals. Prodorsum with 5 pairs of setae, also rostral seta developed. Notogaster with 3 furrows, hence with 4 shields. Shield *Na* with setae *c* (4 pairs), Nm_1 with setae *d* (2 pairs), Nm_2 with setae *e* (2 pairs), and shield *Py* with setae *f*, *h* and *ps* (a total of 7 pairs); notogaster, therefore, with 16 pairs of setae in all. Epimeral setal formula: 3—2—2—4. Seven pairs of genital, no aggenital, 4 pairs of anal and 4 pairs of adanal setae present. All legs monodactyle. Palp without long, ribbon-like hair.

Type-species: *Cosmochthonius (Haplochthonius) simplex* Willmann, 1930.

The two species of the genus is known from many points of the Palaearctic Region.

Key to species

1 (2) All prodorsal and notogastral setae thin, setiform (pl. 6 : A, C); setae h_1 arising about as far from each other as from h_2. — Prodorsal setae short, rostral and lamellar setae turning towards each other, their apices far from meeting, that is, they are shorter than half the distance between their points of insertion. Sensillus elliptical. Chaetom of anogenital region as in pl. 6 : B Length: 270—281 μm, width: 120—131 μm.

Locus typicus: Valkeveen, Netherland. Distribution: Europe, North Africa, Soviet Union (CET, SET, WS, SMA), North America

simplex Willmann, 1930

2 (1) Prodorsal and notogastral setae widening, slightly fusiform or like a knife-blade (pl. 6 : D). Setae h_1 arising considerably nearer to each other than distance between h_1 and h_2; these latter arising immediately near setae ps_3. — Prodorsal setae, long, robust, essentially longer than distance between their points of insertion, their apices crossing each other. Sensillus clavate. Chaetom of anogenital region as shown in pl. 6 : E Length: 255—270 μm, breadth: 120—125 μm.

Locus typicus: Basiluzzo, Isole Eolie, Italy. Distribution: France, Italy, Algeria, Tunesia

sanctaeluciae Bernini, 1973

Species interpretabiles

1. sanctaeluciae Bernini, 1973 : 360, fig. 3a.
2. simplex Willmann, 1930 : 2, figs. 1—2.
   simplex Grandjean: 1947 : 224, figs. 2—6.
   simplex Bernini: 1973 : 360.

Genus 2: AMNEMOCHTHONIUS Grandjean, 1948

Amnemochthonius Grandjean, 1948 : 30.

Body weakly chitinized, with rather extensive rugose surfaces between shields and especially on ventral side. Notogaster subdivided by 3 furrows into 4 shields. 16 pairs of notogastral setae present: setae *c* (4 pairs) on shield *Na*, setae *d* (2 pairs) on Nm_1, setae *e* (2 pairs) on Nm_2, setae *f*, *h* and *ps* (7 pairs) on shield *Py*; all very short. Rostral setae reduced, thus prodorsum with merely four pairs of setae. Genital plates with four pairs, anal plates with three pairs of setae; aggenital setae absent, adanal setae (3 pairs) originating on the rugose surfaces, adanal plate not differentiated. Epimeral setal formula: 3—2—2—3. Legs monodactyle. Palps apically with long, ribbon-like hairs.

Type-species: *Amnemochthonius taeniophorus* Grandjean, 1948.

Only a single species known.

— An elongated, whitish species. Rostrum widely rounded, rostral seta absent, lamellar seta arising far anteriorly, near rostrum. All body setae short, thin, simple (pl. 6 : F, G, I, K). Sensillus basally thin, terminally heavily clavate, ciliate. Palp with an extraordinarily large and modified seta (pl. 6 : J). Famulus (ε) on tarsus of leg I widening bifurcately (pl. 6 : H). Length: 175—190 μm.

Locus typicus: Gruissan, Aude, France. Distribution: France

taeniophorus Grandjean, 1948

Species interpretabilis

1. taeniophorus Grandjean, 1948 : 30, figs. 1—3.

Superfamily 2: BRACHYCHTHONIOIDEA Thor, 1934

Brachychthoniidae Thor, 1934 : 115.
Brachychthonoidea Grandjean: 1969 : 142.

Among the smallest Oribatids, their length usually fluctuating between 120—240 μm. Nearly always weakly pigmented, light, opalescent to dirty white, yellowish white to lemon or orange yellow. For their identification, a detailed knowledge of their morphology is indispensable.

Prodorsal shield very distinct, its surface emitting four pairs of setae *(ro, le, in, exa)*, but insertion point as pores of pair fifth *(exp)* can be recognized also, laterally near bothridia. Stalk of sensillus and its fusiform or rarely sperical capitate section well differentiated, latter part generally heavily ciliate, cilia often modified into spines and arranged into rows.

Rostrum wide; either rounded and smooth or marginally framed with a row of teeth. Prodorsal sides eventually also dentate, occasionally appearing in groups. Slightly anterior to bothridium, nearly always a robust lateral tubercle *(op)* present, exobothridial seta arising on its anterior rim. Surface of prodorsum (and of notogaster) smooth or with a characteristic sculpture consisting of chitinous laths or of well defined fields; these latter appearing usually paired or occasionally fused along median line in notogaster and thus arranged in a single row. Shield *Na* laterally with an anelliform spot framed with robust chitinous laths, each spot surrounded by these fields like a rosette. If no heavy sculpture present, the interbothridial region with four (rarely three) pairs of light, more or less round, small spots: their number, position and shape represent, like those of the small fields, important specific characteristics. Similar spots discernible also on notogasters without heavy sculpture.

Notogaster subdivided into three shields by two furrows. Anterior notogastral shield *(Na)* corresponds to segments *C* and *D*, median notogastral shield *(Nm)* to segment *E*, pygidium *(Py)* to segments *F*, *H* and *Ps*. The notogastral shields join the unsclerotized pleura which may also bear shields, thus posteriorly pleural shields Pl_1 and Pl_2 and anteriorly prodorsal pleural shields Ppl_1 and Ppl_2. Shield Pl_2, more or less fusing with segments *F*, *H* and *Ps*, composes the pygidium, but on the border of their coalescence a slit, the suprapleural incisure *(IcPy)* is also often discernible. The suprapleural plates are especially characteristic to the family Brachychthoniidae. A maximum of four can be present along the notogastral border, and two on the proterosoma; according to their position, they correspond to parts of the segments *C*, *D*, *E* and *F*, and therefore indicated as *SpC*, *SpD*, *SpE* and *SpF*. The suprapleural plate of the proterosoma is designed as *SpPr*. The suprapleural plates may partly coalesce with the shields *Na* and *Nm*, partly become much reduced. Their number, psition and grade of fusion are important generic features.

In addition to the notogastral sculpture discussed above, longitudinal sutures may be decurrent parallel with the border of the shield *MaNa*, *MaNm*, *MaPy)*. The shields carry 16 pairs of setae, *c* and *d* on shield *Na* (seta c_3 occasionally on plate *SpC*), *e* on shield *Nm*, setae *f*, *h* and *ps* on shield *Py*. Five pairs of lyrifissures *(ia, im, ip, ih* and *ips)* may be recognized in the pleural region, but often on the margins of the suprapleural or pleural plates.

The coxosternal region has some importance in the differentiation of the supraspecific categories. The sejugal furrow bisects it into two well distinguishable parts, but medially a longitudinal unsclerotized sternal stripe also separates the right and left epimeres. The inner borders of the epimeres are often heavily sclerotized. The epimeral setal formula is 3—1—3—4 or 3—1—3—3.

In the anogenital region, usually genital, aggenital, peranal, anal and adanal plates can be distinguished in the sexually mature specimens. The aggenital plate may be reduced, and the aggenital setae then arising on the unsclerotized cuticle. The adanal plates often fuse caudally and then form a single, horseshoe-shaped plate. The peranal plate is always separated from the anal one. In general, there are 7 (—6) pairs of genital, 1 pair of aggenital, 1 pair of peranal, 2 pairs of anal and 3 pairs of adanal setae present. Excepting ad_2 and ad_3 all setae are simple while the two former pairs often widen and take the shape of a blade.

The legs and the chaetom of the legs vary less in the several lower taxa, therefore, they are characteristic only to the family. All legs are monodactyle. Leg chaetotaxy:

I: 0 — 3 — 3+2 — 5+1—16−18+3
II: 0 — 4 — 3+1 — 4+1—12−15+1
III: 2 — 3 — 2−3+1 —3+1—11−13+0
IV: 1 — 2−3 — 2−3+1 — 4+1— 11

An accessory bristle near solenidium φ of tibia always present, but hardly discernible.

The known taxa are relegated to a single family.

Family 1: Brachychthoniidae Thor, 1934

Brachychthoniidae Thor, 1934: 115.
Brachychthoniinae Balogh, 1943: 21.
Brachychthoniidae Grandjean: 1947: 223.
Brachychthoniidae Van der Hammen: 1959: 18 (*Brachychthoniidae* Grandjean, 1947!).
Brachychthoniidae Niedbala: 1974: 1.

The species can be found in the whole world. So far 10 genera have been described, containing about 100 species. Every known genus is represented in the Palaearctic Region, with most of the species described from this area.

Key to genera*

1 (8) Seta d_2 in a marginal position, aligned with the longitudinal row of setae *cp*—e_2. Two or more pairs of suprapleural plates present; if two present, then epimeres 4 emitting four pairs of setae.

2 (3) Four pairs of suprapleural plates present. Longitudinal suture of shield *Na* decurrent within points of insertion of setae c_2 and d_2 (pl. 8 : B)
genus 1: **Eobrachychthonius** Jacot, 1936

3 (2) Only two or three pairs of suprapleural plates present. No longitudinal suture present within points of inseriton of setae c_2 and d_2 on shield *Na*.

4 (5) All three adanal pairs of setae thin, simple, setiform (pls. 19 : Fand 20 : I, L), none hypertrophic, blade-shaped. Two pairs of suprapleural plates *(SpC, SpE)* discernible (pl. 19 : F). Fields of shield *Na* appearing in pairs and two groups also in median line, separated by a transverse stripe
genus 3: **Brachychthonius** Berlese, 1910

5 (4) Two pairs of adanal setae *(ad_2 and ad_3)* widened like a blade. Either three pairs of suprapleural plates present (pl. 10 : K), or merely two pairs, but then fields of shield *Na* fused into a uniform longitudinal zone in median line.

6 (7) Three pairs of suprapleural plates *(SpC, SpE, SpF)* recognizable (pl. 10 : K). Fields of sheild *Na* rather differentiated: mostly paired in median line, but in any case segregated into an anterior and a posterior group
genus 2: **Brachychochthonius** Jacot, 1938

* In the differentiation of the genera, the number of suprapleural plates is of decisive importance, but they are not always easy to recognize. Therefore, we submit two keys: one based mainly on the dorsoventral features and one the pleural and suprapleural plates (p. 54).

7 (6) Only two pairs of suprapleural plates *(SpC, SpE)* recognizable (pl. 21 : E). Fields on shield *Na* fused longitudinally and thus forming a narrow band (pl. 21 : A)

genus 4: **Poecilochthonius** Balogh, 1943

8 (1) Seta d_2 not in a marginal position, therefore, originating considerably more mediad from setae cp—e_2 aligned in a row (pl. 35 : A). Suprapleural plates absent or only one pair present. If exceptionally two suprapleural plates present, then only three pair of setae arising on epimeres 4 *(Neobrachychthonius)*.

9 (16) Seta ad_2 hypertrophic, widening like a blade. 0—2 pairs of suprapleural plates present; if only one discernible, then representing either *SpC*, or if *SpE* then genital plates bearing only six pairs of genital setae.

10 (11) All three shields of notogaster with a heavy and definite ornamentation (pl. 35 : A). Genital plates with six pairs of setae. One pair of suprapleural plate *(SpE)* recognizable; adanal plates not fused posteriorly (pl. 35 : B)

genus 7: **Synchthonius** Van der Hammen, 1952

11 (10) Notogaster without ornamentation, at most some arcuate lines or small pairs of spots occasionally discernible. Genital plates with seven pairs of setae. 0-2 suprapleural pairs of plates present; if only one pair present, then representing *SpC*. Adanal plates fused posteriorly; if separate, then epimeres 4 with three pairs of setae.

12 (15) Adanal plates fused posteriorly therefore anal plates embraced by a U-shaped plate (pl. 38 : B). Suprapleural plates either wholly reduced or 2 pairs recognizable.

13 (14) Two free pairs of suprapleural plates present *(SpC, SpE)*. Epimeral setal formula: 3—1—3—3

genus 5: **Neobrachychthonius** Moritz, 1976

14 (13) All suprapleural pairs of plates reduced or adnate to their shields and forming their lobes. Epimeral setal formula: 3—1—3—4

genus 6: **Liochthonius** Van der Hammen, 1959

15 (12) Adanal plates not fused posteriorly (pl. 37 : C), both discrete clearly discernible. One pair of suprapleural plates *(SpC)* present. Epimeral setal formula: 3—1—3—3

genus 8: **Paraliochthonius** Moritz, 1976

16 (9) All three pairs of adanal setae thin and simple. Only one pair of suprapleural plate *(SpE)* present, but genital plates always with seven pairs of setae.

17 (18) Prosomatic suprapleurite *(SpPr)* well differentiated (pl. 38 : C). Prodorsal and notogastral setae long and elongately ciliated. Shield *Py* with setae ps_1, ps_2 and ps_3 aligned in a transverse line, therefore, distance between setae ps_1 nearly double of that between setae ps_2 and ps_1 (pl. 38 : D)

genus 9: **Mixochthonius** Niedbala, 1972

18 (17) No prosomatic suprapleurite present. Prodorsal and notogastral setae short, never ciliate. Shield *Py* setae ps_1 arising far anterior to setae ps_2, both nearly aligned in a longitudinal line (pl. 39 : E)

genus 10: **Verachthonius** Moritz, 1976

Identification key to genera based on pleural and suprapleural plates and their excrescences

1 (2) Four pairs of suprapleural plates differentiated. Margin of shield *Na* with a longitudinal rib, setae c_2 and d_2 arising beyond it, adjacent to body margin
genus 1: **Eobrachychthonius** Jacot, 1936

2 (1) At most three suprapleural plates differentiated, plate *SpD* invariably absent. Shield *Na* without a longitudinal rib within points of insertion of setae c_2 and d_2.

3 (4) Three pairs of suprapleural plates present. In addition to a heavy dorsal sculpture, two pairs of hypertrophic adanal setae present
genus 2: **Brachychochthonius** Jacot, 1938

4 (3) Merely 0-2 suprapleural plates present.

5 (10) Two pairs of suprapleural plates present.

6 (7) All adanal setae thin, simple. Notogaster with a heavy sculpture, shield *Na* without a longitudinal stripe of fused medial fields
genus 3: **Brachychthonius** Berlese, 1910

7 (6) At least one pair of adanal setae hypertrophic, resembling a blade.

8 (9) Two pairs of adanal setae hypertrophic (ad_2, ad_3). A suprapleural incision present. Notogaster with heavy sculpture, shield *Na* with medial fields fused into a longitudinal stripe
genus 4: **Peocilochthonius** Balogh, 1943

9 (8) Only one pair of adanal setae (ad_2) hypertrophic. No suprapleural incision present. Notogaster smooth, or only with some small spots discernible
genus 5: **Neobrachychthonius** Moritz, 1976

10 (5) No or only one suprapleural plate present.

11 (12) No free suprapleural plate present: either reduced or fused with corresponding sclerite and appearing as its lobe
genus 6: **Liochthonius** Van der Hammen, 1959

12 (11) One suprapleural plate present.

13 (14) Suprapleural plate *SpC* present. Epimeral setal formula: 3—1—3—3. No suprapleural incision
genus 8: **Paraliochthonius** Moritz, 1976

14 (13) Suprapleural plate *SpE* present. Epimeral setal formula: 3—1—3—4. Suprapleural incision present.

15 (16) One pair of adanal setae hypertrophic (ad_2). Genital plates only with six pairs of genital hairs. Adanal plates discrete, not fusing into a U-shaped plate. A heavy dorsal sculpture present
genus 7: **Synchthonius** Van der Hammen, 1952

16 (15) All adanal setae thin and simple. Seven pairs of genital setae present. Adanal plates fused posteriorly into a single U-shaped plate. Notogaster without sculpture.

17 (18) Prosomatic suprapleurite present. Dorsal setae long, heavily ciliate
genus 9: **Mixochthonius** Niedbala, 1972

18 (17) Prosomatic suprapleurite absent. Dorsal setae short, simple, smooth
genus 10: **Verachthonius** Moritz, 1976

A synoptic survey of the Brachychthoniid genera

Genera \ Characteristics	Marginal seta ad_2	Marginal suture	Adanal hypertrophy	Adanal coalescence	Present suprapleural plates	Prosmatic suprapleurite	Rostral group of teeth	Genital setae	Epimeral formula	Suprapleural incision
Eobrachychthonius	+	+	— + —	—	C D E F	+	—	7	3 1 3 4	+
Brachychochthonius	+	—	— + +	—	C E F	—	—, +	7	3 1 3 4	+
Brachychthonius	+	—	— — —	—	C E	—	—	7	3 1 3 4	+
Poecilochthonius	+	—	— + +	—	C E	—	+	7	3 1 3 4	+
Neobrachychthonius	—	—	— + —	+	C E	—	—	7	3 1 3 3	—
Liochthonius	—	—	— + —	+	—	— +	—	7	3 1 3 4	—
Synchthonius	—	—	— + —	—	E	—	+	6	3 1 3 4	+
Paraliochthonius	—	—	— + —	—	C	—	—	7	3 1 3 3	—
Mixochthonius	—	—	— — —	+	E	+	—	7	3 1 3 4	+
Verachthonius	—	—	— — —	+	E	—	—	7	3 1 3 4	+

Genus 1: EOBRACHYCHTHONIUS Jacot, 1936

Eobrachychthonius Jacot, 1936: 24.
Eobrachychthonius Niedbala: 1972: 664.
Eobrachychthonius Van der Hammen: 1959: 18.

Comparatively large-sized, wide and flat, vivid yellow to dark yellow species. Prodorsal margins without teeth (not dentate). Notogaster with sporadic round or oval spots, their number occasionally much reduced. Seta d_2 in marginal position, aligned with setae c_2—e_2. A longitudinal median furrow along insertion points of setae c_2, *cp* and d_2. Notogaster with four suprapleural plates *(SpC, SpD, SpE, SpF)*, also prosomatic suprapleurites discernible. Suprapleural incision *(IcPy)* long, decurrent nearly to points of origin of setae h_3. Adanal plates not fused posteriorly. Epimeral setal formula: 3—1—3—(2—4). Seven pairs of genital setae present. Adanal seta ad_2 considerably longer than its mates, usually widened like a blade.

Type-species: *Eobrachychthonius sexnotatus* Jacot, 1936 = *Brachychthonius latior* Berlese, 1910*.

The species of the genus are distributed in the whole world, but there are many dubious species of uncertain position and several erroneous interpretations and thus a revision is indispensable. With some complementation and corrections, we follow D. Krivolutsky's concepts.

*See Van der Hammen: 1959: 18.

Key to species

1 (2) Notogastral setae extremely long, thin, simple, seta c_1 extending beyond insertion point of seta d_1, while e_1 beyond that of seta f_1 (pl. 7 : A), seta e_1 is 40 μm long. — Sensillus fusiform, densely ciliate. Length: 189 μm, with: 98 μm.

Locus typicus: Szakonyfalu, Hungary. Distribution: only the type-locality known **longisetosus** Csiszár, 1961*

2 (1) Notogastral setae shorter, setae c_1 far from approaching points of insertion of setae d_1.

3 (4) Epimeral setal formula: 3—1—3—2(3). — Rostral seta on prodorsum shorter than lamellar setae; insertions of these latter connected by a chitinous lath. Interbothridial region with four pairs of very large, transversely elliptical spots. Stalk of sensillus short, clavus widely fusiform. Notogastral surface with numerous spots, but discernible in midline only on shield *Py*. Notogastral setae acicular, about equal in length excepting marginal setae (f_2 and h_2) on shield *Py* (pl. 8 : D). Aggenital plate whole, approximately triangular. Shapes and chaetom of ventral plates as in pl. 8 : E. Length: 224—246 μm, width: 144—160 μm.

Locus typicus: Kastritsion, Achaia, Greece. Distribution: only the type-locality known **similis** Mahunka, 1979

4 (3) Epimeral setal formula: 3—1—3—4.

5 (6) A straightly truncate projecting part differentiated from arc of anterior margin of rostrum (pl. 7 : B). — Lamellar seta arising near rostrum on prodorsum, long, projecting far beyond anterior margin of body; rostral and interlamellar setae essentially shorter. Notogaster with few spots. Length: 250 μm, width: 180 μm.

Locus typicus: Sint Pietersberg, Netherlands. Distribution: only the type-locality known **mooseri** (Van der Hammen, 1950)

6 (5) Anterior arc of rostrum contiguously convex, without any projecting feature. Lamellar setae arising far from rostrum and shorter than either rostral or interlamellar setae.

7 (8) Aggenital plate whole, not subdivided (pl. 7 : D). — Rostral, lamellar and interlamellar setae about equal in length. Interbothridial region with four very large spots, meeting obliquely each other. Notogaster with comparatively many spots (pl. 7 : C); setae relatively short, d_1 by far not approaching point of insertion of e_1. Length: 270—335 μm, width: 166—200 μm.

Locus typicus: Lake City, Florida, USA. Distribution: the Holarctic Region** **latior** (Berlese, 1910)

* Unfortunately the type-specimen was annihilated and the species has not been collected again; many characteristics, indispensable in modern diagnoses, are missing in the original description which may result in the reallocation of the species.

** For both this and the next species the distribution data are questionable and need careful checking since in this genus misidentification is especially frequent.

8 (7) Aggenital plate subdivided, aggenital seta originating on the posterior, smaller part (pl. 7 : E).

9 (10) Notogaster with some few spots only (pl. 7 : F): two lateral pairs of shield *Na*, and one median pair of shield *Nm*, absent. — Dorsal setae of body comparatively short, interlamellar seta of prodorsum essentially shorter than distance between interlamellar setae; on notogaster d_1 far from approaching insertion point of seta e_1. Length: 199—233 μm, width: 114—147 μm.

Locus typicus: Svartberget, Sweden. Distribution: Sweden, Finland, Poland, Roumania, Soviet Union (SET, KZ)

borealis Forsslund, 1942

10 (9) Notogaster with many spots, every shield also with median spots (pl. 8 : A). Dorsal setae of body comparatively long, interlamellar setae of prodorsum nearly twice longer than lamellar ones, and a considerable difference in length present also between rostral and lamellar setae (pl. 8 : B). Interlamellar setae longer than distance between their insertion points. On notogaster, seta d_1 reaching point of origin of e_1. Chaetom of ventral side as in pl. 8 : C. Length: 250—335 μm, width: 166—200 μm.

Locus typicus: Burgst, near Breda, Netherlands. Distribution: Holarctic Region, South America, Chile

oudemansi Van der Hammen, 1952

Species interpretabiles

1. borealis Forsslund, 1942: 2, figs. 1—2.
   borealis Niedbala: 1972: 468, fig. 16.
   borealis Niedbala: 1976: 22.
2. latior Berlese, 1910: 220, pl. 19, fig. 38 *(Brachychthonius)*.
   latior Forsslund: 1957: 584, figs. 1—3.
   latior Van der Hammen: 1959: 18.
   latior Niedbala: 1972: 469, fig. 17.
   latior Niedbala: 1976: 23.
   brevis (Michael, 1888) var. *glabra* Thor, 1930: 57 (Niedbala: 1972: 469).
   sexnotatus Jacot, 1936: 24, figs. 1: 3—4. (Van der Hammen: 1959: 18).
   grandis Sellnick, 1944: 40 (Niedbala, 1972: 469).
3. longisetosus Csiszár, 1961: 447, fig. 1.
4. mooseri Van der Hammen, 1950: 130, figs. 1a—b *(Brachychthonius)*.
   mooseri Van der Hammen: 1952: 15, fig. 1a.
5. oudemansi Van der Hammen, 1952: 17, fig. 1b.
   oudemansi Niedbala: 1972: 471, fig. 18.
   oudemansi Niedbala: 1976: 24.
   laetepictus Willmann (nec Berlese) 1931: 101 (Niedbala: 1972: 471).
   argentinensis Hammer, 1958: 19, figs. 12, 12a (Niedbala: 1972: 471).
6. similis Mahunka, 1979: 547, figs. 1—2.

Genus 2: BRACHYCHOCHTHONIUS Jacot, 1938

Brachychochthonius Jacot, 1938: 130.
Brachychochthonius Niedbala: 1972: 664.
Brachychochthonius Moritz: 1976: 275.
Sellnickochthonius D. Krivolutsky, 1964: 395 (Niedbala: 1972: 664).

Yellowish, well chitinized species with a heavy prodorsal and notogastral pattern. Rostral margin minutely dentate, but a lateral group of teeth occurring in only some few species. Three pairs of suprapleural plates *(SpC, SpE, SpF)* discernible on sides of notogaster: prosomatic suprapleurite absent. Suprapleural incision of varying depth, but rarely approaching base of seta h_3. Seta d_2 in marginal position, aligned with setae c_2—e_2. Epimeral setal formula: 3—1—3—4. Genital plates comparatively small, sperated from sternocoxal region by a wide zone; seven pairs of genital setae present. Peranal plates gradually narrowing towards their anterior ends. Adanal plates not fusing posteriorly, but their apices meeting; two pairs of adanal setae (ad_2, ad_3) hypertrophic, widened like blades.

Type-species: *Brachychochthonius jugatus* Jacot, 1938.

The species of the genus have so far been extensively mixed with those of nearly related genera, and their correct interpretation was made possible only by Moritz's revison. All previous distribution data must be used with great precaution; 27 species are known from the Palaearctic Region, some with overlapping borders and exhibiting an enormous range of distribution.

Key to species

1 (40) All notogastral setae thin, at most slightly fusiform, simple; either smooth or sparsely ciliate.

2 (3) Prodorsum and notogaster completely smooth, dorsal setae extraordinarily long (pl. 10 : E); interlamellar setae longer than distance between their points of origin; notogastral seta c_1 extending beyond insertion of d_1, seta d_1 beyond that of e_1, and e_1 beyond that of f_1. — Rostrum finely concave. Proportions according to length of prodorsal setae: $exa < ro < le < in$. Stalk of sensillus (pl. 10 : F) long, terminally elongately fusiform with comparatively long cilia. A well discernible transverse lath behind seta h_1 on notogaster. Setae basally characteristically arcuate (pl. 10 : G). Length: 132 μm, width: 79 μm.

Locus typicus: Kagoshima-ken, Noto Peninsula, Japan. Distribution: only the type-locality known

planus (Chinone, 1974)*

3 (2) At least some field portions or details discernible on prodorsum and notogaster. Setae shorter, not reaching their respective points of insertion.

4 (21) Field-pairs in median line of shield *Na* coalesced, generally forming two larger fields; exceptionally major part of fields evanescent; at most field-pair 6 remaining as a distinct pair.

5 (8) Setae with long bilateral cilia.

6 (7) Notogastral setae uniformly and symmetrically ciliate (pl. 9 : A—B), number of cilia: 5—6 pairs. Sensillus narrowly fusiform, head especially elongated with

* The species has been described on the basis of a single holotype, its generic assignment is problematic. The immature or teratological state of the specimen cannot be precluded!

a longitudinal row of 10—12 spines. — Prodorsal pattern heavy, rostral portion without unpaired median field. Prodorsal setae similar to notogastral ones. Notogastral fields, especially median ones, with sinuous and sharply defined borders. Ring-spot of shield *Na* reduced. Length: 136—153 μm, width: 75—85 μm.

Locus typicus: Bent Creek "Exp. Forest", Buncomb Co., N. C., USA. Distribution: USA, Japan

lydiae Jacot, 1938

7 (6) Notogastral setae of divers shape, partly asymmetrically ciliate, partly with shorter cilia (pl. 9: D, F), number of cilia: 2—4 pairs. Sensillus (pl. 9: E) terminally widely fusiform, considerably shorter than in preceding species, with a longitudinal row of 7—8 spines. — Prodorsal pattern weak, especially in basal half, but fields well delimited anterior to lamellar setae; also an unpaired median field present in front. Prodorsal setae similar to notogastral ones. Notogastral pattern weak, fields with partly obscure but straight borders. Posterior field-group of shield *Na* differentiated, ring-spot well developed. Length: 134—137 μm width: 67—68 μm.

Locus typicus: Mitsukaido, Ibaraki-ken, Japan. Distribution: only the type-locality known

japonicus (Chinone, 1974)

8 (5) Notogastral setae smooth, or only a few with minute cilia.

9 (10) Border of notogastral fields sinuous and sharp (pl. 10: A). Prodorsal fields granulous, notogastral ones glabrous. Margin of posterior half of body with a characteristic, discrete apophysis (pl. 10: D). — Seta *ro* longest of all prodorsal setae; c_1 and d_1 of notogaster equal in length, but shorter than setae e_1 (pl. 10: C), f_1 and h_1. Rostral surface of prodorsum without an unpaired field. Notogastral fields often asymmetrical. Capitate fusiform half of sensillus (pl. 10: B) longer than stalk, densely and verticillately spinose. Length: 150—153 μm, width: 83—91 μm.

Locus typicus: Kosuicho, Ishikawa-ken, Noto peninsula, Japan. Distribution: only the type-locality known

gracilis (Chinone, 1974)

10 (9) Border of notogastral fields not sinuous, often obsolete. Posterior margin of notogaster without a robust, projecting apophysis.

11 (12) Rostral part of prodorsum with two paired and one unpaired, well delimited, field (pl. 11: B), enclosed lath, therefore, Y-shaped. Median member of posterior field-group on shield *Na* cordiform (pl. 11: A), last one not fused, therefore, retaining its paired status. — Prodorsal setae *ro* longer than the others, like seta e_1 on notogaster. Anterior field-group of shield *Na* with the original field-pairs still discernible owing to their constrictions. Sensillus (pl. 11: C) terminally slightly asymmetricall fusiform, bearing 5—6 robust spines per row. Length: 146—163 μm, width: 80—97 μm.

Locus typicus: Ryugahara "swampy plain", Mt. Chokai, Akita-ken, North Japan, Japan. Distribution: only the type-locality known

miyauchii Chinone, 1978

12 (11) Rostral part of prodorsum with at most one paired and one unpaired field, their enclosed lath or line arcuate, not Y-shaped. Also posterior field-group of shield *Na* uniform, whole, or completely reduced.

13 (18) The two large median field-groups of shield *Na* well delimited, their anterior and posterior borders sharp and usually also 1—2 further transverse lines visible within field-groups proper. Median field discernible also on shield *Nm*.

14 (15) Anterior fused field-group of shield *Na* (pl. 10 : I) with one constriction, continuing as a weak transverse line. Rostral line (pl. 10 : H) short, medially far from base of rostral hair. — Prodorsal fields well discernible. Rostral setae nearly twice longer than all other prodorsal setae (pl. 10 : K). Also lateral fields of notogaster, especially those of shield *Na*, well defined. Posterior median field-group of shield *Na* slightly widening posteriorad. Spines of sensillus (pl. 10 : J) comparatively robust. Length: 170—195 µm, width: 72—95 µm.

Locus typicus: Västerbotten, Degersfors, Svartberget, Sweden. Distribution: Holarctic Region

immaculatus Forsslund, 1942

15 (14) Anterior fused field-group of shield *Na* with two conspicuous constrictions (pl. 11 : E). Rostral line long, sigmoid and reaching rostral margin (pl. 11 : G).

16 (17) Anterior field in midline of shield *Na* with one conspicuous bordering line, posterior field considerably narrower in front than at its base (pl. 11 : E). No robust transverse lath present behind points of origin of setae h_1 and h_2. — Prodorsal pairs of fields comparatively weakly developed. Spines of sensillus (pl. 11 : F) very small. Ring-spot of shield *Na* small, laterally open, sharp borders of a part of lateral fields discernible. Length: 193—212 µm, width: 95—103 µm.

Locus typicus: Putsche Moer, Calmpthout, Belgium. Distribution: Belgium, Sweden, GDR

formosus (Cooremann, 1947)

17 (16) Anterior field in midline of shield *Na* with two transverse lines, posterior field essentially broader in front than at its base (pl. 12 : A). A robust transverse lath visible behind insertion points of setae h_1 and h_2. — Prodorsal field-pairs well developed, heavily punctate. Spines of sensillus (pl. 12 : B) considerably larger than in the preceding species. Ring-spot of shield *Na* large, rather irregular in shape, closed. Borders of lateral fields obsolete, usually unrecognizable. Length: 177—197 µm, width: 97—104 µm.

Locus typicus: Mt. Chokai, Akita-ken, North Japan. Distribution: only the type-locality known

aokii (Chinone, 1974)

18 (13) Fields of shield *Na* considerably reduced; posterior border of anterior median field-group and anterior border of posterior field-group absent, or posterior field-group entirely obsolete. No median field present on shield *Nm*.

19 (20) Shield *Na* medially with no more than a distinct and two weaker transverse lines in place of anterior field-groups (pl. 12 : E). Interlamellar hairs comparatively long, longer than half distance between their bases and extend beyond anterior margin of basal group of field-pairs on prodorsum (pl. 12 : H). — Rostral lines enclosing a narrow calyciform space (pl. 12 : G). Notogastral

setae relatively long, seta f_1 reaching insertion point of ps_1. Length: 184—194 μm, width: 96—104 μm.

Locus typicus: "Prophet Elias" Cave, Ossa massif, Thessaly, Greece. Distribution: only the type-locality known

guanophilus Mahunka, 1979

20 (19) Shield *Na* medially with both field-groups recognizable, but posterior border of anterior field-group and anterior one of posterior field-group not visible (pl. 12 : J). Interlamellar hairs shorter, not longer then one-third distance between their bases, and extend only to anterior margin of basal group of field-pairs on prodorsum. — Rostral setae enclosing a wide, pan-shaped spece (pl. 12 : I). Notogastral setae also shorter, seta f_1 not reaching insertion point of ps_1 behind it. Length: 184—193 μm, width: 92—100 μm.

Locus typicus: "Ifri El Kaid", Grand Atlas, Morocco. Distribution: only the type-locality known

maroccanus Mahunka, 1980

21 (4) Most field-pairs, but at least four of the original six separated in midline of shield *Na*, therefore, a longitudinal line or lath present between the two members of the pairs.

22 (25) Midline of shield *Na* with heavily punctate and usually nearly rounded fields, removed from each other, but members of pair 5 closely adjacent and often fused, but their transverse suture still discernible (pl. 13 : A). Prodorsum laterally with a deep incision behind rostrum.

23 (24) Shield *Na* anteriorly with only two pairs of rounded fields behind each other (pl. 13 : A). — All prodorsal setae very short, but considerable differences in length between notogastral setae, f_1 being logest of all (pl. 13 : C). Median fields of shield *Nm* also paired. Length: 175—191 μm, width: 90—93 μm.

Locus typicus: Nadap, Velencei Mts., Hungary. Distribution: Palaearctic Region

hungaricus (Balogh, 1943)

24 (23) Shield *Na* anteriorly with three pairs of rounded fields behind each other (pl. 13 : D). — In all other unequivocally definable characteristics (pl. 13 : E) nearly related to the preceding species. Length: 175—205 μm, width: 90—1.03 μm.

Locus typicus: Bent Creek "Experimental Forest", Asheville, USA. Distribution: Austria, Soviet Union (WS), USA

rostratus (Jacot, 1936)

25 (22) Midline of shield *Na* with nearly adjacent field-pairs, angular in shape, or fused and then exceptionally of other outlines. Members of pair 5 never obliquely meeting. No incision behind rostrum on prodorsal margin.

26 (27) Midline of shield *Na* anteriorly with an unpaired field ringed by other fields (pl. 13 : F). — Similar annulate formations recognizable bilaterally near "shoulders". Field-pair 5 of posterior field-group cordiform, fields of pair 6 very wide. Body setae extremely short, *e* and *f* not longer than setae *c* and *d*. Length: 160 μm.

Locus typicus: Obozersky "forest station", Archangelsk, Soviet Union. Distribution: only the type-locality known

borealis (D. Krivolutsky, 1965)

27 (26) Medial fields of shield *Na* appearing in pairs or fused near each other.

28 (29) Midline of shield *Na* with only field-pair 1, and occasionally field-pair 2, fused, field-pair 5 still distinct (pl. 19 : A). — All prodorsal setae well discernibly ciliate, without any essential difference in length. Sensillus fusiform, densely and verticillately ciliate. Notogastral fields with sinuous borders, especially conspicuous in median fields. Chitinous ring of shield *Na* slightly oval, irregular, laterally open. Notogastral setae weakly but visibly fusiform, ciliate; ciliation recognizable on every seta. Adanal setae (pl. 19 : B) comparatively thin, but difference between ad_1 and ad_2, and ad_3, respectively, well discernible. Length: 207—213 μm, width: 106—113 μm.

Locus typicus: Mt. Verdousia, near Amphissa, Greece. Distribution: only the type-locality known

variabilis Mahunka, 1982

29 (28) In midline of shield *Na* field-pair 1 and also field-pair 5 completely fused; field-pair 2 always distinct.

30 (35) In midline of shield *Na* field-pairs 1 and 5 fused, thus only four field-pairs present.

31 (32) Borders of all notogastral fields straight (pl. 14 : A). Chitinous ring of shield *Na* round. — Prodorsal fields resembling notogastral ones, smooth and with straight borders. Field-pair 3 behind lamellar setae hardly narrower than the other ones. Sensillus capitate (pl. 14 : B), comparatively wide, with 5—6 spines in one row. Ring-spot of shield *Na* round. Setae *ro*, c_2 and c_3 ciliate. Length: 135—155 μm, with: 70—75 μm.

Locus typicus: Fenmoller, Strandkjär, Denmark. Distribution: Denmark, Sweden, GDR, Poland, Soviet Union (TC, SMA)

cricoides Weis-Fogh, 1948

32 (31) Border of notogastral fields sinuous. Chitinous ring of shield *Na* with sinuous borders or irregular in shape.

33 (34) Borders of every field on shield *Na* sinuous, (pl. 14 : E). Field-pairs 1 and 5 fused. Prodorsal fields heavily punctate. — Rostral setae ciliate. Sensillus terminally fusiform (pl. 14 : C), with 7 spines in a row. Field-group near setae c_1 with discrete fields. Notogastral setae e_1 and f_1 especially fusiform (pl. 14 : D). Length: 152—164 μm, width: 82—88 μm.

Locus typicus: Nyíregyháza, Hungary. Distribution: Hungary

oesziae Balogh et Mahunka, 1979

34 (33) Borders of only some of the fields on shield *Na* sinuous, waves composing border not so small as in preceding species (pl. 14 : H). Field-pairs 1, 2 and 5 fused (pl. 14 : F).—Rostral setae smooth. Prodorsal fields impunctate, field-pair 3 behind lamellar setae rapidly attenuating. Sensillus (pl. 14 : G) terminally narrow, but still longer than stalk, with a row of 9—10 spines. Field-group near setae c_1 fused into an elongate band. Notogastral setae thin, c_2 and c_3 weakly ciliate. Lenght: 168—180 μm, widht: 92—100 μm.

Locus typicus: Nyíregyháza, Hungary, Distribution: Hungary

subcricoides Balogh et Mahunka, 1979

35 (30) In midline of shield *Na* every field-pairs distinctly separated in middle, therefore, six pairs of fields present.

36 (37) Prodorsal and notogastral setae with well discernible, long cilia (pl. 15 : A—B).—Field-pair 3 behind lamellar setae on prodorsum rapidly tapering. Ring-spot of notogaster elliptical surrounded by merely a few fields, therefore, no rosette evolved. Fields near setae c_1 also weakly developed. Sensillus (pl. 15 : C) widely fusiform, with thin and also spiniform cilia. Length: 145—155 μm, width: 67—78 μm.

Locus typicus: Fenmoller, Strandkjär, Denmark. Distribution: Denmark, Hungary, GDR, Poland

furcatus Weis-Fogh, 1948

37 (36) Most prodorsal and notogastral setae smooth, only setae c_2 and c_3 as well as some notogastral setae (discernible only in lateral view!) with minute cilia, considerably shorter than in preceding species.

38 (39) Every prodorsal and notogastral fields heavily punctate, their margins sinuous. Minute cilia only on setae c_2 and c_3, all other notogastral setae smooth even in lateral view (pl. 15 : F).—Unpaired field of rostral field-group on prodorsum narrow, anteriorly open (pl. 15 : D). Sensillus (pl. 15 : G) less asymmetrical, with both cilia and spines. No complete rosette evolved around ring on shield *Na* (pl. 15 : E), fields near setae c_1 well delimited. Length: 157—162 μm, width: 75—93 μm.

Locus typicus: Västerbotten, Degerfors, Kulbäcksliden, Sweden. Distribution: Holarctic Region

suecicus Forsslund, 1942

39 (38) Prodorsal fields punctate, those of notogaster smooth, their borders straight. Minute cilia recognizable, at least in lateral view, on all notogastral setae (pl. 15 : J). Median unpaired group of rostral field-group on prodorsum wide, well defined also anteriorly (pl. 15 : H). A complete rosette evolved around ring on shield *Na* (pl. 15 : I). Row of fields near setae c_1 composed of well differentiated fields. Length: 152—160 μm, widht: 70—90 μm.

Locus typicus: Ampthill, South Bedfordshire, England. Distribution: England, Germany, Poland, Hungary

jacoti (Evans, 1952)

40 (1) At least some of the notogastral hairs widened like a blade or phylliform.

41 (46) Midline of shield *Na* with all six field-pairs distinct, or at most field-pairs 1 and 5 coalescing into a single transverse field each.

42 (43) Field-pairs 1 and 5 of shield *Na* fused into single fields (pl. 14 : I). Prodorsal and notogastral setae only dentate, not ciliate.—Prodorsal fields punctate, notogastral ones smooth. Row of fields near seta c_1 fused. Rosette rather incomplete. Only few fields unequivocally discrete on shield *Nm*, especially borders of lateral fields evanescent. Length: 152—160 μm, width: 80—85 μm.

Locus typicus: Leopoldsberg, Vienna, Austria. Distribution: Austria, Hungary

phyllophorus Moritz, 1976

43 (42) Every field-pair of shield *Na* distinct. All prodorsal and notogastral setae ciliate.

44 (45) Body hairs thin (pl. 16 : A), setae *e, f* and *h,* widening, their margins dentately ciliate (pl. 16 : C—D). Sensillus (pl. 16 : B) terminally hardly fusiform, its

surface with dense and long cilia.—All prodorsal setae thin, narrow, also *c* and *d* as well as marginal setae of shield *Py*. Rosette weakly evolved, row of fields near setae c_1 fused. Length: 139—144 μm, width: 81—87 μm.

Locus typicus: without exact locality data, North Carolina, USA. Distribution: Holarctic Region*

jugatus Jacot, 1938

45 (44) Every body seta widely phylliform (pl. 16 : G). Sensillus (pl. 16 : H) widely fusiform, with fine cilia.—Basal fields of prodorsum considerably widened transversely, but their heights remaining low. Configuration of rosette and of the other notogastral fields similar to that of preceding species. Length: 167 μm, width: 90 μm.

Locus typicus: Basiluzzo, Isole Eoile, Italy. Distribution: Italy

meridionalis Bernini, 1973

46 (41) Every opposite field along midline of notogaster fused, usually also forming two extensive areas.

47 (50) Prodorsal and notogastral setae thin, short, eventually also notwidened setae present. Seta e_1 longest of all, half as long as distance between its base and insertion point of f_1.

48 (49) Setal margins ciliate (pl. 17 : B), all prodorsal and notogastral setae slightly widened.—Rostral, lamellar and interlamellar setae approximately equal in length. Sensillus terminally widely fusiform, densely squamate. Ring-spot (pl. 17 : D) of shield *Na* irregularly shaped, elongated. Fields along setae c_1 fused into a longitudinal zone (pl. 17 : A). Length: 171—184 μm, width: 92—100 μm.

Locus typicus: "Valley of Death" near Kato Zakros, Crete, Greece. Distribution: only the type-locality known

hauserorum Mahunka, 1979

49 (48) Margin of dorsal setae smooth (pl. 17 : H—I), several among prodorsal and notogastral setae simple, thin.—Rostral setae essentially longer than lamellar and interlamellar ones. Setae *c* of notogaster (pl. 17 : F) not widened, *d* only slightly resembling a blade. Sensillus (pl. 17 : G) terminally very narrow, densely covered with robust spines. Ring-spot of shield *Na* round, marginally sinuous. No ribbon-like fused row of fields along setae c_1. Length: 140—147 μm, width: 76—87 μm.

Locus typicus: Rio Atuel valley at Arroyo El Obscuro, Argentina. Distribution: Argentina, Japan**

elsosneadensis (Hammer, 1958)

50 (47) Prodorsal and notogastral setae wide, long; e_1 reaching base of posteriorly standing seta f_1.

51 (52) Prodorsal and notogastral fields smooth. Margin of notogastral setae smooth, their widest section attained usually well before half their length, margins for a

* The species is surely often confused with *B. suecicus* Forsslund, 1942, and *B. jacoti* (Evans, 1952) (see Moritz: 1976: 280), therefore, the locality and distribution data should be accepted with reservations! In the interpretation of the species, we follow Chinone and Aoki (1972)

** Identification needs checking!

while parallel (pl. 18 : D).—Rostral setae with serrate to dentate margins (pl. 18 : E), all other prodorsal setae nearly smooth. Sensillus (pl. 18 : B) terminally with rows of robust spines. Anterior and posterior three field-pairs each wholly fused along midline of notogaster, not even transverse lines discernible (pl. 18 : A). Length: 148—171 μm, width: 84—96 μm.

Locus typicus: Zehlaubruch, Königsberg (now Kaliningrad), Soviet Union. Distribution: Holarctic Region

zelawaiensis (Sellnick, 1928)

52 (51) Prodorsal and notogastral fields punctate (pl. 18 : H). Margins of notogastral setae finely serrato-dentate, setae shorter, attaining their greatest width at half their length, then rapidly attenuating (pl. 18 : J).—Sensillus narrower, densely ciliate, not spinose. Median fields on shield *Na* of notogaster also with discernible transverse lines. Length: 167—179 μm, width: 87—90 μm.

Locus typicus: Weisswasser, Oberlausnitz, GDR. Distribution: GDR, England, Soviet Union (NET, CET)

honestus Moritz, 1976

Species interpretabiles

1. aokii Chinone, 1974 : 2, figs. 1—6 *(Brachychthonius)*
2. borealis D. Krivolutsky, 1965 : 707, fig. 2 *(Brachychthonius)*.
3. cricoides Weis-Fogh, 1948 : 269, fig. 25.
   cricoides Evans: 1952 : 236, fig. 7 *(Brachychthonius)*.
   cricoides Niedbala: 1972 : 36, figs. 6—7 *(Brachychthonius)*.
   cricoides Niedbala: 1974 : 473, fig. 20.
   cricoides Moritz: 1976 : 287, fig. 19.
4. elsosneadensis Hammer, 1958 : 20, fig. 3 : 14 *(Brachychthonius)*.
   elsosneadensis Chinone et Aoki: 1972 : 229, figs. 41—47 (Brachychthonius).
   elsosneadensis Chinone: 1978 : 28.
5. formosus Cooremann, 1947 : 9, fig. 4 *(Eobrachychthonius)*.
   formosus Moritz: 1976 : 296, figs. 21—22.
6. furcatus Weis-Fogh, 1948 : 269, fig. 25.
   furcatus Moritz: 1976 : 286, figs. 18a—b.
7. gracilis Chinone, 1974 : 4, figs. 15—20 *(Brachychthonius)*.
   gracilis Chinone: 1978 : 29.
8. guanophilus Mahunka, 1979 : 549, figs. 3—5.
   guanophilus Mahunka: 1981 : 799, fig. 3.
9. hauserorum Mahunka, 1979 : 551, figs. 6—10.
10. honestus Moritz, 1976 : 308, figs. 26a—b.
    zelawaiensis Evans: 1952 : 233, fig. 3 *(Brachychthonius)*.
    zelawaiensis Bulanova-Zachvatkina; 1957 : 118, pl. 2. fig. 19 *(Brachychthonius)*.
11. hungaricus Balogh, 1943 : 22, fig. 4 : 9 *(Poecilochthonius)*.
    hungaricus Evans: 1952 : 236, fig. 6 *(Brachychthonius)*.
    hungaricus Chinone et Aoki: 1972 : 224, figs. 23—31 *(Brachychthonius)*.
    hungaricus Moritz: 1976 : 310, figs. 27a—b.
    rostratus D. Krivolutsky: 1975 : 59, fig. 58 *(Sellnickochthonius)*.
12. immaculatus Forsslund, 1942 : 9, fig. 12.
    immaculatus Niedbala: 1972 : 190, figs. 12—14 *(Brachychthonius)*.
    immaculatus Moritz: 1976 : 291, figs. 20-a-d.
    semiornatus Evans, 1952 : 237, fig. 8 *(Brachychthonius)* (Moritz: 1976 : 291).
    arcticus Hammer, 1952 : 20, fig. 14 *(Brachychthonius)* (Moritz: 1976 : 291).

striatus Willmann, 1956: 249, fig. 19 *(Brachychthonius)* (Moritz: 1976: 291).
gygeri Schweizer, 1963: 67, fig. 1 *(Brachychthonius)* (Niedbala: 1974: 474).
obscurus D. Kirovlutsky, 1966: 126, fig. 1 *(Brachychthonius)* (Moritz: 1976: 292).
obscurus Mahunka: 1979: 139 *(Brachychthonius)*.

13. jacoti Evans, 1952: 235, fig. 5 *(Brachychthonius)*.
jacoti Niedbala: 1974: 475, fig. 23.
jacoti Moritz: 1976: 281, figs. 17a—c.

14. japonicus Chinone, 1974: 4, figs. 7—14 *(Brachychthonius)*.
japonicus Chinone: 1978: 29.

15. jugatus Jacot, 1938: 130, figs. 10—11.
jugatus Chinone et Aoki: 1972: 226, figs. 32—40 *(Brachychthonius)*.
jugatus Niedbala: 1972: 193, fig. 16 *(Brachychthonius)*.

16. lydiae Jacot, 1938: 132, figs. 7—8.
lydiae Reeves et Marshall: 1971: 317, figs. 1—6 *(Bracyhchthonius)*.
hammerae Chinone et Aoki, 1972: 233, figs. 55—59 *(Brachychthonius)* (Chinone: 1978: 29).

17. maroccanus Mahunka, 1980: 798, figs. 1—2.

18. meridionalis Bernini, 1973: 364, figs. 4—5.
meridionalis Moritz: 1976: 300, fig. 23.

19. miyauchii Chinone, 1978: 10, figs, 1—7.

20. oesziae Balogh et Mahunka, 1979: 283, figs. 1—3.

21. phyllophorus Moritz, 1976: 303, fig. 24.

22. planus Chinone, 1974: 9, figs. 26—29 *(Brachychthonius)*.
planus Chinone: 1978: 29.

23. rostratus Jacot, 1936: 250, fig. 20: 8 *(Brachychthonius)*.
rostratus Jacot: 1938: 131, fig. 13: 15.
rostratus Moritz: 1976: 314, figs. 28a-b.

24. subcricoides Balogh et Mahunka, 1979: 283, figs. 4—6.

25. suecicus Forsslund, 1942*(= jugatus* var. *suecica)*: 8, fig. 11.
suecicus Moritz: 1976: 279, figs. 15, 16a-d.
jugatus Niedbala: 1972: 193, fig. 16 *(Brachychthonius)* (Moritz: 1976: 279).
jugatus Niedbala: 1974: 478, fig. 24 (Moritz: 1976: 279).

26. variabilis Mahunka, 1982: 505, figs. 13—14.

27. zelawaiensis Sellnick, 1928: 23 *(Brachychthonius)*.
zelawaiensis Sellnick: 1929: 347, fig. 11 *(Brachychthonius)*.
zelawaiensis Chinone et Aoki; 1972: 231, figs. 48—54 *(Brachychthonius)*.
zelawaiensis Moritz: 1976: 304, figs. 25a-d.

Genus 3: Brachychthonius Berlese, 1910

Brachychthonius Berlese, 1910: 219.
Brachychthonius Niedbala: 1972: 665.
Brachychthonius Moritz: 1976: 259.

Body well chitinized, with a characteristic and heavy prodorsal and notogastral pattern. Rostral margin minutely serrate, usually prodorsal margin also with lateral dentation. Besides notogaster, two free suprapleural plates *(SpC, SpE)* differentiated, prosomatic suprapleurite absent. Suprapleural incision varying in depth, often reaching basis of seta h_3. Notogaster with seta d_2 originating marginally, in a position corresponding to the longitudinal alignment of setae c_2 and e_2. Epimeral setal formula: 3—1—3—4. Genital plates large, approaching sternocoxal region, bearing seven pairs of genital setae. Peranal plates considerably widening anteriorad, their anterior margins nearly straight. Adanal plates well discernibly discrete, separated by a wide free band anterior to end of body. All adanal setae simple, thin, without hypertrophic ones.

Type-species: *Brachychthonius berlesei* Willmann, 1928 (= *brevis* Michael, 1888, sensu Berlese).

This designation is, under Article 70 (a) of the Code, a case of "misidentified type-species". Under the provisions of the article cited, either the nominal genus *Liochthonius* Van der Hammen, 1959 (type-species: *brevis* Michael, 1888) will become a junior subjective synonym of *Brachychthonius* Berlese, 1910, or this latter will be upheld by the designation of another type-species selected from among those originally included in the genus (this action may, however, cause further confusion, because none of the originally included species is unequivocally interpretable: they all appear as "species inquirendae" in our catalogue), or, again, the genus will be upheld by the type-species "named by the designator, regardless of the misidentification" (= *brevis* sensu Berlese, 1910 = *berlesei* Willmann, 1928). The case shall of course be referred to the Commission for decision, but, since the purpose of Article 70 (as that of the entire Code) is the serving "of stability and uniformity of nomenclature", we hereby, and only for the purposes of this book, retain "the species named by the designator" and thus render both nominal genera available for immediate and future use — even though we aware of the additional complication of the type-species having been *undescribed* during the period 1910—1928! Another possible solution would be the designation of a type-species selected from among those not originally included in the genus.

Until recently, the genus was not satisfactorily delimited and several species of nearly related genera had also been relegated to it, and, in addition, some highly similar species erroneously interpreted. Accordingly, the majority of data appears uncertain and should be re-evaluated. At present, five species are known from the Palaearctic Region, but some of them occur also in other faunal regions.

Key to species

1 (4) Prodorsal and notogastral setae, but at least lamellar and rostral ones, well discernibly ciliate (pl. 19 : C).

2 (3) In addition to prodorsal setae, also all notogastral ones elongately ciliated bilaterally (pl. 19 : D). — Sensillus widely clavate and with well discernible cilia arranged in rows. Middle of prodorsum with a pair of fields each behind lamellar and rostral setae. No lateral fields attached to chitinous ring on notogaster (pl. 19 : C). Shield Py posteriorly with an obtuse tubercle behind each seta h_1, bearing setae ps_1. Median fields on shields *Nm* and *Py* extending to anterior margin of the respective shield. Length: 157—173 μm, width: 70—80 μm.

Locus typicus: "Meadow near" Hundsfeld ("Psie Pole near" Wroclaw), Poland. Distribution: Sweden, Denmark, England, GFR, GDR, Poland, Hungary

bimaculatus Willmann, 1936

3 (2) In addition to prodorsal setae only those of shield *Na* ciliate, and even they weakly and occasionally hardly discernibly so (visible mostly on setae c_2 and

c_3) (pl. 19 :). — Sensillus less clavate and with minute, sparse cilia. No pairs of fields behind lamellar and rostral setae in middle of prodorsum. Two fields each attached to notogastral rings (pl. 19 : E). Shield *Py* without tubercle behind setae h_1. Median fields not extending to anterior margins of their shields *Nm* and *Py*. Length: 165—171 μm, width: 70—80 μm.

Locus typicus: Feldberg, near Templin, GDR. Distribution: GDR, Poland

hirtus Moritz, 1976

4 (1) All prodorsal and notogastral setae smooth.

5 (6) Caudomedian field *(cm)* among those constituting rosette on shield *Na*, fused with chitinous ring (in front of seta d_1; pl. 20 : A). — Outer side of outer field (opposite seta d_3) closed. Median fields completely fused. Rostral median fields (pl. 20 : C) as wide as long, both together wider than length of one field. Rostral lines in a semicircle around insertion points of rostral setae and reaching to rostral margins. Sensillus (pl. 20 : B) sparsely and asymmetrically ciliate. Notogastral setae comparatively short. Length: 196—212 μm, width: 100—113 μm.

Locus typicus: unknown to us (Italy?). Distribution:* Sweden, Belgium, Germany, Italy, Austria, Hungary

berlesei Willmann, 1928

6 (5) Caudomedian field *(cm)* among those constituting rosette on shield *Na* not fused with chitinous ring. Rostral median fields longer than their combined width.

7 (8) Some median fields of shield *Na* bisected into two halves by a longitudinal line, or paired; fields subdivided also by transverse lines (pl. 20 : K), their borders and to a lesser extent some of their surfaces punctate (pl. 20 : G). — Outer side of lateral fields opposite seta d_3 closed. Rostral fields one and a half times as long as wide (pl. 20 : J). Rostral lines not reaching bases of rostral setae and rostral margins. Sensillus (pl. 20 : H) densely ciliate, cilia long. Length: 175—192 μm width: 87—100 μm.

Locus typicus: Berlin-Buch, GDR. Distribution: Sweden, Poland, GDR, Hungary

impressus Moritz, 1976

8 (7) Median fields of shield *Na* not subdivided by either longitudinal or transverse lines (pl. 20 : F). — Outer side of lateral field opposite seta d_3 open. Rostral fields twice longer than wide. Rostral-lines (pl. 20 : D) extending to insertion points of rostral setae. Sensillus (pl. 20 : E) densely ciliate, cilia short. Length: 175—186 μm, width: 85—102 μm.

Locus typicus: Bad Frankenhausen, GDR. Distribution: Sweden, GDR, Poland, Hungary

pius Moritz, 1976

* The known locality data refer to three nearly related and hitherto not distinguished species, therefore, those revised by Moritz — or those published in the knowledge of his work — can be accepted.

Species interpretabiles

1. berlesei Willmann, 1928: 160, fig. 11 (nom.nov.! = *B. brevis* Berlese: 1910, nec Michael, 1888).
   berlesei Evans: 1952: 234, fig. 4.
   berlesei Van der Hammen: 1959: 20.
   berlesei Moritz: 1976: 262, figs. 10a—c.
2. bimaculatus Willmann, 1936: 290, fig. 22.
   bimaculatus Niedbala: 1974: 485, fig. 31.
   bimaculatus Moritz: 1976: 270, figs. 13a—b.
   helveticus Schweizer, 1956: 230, figs. 145a—b (Niedbala: 1974: 485).
3. hirtus Moritz, 1976: 274, figs. 14a—b.
4. impressus Moritz, 1976: 264, figs. 9a—c, 11a—d.
   berlesei Niedbala: 1974 (nec Willmann, 1928): 484, fig. 30.
5. pius Moritz, 1976: 268, figs. 12a—c.
   berlesei Pérez-Iñigo: 1968: 195, fig. 13 (nec Willmann, 1928).

Genus 4: Poecilochthonius Balogh, 1943

Poecilochthonius Balogh, 1943: 22.
Poecilochthonius Chinone et Aoki: 1972: 235.
Poecilochthonius Niedbala: 1972: 665.
Poecilochthonius Moritz: 1976: 248.

Vivid yellow species with a heavy pattern. Rostrum bilaterally with a group of robust teeth each, anterior margin with minute, sparse teeth. Median line of shield *Na* on notogaster with a longitudinally coalescent narrow ribbon of spots. Seta d_2 in marginal position, aligned with setae c_2—e_2. No longitudinal furrow within insertion points of setae c_2, *cp* and d_2 on shield *Na*. Two pairs of suprapleural plates *(SpC, SpE)* present. No prosomatic suprapleurite. Suprapleural incision present, very short, hardly effecting shield. Adanal plates not fused posteriorly. Epimeral setal formula: 3—1—2—4. Seven pairs of genital setae present, adanal ones ad_2 and ad_3 widened like a blade.

Type-species: *Brachychthonius brevis* var. *italicus* Berlese, 1910.

Only two species of the genus have been described so far;* both inhabit the Palaearctic Region, but data indicate their occurrence also in other zoogeographic regions.

Key to species

1 (2) Notogastral setae short: c_1 and d_1 not or only slightly longer than half distance between setae c_1—c_1 and d_1—d_1 (pl. 21 : A). Chitinous ring of shield *Na* open posteriorad, and also posterolateral field open (pl. 21 : B). — Seta e_1 and especially f_1 short latter not even approaching insertion point of h_1 (pl. 21 : C). Chaetom of anogenital region as in pl. 21 : D, E. Length: 175—207 μm, width: 98—115 μm.

* On the basis of description Moritz (1976 : 255) had rightly doubted the identity of the form published as *P. italicus* (Berlese, 1910) by Chinone and Aoki with both *P. italicus* and *P. spiciger* (Berlese, 1910), but Chinone (1978 : 29), already with full knowledge of Moritz's work, has repeatedly identified it with *spiciger* and thus we accept his opinion.

Reported from a number of areas in the Holarctic Region, and found also in Argentina; in Europe, it is the commonest species of the family. However, the distribution data have not yet been critically revised with respect to the following highly similar species: the majority of data published for this latter will prove to refer to the present species. Locus typicus: Lake City, Florida, USA. Distribution: Holarctic Region, Argentina

spiciger (Berlese, 1910)

2 (1) Notogastral setae longer: c_1 and d_1 as long as two-thirds to three-fourths distance between setae c_1—c_1 and d_1—d_1. Chitinous ring of shield *Na* closed (pl. 21 : F), as also posteromarginal field. — Also setae e_1 and f_1 longer, both extending to point of origin of the corresponding seta behind them. Chaetom of rostral and pygidial parts as in pl. 21 : G-J. Length: 175—187 μm, width: 95—105 μm.

Known from several points in Europe, but essentially rarer than the preceding species. Most known data are uncertain, that is, they might refer to the preceding species. A thermophilous species, inhabiting principally the litter and grassy sites of hillsides exposed to hot sunshine. Locus typicus: Florence, Italy. Distribution: Italy, Austria, Hungary, GDR

italicus (Berlese, 1910)

Species interpretabiles

1. italicus Berlese, 1910: 220, fig. 40 (*Brachychthonius brevis* (Michael, 1888) var. *italicus*).
   italicus Van der Hammen: 1959 : 20 (*Brachychthonius*).
   italicus Moritz: 1976: 251, figs. 7a—d.
2. spiciger Berlese, 1910: 220 (*Brachychthonius brevis* (Michael, 1888) var. *spiciger*).
   italicus spiciger Jacot: 1936 : 249, fig. 7 (*Brachychthonius*).
   spiciger Moritz: 1976 : 254, figs. 8a—e.
   rapaporti (Balogh et Mahunka, 1968): 318, fig. 1 (*Brachychthonius*) (Moritz: 1976: 255).

Genus 5: Neobrachychthonius Moritz, 1976

Neobrachychthonius Moritz, 1976 : 233.

Light yellow, well chitinized animals with very weak sculpture consisting mostly of arcuate lines only. No teeth present either anteriorly or laterally on rostrum. Seta e_2 not in a marginal position on notogaster, arising considerably further inward from aligned setae c_2—e_2. Shield *Na* without a longitudinal furrow inward of insertion points of setae c_2 and d_3, but outwardly with a long suture extending even on to shield *Py*. Two pairs of suprapleural plates (*SpC, SpE*) present, prosomatic suprapleurite absent. Suprapleural incision also absent. Epimeral setal formula: 3—1—3—3. Seven pairs of genital setae. Adanal plates fused posteriorly into a single U-shaped plate; seta ad_2 widened like a knife-blade.

Type-species: *Brachychthonius marginatus* Forsslund, 1942.

Two species, or in the opinion of some authors merely one species represented by two subspecies, are known; both have been collected only in Europe.

Key to species

1 (2) Interlamellar setae as well as those of shield *Na* simple, thin; even all other notogastral setae only slightly incrassate (pl. 22: E, F). — Distance between setae d_1—d_1 much greater than that between d_2—d_2. Prodorsal and notogastral surfaces smooth, without any discernible sculpture. Length: 200—222 µm, width: 102—113 µm.

Collected in acidic forest litter. Locus typicus: Svartberget, Degerfors, Västerbotten, Sweden. Distribution: Sweden, GDR

marginatus (Forsslund, 1942)

2 (1) Every seta of body widened like a willow-leaf, phylliform (pl. 22: D), also longer than in preceding species. — Distance between setae d_1—d_1 equalling that between d_2—d_2. Prodorsal and notogastral surfaces with a weak sculpture consisting of arcuate lines (pl. 22: A) Length: 230—265 µm, width: 128—150 µm.

Occurring in cold forests, mainly in West, North and Central Europe; so far confused with the preceding species, but more frequent than that; the majority of the occurrence data probably refer to the present species. Locus typicus: Bad Frankenstein, GDR. Distribution: Palaearctic Region

magnus Moritz, 1976

Species interpretabiles

1. magnus Moritz, 1976: 238, figs. 4a—c *(marginatus magnus)*.
2. marginatus (Forsslund, 1942): 6, fig. 7 *(Brachychthonius)*.
   marginatus marginatus Moritz: 1976: 234, figs. 2, 3a—b.

Genus 6: Liochthonius Van der Hammen, 1959

Liochthonius Van der Hammen, 1959: 19.
Liochthonius Niedbala: 1972: 666.
Liochthonius Chinone et Aoki: 1972: 241.
Liochthonius Moritz: 1976: 38.

Weakly chitinized animals of light coloration. Body without sculpture, at most some round dorsal spots. Rostrum not dentate or serrulate, no group of teeth even laterally. Seta d_2 of notogaster not in marginal position, arising inwards from aligned c_2 and e_2; nor any longitudinal furrow in this place. No free suprapleural plate present, but *SpC* — carrying also seta c_3 — recognizably fused with anterior margin of shield *Na*, and *SpE* with *Nm*, so that they are occasionally separated by a sharp line. Suprapleural incision absent, replaced eventually by a weak emargination. Epimeral setal formula: 3—1—3—4. Seven pairs of genital setae present. Adanal plates fused into a single U-shaped plate. One pair of adanal setae widened like a knife-blade, the two others simple.

Type-species: *Brachychthonius perpusillus* Berlese, 1910 (see under Species inquirendae!).

The genus is distributed in the whole world, represented by species relegable to several subgroups. There are 36 species known from the Palaearctic Region.

1 (18) All dorsal setae uniformly, strongly widened, usually phylliform or resembling a blade; occasionally inflated; attenuating rapidly in preapical fourth only;* marginally often serrate to fimbriate.

2 (11) Interlamellar setae arising far (more than 17.5 μm) from each other, near bothridia, their distance considerably greater than that between their points of insertion and bothridia (pl. 23 : A).

3 (6) Dorsal setae relatively short, c_1 at most half as long as distance between their bases, while setae d_1 not longer than distance between points of insertion of d_1—d_2.

4 (5) Sensillus (pl. 23 : B) with asymmetrical clavus bearing long spines. Notogastral setae rather wide as compared to their length, marginally completely smooth (pl. 23 : C, D). Rostral setae longest of all prodorsal setae. Spots of interlamellar region situated near each other, excepting first pair. Notogastral surface (pl. 23 : A) also with some recognizable spots. Ventral side shown in pl. 23 : E. Length: 180—195 μm, width: 106—120 μm.

Locus typicus: Motobu peninsula, Okinawa-ken, Japan. Distribution: Japan

penicillus Chinone, 1978

5 (4) Sensillus symmetrical (pl. 23 : G), spines of clavus normal in length. Notogastral setae narrower, their margins serrate or ciliate (pl. 23 : F).—No essential difference in length between prodorsal setae, but interlamellar one considerably thinner than lamellar or exobothridial setae (pl. 23 : F). Neither prodorsum nor notogaster with discernible spots. Length: 207—213 μm, width: 106—113 μm.

Locus typicus: Mt. Verdousia, near Amphissa, "Phocis", Greece. Distribution: only the type-locality known

phitosi Mahunka, 1982

6 (3) Dorsal setae long, setae c_1 at least as long as distance between their bases; d_1 one and a half times as long as distance between d_1—d_2.

7 (8) Sensillus (pl. 23 : J) fusiform, uniapical, clavus evenly covered with spines.—All body (pl. 23 : H) setae phylliform, like a willow leaf (pl. 23 : I), marginally completely smooth. Bases of lamellar setae connected by a robust chitinous lath. Interbothridial region with four pairs of spots, all adjoined by their inner margins. No essential difference in length between notogastral setae (c_1 : 31 μm, d_1 : 32 μm, e_1 : 34 μm). Length: 168—186 μm, width: 97—110 μm.

Locus typicus: Mitsukaido, Ibaraki-ken, Japan. Distribution: Japan

intermedius Chinone et Aoki, 1972

8 (7) Sensillus wider, asymmetrical, biapical, unevenly covered with spines.

9 (10) Notogastral setae mostly smooth, their vela whole, only exceptionally with a slit each, shaped like a knife-blade (pl. 24 : A).—Dorsal setae relatively short,

* In the case of some short-haired species this is not without difficulties (e.g. the species *L. leptaleus* Moritz, 1976, and *L. dilutus* Moritz, 1976, against *L. penicillus* Chinone, 1978), but the symmetrical or asymmetrical state of the sensillus as well as the differences or — in the latter case — the similarities in the prodorsal and notogastral setae give safe points of judgement in the process of identification.

rostral setae hardly longer than distance between their bases, also setae c_1 essentially shorter than distance between their insertion points, and they merely approach but not reach points of origin of setae d_1. Sensillus with two apical tips of nearly equal size. Length: 143—183 µm, width: 79—104 µm.

Locus typicus: Mossiberg, Älvdalen, Dalarna, Sweden. Distribution: England, Sweden, GDR, Poland, Hungary, Japan

evansi (Forsslund, 1958)*

10 (9) Notogastral setae serrately dentate, each velum cleft, especially well discernible on c_2 and c_3 (pl. 24 : C—D).—Dorsal setae longer, rostral setae one and a half times longer than distance between their bases, c_1 nearly as long as distance between their insertion points, and always projecting beyond bases of setae d_1. Sensillus (pl. 24 : B) with one shorter and one longer apical tip. Length: 156—159 µm, width: 75—93 µm.

Locus typicus: Ásotthalom, Hungary. Distribution: Hungary, Poland

forsslundi Mahunka, 1969**

11 (2) Interlamellar setae originating comparatively near (nearer than 17 µm) to each other, far from bothridia; distance between their bases and that between them and bothridia approximately equal.

12 (15) Margins of notogastral setae smooth or sparsely serrulate, but never densely ciliate.

13 (14) Interlamellar setae very wide, their entire surface slightly penicillately ciliate (pl. 24 : F), wholly deviating from structure of notogastral setae (pl. 24 : E).— In these latter, serration (2 rows) decurrent apparently on posterior surface, owing to incurving margins. Sensillus long, clavate section very large, symmetrical, densely ciliate. Interlamellar region without small spots (or hardly discernible). Rostrum narrow, wedge-shaped, with straight sides. Length: 147 µm, width: 103 µm.

Locus typicus: Scanfs, Unterengadin, Graubünden, Switzerland. Distribution: only the type-locality known

gisini (Schweizer, 1948)

14 (13) Interlamellar setae narrow, resembling notogastral setae (pl. 24 : G—H). Serration (2 rows) of setae in normal position: decurrent along margins. Sensillus shorter, clavus smaller, slightly asymmetrical. Interlamellar region with four pairs of well discernible small spots. Rostrum wide, rounded. Length: 166—195 µm, width: 92—105 µm.

Locus typicus: Lötzen* (now Gizycko), Poland. Distribution:*** Sweden, GDR, Poland, Hungary, Bulgaria, Greece

horridus (Sellnick, 1928)

* According to Moritz (1976 : 108), this feature, the same as the length of the setae and the shape of the sensillus, vary considerably, and thus does not consider them as two distinct species. In our view, however, the problem needs further study, because recently discovered specimens of *forsslundi* (in Barcs, Hungary) differ unequivocally and uniformly from all *evansi* specimens we have studied. Accordingly, we still uphold its specific status.

** See Moritz, 1976: 100.

*** Owing to confusions with several other highly similar species, many locality data are to be checked. The species occurs probably in the whole European subcontinent!

15 (12) At least a part of notogastral setae with long and dense marginal cilia (pl. 25 : A).

16 (17) All dorsal setae densely ciliate (pl. 25 : C).—Interlamellar region with four pairs of well discernible spots, members of the two median pairs meeting along their inner margins. Sensillus (pl. 25 : B) widely fusiform with short spines. Notogastral setae (pl. 25 : A) comparatively short, d_1 not projecting beyond insertion point of seta e_1. Length: 165—198 μm, width: 87—100 μm.

Locus typicus: Tihany, Hungary. Distribution: Hungary, Poland, GDR, Korean PDR (?)

plumosus Mahunka, 1969

17 (16) Margins of dorsal setae on shield *Py* partly (f_1,f_2) ciliate, but those of marginal setae (h_1—h_3, ps_1—ps_2) completely smooth (pl. 25 : G).—Interlamellar region without discernible spots. Sensillus (pl. 25 : H) with long clavus, narrowly fusiform, its spines longer than in the preceding species. Notogastral setae (pl. 25 : I—K) generally longer, d_1 projecting far beyond insertion point of seta e_1. Length: 153—165 μm, width: 84—96 μm.

Locus typicus: Mawaba-mura, Gumma-ken, Japan. Distribution* only the type-locality known

alius Chinone, 1978

18 (1) Dorsal setae not widened, or only partly, or only to a small extent; if a more or less narrow marginal membrane present, it is gradually attenuating from base upwards, or only a part of setae wide and essential differences in shape and length among notogastral setae present.

19 (42) Clavus of sensillus fusiform, terminally slightly rounded or terminating in a single tip.

20 (29) Notogastral setae simple, round in cross-section, without a marginal velum**, long, e.g. c_1 not or hardly longer than distance between setae c_1—d_1.

21 (24) Notogastral setae extraordinarily long (setae e_1 50 μm long!), and seta c_1 projecting considerably beyond insertion point of d_1.

22 (23) Notogastral setae straight, rigid, thin, acicular (pl. 25 : E). Bases of lamellar setae not connected by a transverse lath.—Sensillus (pl. 25 : D) visibly expanding, clavus unequivocally differentiated from stalk, bearing sparsely arranged robust spines. Notogastral surface (pl. 25 : F) with 5—6 small round spots. Length: 175—218 μm, width: 92—119 μm.

Locus typicus: Kulbäcksliden, Västerbotten, Sweden. Distribution: Holarctic Region

hystricinus (Forsslund, 1942)

23 (22) Notogastral setae arcuate or flagellately curved, robust (pl. 26 : C). Bases of lamellar setae connected by an arcuate transverse lath (pl. 26 : A).—Clavus of sensillus (pl. 26 : B) hardly widening, transition between stalk and clavus gradual; spines thinner but longer. Notogastral surface with a sculpture more

* Niedbala's (1977: 72) Korean datum may rather refer to this species than to *plumosus* Mahunka, 1969!

** To establish the presence or absence of the velum, the hairs must be examined from every possible plane.

complex than merely round spots, but hardly recognizable, and very fine. Length: 202 µm, width: 89 µm.

Locus typicus: Kamioboro, Hokkaido, Japan. Distribution: only the type-locality known

lentus Chinone, 1974

24 (21) Notogastral setae shorter (seta e_1 at most 35—37 µm long), c_1 at most approaching insertion point of seta d_1.

25 (26) Notogastral setae equal in thickness from base to nearly apex, hence bacilliform (pl. 27 : A); apical third very finely ciliate above, but cilia occasionally discernible only in lateral view (pl. 27 : B).—Interbothridial region with three pairs of hardly recognizable spots, nearly equal in size, occasionally coalescent and arranged in a longitudinal row. Sensillus medium incrassate, with relatively many spines. Notogastral setae generally shorter than in preceding species, but a difference appears only in averages; e_1 reaching insertion point of seta f_1. Length: 185—205 µm, width: 105—110 µm.

Locus typicus: Hohnekamm, Harz range, GDR. Distribution: Sweden, GDR, Poland, Norway

neglectus Moritz, 1976

26 (25) Notogastral setae gradually tapering from base to apex, terminally acicular and always without fine cilia. Interlamellar region with spots not arranged in a longitudinal row.

27 (28) Notogastral setae very thin, acicular (pl. 27 : C). A large distance (20—22 µm) between interlamellar setae.—Clavus of sensillus (pl. 27 : D) comparatively wide, spines shorter on one side, hence asymmetrical. Interlamellar region with spots approaching each other, members of first three pairs nearly in a chevron formation (V), but even those of pair 3 still removed from each other. Notogastral setae generally longer than in following species, also dimensions bigger: length: 192—205 µm, width: 112—120 µm.

Locus typicus: Skidastador, Skagafjördur, Iceland. Distribution: Iceland, Sweden, Germany

tuxeni (Forsslund, 1957)

28 (27) Notogastral setae thicker, more robust (pl. 26 : E). Interlamellar setae arising much nearer (16 µm) to each other than in the preceding species.—Sensillus hardly clavate (pl. 26 : F), spines equal in length, hence appearance symmetrical. Interlamellar region with spots arranged in complete chevron formation, that is, members of pair 3 touching medially. Notogastral setae generally shorter than in the preceding species, but body dimensions also smaller: Length: 158—191 µ, width: 97—105 µ.

Locus typicus: Mt. Tsukuba, Ibaraki-ken Japan. Distribution: only the type-locality known

moritzi nom. nov.*

29 (20) Notogastral setae with a narrower or wider marginal velum, hence cross-section flattened or of a flat V-shape. Setae much shorter than distance between c_1—d_1.

* The taxa *L. tuxeni* Chinone, 1974, and *L. hystricinus* var. *tuxeni* Forsslund, 1957, are obviously not identical. Moritz's inferences (p. 62) are wholly pertinent and we dedicate the species to him.

30 (35) Notogastral setae comparatively long, c_1 longer than half distance between setae c_1—d_1.

31 (32) Interlamellar setae long, longer than distance between their bases (pl. 27 : E).—Members of first pair of interbothridial spots touching medially; imaginary lines connecting centres of four spots on each side straight and parallel. Dorsal setae thin, acicular, with a narrow membrane developed usually only in their basal part (pl. 27 : F). Clavus of sensillus large, with relatively long spines arranged in rows. Length: 160—223 μm,width: 87—116 μm.

Locus typicus: Epping Forest, England. Distribution: Holarctic Region

brevis (Michael, 1888)

32 (31) Interlamellar setae shorter than distance between their bases. Interbothridial spots arranged otherwise, behind pair 1 a distance equal at least to their diameter present. Notogastral setae wider.

33 (34) Interbothridial spots situated at divers distances from each other: the imaginary lines connecting their centres forming an X (pl. 28 : A), because members of pair 3 meeting medially.—Clavus of sensillus (pl. 28 : B) with 6—7 spines in several rows, in a comparatively dense formation. Setae longer since body dimensions also bigger than those of following species. Length: 180—187 μm, width: 110—117 μm.

Locus typicus: Katenburg, Bad Frankenhausen, GDR. Distribution: GDR, Spitsbergen Islands

dilutus Moritz, 1976

34 (33) Interbothridial spots not reaching each other, hence the straight lines connected their centres meeting in a V (pl. 28 : C); their smallest included distance is still larger than diameter of a spot.—Clavus of sensillus (pl. 28 : D) with 4—5 spines in a row, rather removed from each other. A smaller species with proportionally shorter hairs. Length: 160—175 μm, width: 84—100 μm.

Locus typicus: Elisenhain near Eldena, Greifswald, GDR. Distribution: GDR, Hungary, Greece

leptaleus Moritz, 1976

35 (30) Notogastral setae shorter; c_1 always shorter than half distance between setae c_1—d_1.

36 (37) Rostral setae short, always shorter than distance between their bases (pl. 28 : E).—Stalk of sensillus shorter (pl. 28 : F), clavus widely elliptical, with minute, triseriate spines. No spots discernible in interlamellar region. Prodorsal setae minute, but also notogastral ones extraordinarily short; those on shield *Na* thinner than those on shields *Nm* and *Py*. Length: 147—160 μm, width: 75—80 μm.

Locus typicus: Csévharaszt, Hungary. Distribution: only the type-locality known

perelegans Moritz, 1976,

37 (36) Rostral setae considerably longer than distance between their bases. Sensillus with longer clavus and thereon with larger spines.

38 (39) Clavus of sensillus (pl. 29 : C) very narrow, hardly expanding, with long, thin and densely arranged spines.—All dorsal setae with a wide velum (pl. 29 : D, E), seta c_1 only slightly narrower than seta f_1 (pl. 29 : A). Interbothridial spots

well discernible, pair 4 larger than pair 3, their members removed much further from each other than those of pairs 1 and 2. Also notogastral shields with round spots. Length: 166—186 µm, width: 87—115 µm.

Locus typicus: Senjogahara, Nikko, Tochigi-ken, Japan. Distribution: Japan

asper Chinone, 1978

39 (38) Clavus of sensillus wide, well differentiated from stalk, with merely sparsely arranged short spines.

40 (41) Notogastral setae wide, striation of velum discernible, hence setae phylliform, especially e_1 and f_1 (pl. 29: F). Prodorsum with well ascertainable interbothridial spots. Distance between members of pair 1 nearly twice greater than next one. Spines of sensillus (pl. 29: G) smaller. Length: 162—179 µm, width: 82—98 µm.

Locus typicus: Swinia Gora, Kielce, Poland. Distribution: Poland, Hungary

propinquus Niedbala, 1972

41 (40) Notogastral setae narrower, velum less striated, especially seta e_1 hardly phylliform (pl. 29: H). Interbothridial spots of prodorsum absent or not discernible. Spines of sensillus (pl. 29: I) larger and longer. Length: 150—175 µm, width: 85—95 µm.

Locus typicus: Kulbäcksliden, Degerfors, Västerbotten, Sweden. Distribution: Palaearctic Region

simplex (Forsslund, 1942)

42 (19) Apex of clavate sensillus truncate or excised and thus biapical (pl. 30: D, G).

43 (50) Notogastral setae simple, round in cross-section.

44 (47) Body setae extraordinarily long, interlamellar setae longer than distance between their insertion points, notogastral seta c_1 reaching or projecting beyond base of d_1.

45 (46) Seta c_1 (more than 50 µm) longer than diameter of shield *Na*, reaching insertion point of seta e_1, also seta c_2 not shorter.—Body setae rigid, hence lamellar setae directed anteriorad, exobothridial ones straightly (laterally) exclinate (pl. 30: A). Rostral setae (27 µm) essentially shorter than lamellar setae. Clavus of sensillus biapical, tips equal in size, spines marginally well differentiated. Interbothridial region with three pairs of spots. Length: 180—187 µm, width: 111—119 µm.

Locus typicus: Kawaba-mua, Gumma-ken, Japan. Distribution: only the type-locality known

pseudohystricinus nom. nov.*

46 (45) Seta c_1 (at most 40 µm) essentially shorter than diameter of shield *Na*, extending only to insertion point of seta d_1 (pl. 30: C), seta c_2 considerably shorter.—Body setae elastic, thus lamellar and exobothridial setae terminally reclinate or inclinate. Rostral and lamellar setae equal in length. Clavus of sensillus (pl. 30: D) with dorsal tip essentially longer, its margin with less

* *Liochthonius pseudohystricinus* **nom. nov.** pro *L. hystricinus* (Forsslund, 1942) sensu Chinone et Aoki, 1972.

differentiated spines. Spots of interbothridial region as in preceding species. Length: 162—179 µm, width: 85—103 µm.

Locus typicus: "Urwald" Weisswasser, GDR. Distribution: Sweden, Germany

perfusorius Moritz, 1976*

47 (44) Body setae considerably shorter, distance between interlamellar setae one and a half times to twice longer than interlamellar seta; notogastral seta c_1 not even approaching insertion point of d_1.

48 (49) Notogastral setae longer, apex of seta c_1 reaching beyond half distance between c_1—d_1, but never attaining insertion point of seta d_1 (pl. 30 : H), nor d_1 reaching base of e_1. Clavus of sensillus (pl. 30 : G) with arcuately excised apex, dorsal margin convex, ventral margin concave. Interbothridial region with four pairs of spots (pl. 30 : E), distance between members of pair 4 equalling that of pair 3. Length: 200—210 µm, width: 115—123 µm.

Locus typicus: Salemer Moor, Holstein, GFR. Distribution: Germany

peduncularis (Strenzke, 1951)

49 (48) Notogastral setae shorter, c_1 considerably shorter than half distance between setae c_1—d_1 (pl. 31 : A). Sensillus (pl. 31 : B) short, stalk shorter than clavus, terminally straightly truncate, laterally with two minute tips, dorsal and ventral margins equally convex. Interbothridial spots diminishing posteriorad. Distance between members of pair 4 greater than that of pair 3. Prodorsum wider than long, therefore, body conspicuously squat. Chaetom of pygidium, prodorsum and ventral side as in pl. 31 : C—E Length: 185—195 µm, width: 102—110 µm.

Locus typicus: Långfjället, Idre, Dalarna, Sweden. Distribution: Sweden, Germany, Czechoslovakia, Poland

alpestris (Forsslund, 1958)

50 (43) Notogastral setae partly widened, resembling blades, or laterally compressed, eventually with marginal velum and cross-section flat (unequivocal occasionally only in lateral view!).

51 (52) Notogastral setae long, c_1 approaching insertion point of d_1, seta e_1 actually reaching that of f_1 (pl. 31 : F). Interlamellar setae long, longer than distance between their bases. Notogastral setae laterally flattened, widened like a knife-blade (appreciable mainly in lateral view). Interbothridial region with four pairs of large spots, none meeting medially. Sensillus (pl. 31 : G) long, asymmetrically concavely excised, dorsal tip essentially longer than ventral one. Body dark yellow, notogaster with several light spots. Length: 232—243 µm, width: 140—160 µm.

Locus typicus: Dage, near Grossenkneten, Oldenburg, GFR. Distribution: Germany, Austria, Poland

furcillatus (Willmann, 1942)

52 (51) Notogastral setae shorter, c_1 reaching at most to half distance between setae c_1—d_1 (pl. 31 : H). Interlamellar setae shorter than their enclosed distance.Notogastral setae with marginal velum.

* Identification problematic, since easily confusable (cf. Moritz, 1976 : 97) with *L. furcillatus* (Willmann, 1942), but this latter with slightly widening setae and a differently shaped sensillus.

53 (54) Body heavily chtinized, prodorsum with two robust transverse ribs, shield *Py* with a pair of oppositely situated semicircles, a partly covered and framed hollow near seta h_1 (pl. 31 : H).—All dorsal setae wide, lengthening towards end of body. Sensillus with densely arranged spines. Length: 205 μm, width: 136 μm.

Locus typicus: "High mountains", Terskei-Alatau, Kirgiz S.S.R., Soviet Union. Distribution: Soviet Union, Mongolia

kirghisicus D. Krivolutsky, 1971

54 (53) Body without any specially heavy chitinous shield, thus *Py* at most with tubercles and short laths, no framed and deep hollow present.

55 (56) Pygidium with two pairs of tubercles emitting setae f_1 and h_1 (pl. 32 : A).—Interbothridial region with members of fourth pair of spots removed far from each other, enclosed distance much greater than that between spot and interlamellar seta. Body golden yellow, notogaster with several recognizable spots. Sensillus (pl. 32 : C) asymmetrical, margins of clavus approximately straight; dorsally with densely arranged spines. Length: 175—195 μm, width: 95—113 μm.

Locus typicus: Trammer-See, Ost-Holstein, GFR. Distribution: Palaearctic Region

strenzkei Forsslund, 1963

56 (55) Pygidium without round tubercles, notogastral setae longer. Interbothridial region with spots nearer to each other, or shape of sensillus different.

57 (60) Clavus of sensillus (pl. 32 : I) short, wide, on both sides convex, anteriorly straightly truncate. Spots of interbothridial region removed far from each other.

58 (59) Notogastral setae rather narrow, a thin velum recognizable only on those originating on shields *Nm* and *Py* (pl. 32 : J). Sensillus (pl. 32 : I) only with minute, hardly differentiated cilia.—All four pairs of spots of interbothridial region large. Length: 205—215 μm, width: 120—140 μm.

Locus typicus: Kulbäcksliden, Degerfors, Västerbotten, Sweden. Distribution: Sweden, Spitsbergen Islands, Soviet Union (ES)

clavatus (Forsslund, 1942)

59 (58) Notogastral setae wider, vela of setae *d* and *e* recognizable also in superior view (pl. 32 : G—H). Sensillus (pl. 32 : E) with well differentiated spines, considerably larger than in preceding species.—Interbothridial region with merely three pairs of spots of various sizes (pl. 32 : F). Length: 182—191 μm, width: 95—105 μm.

Locus typicus: Mossap, Hokkaido, Japan. Distribution: known only from the type-locality

ohnishii Chinone, 1978

60 (57) Clavus of sensillus long, narrow (pl. 33 : B), at least one side concave or straight, anteriorly concave or obliquely truncate, usually essentially longer than stalk.

61 (62) Prodorsal surface with a chitinous rib between lamellar setae; a similar formation present also behind setae f_1 on pygidium (pl. 33 : A).—Centres of interbothridial spots on prodorsum connectable by a straight imaginary line, a

distance between every pair equalling at least their diameter. Apex of sensillus (pl. 33 : B) straightly truncate, lateral tips equal in size. Considerable differences in length among notogastral setae (pl. 33 : C): seta c_1 hardly more than half as long as e_1. Body brownish golden yellow. Length: 200—237 μm, width: 125—160 μm.

Locus typicus: Tejstfjället, Spitzbergen Islands, Sweden. Distribution: Palaearctic Region

sellnicki (Thor, 1930)

62 (61) No transverse rib present either on prodorsal surface or behind seta f_1. Interbothridial spots arranged otherwise.

63 (68) No straight or arcuate chitinous line present behind setae h_1—h_2. Interbothridial region with well discernible spots.*

64 (65) Clavus of sensillus (pl. 33 : D) with well differentiated spines. Between setae f_1 a larger field of coalescent spots (pl. 33 : F).—Body setae comparatively short (*le:* 20 μm, e_1 : 23 μm). Rostral seta one-fifth longer than lamellar one. Members of pair 4 of interbothridial spots touching medially. Length: 204—223 μm, width: 117—138 μm.

Locus typicus: Mt. Meakandake, Hokkaido, Japan. Distribution: Japan

galba Chinone, 1978

65 (64) Clavus of sensillus (pl. 33 : H) with very small spines, rather cilia, arranged extremely densely along margin. Between setae f_1 distinct spots present or absent.

66 (67) Distance between setae c_1 shorter than, or at least as great as, that between setae e_1 (pl. 33 : G). Distance between anterior members of interbothridial spot-pairs considerably greater than that between pairs 3 and 4. Clavus of sensillus (pl. 33 : H) very long, considerably longer than stalk, thrice longer than wide. Spots present between and beside setae f_1. Length: 200—228 μm, width: 117—134 μm.

Locus typicus: Kulbäcksliden, Degerfors, Västerbotten, Sweden. Distribution: Sweden, Germany, Poland, Hungary, Japan

muscorum Forsslund, 1964

67 (66) Setae c_1 conspicuously farther from each other than setae e_1 (pl. 34 : A). Interbothridial spots as far removed from each other anteriorly as posteriorly. Clavus of sensillus (pl. 34 : C) shorter, hardly longer than stalk, only two and a half times longer than wide. Notogastral setae (pl. 34 : D) slightly shorter (pl. 34 : B) than in preceding species. Length: 175—200 μm, width: 92—110 μm.

Locus typicus: Lule lappmark, Sarek Mts., Sweden. Distribution: Sweden, Germany, Poland, Hungary, Spain, North America

lapponicus (Trägårdh, 1910)

68 (63) A straight or arcuate chitinous line behind aligned setae h_1—h_2 of pygidium. Interbothridial region without spots.

* For the safe distinction of these "dichotomic" features or rather that of the species groups segregated by their means further painstaking and thorough comparative studies are needed. The species involved stand very near each other.

69 (70) Clavus of sensillus (pl. 34: I) narrow, at least thrice longer than wide. Distance between setae c_1 only slightly greater (28 µ) than that between d_1 (26 µ) (pl. 34: E).—Body setae (pl. 34: F, H) slightly longer than in following species. Length: 166—178 µm, width: 101—115 µm.

Locus typicus: Mt. Tomuraushi, Hokkaido, Japan. Distribution: only the type-locality known

pusillus Chinone, 1978

70 (69) Clavus of sensillus (pl. 34 : K) wider, only about twice longer than its greatest width. A considerably greater distance (36 µm) between setae c_1 than between setae d_1 (29 µ) (pl. 34 : J).—Body setae (pt. 34 : J) comparatively longer. Length: 184—192 µm, width: 95—105 µm.

Locus typicus: Beppo, Hokkaido, Japan. Distribution: only the type-locality known

crassus Chinone, 1978* **stat. nov.**

Species interpretabiles

1. alius Chinone, 1978 : 30 **(stat. nov.!)** (*plumosus* ssp. *alius*).
   plumosus Aoki et Chinone (nec Mahunka, 1969) 1972 : 246, figs. 81—87.
2. alpestris Forsslund, 1958 : 78, figs. 5—7 *(Brachychthonius)*.
   alpestris Niedbala: 1974 : 489, fig. 35.
   alpestris Moritz: 1976 : 87, figs. 20b, 21a—b, 30b
3. asper Chinone, 1978 : 15, figs. 21—28.
4. brevis Michael, 1888 : 539, fig. 49 : 14 *(Hypochthonius)*.
   brevis Moritz: 1976 : 40, figs. 6a—b.
   perpusillus Berlese, 1910 : 220, fig. 19 : 41 *(Brachychthonius)* (Moritz: 1976 : 40).
   perpusillus Grandjean: 1963 : 137, figs. 8—12.
   perpusillus Chinone: 1974 : 22, figs. 64—69.
5. clavatus Forsslund, 1942 : 6, fig. 6 *(Brachychthonius)*.
   clavatus Moritz: 1976 : 90, figs. 20c, 22a—b.
6. crassus Chinone, 1978 : 27, figs. 56—60 **(stat. nov.!)** (*alpestris* ssp. *crassus*).
7. dilutus Moritz, 1976 : 44, figs. 7a—b.
8. evansi Forsslund, 1958 : 80, figs. 8—9 *(Brachychthonius)*.
   evansi Mahunka: 1969 : 23, figs. 1—2.
   evansi Niedbala: 1974 : 491, fig. 37.
   evansi Niedbala: 1977: Table I, fig. 17.
   evansi Moritz: 1976 : 108, fig. 28 (nec Sellnick, 1928).
   horridus Evans: 1952 : 279, fig. 1 *(Brachychthonius)*.
   forsslundi levis Chinone, 1974 : 19, figs. 59—63 (Moritz: 1976 : 108).
9. forsslundi Mahunka, 1969 : 23, figs. 3—5.
   forsslundi Niedbala: 1972 : 183, fig. 1.
   forsslundi Niedbala: 1974 : 492, fig. 38.
   forsslundi Niedbala: 1977: Table I, fig. 10.
10. furcillatus Willmann, 1942 : 170, fig. 1 *(Brachychthonius)*.
    furcillatus Moritz: 1976 : 96, fig. 24.
    ensifer Sternzke, 1951 : 238, fig. 4 *(Brachychthonius)* (Moritz: 1976 : 96).
    ensifer Niedbala: 1973 : 62, fig. 2b.
    ensifer Niedbala: 1974 : 489, fig. 36.

* *Liochthonius crassus* Chinone, 1978 **stat. nov.**—*Liochthonius alpestris crassus* Chinone, 1978. Owing to the incrassate dorsal setae, it has in our opinion nothing to do with *L. alpestris* (cf. Chinone, 1978: fig. 60).

11. galba Chinone, 1978: 22, figs. 43—45.
12. gisini Schweizer, 1948: 28, fig. 6 *(Brachychthonius)*.
 gisini Mahunka: 1969: 25, figs. 6—9.
 gisini Niedbala: 1974: 492, fig. 39.
 gisini Moritz: 1976: 105, fig. 27.
13. horridus Sellnick, 1928: 23 *(Brachychthonius)*.
 horridus Mahunka: 1969: 25, figs. 10—11.
 horridus Niedbala: 1974: 494, fig. 41.
 horridus Moritz: 1976: 99, figs. 25a—b.
14. hystricinus Forsslund, 1942: 4, fig. 5 *(Brachychthonius)*.
 hystricinus Moritz: 1976: 66, fig. 14.
 ocellatus Hammer, 1952: 19, fig. 10 *(Brachychthonius)* (Moritz: 1976: 66).
15. intermedius Chinone et Aoki, 1972: 244, figs. 76—80.
16. kirghisicus D. Krivolutsky, 1971: 939, fig. 1.
 kirghisicus Mahunka: 1979: 141, fig. 2.
17. lapponicus Trägårdh, 1910: 549, fig. 352 *(Hypochthonius brevis* var. *lapponicus)*.
 lapponicus Forsslund: 1964: 237, figs. 6—7.
 lapponicus Moritz: 1976: 69, figs. 15a—b.
18. lentus Chinone, 1974: 22, figs. 70—74.
19. leptaleus Moritz, 1976: 45, figs. 8a—b.
20. moritzi nom. nov. (= *hystricinus tuxeni* Chinone, 1974 nec Forsslund, 1957).*
21. muscorum Forsslund, 1964: 236, figs. 1,5.
 muscorum Niedbala: 1974: 498, fig. 45.
 muscorum Moritz: 1976: 72, figs. 16a—b.
22. neglectus Moritz, 1976: 57, figs. 12a—b.
 tuxeni Niedbala: 1974: 505, fig. 56 (nec *tuxeni* (Forsslund, 1957)).
23. ohnishii Chinone, 1978: 24, figs. 46—49.
24. peduncularis Strenzke, 1951: 237, figs. 2—3 *(Brachychthonius)*.
 peduncularis Moritz: 1976: 86, figs. 19a—c, 20a.
25. penicillus Chinone, 1978: 13, figs. 14—20.
26. perelegans Moritz, 1976: 56, figs. 11a—b.
27. perfusorius Moritz, 1976: 93, fig. 23.
 furcillatus Strenzke: 1951: 236 (nec Willmann, 1942).
 furcillatus Niedbala: 1974: 494, figs. 42—43 (nec Willmann, 1942).
28. phitosi Mahunka, 1982: 507, figs. 15—16.
29. plumosus Mahunka, 1969: 25, figs. 12—14.
 plumosus Niedbala: 1974: 500, figs. 49—50.
 plumosus Moritz: 1976: 103, fig. 26.
30. propinquus Niedbala, 1972: 666, fig. 3.
 propinquus Niedbala: 1974: 501, figs. 51—52.
 propinquus Moritz: 1976: 51, figs. 10a—b.
31. pseudohystricinus **nom. nov.** [= *hystricinus* (Forsslund, 1942) sensu Chinone et Aoki, 1972: 248, figs. 88—93].
32. pusillus Chinone, 1978: 25, figs. 50—55.
33. sellnicki Thor, 1930: 58, fig. 7: 2 *(Brachychthonius)*.
 sellnicki Weiss-Fogh: 1947: 267, figs. 40—43 *(Brachychthonius)*.
 sellnicki Niedbala: 1974: 501, fig. 53.
 sellnicki Chinone et Aoki: 1972: 242, figs. 71—75.
 sellnicki Moritz: 1976: 76, figs. 17a—b.
 scalaris Forsslund, 1942: 2, fig. 2 *(Brachychthonius)* (Niedbala: 1972: 187).
 nodosus Willmann, 1952: 170, figs. 24a—b *(Brachychthonius)* (Moritz: 1976: 76).
34. simplex Forsslund, 1942: 7, fig. 8 *(Brachychthonius)*.
 simplex Niedbala: 1974: 502, fig. 54.

* See p. 75.

simplex Chinone: 1974: 17, figs. 53—58.
simplex Moritz: 1976: 50, figs. 9a—b.
35. strenzkei Forsslund, 1963: 282.
strenzkei Niedbala: 1974: 504, fig. 55.
strenzkei Moritz: 1976: 80, figs. 18a—b.
sellnicki Strenzke: 1951: 235, fig. 1 *(Brachychthonius)* (nec Thor, 1930).
36. tuxeni Forsslund, 1957: 590, fig. 13 *(Brachychthonius hystricinus* var. *tuxeni)*.
tuxeni Moritz: 1976: 62, figs. 13a—c.

Species inquirendae

— expolitus Berlese, 1910: 220 *(Brachychthonius brevis* (Michael, 1888) var. *expolitus*).
expolitus Van der Hammen: 1959: 21 *(Brachychthonius)*.
expolitus Moritz: 1976: 316.
— laetepictus Berlese, 1910: 380 *(Brachychthonius)*.
laetepictus Van der Hammen: 1959: 19.
laetepictus Moritz: 1976: 316. *(Brachychthonius)*.
— perpusillus Berlese, 1910: 220, fig. 19: 41 *(Brachychthonius)*
perpusillus Van der Hammen: 1959: 19.

Genus 7: SYNCHTHONIUS Van der Hammen, 1952

Synchthonius Van der Hammen: 1952: 19.
Synchthonius Chinone et Aoki: 1972: 238.
Synchthonius Niedbala: 1972: 665.
Synchthonius Moritz: 1976: 239.

Body heavily chitinized, but of a light colour. Prodorsum and notogaster with a heavy, irregular sculpture, fundamentally different from that of all other Brachychthoniid genera. Rostral part anteriorly and laterally with a group of teeth each, lateral ones composed of irregular and closely adjacent teeth, anterior ones of three uniform and regularly shaped large teeth separated by 2—3 minute ones from each other. Rostral margin serraeto-dentate also between these groups. Notogastral setae simple, d_2 not marginal in position, originating considerably more inwards from aligned setae c_2—e_2. A single suprapleural pair of plates present (*SpE*), situated partly near shield *Na* and partly near shield *Nm*; plates bisected by a transverse line but not discrete. No prosomatic suprapleurite present. Suprapleural incision very deep. Adanal plates not fused posteriorly. Adanal setae ad_2 heavily incrassate.

Type-species: *Synchthonius boschmai* Van der Hammen, 1952 = *Brachychochthonius crenulatus* Jacot, 1938.

Two, widely distributed species of the genus are known.

Key to species

1 (2) Setae long, apex of c_1 reaching insertion point of d_1 and as long as distance between bases of setae c_1—c_1 (pl. 35: D, F). Rostral teeth comparatively short, lateral ones resembling a fleur-de-lis (pl. 36: A—B).—Weakly delimited fields in median zone of shield *Na* connected also by longitudinal lines. Stalk of

sensillus short, clavus long with relatively few long and thin spines. Length: 205—230 µm, width: 120—140 µm.

Locus typicus: Siljansfors "Experimental Park", Mora, Dalarna, Sweden. Distribution: Palaearctic Region

elegans Forsslund, 1957

2 (1) Setae shorter, apex of c_1 only slightly extending beyond half distance between c_1—d_1, and considerably shorter than distance between insertion points of c_1—c_1 (pl. 35 : A, C). Rostral teeth longer, lateral ones serriform (pl. 36 : C, E).—Median zone of shield *Na* with four distinct field-groups unconnected by longitudinal lines, border of fields never contiguous. Chaetom of pygidium and ventral side as in pl. 35 : B, 36 : D. Clavus of sensillus also long, narrow, cilia slightly shorter. Length: 195—206 µm, width: 112—125 µm.

Locus typicus: Bent Creek Exp. Forest, Buncombe Co., North Carolina, USA. Distribution: Holarctic Region

crenulatus (Jacot, 1938)

Species interpretabiles

1. crenulatus Jacot, 1938 : 133, fig. 16 *(Brachychochthonius)*.
   crenulatus Niedbala: 1974 : 482, fig. 28.
   crenulatus Moritz: 1976 : 242, figs. 5a—f.
   boschmai Van der Hammen, 1952 : 20, figs. 1c—f. (Moritz: 1976 : 243).
2. elegans Forsslund, 1957 : 210, figs. 1—3.
   elegans Grandjean: 1963 : 122, figs. 1—7.
   elegans Niedbala: 1974 : 483, fig. 29.
   elegans Moritz: 1976 : 248, figs. 6a—d.

Genus 8: PARALIOCHTHONIUS Moritz, 1976

Paraliochthonius Moritz, 1976 : 122.

Weakly chitinized, nearly white animals. Prodorsum and notogaster without sculpture. Rostrum smooth anteriorly and laterally, no teeth or serration present. Sensillus characteristically spherical. Setae d_2 on shield *Na* not marginal in position, originating considerably mediad from aligned setae c_2—e_2. No longitudinal furrow inside of insertion points of setae c_2—d_3. Suprapleural incision completely absent, a single pair of suprapleural plates (*SpC*) present, no prosomatic suprapleurite. Epimeral setal formula: 3—1—3—3. Genital plates emitting seven pairs of setae. Peranal pairs of plates very long, extending nearly to apices of anal plates. Adanal plates not fused posteriorly; adanal seta ad_2 widened like a knife-blade, while ad_1 and ad_3 setiform.

Type-species: *Brachychthonius piluliferus* Forsslund, 1942.

Only three species of the Holarctic genus is known; they all occur also in Europe.

Key to species

1 (2) Interlamellar seta short (pl. 36 : F), considerably shorter than clavus of sensillus. This latter (pl. 36 : G) very wide, proportion of length to width as 3 : 2. Notogastral setae not widening basally, also those of shield *Py* short, f_1 not even approaching insertion point of seta h_1 (pl. 36 : H). Length: 123—166 µm, width: 56—77 µm.

Locus typicus: Kulbäcksliden, Degerfors, Västerbotten, Sweden. Distribution: Greenland, Great Britain, Sweden, GDR, Poland, Hungary, Japan, USA

piluliferus (Forsslund, 1942)

2 (1) Interlamellar seta long, longer than clavus of sensillus. Larger species, at least 166 µm long.

3 (4) Setae of shields *Nm* and *Py* basally widened, phylliform (pl. 37 : A—D), arising on minute round tubercles. Interlamellar seta hardly longer than clavus of sensillus (pl. 37 : B) and essentially shorter than seta c_1. Chaetom of anogenital region as in pl. 37 : C. Length: 182—210 µm, width: 92—110 µm.

Locus typicus: Pluss-See, Holstein, GFR. Distribution: GFR, GDR, Poland

globuliferus (Strenzke, 1951)

4 (3) Setae of shields *Nm* and *Py*, like those of notogaster, not widening, with some weak tubercles at their bases at most on shield Py (pl. 37 : E, F). Interlamellar setae long, considerably longer than clavus of sensillus, as long as seta c_1. Length: 166—188 µm, width: 100—105 µm.

Locus typicus: "Rezerwat" Pluznica, pow. Strelce Opolskie, Poland. Distribution: Poland, Sweden

occultus (Niedbala, 1971)

Species interpretabiles

1. globuliferus Strenzke, 1951: 243, fig. 8 *(Brachychthonius)*.
   globuliferus Niedbala: 1974 : 493, fig. 40 *(Liochthonius)*.
   globuliferus Moritz: 1976 : 124, figs. 33, 34a—c.
2. occultus Niedbala, 1971 : 677, fig. 1 *(Liochthonius)*.
   occultus Niedbala: 1974 : 497, fig. 46 *(Liochthonius)*.
   occultus Moritz: 1976 : 128, figs. 35a—b.
3. piluliferus Forsslund, 1942 : 8, fig. 9 *(Brachychthonius)*.
   piluliferus Chinone: 1974 : 16, figs. 47—52 *(Liochthonius)*.
   piluliferus Moritz: 1976 : 130, figs. 36a—c.

Genus 9: Mixochthonius Niedbala, 1972

Mixochthonius Niedbala, 1972 : 666.
Mixochthonius Niedbala: 1974 : 487.
Mixochthonius Moritz: 1976 : 228.

Weakly sclerotized, light animals. Prodorsum and notogaster without sculpture. Rostrum without teeth, either frontally or laterally. Seta d_2 of shield *Na* originating considerably inwards from line of setae c_2—e_2. No longitudinal furrow mediad from insertion points of setae c_2—d_2. A single suprapleural pair of plates *(SpE)* present; plate *SpC* adnate to side of shield *Na*, forming its lobe. A prosomatic suprapleurite present. Shield *Py* cleft by a deep suprapleural incision. Epimeral setal formula: 3—1—3—4. Seven pairs of genital setae present. Peranal plates long, reaching to end of anal plates posteriorly. Adanal plates fused behind, forming a U-shaped plate. All three pairs of adanal hairs thin and simple. All prodorsal and notogastral setae thin but densely ciliate.

Type-species: *Brachychthonius pilososetosus* Forsslund, 1942.

Two species are known, but the correctness of generic assignment of the second one: *M. concavus* (Chinone, 1974) needs further corroboration.

Key to species

1 (2) Prodorsal setae (pl. 38 : A) long, arcuate, interlamellar ones nearly reaching bases of lamellar setae. Cilia of setae twice longer than diameter of shaft. Basis of prodorsum with three pairs of weak foveolae. Clavus of sensillus fusiform, lateral margin throughout concave. Chaetom of ventral side and shield *Py* as in pl. 38 : B—D. Length: 195—230 µm, width: 90—100 µm.

Locus typicus: Kulbäcksliden, Degerfors, Västerbotten, Sweden. Distribution: Sweden, GFR, GDR, Poland, Hungary

pilososetosus (Forsslund, 1942)

2 (1) Prodorsal setae essentially shorter, more rigid (pl. 38 : E), interlamellar setae only half as long as distance between their insertion points and those of lamellar ones.—Cilia of notogastral setae (pl. 38 : G) 5—6 times longer than diameter of shaft. Clavus of sensillus (pl. 38 : F) fusiform, marginally sinuous. Chaetom of ventral side as in pl. 38 : H. Length: 173—180 µm, width: 90—92 µm.

Locus typicus: Hosooka, Kusharo "Swampy Plain", Hokkaido, Japan. Distribution: Japan

concavus (Chinone, 1974)

Species interpretabiles

1. concavus Chinone, 1974 : 12, figs. 35—41 *(Liochthonius)*.
   concavus Chinone: 1978 : 29.
2. pilososetosus Forsslund, 1942 : 8, figs. 10a—b *(Brachychthonius)*.
   pilosetosus Niedbala: 1972 : 189, fig. 11 (lapsus calami) *(Liochthonius)*.
   pilososetosus Niedbala: 1974 : 488, fig. 34.
   pilososetosus Moritz: 1976 : 229, figs. 1a—c.
   perpusillus Willmann: 1928 : 162, fig. 13 (nec Berlese, 1910) *(Brachychthonius)*.

Genus 10: Verachthonius Moritz, 1976

Verachthonius Moritz, 1976 : 112.

Weakly sclerotized, light animals. Anterior and lateral margins of rostrum smooth. Prodorsum and notogaster without sculpture, at most a few scattered round spots recognizable. Seta d_2 of shield *Na* originating far inward from aligned setae c_2—e_2. No longitudinal furrow present inwards of insertion points of setae c_2—d_2. A single free suprapleural plate present *(SpE)*, plate *SpC* fused to shield *Na*, forming its lateral lobe. No prosomatic suprapleurite present. A short and comparatively wide suprapleural incision on shield *Py*, continued in a rib near setae h_2 and h_3. Epimeral setal formula: 3—1—3—4. Seven pairs of genital setae. Adanal plates fused posteriorly, forming a U-shaped plate. All three pairs of adanal setae simple and thin. Prodorsal and notogastral setae also simple.

Type-species: *Brachychthonius laticeps* Strenzke, 1951.

The species is distributed in the Palaearctic Region; so far three species have been described, but *L. pseudolaticeps* Subias, 1977, belongs also probably here.

Key to species

1 (2) Clavus of sensillus comparatively wide and short, proportion of its length to width as 3 : 1, its distal end without a sharp apex (pl. 39 : G), its surface evenly covered with minute cilia.—Rostral setae originating near each other, their included distance at most one-third of that between interlamellar setae. Rostral setae arising on an obtusely projecting excrescence. Interbothridial region with minute spots, removed from each other by a distance equalling their diameter; among them one pair elongate, two pairs transversely oval. Shield *Na* also with 3—4 spots outwards from setae c_1 and d_1. Setae comparatively long (pl. 39 : F), d_1 about as long as half distance between bases of d_1—d_1. Length: 200—220 µm, width: 105—118 µm.

Locus typicus: Bad Frankenhausen, Katenburg, GDR. Distribution: only the type-locality known

congruus Moritz, 1976

2 (1) Clavus of sensillus comparatively long and narrow, proportion of length to width as 3.8 : 4.1, apically acute, its surface with longitudinal rows of cilia. Rostral setae not on projecting tubercles. Interbothridial spots large, all transversely oval.

3 (4) A small species: at most 208 µm long. Sensillus (pl. 39 : B) symmetrical, with eight longitudinal rows of 8—11 spines per row.—Spots of interbothridial region nearly meeting medially (pl. 39 : A), nearly filling interbothridial area, middle spots the largest of all. Chaetom of pygidium and anogenital region as in pl. 39 : C—E. Length: 180—208 µm, width: 95—110 µm.

Locus typicus: Plön, GFR. Distribution: Germany, Czechoslovakia, Poland, Japan

laticeps (Strenzke, 1951)

4 (3) A larger species: longer than 215 µm. Sensillus asymmetrical (pl. 39 : H), above with 3—4 rows of spines, each composed of 12—15 minute spines, somewhat shorter lower portion with some few rows of 5—7 considerably larger spines.—Interbothridial spots removed further away from each other medially (pl. 39 : J) median pair not bigger than the other ones. Length: 215—228 µm, width: 110—125 µm.

Locus typicus: Jonastal, Arnstadt, Thüringen, GDR. Distribution: only the type-locality known

diversus Moritz, 1976

Species interpretabiles

1. congruus Moritz, 1976: 118, figs. 32a—c.
2. diversus Moritz, 1976: 117, figs. 31a—b.
3. laticeps Strenzke, 1951: 240, fig. 5 *(Brachychthonius)*.
laticeps Niedbala: 1974: 487, fig. 33 *(Mixochthonius)*.
laticeps Chinone: 1974: 14, figs. 42, 46 *(Liochthonius)*.
laticeps Moritz: 1976: 113, figs. 29a—b, 30a.

Species inquirenda

— pseudolaticeps Subias, 1977: 65, fig. 23 *(Liochthonius)* **(comb. nov.)**.

Superfamily 3: HYPOCHTHONIOIDEA Berlese, 1910*

Hypoctoniidae Berlese, 1910: 218 (lapsus calami).
Hypochthoniidae Van der Hamman: 1959: 13.
Hypochthonioidea Balogh: 1961: 246.

Characteristically angular, dorsoventrally flattened, light or dark yellowish animals. Notogaster subdivided by a furrow into two shields, furrow with insertion points of two pairs of notogastral setae. These setae reduced, hence only a total of 14 pairs of true notogastral setae present: setiform or ensiform, but never palmate or T-shaped.

Representing only a single family.

Family 1: HYPOCHTHONIIDAE Berlese, 1910

Characterizable by the features of the superfamily. The family includes two genera.

Key to genera

1 (2) Genital plate subdivided by a transverse suture into two parts each, every one bearing five pairs of genital setae. Anal and adanal plates well differentiated; two pairs of anal setae present
genus 1: **Eohypochthonius** Jacot, 1938

2 (1) Genital plates whole, without a transverse suture. Anal plates completely reduced, whereas adanal plates well developed. Also anal setae reduced
genus 2: **Hypochthonius** C. L. Koch, 1836

Genus 1: EOHYPOCHTHONIUS Jacot, 1938

Eohypochthonius Jacot, 1938: 133.
Eohypochthonius Beck: 1962: 231.
Eohypochthonius Aoki: 1977: 117.

* Schweitzer (1956) described nine new species in a new genus *Alphypochthonius* and placed them in the family of *Hypochthoniidae*, however, they are nymphs of higher Oribatids *(Ceratozetidae?)*. See also Van der Hammen: 1959: 14.

Body light, pale yellow, dorsally convex. Margins smooth. Notogaster subdivided into two shields by a transverse suture: setae *c* and *d* arising on shield *Na*, seta *e* in furrow, all other setae on shield *Py*. Setae ps_2 and ps_3 shifted to ventral side. Prodorsal and notogastral setae simple, setiform, or more or less widened: ensiform to phylliform (like a willow leaf). 16 pairs of notogastral setae present, but two pairs represented by alveoli in transverse furrow. Genital plates transversally subdivided, each part bearing five pairs of genital setae. A short aggenital plate present, not extending beyond furrow of genital plate; aggenital setae absent. Anal and adanal plates well evolved. two pairs of anal and three pairs of adanal setae well recognizable. Epimeral setal formula: 3—1—2—4.

Type-species: *Hypochthonius gracilis* Jacot, 1936.

The species are frequent principally in the tropics and occur in the Palaearctic Region only in Japan, on the border of the Oriental Region.

Key to species

1 (2) Notogastral setae slightly expanded, fusiform (pl. 40: K). Interlamellar setae also thick (pl. 40: M), with minute surface squamae.—Interlamellar setae shaped like the notogastral ones, as long as distance between insertion points and bothridia. Sensillus (pl. 40: L) with 8—10, exceptionally more, lateral branches, also its other side with some small spines. Chitinous tubercle on prodorsal side large and wide (fig. 40: K). Length: 340—388 µm, width: 160—170 µm.

Locus typicus: Shiraishi, Kitagawa-Mura, Japan. Distribution: Japan, Korean PDR

crassisetiger Aoki, 1959

2 (1) Notogastral setae simpler, thin, also interlamellar seta setiform (pl. 40: H).

3 (4) Sensillus with 10—11 long lateral branches (pl. 40: B). Lateral chitinous tubercle of prodorsum (pl. 40: A) very convex, large, wide.—Lamellar setae (pl. 40: D) very short, not longer than half distance between their insertion points and bothridia. Notogastral setae also comparatively short, their proportions shown in pl. 40: A. Length: 268—300 µm.

Locus typicus: Kamikurihara, in Zama-machi, Kanagawa, Central Japan. Distribution: Japan

parvus Aoki, 1977

4 (3) Sensillus with merely 7—8 long lateral branches (pl. 40: G). Lateral chitinous tubercle of prodorsum narrow, elongated (pl. 40: F). Prodorsal setae (pl. 40: H, I, J) longer, lamellar ones longer than half distance between their insertion points and bothridia. Notogastral setae, also longer than in preceding species, their proportions shown in pl. 40: F. Length: 355—410 µm.

Locus typicus: Mt. Myojo, in Hakone, Kanagawa, Central Japan. Distribution: only the type-locality known

magnus Aoki, 1977

Species interpretabiles

1. crassisetiger Aoki, 1959: 3, fig. 2 *(gracilis crassisetiger)*.
   crassisetiger Aoki: 1977: 121, figs. 2C, 3C, 4C.
2. magnus Aoki, 1977: 119, figs. 2B, 3B, 4B.
3. parvus Aoki, 1977: 118, figs. 2A, 3A, 4A.

Genus 2: HYPOCHTHONIUS C. L. Koch, 1836

Hypochthonius C. L. Koch, 1836: 3, 19—20.
Hypochthonius D. Krivolutsky: 1965: 934.
Hypochthonius Van der Hammen: 1959: 14.

Body pentagonal, conspicuously flattened, invariably vivid orange yellow or dark brown. Rostral margin with teeth. Prodorsal and notogastral surface with minute fields of polygonal sculpture. Setae *c* and *d* arising on shield *Na*, setae *f*, *h* and *p* on shield *Py*, setae p_1 and p_2 shifted to ventral side. Seta *e* indicated by insertion points only, in transverse furrow. Genital plates whole, without transverse furrow. Ten pairs of genital setae. Aggenital plates widely fused along midline anteriorly, not delimited also from epimeral region. Aggenital setae absent. Anal plate and anal setae reduced, entire anal region occupied by adanal plates bearing three pairs of adanal setae. Epimeral setal formula: 3—1—3—4.

Type-species: *Hypochthonius rufulus* C. L. Koch, 1836.

The two Palaearctic species of the genus have been subdivided into several subspecies; their separation is uncertain and needs further comparative investigations. We accept only the two nominate subspecies. Necrophagous.

Key to species

1 (2) Notogastral setae smooth or finely ciliate and distally gradually acicular. Apex of seta c_1 considerably projecting beyond insertion point of seta d_1 (pl. 41 : B). Sensillus with 6—8 long lateral branches. Body wide especially notogaster widening posteriorad; colour darker, often rufous. Length: 650—700 µm, width: 370 µm.
Locus typicus: Regensburg, GFR. Distribution: Holarctic Region
rufulus C. L. Koch, 1836

2 (1) Notogastral setae serrato-dentate or ciliate on one side, apically obtuse; shorter, seta c_1 extending at most to insertion point of d_1. Sensillus with 13—15 lateral branches. Body (pl. 41 : A) more elongated, narrower than in preceding species. Colour lighter, yellowish. Length: 610—650 µm, width: 320 µm.
Locus typicus: Valkenburg, Netherlands. Distribution: Palaearctic Region
luteus Oudemans, 1917

Species interpretabiles

1. luteus Oudemans, 1917 : 343.
   luteus Oudemans: 1917 : 21.
   luteus Pérez—Iñigo: 1968 : 192, figs. 11—12.
2. rufulus C. L. Koch, 1836 : 3, 19, fig.
   rufulus Oudemans, 1917: 22, figs. 45—50.
   rufulus Aoki: 1965 : 289, figs. 1—4.

Species inquirendae

— latirostris Schweizer, 1956 : 224, fig. 134.
— luteus rectosetosus D. Krivolutsky, 1965 : 936, figs. a—g.
— rufulus europaeus D. Krivolutsky, 1965 : 935, figs. a—b, d.

Superfamily 4: ENIOCHTHONIOIDEA Grandjean, 1947

Eniochthoniinae Grandjean, 1947 : 223.
Eniochthonioidea Balogh et Mahunka: 1979 : 282.

Body cylindrical, hardly flattened. Notogaster subdivided by a furrow into two shields; anterior shield with a pseudosuture interrupted medially.

All 16 pairs of notogastral setae evolved, furrows without setal insertion points. Setae *c* and *d* arising on shield *Na*, setae *e, f* and *h* on shield *Py*.

Based on a single family.

Family 1: ENIOCHTHONIIDAE Grandjean, 1947

Eniochthoniinae Grandjean, 1947 : 223.
Eniochthoniidae Grandjean: 1954 : 429.

As characterized above.

Only a single, cosmopolitan genus known.

Genus 1: HYPOCHTHONIELLA Berlese, 1910

Hypochthoniella Berlese, 1910 : 218.
Hypochthoniella Pérez-Iñigo: 1968 : 194.
Arthrochthonius Ewing, 1917: 130.
Eniochthonius Grandjean, 1947 : 223.
Eniochthonius Van der Hammen: 1959 : 16.

Besides the features enumerated in the family diagnosis above, the genus is characterized by the genital plates subdivided by an obsolete transverse suture; plates with 10 pairs of genital setae grouped in 5 + 5. Aggenital plates also subdivided, a single pair of aggenital setae present, originating on anterior part of plate. Epimeral setal formula: 3—1—3—4.

Type-species: *Hypochthonius minutissimus* Berlese, 1904.
Only a single species known.

— Prodorsal surface smooth, its setae simple. Notogastral setae (pl. 41 : E) also thin and setiform, excepting slightly fusiform setae f_1 and h_1. Sensillus long, thin, asymmetrical: with 14—15 long cilia on one side and some few cilia on the other. Shield *Na* displaying some lighter spots (pl. 41 : C). Configuration and chaetom of anogenital region as in pl. 41 : D Length: 370—380 µm, width: 180—195 µm.

Locus typicus: "Giardino di Boboli", Florence, Italy. Distribution: Cosmopolitan (?)

minutissima (Berlese, 1904)

Species interpretabilis

1. minutissimus Berlese, 1904 : 252 *(Hypochthonius)*.
minutissimus Van der Hammen: 1959 : 17 *(Eniochthonius)*.
minutissimus Pérez-Iñigo: 1968 : 194.
pallidulus Berlese: 1896 : 25 (nec C. L. Koch, 1836) *(Hypochthonius)*.
pallidulus Grandjean: 1933 : 49, figs. 9—16 *(Eniochthonius)*.
grandjeani Van der Hammen, 1952 : 13 *(Eniochthonius)*.

Superfamily 5: Cosmochthonioidea Grandjean, 1947

Cosmochthoniidae Grandjean, 1947 : 224.
Cosmochthonioidea Grandjean; 1969 : 142.

Body light, yellow or white, weakly chitinized, bearing long and partly modified setae. Notogaster with 2—4 furrows, therefore, subdivided into 3, 4 or 5 shields. Setae *e* and *f* originating in these furrows, elongately phylliform or plumose, movable. Prodorsal setae plumose or ciliate, occasionally phylliform; all other notogastral setae shorter, mostly setiform, ciliate, plumose or eventually phylliform, but never umbellately widening.

Composed of a single family.

Family 1: Cosmochthoniidae Grandjean, 1947

Cosmochthoniidae Grandjean, 1947 : 224.
Cosmochthoniidae Gordeeva: 1980 : 838.

Characterizable as in the above diagnosis.

Six genera are relegated to the family, five occurring also in the Palaearctic Region.

Key to genera

1 (6) Notogaster subdivided by 3 or 4 furrows into 4 or 5 shields. Tarsi with two or three claws.

2 (3) Notogaster subdivided by 4 furrows into 5 shields. Setae *e* and *f* widely phylliform. Tarsi of legs I—IV with claw formula 2—2—2—3
genus 3: **Phyllozetes** Gordeeva, 1978

3 (2) Notogaster subdivided by 3 furrows into 4 shields: number of claws different.

4 (5) Also setae *c* and *d* of notogaster widened, phylliform. Tarsi of all legs bidactylous

genus 2: **Krivolutskiella** Gordeeva, 1980

5 (4) Notogastral setae setiform, with more or less elongate branches, setae *e* and *f* resembling palm-leaves. Tarsi with claw formula 2—3—3—3

genus 1: **Cosmochthonius** Berlese, 1910

6 (1) Notogaster subdivided by 2 furrows into 3 shields. All tarsi monodactylous.

7 (8) Notogastral setae *e* and *f* thin, arcuate, flagelliform. Intercalar sclerites nearly reduced. Anogenital region with postanal sclerite present, aggenital one absent

genus 5: **Nipponiella** Gordeeva, 1980

8 (7) Setae *e* and *f* widened phylliform. Anogenital region with aggenital sclerite present, adanal one absent

genus 4: **Gozmanyina nom. nov.***

Genus 1: Cosmochthonius Berlese, 1910

Cosmochthonius Berlese, 1910 : 218.
Cosmochthonius Grandjean: 1931 : 659.
Cosmochthonius Van der Hammen: 1959 : 21.
Cosmochthonius Beck: 1962 : 235.
Cosmochthonius Gordeeva, 1980 : 843.

Slightly convex animals of divers yellowisch hues. Notogaster with three furrows, thereby subdivided into four shields. Setae *c* arising on shield *Na*, seta *d* on shield Nm_1, seta *e* in furrow 2, setae *f* in furrow 3, setae *h* and *ps* on shield *Py*. Prodorsal setae as well as setae *h* and *ps* short or medium long, densely (penicillately) ciliate or plumose. Setae *c* and *d* simpler though still densely ciliate, setae *e* and *f* very long and usually rigid (erect). 16 pairs of notogastral setae present, seta *ps* shifted to ventral side. Ten pairs of genital, four pairs of anal, and four pairs of adanal setae present. Epimeral setal formula: 3—2—3—4. Claw formula of legs: 2—3—3—3.

Type-species: *Hypochthonius lanatus* Michael, 1885.

Key to species

1 (8) Seta *d* of notogaster arising in anterior furrow or on its rim, hence their bases connected by it.

2 (3) Seta d_1 short, terminating before middle of segment 3 (pl. 42 : A).—Rostral setae widended phylliform, their cilia comparatively short, sensillus very slightly thickened, verticillately ciliate. Notogastral surface, especially on pygidium, with relatively large, bean-shaped foveolae. Seta *c* essentially longer

* *Gozmanyina* **nom. nov.** pro *Marshallia* Gordeeva, 1980, nec *Marshallia* Zittel, 1887, Spongia; nec *Marshallia* Doherty, 1889, Lepidoptera. We dedicate the new genus to Dr. L. Gozmány, our friend, who extensively helped us in solving several nomenclatorial problems.

than seta *d*. Cilia of setae *e* and *f* very short, only basal halves of these setae ciliate. Posteromarginal setae short, widened, minutely ciliate. Anogenital region as in pl. 42 : B. Length: 290 µm, width: 200 µm.

Locus typicus: Homutovskiy steppe, Donets region, Soviet Union. Distribution: Soviet Union (SET)

ponticus Gordeeva, 1980

3 (2) Seta *d* longer, d_1 reaching or extending beyond posterior margin of segment 3 (p. 42 : E).

4 (5) Colourless animals, pygidium with sparsely punctate, but partially smooth (pl. 42 : F).—Rostral and posteromarginal setae widened, with elongate cilia, notogastral setae also elongately ciliate. Erectile setae *e* and *f* (pl. 42 : G, H) also with short and long cilia, hence these hairs seemingly more compact. Length: 280—300 µm, width: 160—180 µm.

Locus typicus: Mt. Giovi, Mugello, Italy. Distribution: France, Italy, Hungary, Greece, Soviet Union (SET, SMA)

plumatus Berlese, 1910

5 (4) Yellowish brown, more heavily chitinized animals. Pygidium entirely smooth or with a uniform field of minute, round impressions.

6 (7) Distance between setae d_1 equalling that between setae d_1 and d_2 (pl. 42 : I). Notogaster completely smooth.—Prodorsal hairs extraordinarily long, especially rostral hairs which also bear long cilia. Sensillus setiform with short cilia. Notogastral seta *c* longer than *d*, a considerable difference in length extant especially between setae c_1 and d_1. Posteromarginal setae conspicuously long, thin, elongately ciliate. Length: 245 µm, width: 150 µm.

Locus typicus: Foros, Crimea, Soviet Union. Distribution: only the type-locality known

asiaticus Gordeeva, 1980

7 (6) Distance between setae d_1 essentially greater than that between setae d_1—d_2. Pygidium with round, uniformly placed minute impressions (pl. 42 : J).—Prodorsal setae widely phylliform, marginally with short cilia; latter short also on other prodorsal and posteromarginal setae. Notogastral setae conspicuously shorter than setae *d*, while *e* longer than seta *f*. Cilia of erectile setae *e* and *f* very long and robust. Length: 319 µm, width: 148 µm.

Locus typicus: Dusanbe, Soviet Union. Distribution: only the type-locality known

tenuisetus Gordeeva, 1980

8 (1) Bases of notogastral setae *d* not connected by furrow or its rim, arising considerably behind it.

9 (10) Notogastral surface ornamented with foveolae under cerotegument (pl. 43 : E).—Among rigid and erect notogastral seta *e* considerably longer and more robust than *f*; a similar difference existing also setae *c* and *d*. Erectile setae with long and sparse, while setae *c* and *d* with short but dense, cilia (pl. 43 : F). Prodorsal setae hirsutely and densely ciliate, also notogastral setae *h* and *ps* heavily ciliate. Length: 288—330 µm, width: 168—180 µm.

Locus typicus: Land's End, England. Distribution: Holarctic Region

lanatus (Michael, 1885)

10 (9) Notogastral surface with a heavy reticulation, or only finely and irregularly puctate without larger foveolae, under cerotegument.

11 (12) Notogaster with polygonal reticulation (pl. 43: A).—Prodorsum with large foveolae medially, rostral seta incrassate, bacilliform, lamellar seta branching T-shaped (pl. 43 : C). Cilia of erectile setae of notogaster long and uniform. Setae *h* and *ps* at body and widened like rostral setae. Length: 300—340 µm.
Locus typicus: Périgueux, France. Distribution: France, Greece, Morocco, Tunesia, Egypt **reticulatus** Grandjean, 1947

12 (11) Notogaster with a stellate sculpture composed of foveolae, but still lacking a true reticulation (pl. 43 : G).—Rostrum obtuse, without a ridgelike elevation. Rostral seta furcate. Sensillus narrow but still visibly widened, long. Notogastral setae *c* essentially longer than setae *d*. Shields and chaetotaxy of anogenital region as in pl. 43 : H. Length: 290 µm, width: 216 µm.
Locus typicus: Bactandik, Tadzhikistan, Soviet Union. Distribution: only the type-locality known **ugamaensis** Gordeeva, 1980

Species interpretabiles

1. asiaticus Gordeeva, 1980: 846, fig. 2 : 3.
2. lanatus Michael, 1885 : 396, pl. 7, fig. 11 *(Hypochthonius)*.
   lanatus Michael; 1888 : 541, pl. 49, figs. 15—22 *(Hypochthonius)*.
   lanatus Berlese: 1910 : 221.
   domesticus Grandjean, 1947 : 354, fig. 34 (Grandjean: 1950 : 80).
3. plumatus Berlese, 1910 : 221, pl. 20, fig. 48.
   plumatus Grandjean: 1950 : 78, figs. 2A—D.
   plumatus Van der Hammen: 1959 : 22.
4. ponticus Gordeeva, 1980 : 844, fig. 2 : 1.
5. reticulatus Grandjean, 1947 : 354, fig. 30.
   reticulatus Grandjean: 1962 : 404, figs. 3A—D.
6. tenuisetus Gordeeva, 1980 : 844, fig. 2 : 2.
7. ugamaensis Gordeeva, 1980 : 846, fig. 2 : 4.

Genus 2: Phyllozetes Gordeeva, 1978

Phyllozetes Gordeeva, 1978 : 1099.
Phylozetes Gordeeva: 1980 : 847, unjustified subsequent emendation.

Light, yellowisch white to dirty white animals. Configuration and position of setae of notogaster agreeing with those of *Cosmochthonius* Berlese, 1910. However, erectile setae *e* and *f* widened, phylliform, all other setae ciliate. Claw formula of legs: 2—2—2—3.

Type-species: *Cosmochthonius emmae* Berlese, 1910.

Key to species

1 (4) Margin of phylliform setae *e* and *f* serrately ciliate, cilia short and comparatively sparse, especially not denser towards setal apices (pl. 44 : D).

2 (3) Setae *e* and *f* distally rapidly acute, apex proper obtuse (pl. 44 : G—H), distally not filiform, transverse venation of blades well discernible, margins densely

ciliate. — Setae *c* and *d* simple though very densely ciliate, setae *c* considerably longer than *d* (pl. 44 : F). Setae *h* and *ps* widened densley ciliate. Length: 230—250 µm, width: 110—115 µm.

Locus typicus: S. Vincenzo, Pisa, Italy. Distribution: Austria, Italy, Hungary, Greece, Soviet Union (SET)

emmae (Berlese, 1910)

3 (2) Setae *e* and *f* distally elongately filiform, finely attenuating (pl. 44 : D—E); transverse venation of blades obsolete, margins with routh, nearly spiniform cilia. Otherwise very similar to preceding species; setae *h* and *ps* short, with spiniform cilia (fig. 44 : A). Length: 280—290 µm, width: 120—130 µm.

Locus typicus: Crimea, Soviet Union. Distribution: only the type-locality known

tauricus Gordeeva, 1978

4 (1) Margins of phylliform setae *e* and *f* of notogaster with very long and apically ever more densely arranged cilia (pl. 45 : H).

5 (6) Bases of setae *d* meeting, both setae originating immediately near each other (pl. 45 : E). — Prodorsum elongate, anteriorly strongly convex. Prodorsal setae (pl. 45 : F) robust, thick, with spiniform ciliation. Phylliform setae as in pl. 45 : G—H. Length: 160 µm, width: 60 µm.

Locus typicus: Homutovskiy steppe, Donets district, Soviet Union. Distribution: only the typelocality known

osithchnjukovi Gordeeva, 1980

6 (5) Setae d_1 arising removed from each other (pl. 45 : D), their enclosed distance hardly shorter than that between setae d_1 and d_2. Prodorsal and other body setae highly similar to those of preceding species (pl. 45 : A). Phylliform setae *e* and *f* as in pl. 45 : B—C. Length: 150 µm, width: 60 µm.

Locus typicus: Tarhankut, Soviet Union. Distribution: Canary Islands, Soviet Union (SET, CET)

latifolius Gordeeva, 1980

Species interpretabiles

1. emmae Berlese, 1910: 222, p. 20, fig. 49 *(Cosmochthonius)*.
   emmae Van der Hammen: 1959 : 22.
   emmae Mahunka: 1980 : 110, figs. 11—12 *(Ovochthonius)*.
2. latifolius Gordeeva, 1980 : 848, figs. 3 : 1—4.
3. osithchnjukovi Gordeeva, 1980: 848, figs. 3 : 7—10.
4. tauricus Gordeeva, 1978 : 1100, figs. 1—5.

Genus 3: Nipponiella Gordeeva, 1980

Nipponiella Gordeeva, 1980 : 839.

Large-sized, well sclerotized animals. Notogaster with two transverse furrows separating three shields, but traces of third furrow also recognizable. Setae *c* and *d* originating on shield *Na* (setae *d* on the obscure line of a furrow mentioned above), setae *e* in furrow 1 and setae *f* in furrow 2; setae *h* and *p* on shield *Py*, but setae *p* shifted to ventral side. Prodorsal and notogastral setae setiform, finely ciliate. 16 pairs of notogastral setae; two pairs of anal, three pairs of adanal, and 10 pairs of genital setae present. Aggenital setae absent. Epimeral setal formula: 3—1—3—3. Legs monodactyle.

Type-species: *Trichthonius simplex* Aoki, 1966.

Only a single species known from the Palaearctic Region.

— Interlamellar setae longest of all prodorsal setae, extending far beyond rostral apex: rostral and lamellar setae about equal in length, exobothridial setae not much shorter. Sensillus (pl. 46 : H) filiform, not incrassate in any respect, bearing 7—8 long and some minute cilia on one side and several minute ones on the other. Notogastral setae of various lengths (pl. 46 : G), e_1 and f_1 longest of all. Distance between setae e_1 equalling that between setae e_1 and e_2, but setae f_1 and f_2 originating laterally and immediately near each other so that distance between setae f_1 multiply that between f_1–f_2. Genital plate whole, with 10 (7 + 3) pairs of setae. Length: 735—790 μm, width: 433—480 μm.

Locus typicus: Sanshiro, SE of Utsukushigahara, Nagano, Japan. Distribution: only the type-locality known

simplex (Aoki, 1966)

Species interpretabilis

1. simplex Aoki, 1966: 1, fig. 8. *(Trichthonius)*.

Genus 4: Gozmanyina nom. nov.

Marshallia Gordeeva, 1980: 840 (see footnote on p. 93).

Notogaster subdivided by two furrows into three shields, no other furrows recognizable, all traceless. Setae *c* and *d* arising on short chitinous pegs on shield *Na*. Setae *e* originating in furrow 1, setae *f* in furrow 2 or on the enclosed sclerites. Three pairs of prodorsal setae minute, simple, fourth interbothridial pair long and simple, sensillus slightly widened, with a peculiar marginal serration. Epimeral setal formula: 3—1—2—3. Ten pairs of genital, two pairs of anal, and three pairs of adanal, setae present; aggenital setae absent. Lengs monodactylous.

Type-species: *Trichthonius majestus* Marshall, 1970.

Only one species known from the Palaearctic Region.

— Rostrum anteriorly rounded, without any sculpture. Prodorsal surface with a reticulation composed of longitudinal and transverse chitinous laths. Rostral, lamellar and exobothridial setae minute, simple (pl. 46 : C) interlamellar setae arising on chitinous pegs (pl. 46 : D), long and smooth. Sensillus setiform, marginally serrately dentate. Notogastral setae *c* and *d* simple, smooth, originating also on chitinous pegs. Phylliform setae *e* and *f* (pl. 46 : E—F) with a peculiar, squamose marginal ornamentation. Length: 155 μm, width: 95 μm.

Locus typicus: South Kurilsk, Soviet Union. Distribution: only the type-locality known

golosovae (Gordeeva, 1980)

Species interpretabilis

1. golosovae Gordeeva, 1980: 840, figs. 1:1—4 *(Marshallia)*.

Genus 5: Krivolutskiella Gordeeva, 1980

Krivolutskiella Gordeeva, 1980 : 841.

Notogaster subdivided by three furrows into four shields. Seta *c* arising on shield *Na seta d* on shield Nm_1, insertion points of latter ones outside of furrow, setae *e* and *f* within furrows. Rostrum perforated. Prodorsal setae modified, all notogastral setae phylliform, excepting fibriate f_1 (ornamented with long lateral branches). Ten pairs of genital, four pairs of anal and four pairs of adanal setae present.

Type-species: *Krivolutskiella pubescens* Gordeeva, 1980.

Only the type-species known.

— Rostrum concave, surface punctate, arranged in four rows. All prodorsal setae widened, spinosely ciliate. Stalk of sensillus thin, distally considerably widened. Notogastral setae *c* and *d* short, distally triangular with a deep incision or cordiform. Setae *e* and *f* erectile, *e* widely phylliform, f_1 with long ramifying, arcuate, flagellately attenuating lateral branches; hairs f_2 phylliform (pl. 46 : A). Posteromarginal setae very wide, ciliate. Chaetom of anogenital region as in pl. 46 : B. Length: 180 µm, width: 110 µm.

Locus typicus: Canary Islands, Spain. Distribution: only the type-locality known

pubescens Gordeeva, 1980

Species interpretabilis

1. pubescens Gordeeva, 1980: 842, figs. 1:5—6.

Superfamily 6: Sphaerochthonioidea Grandjean, 1947

Sphaerochthoniidae Grandjean, 1947 : 224.
Sphaerochthonioidea Balogh et Mahunka, 1979 : 282.

Squat, angular, dorsoventrally very thick, characteristically gray or black animals. Notogaster with merely one furrow subdividing notogaster into two (occasionally overlapping) shields. Seate *c* originating on shield *Na*, setae *d* in the furrow (often covered, hardly discernible), setae *e*, *f* and *h* on shield *Py*: a total of 15 pairs of notogastral setae present. Setae, excepting *d*, big, mostly T-shaped, densely plumose (catkin-like). Notogaster covered with chitinous ribs arranged in a polygonal pattern, but entire animal usually hidden under a thick cerotegument.

Superfamily monotypical.

Family 1: Sphaerochthoniidae Grandjean, 1947

Sphaerochthoniidae Grandjean, 1947 : 224.
Characterized as the superfamily and containing one genus.

Genus 1: SPHAEROCHTHONIUS Berlese, 1910

Sphaerochthonius Berlese, 1910 : 218.
Sphaerochthonius Grandjean: 1932 : 30.
Sphaerochthonius Van der Hammen: 1959 : 25.
Sphaerochthonius Wallwork: 1960 : 377.
Sphaerochthonius Aoki: 1977 : 87.

Besides the characteristics enumerated in the diagnosis of the superfamily, the genus is characterized by the undivided genital plates bearing 6—10 pairs of setae. Aggenital hairs absent. Anal plates with 4—10 pairs, adanal ones with 4—5 pairs, of (partially catkin-like) hairs. Epimeral setal formula: 3—2—3—4. Legs tridactylous, heterodactylous.

Type-species: *Hypochthonius splendidus* Berlese, 1904.

Key to species*

1 (2) Excepting setae *d*, all notogastral setae T-shaped (pl. 47 : A). Prodorsum only with setae *exa* and *exp* not branching, also interlamellar ones T-shaped. Sensillus auriculatel widened. Five pairs of anal, four pairs of adanal and eight pairs of genital, setae present. Anal setae simple, adanal hairs T-shaped. Length: 275—310 μm, width: 165—200 μm.

Locus typicus: "Giardino di Boboli", Florence, Italy. Distribution: Central and South Europe, North Africa, Soviet Union (SET, TC, KZ, SMA)

splendidus (Berlese, 1904)

2 (1) A part of notogastral setae, those on shield *Py*, incrassate and ornamented with papillae, but not branching in a T-shape (pl.47 : D).—Rostral and lamellar setae of prodorsum also T-shaped, but interlamellar ones also unbranched. Sensillus (pl. 47 : B, C) spoon-shaped in posterior view, but nearly resembling dorsal setae in superior view. Eight pairs of simple genital, 7—8 pairs of also simple anal, four pairs of furcate adanal (pl. 47 : F), setae but none with papillae. Notogastral sculpture as in pl. 47 : G. Length: 310 μm, width: 180 μm.

Locus typicus: Mt. Yumishi-dake, Kakeroma Island, Amami Islands, Japan. Distribution: only the type-locality known

suzukii Aoki, 1977

Species interpretabiles

1. splendidus Berlese, 1904 : 26, pl. 2. fig. 37 *(Hypochthonius)*.
   splendidus Van der Hammen: 1959 : 25.
2. suzukii Aoki, 1977 : 85, figs. 1—5.

* The occurrences of two further *Sphaerochthonius* species have been reported (*S. gemma* (Oudemans, 1909) and *S. transversus* Wallwork, 1960) from the Soviet Union (Crimea and the Caucasus). The first is known only from Java in the Oriental Region, the second from Ghana, Africa. Their occurrences in the above localities are improbable (see D. Krivolutsky: 1975 : 52), and in our opinion we are confronted with two undescribed species.

Superfamily 7: HETEROCHTHONIOIDEA Grandjean, 1954

Heterochthoniidae Grandjean, 1954 : 429.
Heterochthonioidea Grandjean, 1969 : 142.
Heterochthoniidae Ryabinin et D. Krivolutsky: 1977 : 1729.

Convex, rounded, light, dirty whitish to yellowish white animals. Prodorsum with lentil-shaped dark eyespots. Notogaster with six pairs of long and rigid setae, among these d_1 and d_2 reflexed to notogaster, while setae e_1, e_2, f_1 and f_2 rigid, erect; all other setae short and simple.

Representing a single family.

Family 1: HETEROCHTHONIIDAE Grandjean, 1954

Heterochthoniidae Grandjean, 1954 : 429.

Characterized as the superfamily, containing two genera.

Key to genera

1 (2) Anterior half of prodorsum with two-eyespots medially. No proclinate spines present between setae *e* and *f* on notogaster
genus 2: **Ovochthonius** Ryabinin, 1977

2 (1) Anterior half of prodorsum with a single eyespot medially. Proclinate spines present between setae *e* and *f* on notogaster
genus 1: **Heterochthonius** Berlese, 1910

Genus 1: HETEROCHTHONIUS Berlese, 1910

Heterochthonius Berlese, 1910:221 *[Cosmochthonius (Heterochthonius)]*.
Heterochthonius Grandjean: 1928 : 235.
Heterochthonius Van der Hammen: 1959 : 23.
Heterochthonius Ryabinin et D. Krivolutsky: 1977 : 1729.

Prodorsum very widely rounded; anterior third with a median unpaired black eyespot; beside bothridia a smaller spot each. Notogastral setae *c, d, e* and *f* originating behind each other on transverse laths in anterior half of notogaster, long and rigid; all other posteromarginal setae thin, simple. Three proclinate robust spines starting from line of setae *f*. Genital opening considerably larger than anal one, transversely subdivided, each portion bearing five pairs of genital setae. Anal plates emitting only one pair of setae; three pairs of adanal hairs present

Type-species: *Cosmochthonius (Heterochthonius) gibbus* Berlese, 1910.

Key to species

1 (2) Sensillus (pl. 48 : C) weakly but well discernibly widened, marginally with small teeth but not ciliate. Rostrum protruding from rounded anterior margin of prodorsum. Lamellar setae not reaching anterior margin of prodorsum (pl.

48 : A), not even half as long interlamellar setae. Spines proceeding from line of seta *f* on notogaster short, narrow, even unpaired median spine reaching only to transverse line of setae *e*. Long and rigid notogastral setae finely ciliate (pl. 48 : B). Ventral side as in pl. 48 : D. Length: 376—395 μm, width: 232—244 μm.

Locus typicus: Mte. Falterona, Tuscany, Italy. Distribution: France, Italy

gibbus Berlese, 1910

2 (1) Sensillus setiform, bilaterally with long cilia. Lamellar setae considerably projecting beyond rostral apex. Notogastral spines different.

3 (4) Median spine of notogaster long, narrow, nearly reaching transverse line of setae *d;* the two lateral spines wide, approximately triangular, only half as long as median spine (pl. 48 : E).—Rostrum semicirculately rounded, medially without projecting part. Exobothridial setae considerably longer than rostral ones; lamellar pair also relatively long, interlamellar seta only one and a half times longer. Length: 430 μm, width: 240 μm.

Locus typicus: Kemerovo, Salarskiy range, Soviet Union. Distribution: only the type-locality known

byzovae D. Krivolutsky, 1977

4 (3) Median and lateral notogastral spines uniform: all three narrow and long, but median spine still longer (fig. 48 : F).—Exobothridial hairs shorter than rostral setae; all other prodorsal setae shorter than those of preceding species, lamellar seta only half as long as interlamellar setae. Length: 420 μm, width: 240 μm.

Locus typicus: Botanical Garden, Batumi, Soviet Union. Distribution: only the type-locality known

caucasicus D. Krivolutsky, 1977

Species interpretabiles

1. byzovae D. Krivolutsky, 1977 : 1731, fig. zh.
2. caucasicus D. Krivolutsky, 1977 : 1929, fig. e.
3. gibbus Berlese, 1910 : 222, pl. 20, fig. 50 */(Cosmochthonius) Heterochthonius/.*
   gibbus Grandjean: 1928 : 235, figs. 1—4.
   gibbus Van der Hammen: 1959 : 23.
   gibbus Mahunka: 1980 : 108, figs. 9—10.

Genus 2: Ovochthonius Ryabinin, 1977

Ovochthonius Ryabinin, 1977 : 1729.

Anterior third of prodorsum with a pair of black eyespots, those beside bothridia absent. Lamellar setae longer and more robust than interlamellar ones. Spines originating in line of setae *f* absent on notogaster. Proportion and position of genital and anal openings as in preceding genus.

Only a single species known.

— Rostrum widely rounded. Rostral and exobothridial setae equally short. Lamellar seta longest among prodorsal setae (pl. 48 : I). Notogastral seta d_1 essentially shorter than d_2 (pl. 48 : G), latter densely ciliate, the others less so. Sensillus slightly widened pl. 48 : H, marginally with densely arranged long cilia. Ventral side as in fig. 48 : J. Length: 420 µm, width: 240 µm.

Locus typicus: Great Khekhcir, Khrebet Khekhcir, Khabarovsk, Soviet Union. Distribution: only the type-locality known

rossicus Ryabinin, 1977

Species interpretabilis

1. rossicus Ryabinin, 1977 : 1729, figs. a—g.

Superfamily 8: ATOPOCHTHONIOIDEA Grandjean, 1948

Atopochthoniidae Grandjean, 1948 : 53.
Atopochthonioidea **stat. nov.**

Notogaster subdivided, in addition to primary furrows also further hollows and elevated parts present, shields proper difficult to recognize; two furrows differentiating three shields. Setae *c* originating on shield *Na*, setae *d* reduced, *e*, *f* and h_1 on dorsal side of shield *Py*, all other setae *h* and *ps* shifted to ventral side. Most prodorsal and notogastral setae widened, fungiform or phylliform, but also some blade-like setae present. A total of 14 pairs of notogastral setae. Ventral plates with six pairs of genital, three pairs of anal and four pairs of adanal setae present; aggenital and peranal ones absent.

A single family and a single genus known.

Family 1: ATOPOCHTHONIIDAE Grandjean, 1948:

Atopochthoniidae Grandjean, 1948 : 53.

Characterizable by the features of the superfamily.

Genus 1: ATOPOCHTHONIUS Grandjean, 1948

Atopochthonius Grandjean, 1948 : 43.

Besides the features given in the diagnosis of the superfamily characterized by the epimeral setal formula: 3—1—3—4, and 2-clawed legs.

Type-species: *Atopochthonius artiodactylus* Grandjean, 1948.

Key to species

1 (2) A twofold difference in length between notogastral setae *c* and *e* (fig. 49 : A); seta *c* cordiform, *e* long, lanceolate.—Body setae composed of foveolate phylliform lamellae attached to short, spiniform bases (fig. 49 : B); bases of setae *c* and *e* equally short, those of *f* multiply longer. Prodorsal setae also

extremely widened, very large, c_3 and *cp* umbellate or fungiform*, *f* narrow, distal half rapidly attenuating, concave. Ventral setae simple, setae of palp also simple, but legs also with some umbellate setae, especially ventrally. Length: 200 μm.

Locus typicus: Mugello, near Florence, Italy. Distribution: Sweden, Italy, Yugoslavia

artiodactylus Grandjean, 1948

2 (1) Notogastral setae *c* and *e* nearly equal in length and uniform in shape (fig. 49: C—D).—Prodorsal setae as in preceding species, notogastral setae *f* not attentuating distally. Otherwise very similar to its congener (no other difference mentioned even by its author). Length: 191 μm, width: 101 μm.

Locus typicus: Mayma, Mts. Altai, Soviet Union. Distribution: only the type-locality known

maimaensis Grishina, 1971

Species interpretabilis

1. artiodactylus Grandjean, 1948: 43, figs. 4—7.
   artiodactylus Forsslund: 1956: 212, figs. 4—6.
2. maimaensis Grishina, 1971: 71, figs. 1—2.

Superfamily 9: PTEROCHTHONIOIDEA Grandjean, 1950

Pterochthoniidae Grandjean, 1950: 105.
Pterochthonioidea **stat. nov.**

Notogaster subdivided by three furrows into four shields. All 16 pairs of notogastral setae evolved, all modified. Setae *c* originating on shield *Na*, setae *d* on Nm_1, setae *e* on Nm_2, setae *f* on shield *Py*, on anterior margins of shields, e_2 and f_2 arising in the furrows. Setae *h* and *ps* shifted to ventral side, circularly around anal and adanal setae. A peranal plate differentiated, bearing one pair of peranal setae. Eight pairs of genital, three pairs of anal and four pairs of adanal, setae present; aggenital setae absent.

The superfamily includes a single family with a single genus.

Family 1: PTEROCHTHONIIDAE Grandjean, 1950.

Pterochthoniidae Grandjean, 1950: 105.

Characterized by the features of the superfamily.

Genus 1: PTEROCHTHONIUS Berlese, 1913

Pterochthonius Berlese, 1913: 104 *[Cosmochthonius (Pterochthonius)]*.
Pterochthonius Grandjean: 1950 : 85.

In addition to the features given in the diagnosis of the superfamily, the genus can be characterized by the epimeral setal formula: 3—1—3—3 and the monodactyle legs.

* Lamellate portion of setae frequently broken off, only the stalk or base of hair visible.

Type-species: *Cosmochthonius angelus* Berlese, 1910.
Only the type-species known.

— Every prodorsal and notogastral setae expanded: umbellate, fungiform or phylliform (pl. 49 : F), no elongated and attenuating setae present, hence modified setae covering nearly entire body; laminate setac ornamented with large foveolae (pl. 49 : H). Peranal, anal and adanal setae in a concentrical arrangement. Ventral setae simple. End of palp (pl. 49 : I) with several modified, ribbon-shaped setae. Legs with modified setae dorsally. Length: 353—450 μm.

Locus typicus: Vallombrosa, Italy. Distribution: France, Italy, Yugoslavia, Soviet Union (FE), Mexico

angelus (Berlese, 1910)

Species interpretabilis

1. angelus Berlese, 1910 : 338 *(Cosmochthonius)*.
angelus Berlese: 1913 : 104, pl. VIII, fig. 93 *[Cosmochthonius (Pterochthonius)]*.
angelus Grandjean: 1950 : 85, figs. 1—7.
angelus Van der Hammen: 1959 : 24.

Cohort 2: Arthroptyctima Grandjean, 1967

Arthroptyctima Grandjean, 1967 : 245.
Arthroptyctima Balogh et Mahunka: 1979 : 279.

Body ptychoid. Notogaster subdivided: unequivocal in case of several differentiated segments or well discernible furrows, but possibly also concealed and then original segments identifiable only with recourse to chaetotaxy. In this latter case segments *F*, *H* and *PS* fused with anal one, thus a furrow present actually between segments *E* and *F*, separating two large shields.

About 12 genera of the two monotypical superfamilies are known, five of them occurring in the Palaearctic Region.

Key to superfamilies

1 (2) Anal and genital openings elongated, touching (pl. 50 : C); pygidial part of notogaster with several well discernible furrows (pl. 50 : A)
superfamily 1: **Protoplophoroidea** Ewing, 1917

2 (1) Anal opening round, separated from genital opening and situated on a distinct ventral plate (pl. 52 : F); notogaster without well discernible furrows, its segments *(F—PS)* fused with anal plate
superfamily 2: **Mesoplophoroidea** Ewing, 1917

Superfamily 1: Protoplophoroidea Ewing, 1917

Protoplophoridae Ewing, 1917 : 119.
Protoplophoridae Jacot: 1923 : 161.
Protoplophoridae Grandjean: 1932 : 10.

Protoplophoridae Van der Hammen: 1959 : 28.
Protoplophoroidea Grandjean, 1965 : 101.

Minute, mostly weakly, only exceptionally well pigmented species; in case of danger rolling into a pill, so that also genital opening covered besides legs and mouth organs.—The only segmented group of the *Ptyctima* and thus apparently related to the supercohort *Arthronota**. The hysterosoma is covered by four shields; the anterior one is usually the largest: the unpaired pronotaspis, the posterior shield is also unpaired: the metanoptaspis (= pygidium), and between them a lateral pseudaspis on each side. The pygidium displays two sutures or a respective lath (= carina) emitting the setae *e* and *f*.

Family 1: PROTOPLOPHORIDAE Ewing, 1917

Protoplophoridae Ewing, 1917 : 119.

The majority of the known 9 genera and 14 species of the family have been described from the Palaearctic Region, but a datum each is known from every zoogeographical area. Owing to their hidden habits and rarity, information is meagre in every respect. Most species are known by one or two specimens only.

Key to genera

1 (6) Anal and adanal plates distinct, separated by a well discernible furrow (pl. 50 : E).

2 (5) Anal plates small, adanal plates considerably larger, extending along and up to base—and thus enclosing — anal plates (pl. 50 : I).

3 (4) Legs monodactyle. Notogaster smooth or punctate. Rostrum smooth in front, without teeth. Notogastral setae simple, thin. Anal plate essentially longer than wide
genus 1: **Protoplophora** Berlese, 1910

4 (3) Legs tridactylous. Notogastral surface with a rough and heavy polygonal sculpture consisting of chitinous laths. Notogastral hairs mostly T-shaped, branching (pl. 50 : G)
genus 2: **Hauseroplophora** Mahunka, 1977

5 (2) Anal plate large, essentially larger than narrow adanal plates, these latter extending only to half length—and therefore not enclosing—anal plates (pl. 50 : E)
genus 3: **Grandjeanoplophora** Balogh et Mahunka, 1979

6 (1) Anal and adanal plates fused (pl. 51 : H).

7 (8) Ano-adanal plates with four aligned setae on each plate, no deep hollow on their surfaces present. Sensillus spherically widended, densely ciliate
genus 4: **Cryptoplophora** Grandjean, 1932

8 (7) Ano-adanal plates with six (4 + 2) setae in two rows each; plates at half their length with a deep hollow (bursa). Sensillus elongately spatulate (pl. 51 : K)
genus 5: **Bursoplophora** Subias et Pérez-Iñigo, 1978

* See Grandjean: 1932 : 30, 1954 : 334; and Mahunka, 1977 : 467.

Genus 1: Protoplophora Berlese, 1910

Protoplophora Berlese, 1910 : 217.
Protoplophora Van der Hammen: 1959 : 28.
Protoplophora Grandjean: 1932 : 24.

Propodosoma and notogaster without sculpture, their surfaces smooth. Rostral apex without teeth. Sensillus clavate. Prodorsal and notogastral setae simple, short. Anal and adanal plates not fused, adanal plates large, enclosing small and narrow anal plates. Legs monodactyle.

Type-species: *Protoplophora palpalis* Berlese, 1910.

A single Palaearctic species known.

— Rostral setae arising on prodorsum, far from rostrum, considerably nearer to lamellar setae than to anterior margin of body (pl. 50 : B). Sensillus clavate, outer margin densely ciliate. Two pairs of long exostigmatal setae present; all other prodorsal setae very short. Notogaster entirely smooth, even pygidium impunctate. Notogastral setae also simple, *f* comparatively long, apically reaching posterior margin of pygidium (pl. 50 : A). Adanal plates extremely large, wide, smooth, with three pairs of setae. Anal plates narrow, with four pairs of setae (pl. 50 : C). Length: 182 μm, width: 96 μm.

Locus typicus: Palermo, Sicily, Italy. Distribution: Spain, Italy, Soviet Union (Crimea)

palpalis Berlese, 1910*

Species interpretabilis

1. palpalis Berlese, 1910 : 217, figs. 53, 63a.
   palpalis Grandjean: 1932 : 24, figs. 5A—C.
   palpalis Van der Hammen: 1959 : 29.

Genus 2: Hauseroplophora Mahunka, 1977

Hauseroplophora Mahunka, 1977 : 467.

Prodorsum and notogaster with a rough and heavy polygonal sculpture composed of chitinous laths. Colour dark, occasionally black. Dorsal setae bi- or trifurcate, usually T-shaped, their surfaces aciculate and covered with secretion granules. Interlamellar setae phylliform, hence highly similar in habitus to *Sphaerochthonius* Berlese, 1910. Genital plate small, indiscernible in "rolled" state of specimen. Adanal plate very large, twice longer than anal plate, both adanal plates enclosing anal ones. Ventral side and anogenital plates with a rough sculpture similar to that of notogastral surface.

Type-species: *Hauseroplophora soniae* Mahunka, 1977.

Only two species of the genus are known, one from the Palaearctic Region.

* The above description is based on Grandjean's redescription: its identity with Berlese's type needs corroboration!

— Rostrum wide. Entire prodorsal and notogastral surfaces with a heavy polygonal sculpture composed of chitinous laths (pl. 50 : G). Lamellar and notogastral setae ramifying, T-shaped, covered with granules. Sensillus (pl. 50 : H) large, auricular, also covered with granules of secretion. Anal and adanal plates ornamented with large foveolae. Setae very short, hardly discernible; five pairs of anal and four pairs of adanal setae present (pl. 50 : I). Length: 217 µm, with: 191 µm.

Locus typicus: Krestena, Peloponnes, Greece. Distribution: only the type-locality known

phitosi Mahunka, 1977

Species interpretabilis

1. phitosi Mahunka, 1977 : 545, figs. 1—3.

Genus 3: Grandjeanoplophora Balogh et Mahunka, 1979

Grandjeanoplophora Balogh et Mahunka, 1979 : 288.

Body surface smooth. Rostral apex serrate. Sensillus clavate. Prodorsal and notogastral setae simple, short, setae *e* (e_1, e_2) absent, indicated only by their alveoli. Anal and adanal plates not fused, but anal plate very large, adanal plates considerably smaller, narrower, not enclosing anal plate posteriorly. Two pairs of adanal setae. All legs tridactylous, heterodactylous.

Type-species: *Cryptoplophora mauritanica* Grandjean, 1932.

A single Palararctic species known.

— Body smooth, without sculpture. Rostral setae longest of all prodorsal setae, originating in anterior fourth of prodorsum (pl. 50 : D); two pairs of minute exostigmatic setae present. Sensillus long, clavate, outer margin with long and dense cilia. The two pygidial sutures far from each other, neither one reaching lateral margin of pygidium (pl. 50 : F). Setae *f* comparatively short, not extending to posterior margin of pygidium. Adanal plates with two pairs, anal plates with four pairs of minute setae (fig. 50 : E). Length (when extended): 160 µm, width: 80 µm.

Locus typicus: Agadir, Morocco. Distribution: only the type-locality known

mauritanica (Grandjean, 1932)

Species interpretabilis

1. mauritanica Grandjean, 1932 : 21, figs. 3A—D, 4A *(Cryptoplophora)*.

Genus 4: Cryptoplophora Grandjean, 1932

Cryptoplophora Grandjean, 1932 : 18.
Cryptoplophora Grandjean: 1954 : 122.

Body surface smooth. Rostral apex dentate. Notogastral setae simple; sensillus extremely wide, short, spherical. Anal and adanal plates fused, emitting only four pairs of setae. All legs tridactylous, heterodactylous.

Type-species: *Cryptoplophora abscondita* Grandjean, 1932.

A single species of very extensive (?) distribution.*

— Prodosal setae short; rostral pair arising on prodorsal surface, far from rostrum (pl. 51 : B), considerably nearer to each other than lamellar setae. Sensillus (pl. 51 : B) heavily clavate, densely ciliate. Notogastral setae simple, relatively long, *f* reaching to posterior margin of body (pl. 51 : A). The two pygidial sutures (bearing also setae) comparatively near each other, their enclosed distance smaller than that between second suture and posterior margin of pygidium. Anogenital and epimeral regions, configuration of leg I and palp, as well as their chaetotaxy, given in pl. 51 : C—E. Length (when extended): 180 µm, width: 100 µm.

Locus typicus: Algiers, Algeria. Distribution: Algeria, Panama

abscondita Grandjean, 1932

Species interpretabilis

1. abscondita Grandjean, 1932: 18, figs. 2C—G, 4B.
   abscondita Grandjean: 1954: 332, figs. 5—7.

Genus 5: Bursoplophora Subias et Pérez-Iñigo, 1978

Bursoplophora Subias et Pérez-Iñigo, 1978 : 387.

Body with a weak polygonal ornamentation, discernible rather on notogaster. Body setae simple, smooth. Sensillus spatulate. Anal and adanal plates fused, surface with six recognizable pairs of setae; surface of plates deeply impressed medially, this area framed by chitinous rugae. All legs tridactylous.

Type-species: *Bursoplophora iberica* Subias et Pérez-Iñigo, 1978.

Only two Palaearctic species known.

Key to species

1 (2) Hollow (bursa) of ano-adanal plate transversely extended (pl. 51 : H). Sensillus with an also laterally extending long row of cilia (pl. 51 : I),—Prodorsum and segments *C—E* smooth, pygidium heavily punctate (pl. 51 : F). Body setae short. Rostral seta (pl. 51 : G) longest on prodorsum; e_1 longest on notogaster, nearly twice longer than seta f_1; this latter not even approaching posterior margin of body. Length: 250 µm, width: 125 µm.

Locus typicus: Agadir, Morocco. Distribution: Morocco, Greece

bivaginata (Grandjean, 1932)

2 (1) Hollow (bursa) of ano-adanal plate longitudinal, not widened transversely at all (pl. 51 : L). Sensillus (pl. 51 : K) with minute dentation, discernible only apically.—Body surface with irregular and fine polygonal sculpture, (pl. 51 : J),

* Grandjean suggested its introduction to North Africa from Central America. A situation requiring further investigation.

wholly impunctate. Body setae also short, rostral and all other prodorsal setae equal in length; notogaster only with setae e_1 and f_1 elongate, without any essential difference in length; f_1 nearly reaching posterior margin of body. Length: 200—235 μm.

Locus typicus: "Pantano del la Jarosa", Madrid, Spain. Distribution: Spain

iberica Subias et Pérez-Iñigo, 1978

Species interpretabiles

1. bivaginata (Grandjean, 1932): 26, figs. 6a—d *(Protoplophora)*.
   bivaginata Subias et Pérez-Iñigo: 1978: 392.
2. iberica Subias et Pérez-Iñigo, 1978: 388, figs. 1—6.

Superfamily 2: MESOPLOPHOROIDEA Ewing, 1917

Mesoplophorinae Ewing, 1919: 119.
Mesoplophoridae Jacot: 1923: 161.
Mesoplophoroidea Van der Hammen: 1959: 31.
Mesoplophoroidea Grandjean: 1965: 91.

Notogaster apparently entire, without transverse furrows, but this notogaster actually composed merely of enlarged segments *C—E,* while segments *F—Ps,* sharply delimited, are united with anal plate. Thus a distinct "ventral" plate present, bearing genital and anal openings far removed from each other. "Notogaster" with 7—8 pairs of setae, "ventral" plate with 9—10 pairs. Genital plate subdivided by one furrow into two parts, anterior smaller part bearing 1—2 pairs of setae, posterior larger part with 5—6 pairs. Anal plate emitting 2—3 pairs of setae. Legs monodactyle.

The tritonymphs of the group resemble extraordinarily the adults of the *Euphthiracarides:* the notogaster carries 15 pairs of setae, and the anogenital region shows also a similar construction. The inference that only the segments *C, D* and *E* partake in the composition of the notogaster of the *Mesoplophorides,* while segments *F, H* and *Ps* coalesce with the anal segment, is based on the configuration of the tritonymphs. Thus the "ventrale" is to be considered the pygidioventrale, separated from the notogaster by the suture between the segments *E* and *F,* that is, suture ar_3. Accordingly, the setae of the notogaster are as follows: c_1, c_2, c_3, d_1, d_2, d_3, e_1, e_2 (in the species *M. japonica* Aoki, 1970, seta c_3 is absent). It also follows that the setae of the segmentum pygidioventrale represent the setae *f, h* and *ps,* but they have not been homologized so far and are given serial numbers in the descriptions.

According to Grandjean, the superfamily *Mesoplophoroidea* contains two families, the *Mesoplophoridae* Ewing, 1917, and the *Archoplophoridae* Grandjean, 1965. This latter does not occur in the Palaearctic Region and has not been considered in the diagnosis of the superfamily.

One family and a single genus occur in the Palaearctic Region.

Family 1: Mesoplophoridae Ewing, 1917

Mesoplophorinae Ewing, 1917: 119.
Mesoplophoridae Grandjean: 1933: 308.
Based on a single genus.

Genus 1: Mesoplophora Berlese, 1904

Mesoplophora Berlese, 1904:23.
Mesoplophora Van der Hammen: 1959:31.
Phtiracarulus Berlese, 1920:149 (Van der Hammen: 1959:31).

Aspic acute, notogaster convex. Sensillus setiform or filiform, distally occasionally fusiform, smooth or ciliate. Aspis with four pairs of smooth or finely ciliate setae. "Notogaster" with 7—8 pairs, "ventral" plate with 9—10 pairs, genital plates with 5—7 pairs, anal plates with 2—4 pairs, of setae. Anal and genital openings round, separated by a wide bridge from each other.

Type-species: *Mesoplophora michaelina* Berlese, 1904.

The species of the genus occur also in the Holotropic zone in addition to the Holarctic Region. Five species are so far known from the Palaearctic Region.

Key to species

1 (4) Sensillus smooth, setiform or slightly incrassate.

2 (3) Distal end of sensillus gradually thickening: fusiform or clavate (pl. 52:A).—Setae of aspis and "notogaster" long, finely ciliate. "Ventral" setae thinner and smooth; no essential differences in length. "Notogaster" with 8, "ventral" plate with 10, pairs of setae (pl. 52:B). Genital plate with 7 (2+5) or (3+4), anal plate with three pairs of setae. Length of aspis: 180—200 μm, length of notogaster: 255—280 μm, height of notogaster: 190 μm.
Locus typicus: in original description not stated. Distribution: Europe, North Africa, Soviet Union (CET, SET)
pulchra Sellnick, 1928

3 (2) Sensillus (pl. 53:B) gradually thinning apicad or throughout evenly thin, never conspicuously fusiform or clavate. "Notogastral" setae short (pl. 53:D), shorter than length of anal opening.—Seven pairs of "notogastral" setae present, c_3 absent. Ventral plate with 10, genital plate with 1+5, anal plate with three pairs of setae (pl. 53:C).—The measurement data given by Aoki should be checked.
Locus typicus: Mt. Mokkoku, Tsutsu, Japan. Distribution: Japan
japonica Aoki, 1970

4 (1) Sensillus gradually tapering apicad, setiform, with at least 2—3, occasionally spiniform, cilia.

5 (6) "Notogastral" setae very long, *c* as long as sensillus (pl. 53:A). Sensillus with merely 2—3 minute spines.—Prodorsal setae—excepting exoboth-

ridial seta—also very long, finely ciliate, e_1 and e_2 of "notogaster" shorter than seta *c*, very finely ciliate. Length: 420 µm, width: 250 µm.

Locus typicus: Florence, Italy. Distribution: Italy

michaeliana Berlese, 1904

6 (5) "Notogastral" setae essentially shorter, *c* much shorter than sensillus; latter bearing at least 5—6 setae or spines.

7 (8) Sensillus at most with six cilia of divers lengths, first lateral cilium appearing approximately at midlength of sensillus (pl. 53 : E). Prodorsum with an acute rostrum, setae long and arcuate excepting exobothridial ones. "Notogastral" setae comparatively short (pl. 53 : F), c_1 shorter than c_2. Ventral plate with short setae (pl. 53 : G), genital plates with seven (1 + 6), anal plates with two pairs of setae. Measurements unknown.

Locus typicus: Thrace, Greece. Distribution: Greece

graeca Walzl, 1974

8 (7) Sensillus at least with eight cilia equal in length, appearing already in first (basal) third of sensillus (pl. 52 : E). Rostrum acute, setae long (pl. 52 : C). "Notogastral" setae longer than in preceding species, no difference in length between c_1 and c_2. Ventral plate (pl. 52 : F) with 10 pairs of setae, longer than in preceding species: genital plate with seven pairs of setae (first pair not conspicuously distinct!), anal plate with two pairs of setae. Length of aspis: 150—157 µm, length of notogaster: 184—190 µm, height of notogaster: 102—105 µm.

Locus typicus: Kastritsi, Peloponnes, Greece. Distribution: only the type-locality known

pectinata Mahunka, 1979

Species interpretabiles

1. graeca Walzl, 1974 : 534, figs. 1—3.
2. japonica Aoki, 1970 : 397, figs. 2—4.
3. Michaeliana Berlese, 1904 : 23, pl. 2, fig. 51.
   michaeliana Van der Hammen: 1959 : 32.
   michaeliana Mahunka: 1980 : 108, fig. 5.
   perexiguus Berlese, 1920: 149 *[Phtiracarus (Phtiracarulus)]*.
4. pectinata Mahunka, 1979 : 551, figs. 11—14.
5. pulchra Sellnick, 1928 : 38, fig. 88.
   pulchra Grandjean: 1933 : 310, figs. 1—5.
   pulchra Grandjean: 1965 : 93, figs. 1—2.

Species inquirenda

— caucasica D. Krivolutsky, 1975 : 368*.

Supercohort 3: Parhyposomata Balogh et Mahunka, 1979

Parhyposomata Balogh et Mahunka, 1979 : 280.

Weakly chitinized, colourless animals. Mouthparts well visible in superior view. Notogaster subdivided by two obsolete transverse furrows, decurrent ventrally

* The description is unknown to us; we have no other information than D. Krivolutsky's identification key (in: Ghilarov et D. Krivolutsky, 1975:368, fig. 923); in this form a nomen nudum.

between either adanal and anal, or pseudanal and anal, setae. 16 pairs of notogastral setae present, f_1 invariably developed, f_2 occasionally absent. Oil glands conspicuous. Slit organs present. Peranal plate present (absent in tropical forms). One femur present, tarsi invariably 3-clawed, median claw shorter and more arcuate than lateral ones.

A single superfamily with two families; both occur in the Palaearctic Region.

Superfamily 1: PARHYPOCHTHONIOIDEA Grandjean, 1932

Parhypochthoniidae Grandjean, 1932: 426.
Parhypochthonioidea Van der Hammen: 1959: 10.
Parhypochthonoidea Strenzke: 1963: 251.

The diagnosis of the supercohort refers also to the superfamily.

Composed of two families.

Key to families

1 (2) Glandula opening on end of a long, tubuliform appendage laterally on notogaster (pl. 54: 17). Epimeral setal formula: 3—1—3—4. Peranal plate present. Nine pairs of genital and one pair of aggenital setae present
family 1: **Parhypochthoniidae** Grandjean, 1932

2 (1) Glandula opening on surface of notogaster, no tubiform appendage present. Epimeral setal formula: 3—2—3—4. Peranal plate absent. 9—10 pairs of genital and two pairs of anal setae present
family 2: **Gehypochthoniidae** Strenzke, 1963

Family 1: PARHYPOCHTHONIIDAE Grandjean, 1932

Parhypochthoniidae Grandjean, 1932: 426.
Parhypochthoniidae Van der Hammen: 1959: 10.
Parhypochthoniidae Strenzke: 1963: 250.

Glandulae opening distally on tubuliform appendages, placed postero-laterally on notogaster.

A single genus known.

Genus 1: PARHYPOCHTHONIUS Berlese, 1904

Parhypochthonius Berlese, 1904: 25.
Parhypochthonius Grandjean: 1934: 423.

Narrow, elongated animals. Prodorsum separated from notogaster by a soft cuticular zone, body there constricted. Peranal plate present, with one pair of peranal setae: nine pairs of genital, one pair of aggenital, four pairs of anal and four pairs of adanal, setae present. Epimeral setal formula: 3—1—3—4.

Type-species: *Parhypochthonius aphidinus* Berlese, 1904.
Represented by a single species in the Palaearctic Region.

— Prodorsum angular, rostrum hardly projecting medially, rostral hairs arising there near each other. Sensillus with long cilia on one side, with short on the other one (pl. 54 : A—B). Notogaster subdivided into two parts by a furrow. Setae *c* and *d* originating on shield *Na*, all other notogastral as well as the three adanal setae on shield *Py*. Seta f_2 arising on tubiform opening of oil gland. Length: 350—470 μm, width: 200—225 μm.
Locus typicus: Florence, Italy. Distribution: Southern Europe, North Africa, Soviet Union (SET, WS, FE), Japan

aphidinus Berlese, 1904

Species interpretabilis

1. aphidinus Berlese, 1904 : 25, pl. 2, fig. 42.
aphidinus Grandjean: 1932 : 425.
aphidinus Van der Hammen: 1959 : 10.
aphidinus Aoki: 1969 : 117, figs. 1—2.

Family 2: Gehypochthoniidae Strenzke, 1963

Gehypochthoniidae Strenzke, 1963 : 251.
Notogaster without tubiform appendage, oil gland opening directly on notogastral surface.
Composed of a single genus.

Genus 1: Gehypochthonius Jacot, 1936

Gehypochthonius Jacot, 1936 : 22.
Gehypochthonius Strenzke: 1963 : 231.
Notogaster with one furrow and thus subdivided into two shields. Setae *c* and *d* arising on shield *Na*, all other notogastral setae on shield *Py*. A total of 15 pairs of notogastral setae present (seta f_2 absent). No peranal segment and peranal seta present. Two pairs of anal, three pairs of adanal, 1—10 pairs of genital and two pairs of aggenital setae. Glandula without a tubiform opening. Epimeral setal formula: 3—2—3—4.
Type-species: *Gehypochthonius rhadamanthus* Jacot, 1936.
Two Palaearctic species known.

Key to species

1 (2) Posteromarginal setae of notogaster widened, phylliform (pl. 54 : G). Sensillus throughout evenly thin, setiform, with extremely long cilia (pl. 54 : F).—Great differences in length between setae of shield *Na* (pl. 54 : H), *cp* more than four

times as long as seta d_2. Shield *Py* with setae e_1 and ps_3 simple, all other hairs phylliform with dentate margins. 10 pairs of genital setae present. Length: 290 µm, width: 142 µm.

Locus typicus: Sengen-jinja, on eastern slope of Mt. Fuji, Japan. Distribution: only the type-locality known

frondifer Aoki, 1975

2 (1) All notogastral setae simple, setiform (pl. 54: D). Sensillus fusiform (pl. 54: C), its cilia shorter.—No essential difference in length among setae of shield *Na*, setae *cp* and d_2 equal in length. Nine pairs of genital hairs present (pl. 54: E). Length: 262—275 µm, width: 105—112 µm.

Locus typicus: "Pisgah Forest", Bent Creek, North Carolina, USA. Distribution: USA, Japan

rhadamanthus Jacot, 1936

Species interpretabiles

1. frondifer Aoki, 1975: 57, figs. 4—6.
2. rhadamanthus Jacot, 1936: 22, figs. 1—2.
   rhadamanthus Strenzke: 1963: 248.
   rhadamanthus Aoki: 1975: 55, figs. 1—2.

Supercohort 4: Mixonomata Grandjean, 1969

Mixonomata Grandjean, 1969: 143.
Mixonomata Balogh et Mahunka: 1979: 280.

A group of weakly or well chitinized animals Body either dichoid or ptychoid, but notogaster invariably holonotic, bearing 14—16 pairs (exceptionally fewer) pairs of setae; seta f_1 still developed in some families. A gastronotic or ventral neotrichy also occurring occasionally. Oil gland present in some families. Pseudodigastria and schizogastria may also occur. Legs 5-jointed, only one femur present.

A group hard to characterize unequivocally and its further splitting will probably be inevitable, but the superfamilies proper are well characterizable. It can be subdivided into two cohorts, the dichoid *Dichosomata* Balogh et Mahunka, 1979, and the ptychoid *Euptyctima* Grandjean, 1969.

Key to cohorts

1 (2) Body articulated between proterosoma and hysterosoma, therefore, along sejugal furrow, but the two parts not or hardly movable between themselves, dichoid

cohort 1: **Dichosomata** Balogh et Mahunka, 1979

2 (1) Body articulated between prosoma and opisthosoma, therefore, ventrally along disjugal furrow, the two parts able to fold together like a penknife

cohort 2: **Euptyctima** Grandjean, 1969

Cohort 1: Dichosomata Balogh et Mahunka, 1979

Dichosomata Balogh et Mahunka, 1979 : 281.

Notogaster holoid, that is, whole, without transverse furrows. Body dichoid, with a less sclerotized region between legs II and III, therefore, proterosoma and hysterosoma somewhat movable with reference to each other. Anogenital region macropilic; if rarely brachypilic, then always occurring as pseudodiagastria or schizogastria.

The group contains five well differentiated superfamilies. All are monotypical in the Palaearctic Region, therefore, the families will not be discussed separately.

Key to superfamilies

1 (2) A wide or narrow but invariably well discernible praeanal plate present (pl. 55 : B) between anal and genital plates. Anogenital region with 10 pairs of genital setae, arranged longitudinally as 6 + 4, transversely as 5 + 5, two pairs of anal and four pairs of adanal setae present. Anal and adanal plates occasionally fused, hence number and position of setae varying
superfamily 1: **Lohmannioidea** Berlese, 1916

2 (1) No praeanal plate present between anal and genital plates. 7—9 pairs of genital, 3—4 pairs of anal and 3—4 pairs of adanal setae present.

3 (6) Ventral region macropilic: genital and anal plates large, meeting, filling ventral area.

4 (5) Genital plate subdivided by a transverse suture, anterior part emitting six, posterior part two pairs of setae. Body dorsoventrally flattened. An unpaired pygidial plate present behind anal plate
superfamily 3: **Perlohmannioidea** Grandjean, 1958

5 (4) Genital plate whole, bearing eight pairs of setae in one row. Body slightly flattened laterally, but always strongly convex. No unpaired distinct plate present behind anal plate
superfamily 5: **Collohmannioidea** Grandjean, 1958

6 (3) Ventral region brachypilic, anal and genital plates discrete, situated on the "ventral" plate.

7 (8) Dividing line between anal and genital plates parabolic (pl. 62 : H). 9 pairs of genital setae present
superfamily 2: **Eulohmannioidea** Grandjean, 1931

8 (7) Dividing line between anal and genital plates straight (pl. 65 : M) or exceptionally absent. 7—8 pairs of genital setae present
superfamily 4: **Epilohmannioidea** Oudemans, 1923

Superfamily 1: Lohmannioidea Berlese, 1916

Lohmannini Berlese, 1916 : 176.
Lohmanniidae Grandjean, 1931 : 143.
Lohmanniidae Grandjean: 1950 : 96.

Lohmanniidae Van der Hammen: 1959: 56*.
Lohmannoidea Grandjean: 1969: 143.

Body more or less cylindrical. Notogaster with 16 pairs of setae, but a pygidial neotrichy occasionally also occurring. Prodorsum with two exobothridial setae, both as well or nearly as well developed as the other prodorsal setae. Prodorsum and notogaster frequently with a characteristic sculpture consisting of transverse ribs, pore fields (fossulae vittiformes transversales)—occasionally joined by irregularly arranged areae porosae—and a polygonal ornamentation of the cuticle. Neotrichy often occurring on ventral side, especially on mentum and in epimeral region. Genital plate with 10 pairs of setae (in a 6+4 longitudinal arrangement); genital plates frequently subdivided by a transverse furrow (with 5+5 setae). A preanal plate present between genital and anal plates, occasionally partly covered. Anal and adanal plates discrete, or separated by a suture, or fused. Two pairs of anal and 4 or 5 pairs of adanal hairs, varying considerably. All legs monodactyle, legs short and stout.

The species relegated to the superfamilies are distributed mainly in the tropics; only 21 species of 8 genera are known from the Palaearctic Region. Two related families, about 30 genera and several hundreds of species have been described from tropical areas.

Family 1: Lohmanniidae Berlese, 1916

Lohmannini Berlese, 1916: 176.

The family is characterized by the diagnosis given for the superfamily.

A synoptic survey of the main distinguishing features in the Palaearctic genera

Characteristics / Genera	Transverse suture on genital plate	Preanal plate	Suture between anal and adanal plates	Number of adanal setae	Number of anal setae	Pygidial neotrichy
Meristolohmannia	+	wide	+	4	0	—
Lohmannia	+	wide	+	4	2	—
Papillacarus	+	narrow	+	4	2	+
Thamnacarus	+	wide	+	4	2	+
Cryptacarus	+	narrow	+	4	2	+
Heptacarus	+	wide	—	5	2	+
Javacarus	—	wide	—	4	0	—
Mixacarus	—	wide	+	4	2	—

* A detailed history of the group is also presented by the author.

Key to genera

1 (4) Genital plates whole, without transverse suture (pl. 62 : B).
2 (3) Anal and adanal plates separate, two pairs of anal and four pairs of adanal setae present (pl. 61 : D)
genus 6: **Mixacarus** Balogh, 1958
3 (2) Anal and adanal plates fused, united plate with merely four pairs of setae (pl. 62 : B)
genus 8: **Javacarus** Balogh, 1961
4 (1) Genital plates subdivided by a transverse suture (pl. 55 : B).
5 (12) Anal and adanal plates separated by a conspicuous suture.
6 (7) Praeanal plate narrow, terminally furcate (pl. 57 : I)
genus 2: **Papillacarus** Kunst, 1959
7 (6) Preanal plate wide (pl. 55 : B).
8 (9) Anal setae absent: ano-adanal setal formula (0+4)
genus 7: **Meristolohmannia** Balogh et Mahunka, 1966
9 (8) Anal setae present: ano-adanal setal formula (2+4) or (2+5).
10 (11) No pygidial neotrichy, thus notogaster with merely 16 pairs of setae. Preanal plate not covered
genus 1: **Lohmannia** Michael, 1898
11 (10) Pygidial neotrichy present: notogaster at least with 25 pairs of setae, often considerably more. Preanal plate largely covered (pl. 59 : I)
genus 3: **Thamnacarus** Grandjean, 1950
12 (5) Anal and adanal plates fused, no suture present between them.
13 (14) Preanal plate narrow; setal formula of ano-adanal plate: (2+4)
genus 4: **Cryptacarus** Grandjean, 1950
14 (13) Preanal plate wide; setal formula of ano-adanal plate: (2+5)
genus 5: **Heptacarus** Piffl, 1963

Genus 1: Lohmannia Michael, 1898

Lohmannia Michael, 1898 : 75.
Lohmannia Grandjean: 1950 : 118.
Lohmannia Balogh: 1961 : 20.
Michaelia Haller, 1884 : 234.

Notogaster with 16 pairs of setae, no neotrichy present. Genital plate subdivided by a transverse suture. Adanal and anal plates also conspicuously separated from each other. Preanal plate wide. Four pairs of adanal and two pairs of anal setae present.

Type-species: *Michaelia paradoxa* Haller, 1884*.

* Haller described the type-species from vine-roots in Germany. Unfortunately, the description fails to cover some fundamental features (transverse suturae of genital plate, exobothridial setae, etc.) which would allow a safe identification based on modern diagnoses (by its chaetotaxy the species resembles several congeners). It may stand nearest to *Lohmannia loebli* Mahunka, 1974, but, owing also to the removed localities, their identities may turn on successful collectings in the locus typicus.

The genus is cosmopolitan or subcosmopolitan, the majority of species are known from the Holotropic area. Five safely identifiable species have so far been shown from the Palaearctic Region.

Key to species

1 (4) Posterior exobothridial seta *(exp)* simple, at most basally thickened and thus lanceolate, but never expanding leaf-like (pl. 55: A), and as long as, or longer than, all other similarly thin prodorsal setae.

2 (3) Prodorsal and notogastral setae fusiform, incrassate, verticillately ciliate (pl. 55: A). Some setae of anogenital region also similar, with extremely long cilia (pl. 55: B).—No essential difference in length among prodorsal setae, only lamellar hairs slightly longer than the others. Medial setae of notogaster considerably shorter than marginal and mainly than pygidial setae, but of similar shape. Anal setae longer than enclosed distance of an_1—an_2. Body surface with fine sculpture. Length: 865 μm, width: 320μm.

Locus typicus: Split (Spalato), Yugoslavia. Distribution: only the type-locality known

regalis Berlese, 1923

3 (2) Prodorsal and notogastral setae only partly and considerably less incrassate, bimarginally or unimarginally ciliate (fig. 55: F). Setae of anogenital region not widened at all, their cilia short (fig. 55: G).—A great difference in length among prodorsal setae, rostral seta approximately twice longer than anterior exobothridial one. A similarly significant difference between median and lateral notogastral setae, but pygidial setae (h_1, ps_1) essentially shorter than for instance seta c_3. Anal setae simple, both pairs shorter than distance enclosed between an_1—an_2. Body surface ornamented with a heavy polygonal sculpture. Length: 902 μm, width: 374 μm.

Locus typicus: "Cukalazes", Levkas Island, Greece. Distribution: Greece

reticulata Sellnick, 1931

4 (1) Posterior exobothridial seta *(exp)* phylliform, very considerably expanded, nearly circular and much shorter than all other dorsal setae (pl. 55: C).

5 (8) Notogastral setae widely phylliform, excepting some marginal ones (pl. 55: E); also adanal setae widely phylliform.

6 (7) Body ornamented with anextremely heavy, essentially polygonal sculpture (pl. 55: C). Interlamellar setae, similarly to exobothridial ones, widely phylliform, comparatively short (pl. 55: E).—Nearly all notogastral setae extremely wide, c_3 and c_p narrower and longer. All setae of anogenital region (pl. 55: D) more or less widened, especially ad_1. Anal setae shorter than distance between an_1—an_2. Length: 845 μm, width: 414 μm.

Locus typicus: El Kef, Tunisia. Distribution: only the type-locality known

hungarorum Mahunka, 1980

7 (6) Body surface only with a fine punctate sculpture. Interlamellar setae narrow, long, wholly differing from seta *exp* (pl. 56: G).—Every prodorsal and notogastral seta narrower (pl. 56: I) than in preceding species, but still

phylliform. Anogenital region with some median setae, as well as genital and anal ones not widened. Also adanal setae much narrower than in preceding species. Anal setae longer than distance between an_1—an_2 (pl. 56 : F). Length: 931—980 μm, width: 429—454 μm.

Locus typicus: "Kaligoni", Leucade, Greece. Distribution: Greece

loebli Mahunka, 1974

8 (5) Notogastral setae blade-shaped, marginally serrate, only slightly widened, inner setae not or hardly expanded (fig. 56 : A). Anal and adanal setae similar, uniformly narrow (fig. 56 : E).—Rostral seta (fig. 56 : D) rapidly attenuating preapically, much elongated. Interlamellar and anterior exobothridial *(exa)* setae shaped like a willow-leaf. One side of sensillus (fig. 56 : B) with 14—medially very long—lateral branches, cilia of other side minute. Length: 840—890 μm, width: 410—440 μm.

Locus typicus: Ashabad, Turkmenia, Soviet Union. Distribution: Spain, Hungary, Soviet Union

turcmenica Bulanova-Zachvatkina, 1960

Species interpretabiles

1. hungarorum Mahunka, 1980 : 124, figs. 1—3.
2. loebli Mahunka, 1974 : 572, figs. 1—4.
3. regalis Berlese, 1923 : 261.
   regalis Van der Hammen: 1959 : 58.
   regalis Mahunka: 1980 : 111, figs. 13—14.
4. reticulata Sellnick, 1931 : 703, figs. 9—10. *(regalis* var. *reticulata)*.
   regalis var. reticulata Van der Hammen: 1959 : 58.
   regalis Balogh: 1961 : 36, figs. 3—4.
   reticulata Mahunka: 1980 : 111, figs. 15—17.
5. turcmenica Bulanova-Zachvatkina, 1960: 1842, figs. 5, 7 *(lanceolata turcmenica)*.
   hispaniola Pérez-Iñigo, 1967 : 158, figs. la-e.

Species inquirendae

— paradoxa Haller, 1884:229, pl. 16, figs. 5—8.
— murcioides Berlese, 1896: 78, 7 *(Angelia)*.
  murcioides Van der Hammen: 1959: 57.

Genus 2: Papillacarus Kunst, 1959

Papillacarus Kunst, 1959 : 52.
Papillacarus Balogh: 1961 : 26.

Notagaster densely aciculate, with a medium or heavy pygidial neotrichy, but basic chaetom (f_1, h_1, ps_1) still recognizable among minute neotrichial setae in several species. Genital plate subdivided by a transverse suture. Preanal plate narrow, considerably longer than wide, posteriorly furcate. Anal and adanal plates separated by a suture. Number of epimeral setae varying. Anal plates with two, adanal plates with four, pairs of setae.

Type-species: *Lohmannia murcioides* var. *aciculata* Berlese, 1905.

The majority of the species have been described from the Palaearctic Region, though the genus is represented also in the Ethiopian and Oriental Regions.

Key to species

1 (6) Setae equal in length, basic chaetom of pygidium not recognizable among neotrichial setae.

2 (3) Setae c_1, d_1 and e_1 smooth (pl. 57 : A), all other notogastral setae with long cilia. Neotrichial setae with especially long cilia, yet not stellately branching but rather plumose.—Sensillus ciliate on both sides. Prodorsal setae simple, but also visibly ciliate. Notogastral polygonal reticulation also discernible. Several smooth setae also among epimeral ones (pl. 57 : B). Epimeral setal formula: 7—4—3—4. Genital setae minute. Two pairs of anal and four pairs of adanal setae similar in construction. Length: 500 µm, width: 220 µm.

Locus typicus: Madrid, Spain. Distribution: Spain

chamartinensis Pérez-Iñigo, 1967

3 (2) Setae c_1, d_1 and e_1 stellately ramifying, composed of 5—6 cilia equal in length. Neotrichial setae also stellately branching or penicillate.

4 (5) Body setae (pl. 57 : E) short: in sequence of anal and adanal setae none extending beyond insertion point of next one; seta *exa* not reaching insertion of lamellar seta, this latter not reaching that of rostral seta.—Among notogastral setae only *c* and *d* slightly longer than neotrichial setae. Sensillus (pl. 57 : C) unilaterally ciliate, length of cilia divers, basal 2—3 cilia essentially shorter than the others. Epimeral setal formula: 7—4—3—4. Length: 381—423 µm, width: 141—169 µm.

Locus typicus: Kunitachi, near Tokyo, Japan. Distribution: Japan

hirsutus (Aoki, 1961)

5 (4) Body setae (pl. 57 : F) longer: in sequence of anal and adanal setae (pl. 57 : H) apex of every seta extending beyond insertion point of next one; seta *exa* reaching insertion of lamellar seta, this latter projecting considerably beyond insertion of rostral seta.—Notogastral seta e_1 and some lateromarginal setae longer than neotrichial ones. Sensillus (pl. 57 : G) with 15—16 lateral cilia, basal ones not shorter than the others. Epimeral setal formula: 8—5—3—4. Length: 418—434 µm, width: 169—187 µm.

Locus typicus: Bagyon san, San Chontung, Korean PDR. Distribution: only the type-locality known

koreanus (Mahunka, 1973)

6 (1) Pygidial setae divers in length, at least f_1, h_1 and ps_1 of basic chaetom well distinguishable.

7 (8) Pygidial part of body with merely 5—6 neotrichial setae, all verticillately and evenly ciliate (pl. 57 : J). Only 3—4 neotrichial setae between setae h_1.—No essential difference in length among prodorsal setae, only *exp* longer and slightly thicker than the others. Setae c_1, d_1 and e_1 equal in length; e_2 only half

as long as f_2. Epimeral setal formula: 8—4—3—4. Chaetom of anogenital region as in pl. 57 : I. Length: 516—548 µm, width: 246—263 µm.

Locus typicus: El Kef, Tunisia. Distribution: only the type-locality known

pseudoaciculatus Mahunka, 1980

8 (7) Pygidial part of body with 9—10 pairs of neotrichial setae, between or anterior to h_1 6—7 neotrichial setae present.

9 (10) Neotrichial setae asymmetrically ciliate, cilia usually very long (pl. 58 : A—E).—Prodorsal and notogastral setae (pl. 58 : C—D) sparsely ciliate. Aciculi comparatively sparse. Length of setae c_1, d_1 and e_1 increasing in this sequence. Sensillus (pl. 58 : B) slightly widening, cilia irregularly arranged and divers in length. Length: 502—559 µm, width: 234—283 µm.

Locus typicus: Assos, Kefallinia Island, Greece. Distribution: Greece

ondriasi Mahunka, 1974

10 (9) Neotrichial setae evenly ciliate, cilia short, equal in length (pl. 58 : H).

11 (12) A transverse row of 5—6 pairs of neotrichial setae evolved between setae f_1 and h_1 (pl. 58 : F); three pairs of neotrichial setae between setae h_1.—Only a slight difference in length between median notogastral setae c_1 hardly shorter than c_2, proportion of d_1—d_2 similar: seta f_1 hardly longer than c_1. Epimeral setal formula: 6—6—3—4 (pl. 58 : G). Length: 536 µm, width: 248 µm.

Locus typicus: Kutan-Kanabur, Daghestan, Soviet Union. Distribution: Soviet Union

pavlovskii (Bulanova-Zachvatkina, 1960)

12 (11) No transverse row of neotrichial setae evolved between setae f_1 and h_1, this area with only 2—3 pairs of scattered setae (pl. 58 : H). Essential differences in length among median notogastral setae: c_2 nearly twice longer than c_1, proportion of d_1—d_2 similar; also f_1 considerably longer than c_1. Epimeral setal formula: 8—4—3—4 (pl. 58 : I). Length: 490—630 µm, width: 240—324 µm.

Locus typicus: Florence, Italy. Distribution: * Italy, Yugoslavia, Hungary, Bulgaria, Soviet Union (CET, SET)

aciculatus (Berlese, 1905)

Species interpretabiles

1. aciculatus Berlese, 1905: 24, pl. 2, fig. 39 *(Lohmannia murcioides* Berlese, 1896 var. *aciculata)*.
   aciculatus Kunst; 1959 : 52, figs. 1a—b.
   aciculatus Van der Hammen: 1959 : 59.
2. chamartinensis Pérez-Iñigo: 1967 : 167, figs. 1a—b.
3. hirsutus Aoki, 1961 : 65, figs. 2A—E *(Cryptacarus)*.
4. koreanus Mahunka, 1973 : 53, figs. 6—8 *(Vepracarus)* **(comb. nov.)**.
5. ondriasi Mahunka, 1974 : 574, figs. 5—9.
6. pavlovskii Bulanova-Zachvatkina, 1960: 1844, (pl. 6, figs. 1—2) *(Thamnacarus)* **(comb. nov.)**.
7. pseudaciculatus Mahunka, 1980 : 126, figs. 4—5.

* In view of possible confusions arising from misidentifications of the highly similar species, the locality data need careful revision!

Genus 3: Thamnacarus Grandjean, 1950

Thamnacarus Grandjean, 1950 : 118.
Thamnacarus Balogh: 1961 : 26.
Thamacarus Aoki: 1971 : 127 (lapsus calami).
Asiacarus D. Krivolutsky, 1971 : 35, **syn. nov.***

Notogastral setae densely ciliate; a medium or weak pygidial neotrichy present, but neotrichial setae not differring from basic setae. Body with well recognizable polygonal ornamentation. Genital plates divided by a transverse suture. Seta of preanal plate much wider than long, often covered and then not readily discernible. Two pairs of anal and four pairs of adanal hairs similar to notogastral setae. Epimeral setal formula: 6—(7—8)—3—4.

Type-species: *Lohmannia deserticola* Grandjean, 1934. Four species of the genus are known from the Palaearctic Region.

Key to species

1 (2) Sensillus very wide, clavate (pl. 59 : A), apically rounded, capitate. Notogastral setae accumbent, parallel with body surface; a heavy neotrichy present with at least 32 pairs of notogastral setae (pl. 59 : A).—Both prodorsal and notogastral setae obtuse, rather bacilliform, densely ciliate. Seta *exp* of prodorsum essentially shorter than seta *in*, notogastral setae shorter (excepting c_1), about equal in length. Epimeral setae and those of anogenital region similar in form to dorsal setae. Epimeral setal formula: 6—8—3—4. Length: 504 μ.

Locus typicus: Tachila, Morocco. Distribution: Morocco

deserticola (Grandjean, 1934)

2 (1) Sensillus very finely incrassate, lanceolate, apically pointed (pl. 59 : F).

3 (4) Lamellar seta of prodorsum more than twice longer than rostral one. Medial setae of notogaster also longer than marginal ones (pl. 59 : E).—Notogastral setae arcuate, elevating, a weak pygidial neotrichy with at most 25 pairs of neotrichial setae present. Prodorsal, notogastral and ventral setae uniform, gradually tapering apicad, sparsely ciliate (pl. 59 : G—H). Length or proportions of setae as in preceding species. Epimeral setal formula: 6—(7—8)—3—4 (pl. 59 : I). Length: 530 μm, width: 240 μm.

Locus typicus: Beskentskaya valley, Tadzhikistan, Soviet Union. Distribution: Ethiopia, Soviet Union (SMA)

elongatus (D. Krivolutsky, 1971)

4 (3) No appreciable length difference among setae of prodorsum, and marginal and medial setae of notogaster.

* An examination of the slide containing the holotype showed that—in contrast with the description—the preanal plate is present but largely covered and not easily discernible, and the epimeral setae agree with those of *Thamnacarus*. Accordingly, *Asiacarus* is herewith drawn in as a junior subjective synonym of *Thamnacarus*.

5 (6) Prodorsum convexly rounded in front. Notogastral setae extraordinarily long, c_1 much longer than distance between c_1 and d_1 (pl. 60 : A).—Lamellar seta of prodorsum only weakly longer than rostral one, all bent, twisted, heavily ciliate. Among notogastral setae, the pteromarginal ones especially long, straight, rest of setae strongly bent. Setae of ventral side as depicted in pl. 60 : B. Length: 460—520 µm, width 200—220 µm.

Locus typicus: Balla-Isem, Turkmenia, Soviet Union. Distribution: Soviet Union (TC, SMA)

longisetosus Bulanova-Zachvatkina, 1978

6 (5) Prodorsum concavely rounded in front. Body setae much shorter (pl. 59 : C).—Lamellar and rostral setae of prodorsum of equal length, short, scarcely longer than exobothridial ones. Notogastral setae also shorter, less twisted and bent than in previous species. Ventral side depicted in pl. 59 : D. Length: 480—500 µm, width: 200 µm.

Locus typicus: Balla-Isme, Turkmenia, Soviet Union. Distribution: only locus typicus known

smirnovi Bulanova-Zachvatkina, 1978

Species interpretabiles

1. deserticola Grandjean, 1934 : 247, figs. 4—5 *(Lohmannia)*.
   deserticola Grandjean: 1950 : 118.
2. elongatus D. Krivolutsky, 1971 : , fig. 1 *(Asiacarus)*.
   moribei Aoki, 1971 : 127, figs. 1—6; **syn. nov.*** *(Thamacarus!)*.
3. longisetosus Bulanova-Zachvatkina, 1978 : 919, figs. 3—4.
4. smirnovi Bulanova-Zachvatkina, 1978 : 918, figs. 1—2.

Genus 4: Cryptacarus Grandjean, 1950

Cryptacarus Grandjean, 1950 : 118.
Cryptacarus Balogh: 1961 : 27.

Body relatively flat. Notogaster with very heavy neotrichy, thus notogastral setae in two forms: c_1, d_1, e and f, as well as ventrally originating h and ps long and ciliate, whereas pygidial neotrichial setae ramifying, dendriform to shrub-shaped. Epimeral setal formula: 7—4—3—4. Genital plate with transverse suture. Preanal plate narrow, longer than wide. Anal and adanal plates fused, without traces of a suture, but position of setae indicating location of original plates.

Type-species: *Cryptacarus promecus* Grandjean, 1950.

The species inhabit chiefly the Oriental Region, with one species known also from the Palaearctic Region.

— Rostrum wide, nearly angular, but anteriorly with two small excisions producing a small central apex (pl. 60 : C), hence rostrum actually tricuspidate. Rostral setae longer than the rest; all of them ciliate. Sensillus (pl. 60 : D) gradually incrassate, fusiform, outer side with dense and long cilia, about 30 bristles, inner side with

* *elongatus* D. Krivolutsky: February, 1971. *moribei* Aoki: 30 March, 1971.

merely 15 shorter setae. Prodorsum and notogaster with a polygonal ornamentation. Setae of notogastral neotrychy heavily ramifying (pl. 60 : E). Length: 410—435 µm, width: 160—180 µm.

Locus typicus: Bone, Algeria. Distribution: Algeria, Sudan, Spain, Soviet Union (SMA)

promecus Grandjean, 1950

Species interpretabilis

1. promecus Grandjean, 1950 : 138, figs. 9a—f.
   promecus Pérez-Iñigo: 1967 : 163, figs. 2a—d.
   valdemorica Mihelčic, 1956 : 156, fig. 8 (Pérez-Iñigo: 1967 : 163).

Genus 5: Heptacarus Piffl, 1963

Heptacarus Piffl, 1963 : 1.
Heptacarus Mahunka: 1977 : 268.

Wide, large-bodied species. Genital plates divided by a transverse suture. Preanal plate considerably wider than long. Anal and adanal plates completely fused, without traces of a suture. Two pairs of anal and five pairs of adanal setae present. A medium or heavy dorsal and ventral neotrichy: at least 25 but occasionally considerably more pairs of notogastral setae present; additional setae appearing in rows *c, d. e* and *f,* or pygidially. Notogaster with a polygonal or map-like weak ornamentation and fields of pori.

Type-species: *Heptacarus notoneotrichus* Piffl, 1963.

The species are known from the southern confines of the Palaeartic and Ethiopian and Oriental Regions. So far two species are known from the Palaearctic Region.

Key to species

1 (2) 26—27 pairs of notogastral setae (pl. 61 : E); four setae in row *c,* considerably more posteriorad seta *cp,* rows *d—e, f, h* and *ps* with three setae each, and on ventral side of notogaster (pl. 61 : F) with six pairs of unidentifiable setae *b.* All notogastral setae fine and smooth, those of ventral side hardly discernibly ciliate. Sculpture of notogaster resembling a city map, with irregular triangles, quadrangles, or pentagons, enclosing "streets" of lighter stripes, in some places with well defined areae porosae. Length: 580 µm, width: 300 µm.

Locus typicus: Abu Monquar, Chardaqua, Egypt. Distribution: Egypt, Sudan

notoneotrichus Piffl, 1963

2 (1) Considerably more than 27 pairs of notogastral setae present (pl. 61 : G), with a very heavy pygidial neotrichy, ventral side also with many unidentifiable setae (pl. 61 : H). All notogastral setae heavily ciliate; those of ventral side, including anogenital region, similarly formed. Ornamentation of notogaster polygonal, composed of rather regular pentagons or hexagons; this or-

namentation locally disappearing and replaced by irregular groups of pori. Length: 525 µm, with: 225 µm.

Locus typicus: Iraq. Distribution: Iraq

supertrichus Piffl, 1967

Species interpretabiles

1. notoneotrichus Piffl, 1963 : 1, figs. A—I.
2. supertrichus Piffl, 1967 : 1, figs. A—G.

Genus 6: Mixacarus Balogh, 1958

Mixacarus Balogh, 1958 : 4.
Mixacarus Balogh: 1961 : 27.
Mixacarus Aoki: 1970 : 403.

Wide, large-sized animals. Neither dorsal nor ventral neotrichy. Notogaster with ribbon-like transverse bands. Epimeral setal formula: 3—1—3—4. Genital plates undivided, without a transverse suture. Preanal plate wide. Anal and adanal plates separated by a distinct suture; two pairs of anal and four pairs of adanal setae present.

Type-species: *Mixacarus integer* Balogh, 1958.
The species are known from every faunal region, excepting the Austalian area; two have been described from the Palaearctic Region.

Key to species

1 (2) Outer row of genital setae consisting of, three inner row of five setae (pl. 60 : H); eight pairs nearly equal in length, two pairs considerably longer. Seta an_1 as long, or nearly as long, as an_2. All anogenital setae smooth.—Seta *exp* longer than *exa*. Relative lengths of notogastral setae: $c_1 = c_2 = d_2 = = e_1 = f_1 < h_1$. Notogastral setae very finely ciliate (pl. 60:G). Length: 605–635 µm, width: 300—320 µm.

Locus typicus: Mt. Mokkoku, Tsutsu, Japan. Distribution: Japan

exilis Aoki, 1970

2 (1) Outer row of genital setae consisting of four, inner row of six setae (pl. 61 : D); six pairs short, four pairs long. Seta *exa* twice longer than *exp*. Relative lengths of notogastral setae: of notogastral setae: $c_1 < c_2 = f_1 < d_1 < e_1 < d_2 < h_1$ (pl. 61 : A). Notogastral setae densely ciliate (pl. 61 : C). Length: 650—680 µm, with: 328—346 µm.

Locus typicus: Bagyon san, near Keasong, Korean PDR. Distribution: only the type-locality known

vanhonggui Mahunka, 1973

Species interpretabiles

1. exilis Aoki, 1970 : 401, figs. 9—17.
2. vanhonggui Mahunka, 1973 : 49, figs. 1—5.

Genus 7: Meristolohmannia Balogh et Mahunka, 1966

Meristolohmannia Balogh et Mahunka, 1966: 553.

Large-sized, extremely wide animals. Notogaster with ribbons of fossulae vittiformes. Neither pygidial nor ventral neotrichy observable. Genital plate divided, preanal plate wide. Anal adanal plates separated by a well discernible suture, but setal formula: 4: 0.

Type-species: *Meristolohmannia meristacaroides* Balogh et Mahunka, 1966.

Besides the Australian type-species, only a single Palaearctic species known.

— Prodorsal setae approximately equal in length (pl. 62: C), thin, sparsely ciliate. Sensillus with 4—5 long lateral branches, on opposite side with about an equal amount of short teeth, Notogastral setae long and thin, setae c_1 only slightly shorter than c_2 and c_3. Epimeral setal formula: 3—1—3—4. Form and chaetom of anogenital region as in pl. 62: D. Length: 940—980 µm, width: 620—640 µm.
Locus typicus: Tingusany, China. Distribution: only the type-locality known
chinensis (Bulanova-Zachvatkina, 1960)

Species interpretabilis

1. chinensis Bulanova-Zachvatkina, 1960: 1845, figs. 2: 1—4 *(Meristacarus)* **(comb. nov.).**

Genus 8: Javacarus Balogh, 1961

Javacarus Balogh, 1961 : 31.

Body with heavy sculpture. Neither notogastral nor ventral neotrichy observable. Genital plate undivided, preanal plate considerably wider than long. Anal and adanal plates fused, with merely four pairs of setae. Epimeral setal formula: 3—1—3—4.

Type-species: *Javacarus kuehnelti* Balogh, 1961.

The single known species of the genus have been reported from the Oriental Region and from the southern confines of the Palaearctic Region.

— Rostrum widely rounded. Prodorsal surface ornamented with foveolae and fields of pori, that is, transverse bands of these elements (pl. 62: A). Prodorsal and notogastral setae narrow phylliform or blade-shaped, no essential difference in length among prodorsal setae, on notogaster median setae generally shorter than marginal ones. Ventral setae (pl. 62: B) minute, thin, simple. 10 pairs of genital setae in a longitudinal arrangement of 6+4. Anoadanal plates with simple but considerably longer setae. Length: 628—648 µm, width: 304—311 µm.
Locus typicus: Java. Distribution: Java, Egypt
kuehnelti Balogh, 1961

Species interpretabilis

1. kühnelti Balogh, 1961:31, figs. 39—40.

Superfamily 2: EULOHMANNIOIDEA Grandjean, 1931

Eulohmanniidae Grandjean, 1931 : 144.
Eulohmannoidea Grandjean: 1969 : 143.

Body cylindrical, very weakly chitinized, light yellow or occasionally white. Proterosoma joined to hysterosoma by a wide, rugose, "neck"-like zone, rendering a characteristic form to body. Prodorsum with five pairs (two pairs of exobothridial) setae. Notogaster with 16 pairs of setae. Ventral plate with diagastria, parabolic furrow decurrent between genital and anal openings. Nine pairs of genital setae, aggenital region with neotrichy. Anal and adanal plates differentiated, each with four pairs of setae. Oil gland absent. Legs thin, long, 2-clawed.

Monotypical.

Family 1: EULOHMANNIIDAE Grandjean, 1931

Eulohmanniidae Grandjean, 1931 : 144.
Eulohmanniidae Grandjean: 1956 : 282.

Only a single genus known, occurring in the entire Palaeartic Region.

Genus 1: EULOHMANNIA Berlese, 1910

Lohmannia (Eulohmannia) Berlese, 1910 : 223.
Eulohmannia Grandjean: 1956 : 282.
Eulohmannia Van der Hammen: 1959 : 55.
Arthronothrus Trägårdh, 1910 : 554.

Characterized by the features of the family.
Type-species: *Eulohmannia ribagai* Berlese, 1910.
A single Palaearctic species known*.

— Rostrum (pl. 62 : G) wide, medially with a distinct cuspis; rostral setae arising near each other. Sensillus very long, arcuate, with long cilia on one side (pl. 62 : F). Body surface with polygonal ornamentation. Notogastral setae (pl. 62 : E) simple, thin, epimeral setal formula: 3—1—3—4. Genital plate small, (pl. 62 : H) much smaller than anal plate. Every ventral seta simple. Length: 600—700 μm, width: 220—245 μm.
Locus typicus: Tiarno, Italy. Distribution: Palaearctic Region
ribagai Berlese, 1910

Species interpretabilis

1. ribagai Berlese, 1910 : 223, fig. 20 : 51 *(Lohmannia (Eulohmannia))*.
   ribagai Grandjean: 1956 : 282, fig. A.
   ribagai Aoki: 1975 : 57, figs. 1—4.
   biunguiculatus Trägårdh, 1910 : 554, figs. 342 : 346 *(Arthronothrus)*.

* The Palaearctic forms, at present considered to represent a single species, may eventually be segregated to subspecies.

Superfamily 3: PERLOHMANNIOIDEA Grandjean, 1954

Perlohmanniidae Grandjean, 1954 : 430.
Perlohmannoidea Grandjean, 1958 : 116
Perlohmannoidea Van der Hammen: 1959 : 51.

Large-sized (800—1200 µm), well chitinized animals. Notogaster slightly flattened, otherwise cylindrical. Prodosum with five pairs (two pairs of exobothridial) of setae. Notogaster with 15 pairs of true setae, f_1 recognizable only by their reduced insertion alveoli. Oil gland present. Anogenital region well differentiated, aggenital and genital plates about half as long as anal and adanal plates. Peranal plate absent, but an unpaired pygidial plate present behind anal plates. Genital plates transversely divided. Eight (6 + 2) pairs of genital, two pairs of aggenital, three pairs of anal and three pairs of adanal hairs present. Epimeral setal formula: 3—1—3—4. Legs mono- or tridactylous.

The superfamily is monotypical, composed of a single family and one genus.

Family 1: PERLOHMANNIIDAE Grandjean, 1954

Perlohmanniidae Grandjean, 1954 : 430.
A monotypical family.

Genus 1: PERLOHMANNIA Berlese, 1916

Perlohmannia Berlese, 1916 : 176.
Perlohmannia Grandjean: 1958 : 57.
Perlohmannia Van der Hammen: 1959 : 52.
Perlohmannia Schuster: 1960 : 185.
Apolohmannia Aoki, 1960 : 507 (Aoki: 1969 : 220).
Neolohmannia Bulanova-Zachvatkina, 1960 : 1847 (Bulanova-Zachvatkina in: Ghilarov et D. Krivolutsky, 1975:69).

Characterized by the features of the superfamily.

Type-species: *Lohomannia insignis* Berlese, 1904 (lapsus calami) *(= Lohmannia insignis)*.

Eight species of the genus have been described from the Palaearctic Region so far; some of them are in need of revision.

Key to species

1 (10) Tarsi of all legs with 1 claw (monodactyle).
2 (9) Sensillus with 5—8 lateral branches.
3 (4) Prodorsum gradually attenuating, its margins finally somewhat concave, rostrum elongately acute (pl. 63 : G).Rostral setae arising near each other immediately at rostrum proper; short, nearly as long as lamellar setae. Sensillus with merely 4—5 lateral branches. Length: 1000 µm, with: 400 µm.
Locus typicus: Dublin, Ireland. Distribution: only the type-locality known
insignis (Berlese, 1904)*

* The species is known only from the original description and from a later comparative examination (Hewitt, 1908); a redescription is urgently needed. It was fitted in the key on the basis of the known characteristics, any identification of related new material should be made with great care.

4 (3) Prodorsum abruptly attenuating, rostrum rounded or widely triangular, occasionally concave with a small median cuspis.

5 (8) Rostrum widely triangular or rounded with a convex arc.

6 (7) Rostral setae arising behind each other (pl. 63 : A); lateral rim of notogaster narrow (pl. 63 : C) emitting setae c_3, c_2, cp and e_2.—Lamellar setae considerably shorter than rostral and especially than interlamellar setae: these latter nearly thrice longer. Setae of median row of notogaster shorter than lateral and pygidial ones. Genital plate narrow, long (pl. 63 : B), considerably longer than wide. Length: 860—1000 μm, width: 380—420 μm.

Locus typicus: Withington, Great Britain. Distribution: Great Britain, France, Austria, Hungary, Soviet Union

dissimilis (Hewitt, 1908)

7 (6) Rostral setae arising "symmetrically" beside each other* (pl. 63 : D); notogastral marginal rim wide, emitting 63 : F only setae c_3, cp and e_2, while c_2 situated nearer to midline of body.—Difference in length among prodorsal setae smaller, rostral setae not longer than lamellar ones, also interlamellar setae at most twice longer. Proportions of notogastral setae similar to those of preceding species. Genital plates (pl. 63 : E) wide, together about as wide as long. Length: 1000—1370 μm, width: 407—571 μm.

Locus typicus: Arbas, France. Distribution: France, Soviet Union, Japan

coiffaiti Grandjean, 1961

8 (5) Prodorsal margins rapidly constricted behind rostrum, the two angles convex, rostral part nearly truncate, but medially with a short, nasute appendage (pl. 64 : H). Position of rostral setae varying, often "asymmetrical". Rostral and lamellar setae approximately equal in length (pl. 64 : G). Sensillus with 7—8 lateral branches. Configuration and chaetom of anogenital region as in pl. 64 : I. Length: 626—782 μm, width: 218–308 μm.

Locus typicus: Weizklamm, near Weiz, Steiermark, Austria. Distribution: Austria, Hungary, Yugoslavia

nasuta Schuster, 1960

9 (2) Sensillus with 10—12 lateral branches (pl. 63 : H). Rostral hairs arising slightly "asymmetrically", but near each other, far from rostrum, somewhat longer than lamellar setae.** Notogaster with median setae (along midline of body) shorter than outer setae, excepting seta c_1; seta c_3 shorter than c_1. Genital plates wide and short (pl. 63 : I). Length: 955—966 μm, width: 375—386 μm.

Locus typicus: Gorno-Altaiska, Soviet Union. Distribution: Soviet Union (ES)

altaica Grishina, 1968

10 (1) Legs tridactylous.

11 (12) Lamellar setae—as in most *Perlohmannia* species—standing considerably nearer to interlamellar setae than to rostral setae, and hardly shorter than these letter ones (pl. 64 : A). Setae c_1 essentially shorter than setae c_2 and c_3. Median

* Some small-scale asymmetry is possible (see Fujikawa et Aoki, 1969 : 224).

** The shape and position of the lamellar setae are given erroneously both in Grishina's original publication and in Ghilarov and D. Krivolutsky's comprehensive work by Bulanova-Zachvatkina, 1975. We submit here new drawings made from specimens of the type-series.

setae of notogaster (along midline of body) considerably shorter than marginal ones. Sensillus (pl. 64 : B) with 7—9 lateral branches. Configuration of plates and chaetom of anogenital region as in pl. 64 : E. Length: 1490—1510 µm, width: 790 µm.

Locus typicus: Kare-Numa, Nikko, Japan. Distribution: Japan

gigantea (Aoki, 1960)

12 (11) Lamellar setae arising near rostral ones, in anterior third of prodorsum (pl. 64 : J), much shorter than rostral setae (pl. 64 : L). Notogastral setae approximately equal in length, no essential difference in length extant between c_1 and c_2, or between median and marginal setae. Sensillus (pl. 64 : K) with eight lateral branches. Length: 1240—1400 µm, width: 560—680 µm.

Locus typicus: Ladochski nature reservation, Gruzia, Soviet Union. Distribution: only the type-locality known

zachvatkini (Bulanova-Zachvatkina, 1960)

Species interpretabiles

1. altaica Grishina, 1968 : 297, figs. A—B.
2. coiffaiti Grandjean, 1961 : 604, figs. 1—3.
   coiffaiti Fujikawa et Aoki: 1969 : 221, figs. 1—7.
3. dissimilis Hewitt, 1908 : 1, figs. 1—2 *(Lohmannia insignis* var. *dissimilis)*.
   dissimilis Grandjean: 1958 : 57, figs. 1—6.
4. gigantea Aoki, 1960 : 507, figs. 1—3 *(Apolohmannia)*.
   gigantea Aoki: 1969 : 225.
5. insignis Berlese, 1904 : 23, pl. 2, fig. 41 *(Lohomannia)* (lapsus calami).
6. nasuta Schuster, 1960 : 191, figs. 4, 26.
   nasuta Schuster: 1965 : 214, fig. 4.
7. zachvatkini Bulanova-Zachvatkina, 1960 : 1847, figs. 3,7 *(Neolohmannia)*.

Species inquirenda

— skrjabini Baskirova, 1958 : 204, figs. 1a—b *(Eulohmannia)*; **comb. nov.***

Superfamily 4: EPILOHMANNIOIDEA Oudemans, 1923

Epilohmanniidae Oudemans, 1923 : 79.
Epilohmannioidea Grandjean, 1969 : 144.
Lesseriidae Oudemans, 1917 : 76.

Well chitinized animals with a cylindrical body. Notogaster with 14 pairs of setae, f_1 and f_2 absent. Prodorsum with five pairs of setae two pairs of exobothridial ones). Oil gland developed. Ventral plate with schizogastria, exceptionally with hologastria *(Epilohmannoides)*. Genital plates with 7 or 8 pairs of setae, aggenital plate with 3—16 pairs. Adanal plates also fused medially, with three pairs of setae, anal plates also with three pairs. Epimeral setal formula: 3—1—3—3. Legs monodactyle.

The monotypical superfamily.

* Bulanova-Zachvatkina considers it (in: Ghilarov and D. Krivolutsky, 1975 : 71) erroneously as a synonym of *Epilohmannia cylindrica!*

Family 1: Epilohmanniidae Oudemans, 1923

Epilohmanniidae Oudemans, 1923 : 79.

Characterized as the superfamily and containing two genera.

Key to genera

1 (2) Rostrum rounded, rostral setae situated posteriorly, far from rostral apex. Ventral plate with a transverse furrow between genital and anal plates (schizogastria). Eight (5 + 3) pairs of genital setae

genus 1: **Epilohmannia** Berlese, 1916

2 (1) Rostrum forming an acute nose; rostral setae at base of nose, therefore, far anteriorly. Ventral plate without a transverse furrow (hologastria). Seven (4 + 3) pairs of genital setae

genus 2: **Epilohmannoides** Jacot, 1936

Genus 1: Epilohmannia Berlese, 1916

Epilohmannia Berlese, 1916 : 176.
Epilohmannia Schuster: 1960 : 195.
Epilohmannia Van der Hammen: 1959 : 53.
Epilohmannia Bayoumi et Mahunka: 1976 : 5.
Lesseria Oudemans, 1917 : 78.

Rostrum rounded, rostral setae situated posteriorly, far behind rostral apex, usually with asymmetrical insertions. Schizogastry present. Eight (5 + 3) pairs of genital setae.

The genus is subdivided into two subgenera; representatives are known from every faunal region, except Australia. Eleven species and subspecies have been reported from the Palaearctic Region.

Type-species: *Lohmannia cylindrica* Berlese, 1904.

Key to species

1 (6) Beside solenidium φ_1 of leg IV one or more incrassate, spiniform seta, thus tibia (pl. 65:C) with 4 of 5 setae in addition to solenidium (subgenus 1: **Sinolohmannia** Balogh et Mahunka, 1979).

2 (3) Two spiniform, incrassate setae present beside solenidium φ_1 on tibia of leg IV, therefore, tibia with a total of five setae in addition to solenidium (pl. 65 : C).— Sensillus fusiform (pl. 65 : A). Notogastral setae thin but heavily ciliate (pl. 65 : D). No essential difference in length between setae c_1—c_3. All setae on epimeres and in anogenital region very heavily ciliate (pl. 65 : B). Length: 728 µm, with: 348 µm.

Locus typicus: Brasov, Roumania. Distribution: only the type-locality known

schusteri Calugar et Vasiliu, 1976

3 (2) Only a single spiniform, incrassate seta present beside solenidium φ_1 on tibia of leg IV, therefore, tibia with a total of merely four setae in addition to solenidium. Setae and sensillus different.

4 (5) Notogastral setae ensiform (pl. 65 : H), excepting setae c_3. Sensillus weakly but recognizably fusiform (pl. 65 : G), with long cilia.—Apodemes 1 fused in front of apodemes 2, therefore, a sternal apodeme well recognizable already in this species (pl. 65 : F). Tarsus of leg IV with four, tibia with two spines, one of them ciliate. Length: 680—816 μm.

Locus typicus: "Giardino di Boboli", Florence, Italy. Distribution: Italy, Yugoslavia, Greece, Soviet Union (SET)

gigantea Berlese, 1916

5 (4) All notogastral setae simple, thin (pl. 65 : I), also cilia short. Sensillus (pl. 65 : K) setiform, gradually tapering apicad, also with short cilia.—Configuration of ventral side (pl. 65 : M) similar to that of preceding species. Tarsus of leg IV with four spiniform setae (pl. 65 : J). Length: 650—803 μm, width: 320—344 μm,

Locus typicus: Kolozsvár (Cluj), Roumania. Distribution: only the type-locality known

imreorum Bayoumi et Mahunka, 1976

6 (1) No spiniform setae present beside solenidium φ_1 of leg IV, therefore, tibia with only three pairs of setae in addition to solenidium (pl. 66 : E) (subgenus 2: **Epilohmannia** Berlese, 1910).

7 (8) A heavy aggenital neotrichy, 16 pairs of aggenital setae present (pl. 66 : C)—Ciliate portion of sensillus (pl. 66 : B) slightly incrassate, but not fusiform. Rostral setae minute, arising near each other. Interlamellar and lamellar setae ciliate, without an appreciable difference in length (pl. 66 : A), both essentially shorter than sensillus. Notogastral setae simple. Length: 536—917 μm, width: 248—425 μm.

Locus typicus: Kumitachi, near Tokyo, Japan. Distribution: Japan

ovata Aoki, 1961

8 (7) Only three pairs of aggenital setae present.

9 (10) Tarsus of leg IV without a spiniform seta, setae *"a"* and *"pv"* spatulate (pl. 66 : J—K).—Rostral and lamellar setae minute, interlamellar approximately four times as long. Notogastral setae fine and thin, similar to lamellar ones. Sensillus (pl. 66 : I.) weakly fusiform, densely ciliate. Epimeres 1 not meeting medially, also epimeres 3 and 4 separate (pl. 66 : H). Length: 520—550 μm, width: 230—250 μm.

Locus typicus: Mt. Oboshi, Miheson, Japan. Distribution: only the type-locality known

spathulata Aoki, 1970

10 (9) Tarsus of leg IV at least with one spiniform seta, but without spatulate setae.

11 (12) Seta *"ft"* on tarsus of leg IV spiniform, incrassate, in addition to also three other spines (pl. 66 : E).—Rostral seta minute, no essential difference in length between lamellar and interlamellar setae. Sensillus fusiform. Notogastral setae ciliate, of divers length, anterior ones usually essentially shorter than pygidial ones (pl. 66 : D). Seta c_1 only half as long as ps_1. Apodemes 1 not meeting medially, each one coalescent with corresponding apodemes 2. Lateral margin

of posterior sternal plate irregular (pl. 66 : G). Length: 396—410 μm, width: 166—174 μm.

Locus typicus: Alexandria, Egypt. Distribution: only the type-locality known

pallida aegyptica Bayoumi et Mahunka, 1976

12 (11) Seta *"ft"* on tarsus of leg IV thin, setiform, spiniform setae occasionally present.

13 (18) Distal half of sensillus fusiform.

14 (17) Tarsus of leg IV with at least two (2—4) spiniform, modified setae).

15 (16) Tarsus of leg IV with two *(s, a)* spiniform setae (pl. 67 : C).—A double difference in length between lamellar and interlamellar setae, and approximately the same between setae c_1, c_2 and posterior h_1 and h_2 (pl. 67 : A, B). Longer inner setae ciliate, h_3 and ps_3 thin and smooth. Apodemes 1 not meeting each other nor touching corresponding apodemes 2 posteriorly (pl. 66 : D), therefore, epimeres 1 open. Length: 453—573 μ.

Locus typicus: Florence, Italy. Distribution: Palaearctic Region, Hawaii, Laysan Islands

cylindrica cylindrica (Berlese, 1904)

16 (15) Tarsus of leg IV with four *(s, a, p, "pv")* spiniform setae (pl. 67 : H), otherwise as in nominate subspecies, only smaller. Length: 370—440 μm.

Locus typicus: Stams ("Umgebung"), Inntal, Austria. Distribution: among nominate subspecies, so far: Austria, Czechoslovakia, Italy, France, Hungary

cylindrica minima Schuster, 1960

17 (14) Only one *(s)* spiniform, modified spine on tarsus of leg IV (pl. 66 : M).—Apodemes 1 meeting each other and apodemes 2 at one point (pl. 66 : L). Form and proportions of propodosomal and notogastral setae resembling those of preceding species. Length: 425—520 μm.

Locus typicus: "Etang des Salces", Perpignan, France. Distribution: only the type-locality known

inexpectata Schuster, 1960

18 (13) Sensillus of identical thickness throughout, or tapering apicad; setiform or bacilliform (pl. 67 : K).

19 (20) An essential difference in length between setae c_1 and h_1, latter one more than thrice longer than former, and also considerably thicker (pl. 67 : F). Setae c_1, c_2 and c_3 smooth, h_1 heavily ciliate.—Sensillus, long, setiform, gradually tapering. Apodemes 1 fused anteriorly to apodemes 2 (pl. 67 : G). Length: 635—737 μm, width: 294—325 μm.

Locus typicus: Mt. Spitz, near Recoaro, Italy. Distribution: only the type-locality known

heterotricha Mahunka, 1966

20 (19) No such difference in length between setae c_1 and h_1; setae c_1, c_2, c_3 as thick and cilite as h_1: latter merely twice longer than c_1 (pl. 67 : L).—Apodemes 1 fused, but no sternal apodeme forming behind them anterior to apodemes 2 (pl. 67 : M—N). Sensillus bacilliform (pl. 67 : K). Length: 475—782 μm.

Locus typicus: Weinzödel, Graz, Austria. Distribution: Austria, Belgium, Hungary, Yugoslavia

styriaca Schuster, 1960

Species interpretabiles

1. cylindrica Berlese, 1904: 23, fig. 2: 40 *(Lohomannia)* (lapsus calami).
   cylindrica Van der Hammen: 1959: 53.
   cylindrica Aoki: 1965: 309, figs. 1—3.
   cylindrica Bayoumi et Mahunka: 1976:6, figs. 1—5.
   szanislói Oudemans, 1917:79, figs. 127—139 *(Lesseria).*
   szanislói Schuster: 1960:202, figs. 4—5.
   verrucosa Jacot, 1934:6, pl. 1, figs. 1—6.
2. cylindrica minima Schuster, 1960;205, fig. 56 *szanisloi* forma *minima).*
   cylindrica minima Bayoumi et Mahunka: 1976: 7.
3. gigantea Berlese, 1916: 335.
   gigantea Sellnick: 1931: 707, figs. 11—13.
   gigantea Schuster: 1960: 208, figs. 7a—d.
   gigantea Bayomi et Mahunka: 1976: 7, figs. 10, 12, 14, 16.
4. heterotricha Mahunka, 1966: 372, figs. 1—5.
   heterotricha Bayoumi et Mahunka: 1976: 7, fig. 6.
5. imreorum Bayoumi et Mahunka, 1976: 7, figs. 7—9, 11, 13, 15, 17.
6. inexpectata Schuster, 1960: 207, figs, 6a—c.
7. ovata Aoki, 1961: 64, figs. 1a—c.
8. pallida aegyptica Bayoumi et Mahunka, 1976: 8, figs. 17—21.
9. schusteri Calugar et Vasiliu, 1976: 13, figs. 1—4.
10. spathulata Aoki, 1970: 399, figs. 5—8.
11. styriaca Schuster, 1960: 198, figs. 1—2.
    styriaca Schuster: 1965: 217.

Genus 2: Epilohmannoides Jacot, 1936

Epilohmannoides Jacot, 1936

Epilohmannoides Jacot, 1936: 546.
Epilohmannoides Norton, Metz et Sharma: 1978: 145.

Rostrum an acute nose; rostral setae arising far from rostral cuspis, at base of nose, usually at divers heights. Ventral plate without schizogastria (the sole case in the family *Epilohmanniidae!*). Seven (4+3) paris of genital setae. A single Palaearctic species known.

Type-species: *Epilohmannoides terrae* Jacot, 1936.

— Length: 340—420 μm, the smallest species of the genus (*E. terrae* Jacot, 1936: 365—446 μm; *E. jacoti* Norton, Metz et Sharma, 1978: 587—613 μm, differring from the other two known congeners also in the length of the notogastral setae and some other minor features. Representatives of the genus probably occurring also in Europe.

Locus typicus: Higashiyama Park, Nagoya-shi, Japan. Distribution: only the type-locality known

esulcatus Ohkubo, 1979

Species interpretabilis

1. esulcatus Ohkubo, 1979: 261, figs. 1—10.

Superfamily 5: COLLOHMANNIOIDEA Grandjean, 1958

Collohmanniidae Grandjean, 1958 : 118.
Collohmannoidea Grandjean, 1969 : 144.

Notogaster slightly compressed laterally, with a high convex dorsum, otherwise body characteristically ovoid. Notogaster with 15 pairs of visible setae, f_1 represented only by its alveolus. Prodorsum with five pairs of setae (two pairs of exobothridial ones). Genital and aggenital as well as anal and adanal plates differentiated. Anal plates nearly twice longer than genital plates; eight pairs of uniseriate genital, two pairs of aggenital, three pairs of anal and two pairs of adanal setae present. Epimeral setal formula: 3—1—2—5. Legs tridactylous.

Composed of a single monotypical family; known only from the Palaearctic Region.

Family 1: COLLOHMANNIIDAE Grandjean, 1958

Collohmanniidae Grandjean, 1958 : 118.

Only a single genus known, occurring in the entire Palaearctic Region.

Genus 1: COLLOHMANNIA Sellnick, 1922

Collohmannia Sellnick, 1922 : 18.
Collohmannia Grandjean: 1958 : 131.
Phthiracaroides Storkán, 1923 : 32.

Type-species: *Collohmannia gigantea* Sellnick, 1922.

Characterized by the superfamilial features. Secondary sexual dimorphism is known in the single species.

— Interlamellar seta longest of all prodorsal setae. Sensillus gradually thickening apicad. Essential differences in length between notogastral setae (pl. 68 : A): d_2, h_2 and ps_1 seven to eight times longer than shortest (c_1, c_2, cp) setae. Males and females differing by appendages on legs I and IV (considered as distinct species): male leg I with ribbon-shaped setae, tibia of leg IV (pl. 68 : C) with pustulate excrescences; dorsal seta of genu modified into a large spine, about two-thirds of joint. Female leg (pl. 68 : D) normal. Length: 1450—1804 μm, width: 710—968 μm.

Locus typicus: Comana Vlasca, Roumania. Distribution: Austria, Roumania, Greece, Soviet Union (SMA)

gigantea Sellnick, 1922

Species interpretabilis

1. gigantea Sellnick, 1922 : 18, fig. 1.
gigantea Schuster: 1965 : 216.
gigantea Grandjean: 1966 : 328, figs. 1—8.
gigantea Grandjean: 1969 : 127.
nova Sellnick, 1932 : 704, figs. 7—10.
incertus Storkán, 1923 (*Phthiracaroides*).

Cohort 2: EUPTYCTIMA Grandjean, 1967*

Euptyctima Grandjean, 1967 : 245.

Body ptychoid. Notogaster whole, without furrows or sutures, laterally compressed. Anogenital modified characteristically, in essence macropilic, but genital, aggenital, anal and adanal plates fused in divers variations, resulting in configuration resembling "french windows" or long and V-shaped, divided or undivided. Accordingly, cohort subdivided into two sharply separable superfamilies, with their morphologies similar in many respects yet considerably differing from all other Oribatids. In details see the superfamily *Phthiracaroidea*.

Key to superfamilies

1 (2) Anogenital region wide, with genital and anal plates resembling four wings of a "french window" (pl. 69 : F). Body less compressed laterally
superfamily 1: **Phthiracaroidea** Perty, 1841

2 (1) Anogenital region narrow, V-shaped (pl. 99 : C). Body considerably compressed laterally
superfamily 2: **Euphthiracaroidea** Jacot, 1930

Superfamily 1: PHTHIRACAROIDEA Perty, 1841

Phthiracarea Perty, 1841 : 874.
Phthiracarinae Michael: 1898 : 77.
Phthiracaroidea Grandjean: 1954 : 430.
Phthiracaroidea Van der Hammen: 1959 : 33.
Phthiracaroidea Grandjean: 1969 : 144.

It is the anterior part, the aspis**, of the widely oval ptychoid body which is movable against the opisthosoma (notogaster) and thus encloses the legs. The homologization of the interlamellar and lamellar setae on the aspis is not always safe and, therefore, (in such cases) anterior *(sr)*, median *(ss)* and lateral *(sl)* ones are distinguished according to position, usually, however, the seta near or nearest the bothridium is considered as the interlamellar and the more removed one the lamellar seta. Near the bothridium (anteriorly, posteriorly or laterally) a squamiform appendage, the bothridial squama appears; the insertion point of the exobothridial seta (or this seta itself) lies under (in the lateral view) the bothridium. The aspis may bear a crest (or a robust ridge-like elevation), and one or rarely two sharp lines (carina) start at or near the bothridium. From the basal part of the aspis to the rostrum a marginal line, parallel with the margin, is also generally recognizable.

The notogaster displays, also a marginal band delimited by a sharp line anteriorly and posteriorly, designated in front also as the collar.***

* Unfortunately, three important papers of Aoki: A revision of the Oribatid mites of Japan (1980) could not be incorporated in our work owing to belated postage.

** A term used in ptychoid groups instead of the prodorsum which rather agrees with the prosoma without the legs. The same holds for the superfamily *Euphthiracaroidea*.

*** Willmann (1951 : 171—172, 1954 : 264) made several attempts to separate the *Phthiracarus* species on the basis of the position of setae *c* and their distance from the posterior margin of the collar. Jacot

Lyrifissures are also recognizable *(ia, im, ip)* on its surface, and occasionally the insertion points of reduced setae (f_1, f_2). 15—33 pairs of notogastral hairs may be present, but neotrichy is very rare. The shape of the anogenital region, resembling "french windows" is attributable to the two pairs of fused genito-aggenital and ano-adanal plates; the former bears 7—9, the latter 3—9, pairs of setae. Originally, the ano-adanal plate emitted three pairs of adanal and two pairs of anal setae arranged in two longitudinal rows. In the several genera, the number of setae may vary: decrease or increase, or their positions may (also) alter. The decrease happens by the reduction of setae ad_1 and ad_2, in which cases they are represented virtually by their insertion points only (*Notophthiracarus* Ramsay, 1966; *Phthiracarus* Perty, 1841). The increase in the numbers of setae occurs mainly in certain genera of the southern hemisphere (*Neophthiracarus* Balogh et Csiszár, 1963; *Protophthiracarus* Balogh, 1972, etc.). The position of the setae alters mainly by one or two of the adanal pairs shifting to the inner margins of the ano-adanal plate, and these 3 or 4 setae becoming arranged closely behind each other (*Steganacarus* Ewing, 1917; *Hoplophorella* Berlese, 1923, etc.).

The majority of genera, established principally on the position of setae and other artificial features, cannot be upheld; on the other hand, the revision and redescription of the species of the Phthiracaroid group may eventually involve—following a change in concept—the subdivision of the genera. On the basis of these considerations we find four Palaearctic genera as acceptable, therein assigning those of the hitherto described species which can be keyed out according to their diagnoses or which we know by actual representatives. It seems probable that there are several "good" species among those described incompletely or all too briefly; however, pending studies made on the type-materials or on collections made in the type-localities, their status remains unclarified.

The superfamily contains a single family with five genera.

Family 1: Phthiracaridae Perty, 1841

Phthiracarea Perty, 1841 : 874.
Phthiracarinae Michael: 1898 : 77.
Phthiracaridae Van der Hammen: 1959 : 39.
Hoplophoridae Berlese, 1895 : 14.
Characterizable as the superfamily.

Key to genera

1 (2) Inner margin of anoadanal plate with 3—4 pairs of setae near each other (pl. 69 : F).—Notogaster with a rough sculpture of tubercles and pits. Notogastral setae often phylliform or bacilliform, occasionally incrassate, ciliate
genus 1: **Steganacarus** Ewing, 1917

(1936 : 179—180, fig. 26—27) found, however, that the positions of the setae *c* differ essentially in the males and females of *Phthiracarus ferrugineus* (C. L. Koch). Since all other authors described species without distinguishing between the sexes, the possibility exists that a part of the nominate species represents males and females of certain species. At the present state of knowledge, however, we cannot do but accept the specific interpretations accepted in literature.

2 (1) Inner margin of ano-adanal plate with two, rarely three, pairs of setae removed far from each other (pl. 77 : G).—Notogaster usually smooth, a rough sculpture only exceptionally present. Notogastral setae always setiform, thin and smooth.

3 (6) Interlamellar setae long, rigid, erect (pl. 78 : A), all other prodorsal setae considerably shorter and thinner.

4 (5) Number and position of notogastral setae normal; 15—18 pairs of setae present genus 2: **Hoplophthiracarus** Jacot, 1933

5 (4) Notogaster with a heavy pygidial neotrichy, only setae *c*, *d*, and *e* in normal numbers and positions genus 3: **Hauserophthiracarus** Mahunka, 1980

6 (3) No conspicuous difference among prodorsal hairs. Interlamellar setae short, accumbent.

7 (8) Setae ad_1 and ad_2 reduced, only their insertion points discernible, thus ano-adanal plate with only three pairs of true setae (pl. 81 : B) genus 4: **Phthiracarus** Perty, 1841

8 (7) Also setae ad_1 and ad_2 evolved, ano-adanal plate, therefore, with five pairs of setae (pl. 91 : B) genus 5: **Archiphthiracarus** Balogh et Mahunka, 1979

Genus 1: Steganacarus Ewing, 1917

Steganacarus Ewing, 1917 : 125.
Steganacarus Van der Hammen: 1959 : 41*.
Hoplophora C. L. Koch, 1841 : 2, 9.
Hoploderma Michael, 1898 : 77.
Atropacarus Ewing, 1917 : 125.
Tropacarus Ewing, 1917 : 125.
Trachyhoplophora Berlese, 1923 : 257.
Calhoplophora Berlese, 1923 : 257.

Aspis with a crista (mostly in its anterior section), emitting a chitinous ridge posteriorad. Notogaster with a rough sculpture consisting of pits, laths and tubercles. Setae often bacilliform or phylliform, ciliate or dentate. Inner margin of ano-adanal plate with 3 or 4 pairs of setae near each other, 1 or 2 other pairs removed far from these, on outer section.

Type-species: *Hoplophora anomala* Berlese, 1883.

The genus is probably an artificial unit, to be subdivided after the revision of the doubtful species.

Key to species

1 (20) Inner margin of ano-adanal plate with 3—4 pairs of very short setae, not or hardly longer than removed 1 or 2 pairs of adanal setae (pl. 69 : N), or than distance between insertion points of short setae**.

* A detailed survey of the nomenclatorial history is also presented; in the problem involved we follow this author's solution.

** Borderline cases are keyed out in both variations.

2 (7) Notogastral setae heavily ciliate, or serrately dentate (pl. 69 : L), occasionally penicillate. Sensillus long, arcuate, its outer margin dentate (pl. 68 : G).

3 (6) Inner pair of setae *(si)* of aspis long, straight and ciliate, outer pair *(sl)* only half as long, smooth (pl. 68 : F). Notogastral setae (pl. 68 : E) long, arcuate, heavily ciliate or dentately serrate; 19—20 pairs of notogastral setae present.

4 (5) Basal part of prodorsum with an extraordinarily heavy polygonal sculpture (pl. 68 : F). Notogaster with two pairs of setae (d_1, e_1) originating immediately near each other (in superior view), their enclosed distance smaller than diameter of an insertion point (pl. 68 : G).—Prodorsum with a low crest, its surface with large foveolae, especially laterally. Sensillus thin, long, arcuate. Considerable differences in length among 20 pairs of notogastral setae (pl. 68 : E), but seta *cp* not or hardly shorter than c_1 and c_2. Length of aspis: 184—236 µm, length of notogaster: 384—492 µm, height of notogaster: 236—304 µm.

Locus typicus: Panachaikon massif, near Kastritsion, Greece. Distribution: only the type-locality known **mirabilis** Mahunka, 1979

5 (4) Basal part of prodorsum (pl. 69 : M) without polygonal sculpture, at most some longitudinal laths and lines. In the superior view, the members of all pairs of setae situated at normal distances from each other, their enclosed distance multiply greater than diameter of an insertion point.—Lateral sculpture of prodorsum similar to that of preceding species, as also long, arcuate and thin sensillus (pl. 69 : K). Notogaster with 19 (?) pairs of setae, their length also considerably varying, but seta *cp* essentially shorter than c_1 and c_2 (pl. 69 : J). Length of aspis: 200—272 µm, length of notogaster: 411—570 µm, height of notogaster: 210—350 µm.

Locus typicus: Belin, region Stalin, Roumania. Distribution: Hungary, Roumania **serratus** Feider et Suciu, 1957

6 (3) Setae of apsis (pl. 69 : G) and notogaster equally short. Outer *(sl)* and inner *(si)* hairs of aspis nearly equal in length, both ciliate. Notogastral setae densely, occasionally penicillately ciliate, not arcuate. Only 15 pairs of notogastral setae present.—Aspis medially with only a weak, hardly dentate crest. Notogaster (pl. 69 : H) with very large foveolae. Sensillus long, arcuate, thin. Sculpture of genito-aggenital plate weaker. Length of aspis: 243—247 µm, length of notogaster: 396—413 µm, height of notogaster: 243—256 µm.

Locus typicus: "Giardino di Boboli", Florence, Italy.* Distribution: Hungary, Italy **clavigerus** (Berlese, 1904)

7 (2) Notogastral setae smooth, short or minute, straight or arcuate, apically flagelliform. Sensillus long, never sigmoid, not heavily ciliate or dentate.

8 (11) Notogaster anteriorly with a very large head- or eaves-shaped elevation jutting anteriorad, setae *c* arising on this excrescence.

9 (10) A deep hollow behind projecting notogastral appendage (pl. 69 : D).—Prodorsum with a heavier median and two adjacent weaker crests (pl. 69 : C).

* See Van der Hammen: 1959 : 44.

Stalk of sensillus long, distally fusiform, well differentiated from stalk (pl. 69 : B). Notogastral sculpture normal, ornamented with characteristic tubercles, setae short (pl. 69 : A). Configuration of anogenital region as in pl. 69 : F. Length of aspis : 486—567 µm, length of notogaster : 906—1130 µm, height of notogaster: 648—729 µm.

Locus typicus: Safad, Galilee, Israel. Distribution: Israel

grandjeani Mahunka, 1977

10 (9) No deep hollow present behind projecting appendage of notogaster, at most a saddle-shaped impression between setae c_1—d_1 (pl. 70 : A).—Notogaster with a longitudinal ridge medially. Prodorsal crests weaker (pl. 70 : D), but sculpture of body considerably rougher than in preceding species. All body setae minute. Sensillus thin, also distally hardly incrassate (pl. 70 : B). Length of aspis: 280—340 µm, length of notogaster: 460—700 µm, height of notogaster: 300—470 µm.

Locus typicus: "Colli Euganei", Padova, Italy. Distribution: Hungary, Italy, Roumania, Soviet Union (SET)

pulcherrimus (Berlese, 1887)

11 (8) Notogaster anteriorly without a head- or eaves-shaped excrescence projecting over prodorsum.

12 (19) Notogaster anteriorly with one or two parallel median ridges, at least partly well discernible (pl. 70 : E, 70 : H). Sensillus long.

13 (14) Notogaster anteriorly with two parallel ridges decurrent from setae c_1 to p_1 (pl. 70 : E).—Setae c_1, d_1, e_1, h_1 and ps_1 originating near the two ridges, on inner or outer sides. Notogastral setae (pl. 70 : F) setiform, acute, erect to reclinate, short. Sensillus arcuate, distally flagellately incrassate, pro- and exclinate. Inner setae of aspis *(si)* twice longer than outer ones. Length of aspis: 246—292 µm, length of notogaster: 575—622 µm, height of notogaster: 375—457 µm.

Locus typicus: Gorski Szvor, near Harmanli, Bulgaria. Distribution: only the type-locality known

bicarinatus Jeleva, 1970

14 (13) Notogaster with a single median carina, at least partially well discernible.

15 (18) Crista decurrent along entire length of notogaster. Notogastral setae either minute or very long and flagelliform.

16 (17) Notogastral setae minute (pl. 70 : G), apically straight.—Prodorsal setae very short, sensillus longer than even inner pair. Notogastral sculpture composed of tubercles and foveolae, median ridge often forming a minute tip anteriorly. Setae of ano-adanal plates minute (pl. 70 : I—J). Length of aspis: 209—335 µm, length of notogaster: 418—725 µm, height of notogaster: 293—488 µm.

Locus typicus: Regensburg, Germany. Distribution: Europe, Soviet Union (WS)

carinatus (C. L. Koch, 1841)

17 (16) Notogastral setae very long, fine, thin, distally flagellate (pl. 71 : A). — Inner pair of prodorsal setae *(si)* longer than sensillus (pl. 71 : C). Notogastral sculpture composed of very large foveolae. Setae of ano-adanal plate (pl. 71 : B) slightly longer, hindmost one shortest of all. Length of aspis: 308—

409 µm, length of notogaster: 623—844 µm, height of notogaster: 266—583 µm.

Locus typicus: Lasithi Plateau, Crete, Greece. Distribution: only the type-locality known

lasithiensis Mahunka, 1979

18 (15) Ridge in midline of notogaster extant only in posterior half of body, between setae e_1—p_1 (pl. 71 : E).—Sensillus distally widened: fusiform (pl. 71 : D). Notogastral setae longer than in preceding species, recumbent nearly at right angles to body surface. Chaetom of anogenital region as in pl. 71 : F—G. Lenght of aspis: 290 µm, length of notogaster: 573 µm, height of notogaster: 382 µm.

Locus typicus: "Stazione di Entomologia Agraria", Florence, Italy. Distribution: Italy, Hungary **brevipilus** (Berlese, 1923)

19 (12) Notogaster without any crest above, entirely smooth, of uniform curvature (pl. 71 : K). Notogastral setae distally very fine, flagellately recurving, (pl. 71 : J); sensillus (pl. 71 : I) short, phylliform, widest in its distal third. Lamellar setae minute, hardly discernible, interlamellar ones approximately as long as sensillus, at least twice longer than lamellar setae (pl. 71 : L); both pairs of setae fine. Ano-adanal setae comparatively long (pl. 71 : M). Length of aspis: 258 µm, length of notogaster: 537 µm, height of notogaster: 316 µm.

Locus typicus: Sainte Baume, Var, France. Distribution: only the type-locality known

travei Lions, 1968

20 (1) Ano-adanal plate with 3—4 pairs of setae on inner margins long, essentially longer than far removed 1—2 pairs of adanal setae (pl. 72 : H).

21 (24) Notogastral setae wide: spoon-shaped or phylliform (pl. 72 : D). Collar elevated on anterior margin of notogaster.

22 (23) All setae of ano-adanal plate setiform (pl. 72 : C), thin.—Prodorsal surface with a high crest (pl. 72 : A). Sensillus long, thin, distally weakly ciliate. Collar projecting on anterior margin of notogaster continuous also medially (pl. 72 : B). Notogastral sculpture very rough, with a nearly polygonal ornamentation. Notogaster with 16 pairs of setae, all phylliform, their surfaces finely ciliate. Length of setae arising on ano-adanal plate divers, increasing posteriorad. Length of aspis: 220—250 µm, length of notogaster: 380—415 µm, height of notogaster: 185—210 µm.

Locus typicus: Florence, Italy. Distribution: Austria, Italy, Hungary, Roumania, Bulgaria **phyllophorus** (Berlese, 1904)

23 (22) Setae of ano-adanal plate of divers shape, third seta from behind widened, phylliform (pl. 72 : F).—Setae of prodorsum (pl. 72 : G) and sensillus (pl. 72 : E) resembling those of preceding species. Notogastral collar interrupted medially. (pl. 72 : D); setae wide, spoon-shaped, smooth. Length of aspis: 200—210 µm, length of notogaster: 360—390 µm, height of notogaster: 250 µm,

Locus typicus: Columbia, Mo., USA. Distribution: North America, Japan

cucullatus (Ewing, 1909)

24 (21) Notogastral setae simpler: thin, setiform, clavate or bacilliform, eventually slightly spatulate, but never phylliform.

25 (26) Notogaster (pl. 73 : A) anteriorly with an extensive, projecting eaves-like excrescence, emitting seta c_1.—Sensillus (pl. 73 : B, F) filiform, arcuate; interlamellar setae setiform, very long, proclinate, reaching base of rostral setae (pl. 73 : C). Notogastral setae also long, erect. Entire body densely foveolate. Length of aspis: 375 μm, length of notogaster: 1170 μm, height of notogaster: 550 μm.

Locus typicus: Agri Romani, Frosino, Italy.* Distribution: Great Britain, Italy, Yugoslavia, Hungary

anomalus (Berlese, 1883)

26 (25) Notogaster without an eaves-like projection anteriorad.

27 (28) Notogastral setae equal in length, thin, fine, distally flagellate. Setae of ano-adanal plate short [see couplet 17 (16), if establishing lengths of ano-adanal plate problematic].

28 (27) Notogastral setae different; setae of ano-adanal plate considerably longer.

29 (34) Sensillus short, invariably essentially shorter than interlamellar setae *(si)*; widened, phylliform, marginally always smooth.

30 (31) Notogastral setae short, with a very fine flagellately recurving distal portion [see under 13 (8) if establishing lengths of anogenital setae problematic].

31 (30) Notogastral setae robust, thick, without recurving distal section.

32 (33) Notogastral setae uniform, recurving to surface of body, acute, entirely smooth, reclinate (pl. 73 : H), approximately equal in length; apex of seta c_1 reaching beyond half distance c_1—d_1. — Inner pair of setae *(si)* of aspis long and smooth, outer one *(sl)* very short. Sensillus (pl. 73 : G) crescent-shaped, asymmetrical. Body foveolate. Ano-adanal plate with four aligned setae equal in length. Length of aspis: 250—300 μm, length of notogaster: 500—620 μm.

Locus typicus: "Playa de Magaluf", Palma Nova, Mallorca, Spain. Distribution: only the type-locality known

balearicus Pérez-Iñigo, 1969

33 (32) Notogastral setae of two types: c_2, c_3, *cp* and ps_4 setiform, acute, smooth, reclinate, all others bacilliform, obtuse, apically finely ciliate (pl. 72 : H).—Sensillus (pl. 72 : I) short, phylliform, weakly fusiform, apically obtuse, with smooth margins. Setae *le* setiform and pointed, setae *in* bacilliform and proclinate, apically finely ciliate and obtuse. Body surface foveolate. Ano-adanal plate with setae gradually lengthening posteriorad. Length of aspis: 180—315 μm, length of notogaster: 465—655 μm.

Locus typicus: "El Baillardo", Tenerife, Spain. Distribution: only the type-locality known

tenerifensis Pérez-Iñigo, 1972

34 (29) Sensillus setiform, long, arcuate, distally ciliate or serrato-dentate.

35 (36) Notogaster with 19 pairs of setae (pl. 73 : J).—Setae rigid, erect; c_1, d_1 e_1 and h_1 about equal in length, lateral setae considerably shorter, than *cp*, h_2, h_3 and ps_4; these two latter at most half as long as c_1 or h_1. Aspis (pl. 73 : L) with setae *sl*

* See Van der Hammen: 1959 : 42.

and *si* straight and proclinate. Sensillus (pl. 73 : K) thin, arcuate, ciliate. Ano-adanal setae smooth, about equal in length (pl. 73 : I). Length of aspis: 194 μm, length of notogaster: 348 μm, height of notogaster: 208 μm.

Locus typicus: Mts. Keszthely, near Keszthely, Hungary. Distribution: only the type-locality known

csiszarae Balogh et Mahunka, 1979

36 (35) Notogaster with merely 15 pairs of setae.

37 (40) Notogastral setae e_1 essentially shorter than h_1; latter at least twice longer than h_3 and ps_4.

38 (39) Inner seta *(si)* of aspis (pl. 74 : F) arcuately reclinate, hair c_1 also arcuate, proclinate, almost as long as seta *si* (pl. 74 : E). Crest of aspis underdeveloped, not projecting from body profile in lateral view. Notogastral setae (pl. 74 : H) longer, c_1 as long as distance between c_1—d_2, seta h_3 as long as, or longer than, distance between h_3—ps_4. Body brown. Chaetom of anogenital region as in pl. 74 : G. Length of aspis: 405 μm, length of notogaster: 835 μm.

Locus typicus: Ojo Guarena, Spain. Distribution: a single specimen known from the type-locality

ortizi Pérez-Iñigo, 1969

39 (38) Inner seta *(si)* of aspis proclinate (directed towards rostrum), c_1 erect and slightly reclinate (pl. 74 : D), and at most half as long as seta of aspis.—Crest of aspis developed, arcuately projecting from profile of aspis. Seta c_1 short, halfs as long as distance between c_1—d_1, ratio similar between length of seta h_3 and distance of h_3—ps_4. Body dark, nearly black. Length of aspis: 450 μm, length of notogaster: 825 μm.

Locus typicus: "Gipfel des Unterberges", Voralpen, Austria. Distribution: only the type-locality known

herculeanus Willmann, 1953

40 (37) Notogastral setae about equal in length, if some smaller differences extant, then seta e_1 as long as h_1.

41 (42) Notogastral setae proclinate, rigid, hardly arcuate, apically denticulate. Notogaster long, more than twice longer than its greatest height, its surface irregularly rugulose (pl. 74 : A).—Prodorsum with a low crest (pl. 74 : C), its surface foveolae coalescent into rugae. Sensillus long, setiform, arcuate, pointed (pl. 74 : B), finely ciliate. Setae *in* similar to notogastral ones, also proclinate. Notogastral setae c_1 considerably longer than c_3 or ps_4. Ano-adanal plate with four pairs of setae originating along inner margin about equal in length, arcuate. Length of aspis: 210 μm, length of notogaster: 390 μm, height of notogaster: 200 μm.

Locus typicus: Sanshiro-Weide, Utsukushigaha ra, Nagano district, Japan. Distribution: only the type-locality known

senex Aoki, 1958

42 (41) Notogastral setae at least partly erect or reclinate, of different form. Notogastral surface never rugose.

43 (46) Notogastral setae simple, thin, smooth and acute.

44 (45) Notogastral setae recurving to surface of body, very long, distally often flagelliform (pl. 75 : A).—Sensillus short, hardly incrassate, not shorter than

inner bothridial setae (pl. 75: B). Notogaster ornamented with very large foveolae, therefore, surface locally covered with a polygonal reticulation. Ano-adanal plates (pl. 75: C) with four pairs of setae on their inner margins approximately equal in length, thin, slightly arcuate. Length of aspis: 375—402 µm, length of notogaster: 857—938 µm, height of notogaster: 510—550 µm.

Locus typicus: Kato Zakros, Crete, Greece. Distribution: only the type-locality known

flagellatissimus Mahunka, 1979

45 (44) Every seta erect, thin, straight.—Crest of aspis not high. A polygonal sculpture on anterior part of aspis and a part of notogaster. Body olive brown. Length of aspis: 262—286 µm, length of notogaster: 450—572 µm, height of notogaster: 350—380 µm.

Locus typicus: Gizycko (olim: Lötzen), Poland. Distribution: Yugoslavia, Hungary, Poland, Germany, Soviet Union

spinosus (Sellnick, 1920)

46 (43) Notogastral setae either of divers shape or at least partly not smooth and simple.

47 (50) Notogastral setae partly accumbent to body surface partly rigidly erect.

48 (49) Notogastral setae accumbent to body surface, only last two pairs erect and rigid (pl. 75: E).—Setae mostly smooth, but erect ones distally ciliate. Prodorsal setae of divers lengths, inner pair *(si)* considerably longer than outher one (pl. 75: D). Length of aspis: 387 µm, length of notogaster: 758 µm, height of notogaster: 387 µm.

Locus typicus: St. Blasien, Germany. Distribution: Germany, Poland

applicatus (Sellnick, 1920)

49 (48) Every seta in midline of notogaster (c_1, d_1, e_1, h_1) rigid, straight and erect; several lateral setae of notogaster accumbent to body (pl. 75: H).—Inner pair *(si)* of aspis (pl. 75: I) shorter than outer one. Sensillus (pl. 75: F) hardly incrassate. Chaetom of anogenital region as in pl. 75: G. Length of aspis: 313—550 µm, length of notogaster: 700—1196 µm, height of notogaster: 656—840 µm.

Locus typicus: environment of Paris, France. Distribution: Europe

magnus (Nicolet, 1855)

50 (47) Every notogastral seta rigid and erect.

51 (52) Notogastral setae, but especially those along midline (c_1, e_1, h_1), distally considerably widended (pl. 76: A). Both pairs of setae in interbothridial region short (pl. 76: B).—Seta *si* only half as long as distance between *si*—*ro*. Basal part of prodorsum with robust longitudinal laths. Sensillus heavily ciliate. Setae on inner margins of ano-adanal plate (pl. 76: C) long, arcuate and thin. Length of aspis: 181—330 µm, length of notogaster: 340—484 µm, height of notogaster: 230—370 µm.

Locus typicus: Lasithi Plateau, Crete, Greece. Distribution: only the type-locality known

platakisi Mahunka, 1979

52 (51) Notogastral setae not clavately widened, prodorsal ones longer, at least inner pair essentially longer than half distance between *si*—*ro*.

53 (54) A great, nearly twofold, difference in length between setae originating in interbothridial region (pl. 76: D). Seta *si* heavily ciliate, *sl* short and smooth. Sensillus arcuate, distally dentato-serrately ciliate. Notogastral setae distally similar. Setae of ano-adanal plate (pl. 76: E) straight, not incrassate. Body surface always characteristically squalid owing to adhering grains of soil. Length of aspis: 260—292 µm. length of notogaster: 480—580 µm, height of notogaster: 390—400 µm

Locus typicus: Regensburg, Germany. Distribution: Holarctic Region

striculus (C. L. Koch, 1836)

54 (53) Both pairs of setae in interbothridial region long; even if outer seta slightly shorter, still more than two-thirds as long as inner one.

55 (56) Sensillus weakly fusiform (pl. 76: G) not sigmoid, smooth. Both interbothridial pairs of setae on prodorsum distally heavily ciliate (pl. 76: F), similar to notogastral setae.—Setae of ano-adanal plate slightly thickened, gradually shortening anteriorad. Length of aspis: 370—405 µm, length of notogaster: 680—900 µm.

Locus typicus: Santander, Spain. Distribution: Spain

hirsutus Pérez-Iñigo, 1974

56 (55) Sensillus long, sigmoid, setiform, distally heavily ciliate (pl. 76: H). Both interbothridial pairs of setae long, but only inner one *(si)* distally ciliate, outer pair *(sl)* throughout smooth. Notogastral setae ciliate from base to apex (pl. 76: I).—Hindmost hair of ano-adanal plate considerably longer than the anterior three pairs of about equal length. Length of aspis: 315 µm, lengt of notogaster: 675—750 µm, height of notogaster: 480 µm.

Locus typicus: Rabacal, Madeira, Portugal. Distribution: only the type-locality known

similis Willmann, 1939

Species interpretabiles

1. anomala Berlese, 1883: 6,5 *(Hoplophora)*.
   anomalus Van der Hammen: 1959: 42.
   anomalus Schuster: 1965: 219.
   karamani Tarman, 1962: 141, figs. A—B
2. applicatus Sellnick, 1920: 36 *(Phthiracarus)*.
   applicatus Sellnick: 1960: 128.
   berlesei Oudemans, 1915: 214 *(Phthiracarus)*.
3. balearicus Pérez-Iñigo, 1969: 377, figs. 1—3.
   balearicus Pérez-Iñigo: 1968: 211, fig. 26 (nom. nudum).
4. bicarinatus Jeleva, 1970: 411, figs. 1a—c.
5. brevipilus Berlese, 1923: 257 *(Phthiracarus/Trachyhoplophora)*.
   brevipilus Van der Hammen: 1959: 41 *(Tropacarus)*.
   brevipilus Bernini: 1971: 3, figs. 2—9, pl. I—III.
   applicatum Willmann: 1931: 191, fig. 349 *(Hoploderma)*.
   applicatus Van der Hammen: 1959: 43.
6. carinata C. L. Koch, 1841: fasc. 32, 9, fig. *(Hoplophora)*.
   carinata Berlese: 1887: 36, pl. 1, figs. 1—4 *(Hoplophora)*.
   carinatus Feider et Suciu: 1957: 14, figs. 129—137 *(Tropacarus)*.
   carinatus Bernini: 1971: 15, pl. 7, figs. 1—3.

7. clavigerum Berlese, 1904: 275 *(Hoploderma)*.
clavigerum Berlese: 1913: 104, pl. 8, fig. 104 *(Hoploderma)*.
clavigerus Bernini: 1971 : 15, fig. 8 : a, pl. 6 : 1—2.
clavigerus Balogh et Mahunka: 1979: 286, figs. 10—11.
8. csiszarae Balogh et Mahunka, 1979: 284, figs. 7—9.
9. cucullatum Ewing, 1909: 133, pl. 6, fig. 35 *(Hoploderma)*.
cucullata Jacot: 1933: 249 *(Hoplophorella)*.
cucullata Van der Hammen: 1959: 46 *(Hoplophorella)*.
licnophorum Berlese, 1913: 102, pl. 8, fig. 99 *(Hoploderma)*.
10. flagellatissimus Mahunka, 1979: 554, figs. 15—17.
11. grandjeani Mahunka, 1977: 132, figs. 1—6.
sacculiferus Parry, 1978: 279, figs. 1—14.
12. herculeanus Willmann, 1953: 515, fig. 52.
13. hirsutus Pérez-Iñigo, 1974: 419.
clavigerus Pérez-Iñigo (nec Berlese): 1968: 209, figs. 24—25.
perezinigoi Balogh et Mahunka, 1979: 286 **(syn. nov.)**.
14. lasithiensis Mahunka, 1979: 554, figs. 18—20.
15. magna Nicolet, 1855: 472, pl. 10, fig. 4 *(Hoplophora)*.
magna Berlese: 1892: 67, 9, pl. 15, figs. 1—5 *(Hoplophora)*.
magnum Feider et Suciu: 1957: 32, figs. 91—100.
magnus Pérez-Iñigo: 1968: 208, fig. 23.
16. mirabilis Mahunka, 1979: 556, figs. 21—23.
17. ortizi Pérez-Iñigo, 1969: 144, figs. 1—4.
ortizi Pérez-Iñigo: 1968: 209 (nom. nudum).
18. phyllophorum Berlese, 1904: 265 *(Hoploderma)*.
phyllophorum Berlese: 1913: 103, pl. 8, figs. 102, 102a *(Hoploderma)*.
phyllophorus Schuster: 1957: 97, figs. 1—3.
ropalus Feider et Suciu, 1957: 35, figs. 120—128.
19. platakisi Mahunka, 1979: 557, figs. 24—26.
20. pulcherrima Berlese, 1884: ser. 5* *(Hoplophora)*.
pulcherrima Berlese: 1887: 35,—10, pl. 14, figs. 1—7 *(Hoplophora carinata* K. var. *pulcherrima* n. var.(!)).
pulcherrimus Van der Hammen: 1959: 40.
pulcherrimus Feider et Suciu: 1957: 37, figs. 138—147 *(Tropacarus)*.
21. robertsi Sheals, 1965: 11, figs. 11—19.
22. senex Aoki. 1958: 172, figs. 3A—D.
23. serratus Feider et Suciu, 1957: 34, figs. 111—119.
24. similis Willmann, 1939: 39, figs. 25a—b.
25. spinosus Sellnick, 1920: 37 *(Phthiracarus)*.
spinosum Willmann: 1931: 190, fig. 345 *(Hoploderma)*.
26. stricula C. L. Koch, 1836: 2, 10 *(Hoplophora)*.
striculum Michael: 1888: 563, fig. 7 *(Hoploderma)*.
striculus Feider et Suciu: 1958: 33, figs. 101—110.
27. tenerifensis Pérez-Iñigo, 1972: 190, figs. 2—3.
28. travei Lions, 1968: 500, figs. 1—8.

Species inquirendae

— curtipilus Strenzke, 1952: 152.
curtipilus Van der Hammen: 1959: 41 *(Tropacarus)*.
curtipilus Bernini: 1971: 4, figs. 1a—c *(Tropacarus)*.
— perfectus Sellnick, 1931: 721 *(Hoploderma curtipilus* Berlese var. *perfecta)*.

* See Van der Hammen: 1959: 40.

Genus 2: HOPLOPHTHIRACARUS Jacot, 1933

Hoplophthiracarus Jacot, 1933 : 239.
Hoplophthiracarus Van der Hammen: 1959 : 46.
Hoplophthiracarus Van der Hammen: 1963 : 307.

Interlamellar setae *(si)* of aspis long, rigid, erectile, essentially thicker than all other prodorsal setae. Notogaster at least partly punctate or granulose. Ano-adanal plate with two pairs of setae along its inner margins, the other three pairs also developed and originating in a semicircle removed from inner margin.

Type-species: *Hoploderma histricinum* Berlese, 1908.

Key to species

1 (6) Notogaster with more than 15 pairs of setae.

2 (3) Notogaster with 18 pairs of setae (pl. 77 : A). Sensillus distally hardly perceptibly thickening, its apex obtuse (pl. 77 : C).—The two pairs of anal setae originating near inner margin considerably shorter than adanal seta, even ad_1 situated far from inner margin (pl. 77 : B). Length of aspis: 298—398 µm, length of notogaster: 583—876 µm,

Locus typicus: Tivon, Israel. Distribution: Israel

costai Macfarlane et Sheals, 1965

3 (2) Notogaster with only 17 pairs of setae, sensillus distally considerably widened.

4 (5) All notogastral setae uniform, erect then proclinate (towards prodorsum) (pl. 77 : F); setae ciliate, except basally.—Rostral setae ciliate, longer than entirely smooth *sl* (pl. 77 : E). Stalk of sensillus (pl. 77 : D) long, clavately capitate, ciliate. Ano-adanal plate (pl. 77 : G) with seta ad_3 also short, not longer than anal setae, but originating far from inner margin of plate. Length of aspis: 152—213 µm, length of notogaster: 402—451 µm, height of notogaster: 225—267 µm.

Locus typicus: Mt. Psylodendron, Ossa massif, Thessaly, Greece. Distribution: only the type-locality and its immediate environment known

cretensis Mahunka, 1979

5 (4) Notogastral setae divers, *c* long, thin, arcuate (pl. 77 : I), smooth; all other ones shorter, more rigid, ciliate, resembling those of preceding species.—Rostral and outer *(sl)* setae of prodorsum nearly equal in length, both pairs smooth. Sensillus very short widely clavate or spatulate (pl. 77 : J). Ano-adanal plate (pl. 77 : H) with anal setae equal in length, adanal ones considerably longer, ad_1 longest, ad_3 shortest among them. Length of aspis: 266—295 µm, length of notogaster: 418—576 µm, height of notogaster: 375—422 µm.

Locus typicus: "Valley of Death", Kato Zakros, Crete, Greece. Distribution: only the type-locality known

heterotricha Mahunka, 1979

6 (1) Notogaster with merely 15 pairs of setae.

7 (8) Sensillus setiform, apically tapering, acute, finely ciliate (pl. 78 : B). Ano-adanal setae (pl. 78 : C) with ad_1 also immediately at inner margin of plates,

together with anal ones. Notogastral setae (pl. 78 : A) apically finely ciliate. Length of aspis: 215—290 μm, length of notogaster: 330—500 μm.

Locus typicus: Sandakhpu, Nepal. Distribution: only the type-locality known

nepalensis Sheals, 1965

8 (7) Sensillus distally thickening.

9 (10) Sensillus short, its head nearly round.—Erectile setae of aspis very long, curved. Notogastral setae different in length, d_1 much shorter than *cp* or h_3. All setae thin well ciliated. Anal setae equal, both pairs much shorter than adanal ones. No measurement data available.

Locus typicus: Cazane, region Caras-Severin, Roumania. Distribution: only the type-locality known

cazanicus Feider et Calugar, 1970

10 (9) Sensillus long, its head elongated.

11 (12) Sensillus distally (pl. 78 : E) and notogastral setae entirely smooth.—Rostral setae slightly longer than lateral ones *(sl)*. Body entirely smooth. Essential differences in length among notogastral setae: h_2 and h_3 considerably shorter than e_1 and h_1(pl. 78 : G). Length of aspis: 180—225 μm, length of notogaster: 295—395 μm.

Locus typicus: Tiarno, Italy. Distribution: Germany, Italy, Soviet Union (SET)

pavidus (Berlese, 1913)

12 (11) Sensillus and notogastral setae distally finely ciliate or dentate.

13 (14) Notogastral setae erect, distally weakly proclinate and finely ciliate (pl. 79 : A). Rostral setae comparatively long, originating nearer to rostrum than to insertion points of interlamellar setae (pl. 79 : B). Notogaster densely foveolate. Considerable differences in length among notogastral setae: those arising on midline essentially longer than lateral ones. Sensillus (pl. 79 : C) long, arcuate, distally dentato-serrate. Length of aspis: 300—320 μm, length of notogaster: 590—630 μm, height of notogaster: 410—450 μm.

Locus typicus: Iwakiri, Miyazaki district, Japan. Distribution: only the type-locality known

kugohi Aoki, 1959

14 (13) Notogastral setae conspicuously arcuate, from base towards aspis (pl. 78 : D), distal half of setae roughly dentate. Rostral setae about midway between rostrum and bases of interlamellar ones. — Notogaster sparsely foveolate. Notogastral setae about equal in length. Length of aspis: 140 μm, length of notogaster: 290 μm.

Locus typicus: Blagoveshchensk, on the Amur, Soviet Union. Distribution: Soviet Union (SMA, FE)

minus (D. Krivolutsky, 1965)

Species interpretabiles

1. costai Macfarlane et Sheals, 1965 : 577, figs. 1—9.
2. cazanicus Feider et Calugar, 1970 : 25, figs. 15—18.
3. cretensis Mahunka, 1979 : 558, figs. 27—30.

4. heterotricha Mahunka, 1979: 560, figs. 31—33.
5. kugohi Aoki, 1959: 17, figs. 12A—D.
kugohi Aoki: 1977: 185, fig. 5.
6. minus D. Krivolutsky, 1965: 127, fig. 2 *(Phtiracarus)*.
7. nepalensis Sheals, 1965: 15, figs. 21—29.
8. pavidum Berlese, 1913: 103, pl. 8, figs. 101, 101a *(Hoploderma)*.
pavidus Sellnick: 1929: 40 *(Phthiracarus)*.
pavidus Van der Hammen: 1959: 49 *(Phthiracarus)*.
pavidus Van der Hammen: 1963: 307, figs. 1—8.

Genus 3: HAUSEROPHTHIRACARUS Mahunka, 1982

Hauserophthiracarus Mahunka, 1982: 502.

Both basal pairs of prodorsal setae rigid, erect; inner pair recurving, very long. Sensillus setiform. Notogaster with robust, ciliate setae, *c, d* and *e* normal, a heavy neotrichy posteriorly; 33 pairs of notogastral setae. Ano-adanal plates with five pairs of long setae.

Type-species: *Hauserophthiracarus oenipontanus* Mahunka, 1982.

Only a single species known.

— Carina on prodorsum (pl. 79: E) short, not reaching prodorsal margin. Rostral seta short, hardly longer than exobothridial one. Both pairs of setae in interbothridial region rigidly erect, inner pair considerably longer and reclinate towards notogaster. Sensillus (pl. 79: G) long, thin, setiform, ciliate. Notogastral setae semicircular, arcuate, directed towards prodorsum. Setae *c* divers in length, c_1 more than twice longer than c_3; c_1, c_2 and c_3 originating nearly equidistant from collar line. Neotrichial and normal setae not distinguishable. Nine pairs of short genital and five pairs of ano-adanal setae present (pl. 79: F)), two of these latter pairs considerably shorter than those emitted much farther from inner margins. Length of aspis: 388—495 μm, length of notogaster: 786—1010 μm, height of notogaster: 485—703 μm.

Locus typicus: Ossa massif, near Spilia, Thessaly, Greece. Distribution: only the type-locality known

oenipontanus Mahunka, 1982

Species interpretabilis

1. oenipontanus Mahunka, 1982: 502, figs. 1—3.

Genus 4: PHTHIRACARUS Perty, 1839

Phthiracarus Perty, 1839: column 847.
Phthiracarus Van der Hammen: 1959: 47.
Phthiracarus Parry: 1979: 334.
Hoplophora C. L. Koch, 1836: 2, 9.
Hoploderma Michael, 1898: 77.

Aspis and notogaster smooth, aspis without a high crest (occasionally only very weak chitinous laths observable). Lamellar and interlamellar setae aligned with bothridia in a nearly transverse line. Anterior margin of notogaster often thickened into a "collar". Notogaster with 15 or exceptionally more and generally simple setae. Ad_1 and ad_2 reduced on ano-adanal plate, recognizable only by their insertion points. A secondary sexual dimorphism may also occur.

Type-species: *Phthiracarus contractilis* Perty, 1841 = *Hoplophora laevigata* C. L. Koch, 1841.

The remarks made in the diagnosis of the genus *Steganacarus* hold also for *Phthiracarus;* nor are the redescriptions reliable.* The genus itself should be validated in view of the problematic identity of the type-species.

Key to species**

1 (4) Sensillus long, thin, setiform (pl. 80 : B), at most basally slightly thickened, but always gradually tapering apicad, never fusiform.

2 (3) Setae c_1 and c_3 originating on "collar" of notogaster (pl. 80 : A).—Aspis between interlamellar and rostral setae rather concave than convex in lateral view, no crest present; sinuous anterior to rostral setae; aspis throughout with a rim, also apically well discernible. Notogaster entirely smooth, setae short, d_1 considerably shorter than e_1 and h_1. Distance between setae c_1 and c_2 essentially greater than that between c_2 and c_3. Length: 400—600 µm.

Locus typicus: Regensburg, Germany. Distribution: Germany.

testudineus (C. L. Koch, 1841)

3 (2) Setae c_1 and c_2 originating considerably posterior to notogastral "collar" line (pl. 80 : E).—Notogastral setae fine, distance between c_1 and c_2 hardly greater than that between c_2 and c_3. Setae d_1, e_1 and f_1 equal in length. Length of aspis (pl. 80 : D): 258—279 µm, length of notogaster: 558—588 µm.

Locus typicus: "Higher Kiln Quarry", Buckfastleigh, Devon, Great Britain. Distribution: only the type-locality known

juvenalis Parry, 1979***

4 (1) Sensillus either short, lanceolate, medially greatly widened, or long and then incrassate, eventually fusiform, bacilliform, apically obtuse.

* For instance, some of Feider and Suciu's identifications can be accepted only with great reservations. *Phthiracarus anonymum* Grandjean, 1933, is not identical with *Phthiracarus anonymus* Feider et Suciu, 1957. Grandjean's species has three adanal setae and thus belongs in *Archiphthiracarus* Balogh et Mahunka, 1978, whereas Feider et Suciu's species displays one true adanal seta and, therefore, represents a *Phthiracarus* s. str. taxon. Similarly, *Phthiracarus ligneus* Willmann, 1932, is not identical with *Phthiracarus ligneus* Feider et Suciu, 1957. The former again belongs in *Archiphthiracarus*, and the latter in *Phthiracarus*. Figures made from representative specimens are also often confused (see *Phthiracarus sellnicki*, figs. 86—88).

** The separation of the species is extremely difficult, because features now considered important have not been given in older descriptions; furthermore, some species have also been described recently without a previous redescription of the old ones. Therefore, the identification key is virtually based on diagnoses instead of actual specimes.

*** In the description of the species Parry disregarded *testudineus* (C. L. Koch, 1841) as redescribed by Jacot, a species which, by the form of the sensillus, stands very near to it; indeed, they may even be identical.

5 (14) Sensillus long, usually longer than lamellar seta, if not, still projecting far beyond margin of aspis (in superior view), distally slightly but gradually thickening, apically slightly clavate.

6 (7) Sensillus distally much widened, lanceolate, ciliate (pl. 80 : F—G).—Setae of aspis, especially interlamellar pair, very long, latter reaching insertion points of rostral setae, also exobothridial setae long (pl. 80 : I). Notogastral setae extraordinarily long, thin, arcuate (pl. 80 : H), c_1 essentially longer than distance between c_1—d_1. No conspicuous difference in length among notogastral setae, excepting the slightly shorter setae *ps*. Ano-adanal setae (three pairs) divers in length, ad_3 shortest of all. Length of aspis: 262 μm, length of notogaster: 534 μm, height of notogaster: 394 μm.

Locus typicus: Bogdo ul, Bugijn az achuj, Mongolia. Distribution: only the type-locality known

kaszabi Mahunka, 1981

7 (6) Sensillus smooth or at most marginally slightly dentate, distally widened part rounded.

8 (9) Notogastral seta c_1 originating immediately adjacent to "collar" line, c_2 and c_3 far removed and at nearly equal distances (pl. 81 : F). No essential difference in length between lamellar and interlamellar setae, though latter slightly longer (pl. 81 : E). Sensillus resembling a hockey stick (pl. 81 : D), laterally with some teeth. Majority of notogastral setae directed towards aspis, more robust than those of preceding species. Setae of ano-adanal plate approximately of equal length. Length of aspis: 182—213 μm, length of notogaster: 456—466 μm.

Locus typicus: Wytham Woods Estate, Berkshire, Great Britain. Distribution: only the type-locality known

serrulatus Parry, 1979

9 (8) Setae c_1 and c_3 originating adjacent to "collar" line of notogaster, c_2 much farther removed.

10 (11) Sensillus with a membraniform lamella decurrent from its lower third to apex (pl. 81 : H).—Aspis setae short (pl. 81 : G), interlamellar one about half as long as lamellar seta. Inclination of thin notogastral setae divers (pl. 81 : I). Length of aspis: 213—324 μm, length of notogaster: 314—517 μm.

Locus typicus: "Sitka spruce F., Tintern Forest," Monmouthshire, Great Britain. Distribution: only the type-locality known

membranifer Parry, 1979

11 (10) Sensillus fusiform or bacilliform, without a marginal membrane.

12 (13) Notogastral setae proclinate, directed towards apex of aspis (pl. 81 : A, C). Sensillus straight, basally hardly perceptibly thickened.—A weak crest on aspis well discernible in lateral view. Lamellar and interlamellar setae approximately equal in length, much longer than sensillus. Margin of aspis forming a flat projection at apex in lateral view. Chaetom of anogenital region as in pl. 81 : B. Length of aspis: 330—360 μm, length of notogaster: 500—800 μm, height of notogaster: 500—550 μm.

Locus typicus: Regensburg, Germany, Distribution: West and Central Europe.

lentulus (C. L. Koch, 1841)

13 (12) Notogastral setae erect or reclinate, directed towards end of body (pl. 82: A). Sensillus arcuate or sigmoid, throughout evenly thick, occasionally slightly widened in middle.—Inner seta of interbothridial region considerably shorter than outer one, sensillus as long as, or longer than, these setae. Aspis without crest, but visibly convex between rostral and interlamellar setae in lateral view; its margin not projecting anteriorly in profile. Chaetom of anogenital region as in pl. 82: B. Length of aspis: 225—265 µm, length of notogaster: 470—625 µm, height of notogaster: 380—410 µm.

Locus typicus: Tiarno, Italy. Distribution: Holarctic Region

italicus (Oudemans, 1906)

14 (5) Sensillus short, fusiform, phylliform, or lanceolate, essentially shorter than lamellar seta.

15 (16) An extraordinarily wide species extremely convex and both nearly round in lateral and superior views (pl. 82: E). Aspis with a well developed crista (pl. 82: D).—Sensillus comparatively short, basally widened but with an acute apex. Both pairs of setae in interbothridial region nearly equal in length, short, not even approaching insertion points of rostral seta. Notogastral. setae of divers inclination, c_2 and c_3 originating much nearer to "collar" line than c_1. Chaetom of anogenital region as in pl. 82: C. Length of aspis: 260—330 µ, length of notogaster: 470—700 µm, height of notogaster: 560—680 µm.

Locus typicus: Regensburg, Germany. Distribution: Holarctic Region

globosus (C. L. Koch, 1841)

16 (15) Notogaster narrower, much less convex animals. Aspis without crest.

17 (26) Notogastral setae long, c_1 at least as long as distance of c_1—d_1.

18 (19) Notogastral setae c_1 and c_3 originating far in front, anterior to "collar" cline; c_2 situated far behind "collar" (pl. 83: A).—Notogastral setae proclinate, towards aspis. Setae c_1, c_2 and c_3 about equal in length. Inner setae *(si)* of aspis long, terminating not far from insertion points of rostral setae. Length of aspis: 245 µm, length of notogaster: 375 µm, height of notogaster: 318 µm.

Locus typicus: Ebreichsdorf, Austria. Distribution: Switzerland, Austria

crenophilus Willmann, 1951

19 (18) Setae c_1, c_2 and c_3 originating on, or behind, "collar" line of notogaster.

20 (21) Setae c_1, c_2 and c_3 originating on "collar" line of notogaster, c_1 and c_2 long, c_3 much shorter (pl. 82: G). Notogastral setae reclinate, directed towards end of body, arcuate. Aspis setae shorter (pl. 82: H), inner seta *(si)* terminating far from rostral one. Sensillus narrow, medially incrassate, fusiform, apex acute. Chaetom of anogenital region as in fig. 82: F. Length of aspis: 205—228 µm, length of notogaster: 3360—369 µm, height of notogaster: 279—296 µm.

Locus typicus: Örvényes, valley of Péczeli stream, Hungary. Distribution: only the type-locality known

zicmani nom. nov.*

21 (20) At least one of setae *c* originating far from "collar" line of notogaster; no essential difference in length between them.

* *Phthiracarus zicmani* **nom. nov.** = *P. feideri* Balogh et Mahunka, 1979, junior homonym: *P. feideri* Balogh et Csiszár, 1963.

22 (23) Inner interlamellar pair of setae reaching base of rostral seta (pl. 83 : F), lamellar ones short.—Setae c_1 and c_2 originating far from "collar" of notogaster (pl. 83 : E), c_3 much nearer to it. Sensillus (pl. 83 : D) lanceolate. Notogastral setae of divers directions, thin, arcuate, *c* and *d* proclinate (directed towards aspis). Length of aspis: 230—255 µm, length of notogaster: 560 µm.

Locus typicus: New Forest, Hampshire, Great Britain. Distribution: only the type-locality known

flexisetosus Parry, 1979

23 (22) Setae of aspis considerably shorter, interlamellar ones terminating far from rostral setae.

24 (25) Aspis conspicuously, nasutely convex beyond rostral setae, then again concave before lower margin (pl. 83 : C).—Sensillus finely lanceolate, distally elongated, pointed; occasionally widened part extending to apex but even then narrow and long. Inner pair of setae *(si)* projecting beyond half distance between its insertion point and rostral seta. Aspis in lateral view with a sharp rim from apex to height of bothridia. Chaetom of anogenital region as in pl. 83 : B. Length: 500—800 µm.

Locus typicus: Regensburg, Germany. Distribution: Germany, Hungary, Bulgaria

crinitus (C. L. Koch, 1841)

25 (24) Aspis (pl. 84 : F) with an equally convex curvature before and behind rostral setae, without any conspicuous convexity or concavity in lateral view (pl. 84 : B).—Sensillus (pl. (84 : C) slightly fusiform, distally obtuse. Seta c_1 of notogaster originating further from "collar" line than c_3. Setae of divers inclinations (pl. 84 : A, E). Length of aspis: 223—253 µm, length of notogaster: 406—487 µm.

Locus typicus: West Allendale, Northumberland and Gibside, Durham, England. Distribution: England.

affinis (Hull, 1914)

26 (17) Notogastral setae considerably shorter, c_1 always essentially shorter than distance enclosed by setae c_1—d_1.

27 (30) Setae c_1 originating on posterior margin of "collar" or in its immediate vicinity.

28 (29) Setae c_3 further removed from "collar" than either c_1 or c_2 (pl. 83 : I).—Aspis evenly convex from its base to rostral apex; outer setae originating far in front of inner ones, anterior to bothridium (pl. 83 : G). Notogastral setae c_1, c_2 and d_1 directed towards apex of aspis, all others erect, all short. Sensillus (pl. 83 : J) lanceolate. Chaetom of anogenital region as in pl. 83 : H. Length of aspis: 220—231 µm, length of notogaster: 440—460 µm, height of notogaster: 276—331 µm.

Locus typicus: Bîrnova, region Iași, Roumania. Distribution: only the type-locality known

pallidus Feider et Suciu, 1958

29 (28) Seta c_3 emitted on "collar" line of notogaster *Ph. ferrugineus* C. L. Koch, male specimen, see detailed characterization of female under couplet 36 (37).

30 (27) Seta c_1 originating far behind "collar" line of notogaster.

31 (34) Seta c_1 about twice as far removed from "collar" of notogaster than c_2.

32 (33) Seta c_2 nearest to "collar" line of notogaster, essentially shorter than c_1 and c_3 (pl. 84 : H); c_1 reaching only slightly beyond half distance between c_1—d_1.—Setae originating in interbothridial region about equal in length (pl. 84 : G). Sensillus (pl. 84 : I) conspicously wide, slightly obtuse. Chaetom of anogenital region as in pl. 84 : J. Length of aspis: 312—404 μm, length of notogaster: 625—809 μm, height of notogaster: 515—607 μm.

Locus typicus: Predeal, region Stalin, Roumania. Distribution: only the type-locality known

jacoti Feider et Suciu, 1958

33 (32) Seta c_3 nearest to "collar" line of notogaster, c_1 longer than c_2 and c_3 of approximately equal length. (pl. 85 : G). Seta c_1 longer, hardly shorter than distance of c_1–d_1. Inner pair *(si)* of interbothridial region considerably longer than outer one. Sensillus narrower, also obtuse (pl. 85 : H). Chaetom of anogenital region as in pl. 85 : F. Length of aspis: 258—295 μm, length of notogaster: 460 μm, height of notogaster: 368 μm.

Locus typicus: Babadag, region Constanta, Roumania. Distribution: only the type-locality known

dubinini Feider et Suciu, 1958

34 (31) Seta c_1 originating as far from "collar" as c_2, or even nearer to it.

35 (38) Seta c_1 originating as far from "collar" as width of collar, or even nearer to it.

36 (37) Sensillus basally lanceolate, distally short but acute, thus apical section abruptly attenuating (pl. 85 : B, C). Profile of aspis never straight but sinuous owing to presence of several tubercles and pits.—Notogastral hairs longer in female (pl. 85 : E, with an imaginary triangle connection points of setae c_1, c_2 and c_3. Setae of male shorter (pl. 85 : A), bases of these setae connectable by a nearly straight line. Length of notogaster: 300—500 μm.

Locus typicus: Regensburg, Germany. Distribution: Europe

ferrugineus (C. L. Koch, 1841)

37 (36) Sensillus (pl. 86 : B) wider, distally gradually elongated to a point, not abruptly acute. Surface of aspis without tubercles and pits, hence its profile not sinuous in lateral view, at most finely concave (pl. 86 : A).—Notogastral setae (pl. 86 : D) uniformly longer than in preceding species, only c_3 essentially shorter than the others. Setae c_1—c_2 equidistant from "collar" line, c_3 immediately adjacent to it. Length of aspis: 200—265 μm, length of notogaster: 375—560 μm.

Locus typicus: Västerbotten, Degerfors, Svartbergets, Sweden. Distribution: Sweden, Great Britain

tardus Forsslund, 1956

38 (35) Seta c_1 originating far from "collar", considerably further than width of collar (pl. 86 : F).

39 (40) Notogastral setae very short: c_1 essentially shorter than half length enclosed by c_1—d_1 (pl. 86 : F). Sensillus short, but occasionally with a long, pointed apex (pl. 86 : G). Seta ps_4 nearer to notogastral margin than setae ps_3 and ps_2. Inner pair of setae *(si)* of aspis (pl. 86 : E) considerably longer than outer one.

Chaetom af anogenital region as in pl. 86 : H. Length of aspis: 345—540 μ, length of notogaster: 625—1005 μm, height of notogaster: 465—870 μm.

Locus typicus: Regensburg, Germany. Distribution: Europe

laevigatus (C. L. Koch, 1841)

40 (39) Notogastral setae longer, seta c_1 only slightly shorter than distance enclosed by hairs c_1—d_1.

41 (42) Sixteen pairs of notogastral setae present.*—Lamellar and interlamellar setae short, thin, ending far from insertion points of rostral ones. Sensillus elongated but rounded apically. Notogaster very convex in lateral view, setae c_1—c_p originating very far from collar line. All setae thin and curved. Anal and one pair of adanal setae short. No measurement data available.

Locus typicus: Cazane, region Mehidinti, Roumania. Distribution: only the type-locality known

danubianus Feider et Calugar, 1970

42 (41) Fifteen pairs of notogastral setae present.

43 (44) Setae ps_2, ps_3 and ps_4 originating below lower margin of notogaster (pl. 87 : A, C). An essential difference in length between notogastral setae: those arising in midline much longer than c_2,c_3, cp, d_2 and e_2 (difference respectively more than twofold). Sensillus (pl. 87 : B) narrow, with an acute apex (pl. 87 : B). Inner *(si)* and outer *(sl)* setae of aspis approximately equal in length. Length of aspis: 300—510 μm, length of notogaster: 530—1030 μm, height of notogaster: 390—620 μm.

Locus typicus: Paris, France. Distribution: Europe

nitens (Nicolet, 1855)

44 (43) Setae ps_2, ps_3 and ps_4 originating on lower notogastral margin (pl. 87 : D) Notogastral setae approximately equal in length, only c_1, d and ps_1 longer than the rest (even greatest difference not more than one and a half length). Sensillus with obtuse or a hardly pointed apex. Chaetom of anogenital region as in pl. 87 : E. Length of aspis: 300—514 μm, length of notogaster: 828—1028 μm.

Locus typicus: Karniolie (Karniola), Yugoslavia. Distribution: Palaearctic Region

piger (Scopoli, 1763)

Species interpretabiles

1. affine Hull, 1914 : 287, pl. C : 9 *(Hoploderma)*.
   affinis Parry: 335, figs. 1, 2, 4, 5.
2. crenophilus Willmann, 1951 : 172, fig. 38.
3. crinita C. L. Koch, 1841 : 32, 8 *(Hoplophora)*.
   crinitus Jacot: 1936 : 172, figs. 13—17.
4. danubianus Feider et Calugar, 1970 : 17, figs. 1—5.
   danubianus Feider et Calugar: 1969 : 76, fig. 1 (nom. nudum).
5. dubinini Feider et Suciu, 1958 : 401, figs. 12—19.
6. ferruginea C. L. Koch, 1841: 32, 10 *(Hoplophora)*.
   ferrugineus Jacot: 1936 : 179, figs. 26—33.
7. flexisetosus Parry, 1979 : 340, figs. 7D—F.
8. globosa C. L. Koch, 1841 : 32, 12 *(Hoplophora)*.
   globosus Feider et Suciu: 1957 : 27, figs. 26—34.

* Needing further confirmation.

9. italicum Oudemans, 1900: 170, *(Hoploderma)*.
italicus Feider et Suciu: 1957: 30, figs. 66—73.
italicus Van der Hammen: 1959: 48.
? *dasypus* Dugés, 1834, sensu Berlese, 1883 *(Hoplophora)*.
10. jacoti Feider et Suciu, 1958: 397, figs. 1—11.
11. juvenalis Parry, 1979: 342, figs. 10A—C.
12. kaszabi Mahunka, 1981: 300, figs. 5—8.
13. laevigata C. L. Koch, 1841: 38, 16 *(Hoplophora)*.
laevigatus Jacot: 1936: 167, figs. 1—6.
laevigatus van der Hammen: 1963: 706, figs. 1—6.
laevigatus Parry: 1979: 344, figs. 10D—F.
14. lentula C. L. Koch, 1841: 32, 16 *(Hoplophora)*.
lentulus Jacot: 1936: 175, figs. 18—20.
lentulus Feider et Suciu: 1957: 27, figs. 35—44.
lentulus Feider et Calugar: 1969: 76, fig. 3.
canestrinii Michael, 1898: 81 (Feider et Suciu: 1957: 27).
15. membranifer Parry, 1979: 344, figs. 11A—C.
16. nitens Nicolet, 1855: 472, pl. 10, figs. 6—6a *(Hoplophora)*.
nitens van der Hammen: 1964: 402, figs. 1—6.
nitens Parry; 1979: 346, figs. 12A—C.
17. pallidus Feider et Suciu, 1958: 405, figs. 20—27.
18. piger Scopoli, 1763: 392 *(Acarus)*.
piger Feider et Suciu: 1957: 25, figs. 8—16.
19. serrulatus Parry, 1979: 348, figs. 13A—C.
20. tardus Forsslund, 1956: 216, figs. 11—12.
21. testudinea C. L. Koch, 1841: 32, 11 *(Hoplophora)*.
testudineus Jacot: 1936: 170, figs. 7—12.
22. zicmani **nom. nov.**
feideri Balogh et Mahunka, 1979: 286, figs. 12—14.

Species inquirendae

— boreale Trägårdh, 1910: 547, figs. 347—351 *(Hoploderma)*.
— contractilis Perty, 1841: 874.
— dasypus Dugés, 1834: 47 *(Oribates)*.
— roubali Berlese, 1923: 259.
— rotundus Berlese, 1923: 259.
— sellnicki Feider et Suciu, 1957: 31, figs. 82—90.
— subglobosus Berlese, 1923: 259.

Genus 5: Archiphthiracarus Balogh et Mahunka, 1979

Archiphthiracarus Balogh et Mahunka, 1979: 286.

Habitus, body structure and sculpture agree in general with those of *Phthiracarus* Perty, 1839. Ano-adanal plate with five pairs of well developed setae, therefore, ad_1 and ad_2 evolved to the same degree as ad_3, an_1 and an_2.

Type-species: *Archiphthiracarus variabilis* Balogh et Mahunka, 1979.

Key to species

1 (2) Aspis and notogaster with a rough sculpture, aspis also with three lateral keels (pl. 88 : B, D), entire notogastral surface ornamented with tubercles (pl. 88 : A, E).—Interlamellar seta slightly longer than lamellar one. Sensillus (pl. 88 : C)

short, approximately lanceolate. Notogaster with 14 pairs of robust setae, about equal in length. Ano-adanal plate (pl. 88 : F) with setae ad_1 and ad_3 shorter than the other three pairs. Length of aspis: 242—282 μm, length of notogaster: 464—606 μm.

Locus typicus: Leopoldsberg, Vienna, Austria. Distribution: only the type-locality known

papillosus (Parry, 1979)

2 (1) Aspis and notogaster smooth, at most finely punctate or rugulose, no elevated laths, tubercles or papillae present on body.

3 (4) Notogaster with two pairs of setae (c_1, d_1) long, erect to proclinate; all other setae extremely short, hardly discernible (pl. 87 : G).—Sensillus bacilliform, narrow, comparatively long, falcately arcuate. Interbothridial region with both pairs of setae short, inner one slightly longer, both acute and smooth. Anogenital region (pl. 87 : F) with seta ad_3 essentially shorter than ad_1 or ad_2. Length of aspis: 340—504 μm, length of notogaster: 773—992 μm, height of notogaster: 558—736 μm.

Locus typicus: Vladiceni, region Iasi, Roumania. Distribution: only the type-locality known

baloghi (Feider et Suciu, 1957)

4 (3) No similar extensive difference in length between notogastral setae; c_1 and d_1 longer than e_1 and ps_1 (pl. 88 : G).

5 (6) Anterior margin of notogaster with an obtusely elevated "collar", and its entire length with a wide, lath-like excrescence (pl. 88 : G).—Sensillus very long, thin, longer than also elongated interbothridial setae, inner pair (*si*) longer than half distance between *si* and *ro*. Notogastral setae short, majority of seta—among them c_1, d_1 and e_1 unequivocally—erect to proclinate (towards prodorsum). Length of aspis: 390 μm, length of notogaster: 810 μm, height of notogaster: 555 μm.

Locus typicus: "Levada do Inferno", Madeira, Portugal. Distribution: only the type-locality known

torosus (Willmann, 1939)

6 (5) Anterior margin of notogaster without an elevated "collar", also a longitudinal lath absent.

7 (8) Adanal setae arising near inner margin of plate, nearly aligned with anal setae; all five pairs of ano-adanal setae minute (pl. 89 : A).—Sensillus basally lanceolately widening, then setiform, long, pointed, throughout smooth (pl. 89 : B). All setae of aspis very short and thin. Crista conspicuously projecting from profile of aspis. Notogastral setae also minute, erect, setiform; notogaster with a foveolate sculpture. Length of aspis: 320—480 μm, length of notogaster: 475–830 μm.

Locus typicus: Milke Danra, Nepal. Distribution: only the type-locality known

robertsi (Sheals, 1965)

8 (7) Adanal setae originating much further removed from inner margin of plate than anal setae.

9 (16) Sensillus narrow and long, at least five times longer than its greatest width; never clavate or phylliform.

10 (13) Notogastral setae long; c_1 as long, or nearly as long, as distance enclosed by c_1—d_1.

11 (12) Seta c_1 originating on "collar" line of notogaster, c_3 in same position, while c_2 situated slightly behind marginal line (pl. 89 : C).—Aspis medially with a crest, well visible in lateral view. Setae of aspis long, inner pair *(si)* approaching base of *ro*. Sensillus comparatively long, narrow, lanceolate. Rostral margin very wide, behind it rostrum sinous in lateral view. No essential difference in length between notogastral setae, d_1 not longer than d_2. Anal setae considerably longer than adanal ones. Length of aspis: 225 µm length of notogaster: 420 µm, height of notogaster: 300 µm.

Locus typicus: "Jesuitenbach", near Vienna, Austria. Distribution: Austria, Hungary

peristomaticus (Willmann, 1954)

12 (11) Setae c_1 and c_3 originating behind "collar" of notogaster, c_2 very far removed (pl. 90 : A).—Aspis without crest, its setae much shorter: *si* shorter than half distance between *si*—*ro*. Sensillus (pl. 90 : B) very long, slightly widened only at base, otherwise bacilliform, essentially longer than outer seta of inter-bothridial region. Notogastral setae erect to slightly proclinate (towards apex of aspis), without any considerable difference in length. Anal setae, especially an_1, slightly longer than adanal ones. Length of aspis: 220—250 µm, length of notogaster: 180—250 µm.

Locus typicus: Dummersdorf, Germany. Distribution: Only the type-locality known

ligneus (Willmann, 1931)

13 (10) Notogastral setae short: c_1 at most half as long as distance between c_1—d_1.

14 (15) Notogastral setae very short: c_1 shorter than half distance enclosed by c_1—d_1 (pl. 89 : E); setae p_2, p_3 and p_4 originating below marginal line of notogaster.—Notogastral setae apically flagellate. Sensillus (pl. 89 : D) resembling a knife-blade, basal half hardly perceptibly widening. Base of aspis with an obtuse tubercle in lateral view. Setae short, but exobothridial one strikingly long. Ano-adanal plate (pl. 89 : F) with very short setae, anal ones only slightly longer than adanal setae (equal in length). Length of aspis: 320 µm, length of notogaster: 580—620 µm, height of notogaster: 465—480 µm.

Locus typicus: Tokyo, Japan. Distribution: only the type-locality known

clemens (Aoki, 1963)

15 (14) Notogastral setae longer, c_1 as long as half distance enclosed by c_1—d_1 (pl. 90 : D); setae p_2, p_3 and p_4 originating on marginal line of notogaster.—Sensillus (pl. 90: E) resembling that of preceding species, but comparatively, longer. Basal half of aspis (pl. 90: E) smooth in lateral view. Setae longer, but exobothridial one also long. Ano-adanal plate with long setae (pl. 90 : C), ad_2 not shorter than anal setae. Length of aspis: 368—423 µm, length of notogaster: 754—920 µm, height of notogaster: 605—718 pm.

Locus typicus: Babadag, Constanta, Roumania. Distribution: Roumania

parabotrichus (Feider et Suciu, 1957)

16 (9) Sensillus short, phylliform or distally incrassate.

17 (22) Both pairs of interbothridial setae very long; either both pairs *(si, sl)* extending to base of rostral setae, or outer pair longer than inner one. Notogastral setae also very long, c_1 or c_2 longer than distance between c_1—d_1.

18 (19) A considerable difference in length among notogastral setae: c_2 and cp considerably longer than c_1 or d_1, while c_3 much shorter than even these two latter ones (pl. 90 : I).—Interlamellar pair of setae originating well behind slightly longer lamellar pair (pl. 90 : G, J). A finely aciculated area anterior to bothridium. Sensillus (pl. 90 : H) clavate, apically rounded. Seta ad_3 longest of all ano-adanal setae, ad_2 slightly shorter than ad_1; anal setae considerably shorter, but equal in length. Length of aspis: 254—336 μm, length of notogaster: 467—608 μm, height of notogaster: 342—424 μm.

Locus typicus: Somon Bajanzogt, Central aimak, Mongolia. Distribution: only the type-locality known

gobiensis Mahunka, 1981

19 (18) No essential difference in length among notogastral setae, c_2 never longer than c_1 or d_1, and even c_3 similar to the other setae.

20 (21) Sensillus disciform, apically not pointed (pl. 91 : A).—Notogastral setae extremely long, all more or less equal in length. Setae c_1 and c_2 originating about equally far from "collar" line of notogaster. Ano-adanal plate with ad_3 considerably longer than the other adanal setae, anal ones equal in length but shorter than adanal setae. Length of aspis: 285 μm, length of notogaster: 555 μm.

Locus typicus: Glatz (Kłodzko), Poland. Distribution: only the type-locality known

crinitosimilis (Willmann, 1939)*

21 (20) Sensillus phylliform (pl. 91 : E), apex slightly pointed.—Notogastral setae (pl. 91 : C) similar, but c_2 originating considerably nearer to notogastral "collar" than c_1, but c_3 somewhat further removed from it than in the preceding species. Proportion of setae on ano-adanal plate resembling that of preceding species (pl. 91 : B). Length of aspis: 272—278 μm, length if notogaster: 533 μm, height of notogaster: 368 μm.

Locus typicus: Zetea, Odorhei, Hungarian Autonomous Region, Roumania. Distribution: Roumania

lanatus (Feider et Suciu, 1957)*

22 (17) Aspis with shorter setae, usually even inner one *(si)* not extending beyond half distance enclosed by *si—ro,* but seta *sl* always shorter. Also notogastral setae shorter, c_1 reaching base of d_1 only exceptionally.

23 (24) Notogastral setae extraordinarily short (pl. 92 : A),at least two pairs of setae *c* shorter than width of "collar"; also ano-adanal setae extremely short** (pl. 92: B). Setae *ro, si* and *sl* of aspis approximately equal in length. Sensillus

* The species *Archiphthiracarus crinitosimilis* and *A. lanatus* stand extremely near each other, the differences given here are based on the descriptions and corresponding figures. The identity of the two species is a definite possibility!

** The species may belong in the genus *Phthiracarus* Perty, 1839, because the reduced or very short status of the adanal setae cannot be established unequivocally from the description and the accompanying figure.

lanceolate. Length of aspis: 270 μm, length of notogaster: 520 μm, height of notogaster: 400 μm.

Locus typicus: Utsukushigahara, Nagano, Japan. Distribution: only the type-locality known

japonicus (Aoki, 1958)

24 (23) Notogastral setae longer, every seta *c* longer than width of "collar"; also ano-adanal setae longer and well discernible on their plate.

25 (34) Ano-adanal plate with setae ad_1 and ad_2 as long as anal *(an)* setae, or at most slightly longer.

26 (27) Seta ad_3 considerably shorter than the other two adanal setae (pl. 91 : G). Notogastral setae shorter, erect and proclinate, directed towards apex of aspis (pl. 91 : H). Sensillus (pl. 91 : F) lancolate. A nearly twofold difference in length between setae of aspis (pl. 91 : I). Length of aspis: 305—455 μm, length of notogaster: 600—950 μm.

Locus typicus: Lambridge Wood, Oxfordshire, England. Distribution: England

murphyi (Harding, 1976)

27 (26) Only a small difference in length among adanal seteae; either equal in length, or ad_3 longer than the other two pairs.

28 (29) Notogastral setae very thin, long, arcuate (pl. 92 : D), c_1 longer than distance between setae c_1—d_1. Setae c_1, c_2 and c_3 connectable by an imaginary straight line.—Inner pair of setae *(si)* in interbothridial region of aspis longer than half distance between its insertion point and rostral setae. Great differences in length among notogastral setae; c_3 essentially shorter. Sensillus rounded (pl. 92 : C). Chaetom of ano-adanal plate as in pl. 92 : E. Length of aspis: 227 μm, length of notogaster: 413 μm, height of notogaster: 292 μm.

Locus typicus: Kőszeg, Hungary. Distribution: only the single type-specimen collected in the type-locality is known

hungaricus Balogh et Mahunka, 1979

29 (28) Notogastral setae considerably shorter, c_1 at most as long as half distance between setae c_1—d_1. An imaginary line connecting bases of c_1, c_2 and c_3 forming a triangle.

30 (31) At least a twofold difference in length between setae c_1 and c_3 (pl. 92 : F); both originating at equal distances from "collar" line of notogaster. Setae long, directed towards apex of aspis. Sensillus (pl. 92 : H) short, lanceolate. Chaetom of ano-adanal plate as in pl. 92 : I. Body in superior view in pl. 92 : G. Length of aspis: 346 μ, length of notogaster: 360—540 μm.

Locus typicus: Regensburg, Germany. Distribution: Germany, Poland, Hungary, Soviet Union (SET)

stramineus (C. L. Koch, 1841)

31 (30) Setae c_1, c_2 and c_3 equal in length, c_1 originating much nearer to "collar" line of notogaster than c_3 (pl. 93 : A).

32 (33) Notogastral setae longer, length of c_3 equalling distance between c_1—c_3; seta c_1 emitted on "collar" of notogaster (pl. 93 : A).—Setae ps_2, ps_3 and ps_4 originating below lower margin of notogaster, ps_4 shorter than the other two pairs. Sensillus short, lanceolate. Setae of interbothridial region also short,

inner seta about as long as distance between its insertion point and basis of rostral seta. Length of aspis: 270 μm, length of notogaster: 540 μm, height of notogaster: 250 μm.

Locus typicus: Regensburg, Germany. Distribution: Germany, Czechoslovakia

lucidus (C. L. Koch, 1841)

33 (32) Notogastral setae short, c_3 considerably shorter than distance between c_1—c_3 (pl. 93 : D); seta c_1 originating behind "collar" line.—Setae ps_2, ps_3 and ps_4 aligned on lower marginal line of notogaster. Sensillus (pl. 93 : E) less wide, rather clavate. Setae of interbothridial region longer, inner setae considerably longer than half distance between their insertion points and bases of rostral setae. Chaetom of anogenital region given in pl. 93 : D. Length of aspis: 270—370 μm, length of notogaster: 525—800 μm.

Locus typicus: Navacerrada, Madrid, Spain. Distribution: Spain

montanus (Pérez-Iñigo, 1969)

34 (25) An essential difference in height existing between ad_1, ad_2 and anal setae of ano-adanal plate; at least one of them considerably longer or shorter than the others.

35 (38) Inner seta *(si)* of aspis reaching base of rostral seta.

36 (37) Interlamellar seta extremely long, (pl. 93 : B), projecting far beyond insertion point of rostral seta. Considerable differences in length extant also among notogastral setae (pl. 93 : C).—Sensillus lanceolate. Seta c_1 remote from "collar"*. Setae originating along midline (c_1, d_1, e_1, etc.) essentially longer than lateral ones; all thin and arcuate. Setae ad_1 and ad_2 of ano-adanal plate slightly longer than ad_3. Length of aspis: 243—324 μm, length of notogaster: 437—608 μm, height of notogaster: 240—405 μm.

Locus typicus: "Ördögorom", Farkasrét, Budapest, Hungary. Distribution: only the type-locality known

hammeni Balogh et Mahunka, 1979

37 (36) Interlamellar seta reaching only to base of rostral seta, lamellar one essentially shorter (pl. 93 : F). Notogastral setae about equal in length (pl. 93 : G).—Sensillus (pl. 93 : H) slightly fusiform. Notogastral setae c_1 and c_2 at equal distances from "collar", c_3 situated on collar proper. Length of aspis: 324—365 μm, length of notogaster: 426—553 μm.

Locus typicus: Rydal Water, Westmorland, Great Britain. Distribution: Great Britain

globus (Parry, 1979)

38 (35) Setae of aspis not even approaching bases of rostral setae; notogastral setae considerably shorter.

39 (40) Setae c_1 originating on "collar" line of notogaster, or in front of it (pl. 94 : A). Seta ad_2 very short, shorter than ad_3 and ad_1. Rostral apex densely and heavily striolate (pl. 94 : B), apex with a distinct rim in lateral view. Sensillus (pl. 94 : C) medium long, lanceolate. Great differences extant among notogastral setae:

* Line of notogaster, c_3 at same distance, c_2 far removed from these, considerably posteriorad from "collar" line.

those along midline (c_1, d_1, e_1, h_1) generally longer than those removed from it $(d_2, c_3, e_2, h_2, h_3)$. Length of aspis: 178—218 μm, length of notogaster: 330—420 μm, height of notogaster: 412 μm.

Locus typicus: Périgueux, France. Distribution: France, Hungary, Roumania (?)

anonymus (Grandjean, 1933)

40 (39) Seta c_1 originating behind "collar" line of notogaster.

41 (44) All three pairs of setae c arising near "collar" line of notogaster.

42 (43) Considerable differences in length between notogastral setae (pl. 94 : G): at least c_3 and ps_4 essentially shorter than c_1 and d_1.—Among interbothridial pairs of setae only smaller differences in length present (pl. 94 : I). Notogastral setae arising along midline (c_1, d_1, etc.) generally long, reclinate, all other setae shorter. Sensillus lanceolate, apically acute. Ano-adanal plate with the two anal pairs of setae essentially longer than adanal setae (pl. 94 : H). Length of aspis: 235—251 μm, length of notogaster: 436—446 μm, height of notogaster: 324—344 μm.

Locus typicus: "Tolvajárok", Sopron, Hungary. Distribution: Austria, Hungary

variabilis Balogh et Mahunka, 1979

43 (42) Notogastral setae approximately equal in length, c_3 and ps_4 never significantly shorter than the others.—Aspis setae similar to those of preceding species, (pl. 95 : G), but sensillus (pl. 95 : E) very obtuse, nearly rounded though spatulate. Midline seta e_1 slightly longer than the others, but in essence no significant differences in length among notogastral setae present (pl. 95 : F). Length of aspis: 279—324 μm, length of notogaster: 616—659 μm.

Locus typicus: Meathop Wood, Westmorland, Great Britain. Distribution: Great Britain

clavatus (Parry, 1979)

44 (41) Setae c originating at divers distances from "collar" of notogaster.

45 (46) Notogastral setae c_3, cp and ps_4 very short, considerably shorter than c_1 or ps_2 (pl. 95 : A). Setae c_2 hardly more removed from "collar" than c_1.—Setae of aspis (pl. 95 : C) short, interlamellar setae hardly longer than half distance between its insertion point and base of rostral seta. Majority of notogastral setae proclinate, all short. Sensillus short (pl. 95 : B, E). Anal setae longer than adanal ones. Length of aspis: 312—336 μm, length of notogaster: 657—737 μm, height of notogaster: 510—540 μm.

Locus typicus: Lasithi plateau, Crete, Greece. Distribution: only the type-locality known

tzanoudakisi Mahunka, 1979

46 (45) Notogastral setae about equal in length, c_3 and ps_4 not shorter than the others (pl. 94 : E); c_2 nearly twice as far removed from "collar" than c_1.—Aspis setae (pl. 94 : F) also short and thin. Sensillus slightly asymmetrical (pl. 94 : D), spatulate. Ano-adanal plate with ad_1 and ad_2 essentially shorter than ad_3 and anal setae. Length of aspis: 350—477 μm, length of notogaster: 710—994 μm.

Locus typicus: Burnham Beeches, Buckinghamshire, Great Britain. Distribution: Great Britain

rectisetosus (Parry, 1979)

Species interpretabiles

1. anonymum Grandjean, 1933: 312, figs. 3A—B *(Phthiracarus)*.
anonymum Grandjean: 1934: 51, figs. 1—4 *(Phthiracarus)*.
anonymum Feider et Suciu: 1957: 24, figs. 1—7 *(Phthiracarus)*.
anonymum Parry: 1979: 336, figs. 6A—E *(Phthiracarus)*.
2. baloghi Feider et Suciu, 1957: 28, figs. 45—55 *(Phthiracarus)*.
baloghi Feider et Calugar: 1969: 77, fig. 5 *(Phthiracarus)*.
3. clavatus Parry, 1979: 338, figs. 7A—C *(Phthiracarus)*.
4. clemens Aoki, 1963: 218, figs. 1—4 *(Phthiracarus)*.
5. crinitosimilis Willmann, 1939: 452, figs. 183—185 *(Phthiracarus)*
6. globus Parry, 1979: 341, figs. 8A—C *(Phthiracarus)*.
7. gobiensis Mahunka, 1981: 298, figs. 1—4.
8. hammeni Balogh et Mahunka, 1979: 287, figs. 18—19.
9. hungaricus Balogh et Mahunka, 1979: 288, figs. 20—22.
10. japonicus Aoki, 1958: 171, figs. 1A—D *(Phthiracarus)*.
11. lanatus Feider et Suciu, 1957: 26, figs. 17—25 *(Phthiracarus)*.
12. ligneus Willmann, 1931: 192, figs. 352—353 *(Phthiracarus)*.
ligneus Willmann: 1932: 438, figs. 1—3 *(Phthiracarus)*.
ligneus Feider et Suciu: 1957: 31, figs. 74—81 *(Phthiracarus)*.
13. lucidus C. L. Koch, 1841: 32, 14.
lucidus Willmann: 1955: 268, fig. 22 *(Phthiracarus)*
14. montanus Pérez-Iñigo, 1969: 380, figs. 6, 7, 11, 12 *(Phthiracarus)*.
montanus Pérez-Iñigo, 1968: 204 (nom. nudum).
15. murphyi Harding, 1976: 163, figs 1—22 *(Phthiracarus)*.
murphyi Parry: 1979: 346, figs. 11D—F *(Phthiracarus)*.
16. papillosus Parry, 1979: 73, figs. 1—13 *(Phthiracarus)*
17. parabotrichus Feider et Suciu, 1957: 29, figs. 56—65 *(Phthiracarus)*.
18. peristomaticus Willmann, 1951: 173, fig. 39 *(Phthiracarus)*.
peristomaticus Willmann: 1954: 263, fig. 23 *(Phthiracarus)* (non sp. nov.!).
19. rectisetosus Parry, 1979: 348, figs. 12D—F *(Phthiracarus)*.
20. robertsi Sheals, 1965: , figs. 11—19 *(Phthiracarus)*
21. straminea C. L. Koch, 1841: 32, 13 *(Hoplophora)*.
stramineus Jacot: 1936: 177, figs. 21—25 *(Phthiracarus)*.
22. torosus Willmann, 1939: 40, figs. 26A—B *(Phthiracarus)*.
23. tzanoudakisi Mahunka, 1979: 561, figs. 34—37.
24. variabilis Balogh et Mahunka, 1979: 287, figs. 15—17.

Species inquirenda

— longula C. L. Koch, 1841: 32, 16 *(Hoplophora)*.

Superfamily 2: EUPHTHIRACAROIDEA Jacot, 1930

Euphthiracarinae Jacot, 1930: 214.
Euphthiracaridae Grandjean: 1954: 430.
Euphthiracaridae Van der Hammen: 1959: 35.
Euphthiracaridae Märkel: 1964: 2.
Euphthiracaroidea Grandjean: 1967: 246.

Body in lateral view resembling a much compressed, closed sea-shell. In basic structure similar to preceding superfamily. Aspis without crista, position of bothridial squama and configuration of carinae very characteristic. Exobothridial seta occasionally reduced. Notogaster rarely with any sculpture, mostly smooth, emitting

14—15 pairs of setae, with one or two *(f_1—f_2)* surviving only as pori: 3—4 lyrifissures *(ia, im, ips,* and seldom *ih)* and opening of lateroabdominal gland *(gla)* also discernible. Posteriorly a rounded excision (sinus terminalis) or a narrow and long slit (fissura terminalis) present. Ventral plate originally formed by various rates of coalescence of four pairs of plates, aggenital, genital, anal and adanal ones; aggenital and adanal plates always at least partly fused. On border of genital and anal plates a peculiar formation: a triangular chitinous thickening may be present, and a similar one in midline at end of plates. If genital and anal plates separate, an incision, directed obliquely anteriorly and extending also to aggenital and adanal plates, may also occur as a continuation of the suture (incisura aggenitalis—adanalis). Pori *iad* also recognizable in adanal part of aggenito-adanal plate. Chaetotaxy of ventral plates highly varying: 5—11 pairs of genital, 2—4 pairs of aggenital, 0—6 anal and 2—5 adanal, setae may be present in divers combinations.

In the separation of taxa of various levels the claws of legs (1 or 3) as well as the leg chaetotaxy also play a role. The most important characteristics in the Palaearctic genera of the superfamily are as follows:

Characteristics / Genera	Genito-aggenital suture	Ano-adanal suture	Number of triangles	Aggenital sutures	Bothridial squama	Number of setae				Number of palp joints
						genital	aggenital	anal	adnal	
Oribotritia	+	+	0	+	above	5—10	2—3	1—3	3—4	5
Mesotritia	+	+	0	+	below	5—11	2—4	1—2	3—4	3
Maerkelotritia	+	+	0	—	below	7—10	2	0—1	4	3
Protoribotritia	+	+	0	—	below	7	2	3	3—4	3—4
Indotritia	+, —	+	0	+	above	9	2	2	2	5
Paratritia	—	+	0	—	above	8	2	2	3	3
Euphthiracus	—	—	2	—	below	7—9	2	3—6	3—5	3
Microtritia	—	—	1	—	above	4	2—3	3	3	3
Rhysotritia	—	—	1	—	above	5—9	2	3—9	3	3

The superfamily contains three families, but only two have representatives in the Palaearctic Region.

Key to families

1 (2) Ventral plates not completely fused, at least anal plate differentiated by a well discernible suture. At meeting of anal and genital plates never a triangle, at most the anogenital slit, present
family 1: **Oribotritiidae** Grandjean, 1954

2 (1) Ventral plates completely fused, hence anogenital region formed by a single pair of plates, occasionally enclosing 1 or 2 triangles
family 2: **Euphthiracaridae** Jacot, 1930

Family 1: ORIBOTRITIIDAE Grandjean, 1954

Oribotritiidae Grandjean, 1954 : 430.
Oribotritiidae Van der Hammen: 1959 : 34.
Oribotritiidae Grandjean 1967 : 247.

Anal and adanal plates always, genital and aggenital plates either wholly or partly, separate. A triangle separating anal and genital plates always absent, but an aggenital incision often well developed. 5—11 pairs of genital, 2—4 pairs of aggenital, 0—3 pairs of anal and 2—4 pairs of adanal setae present in various combinations. Palp with 3 to 5 joints.

Six Palaearctic genera are known.

Key to genera

1 (8) Suture between genital and aggenital plates present in its entire length, well visible.

2 (3) Bothridial squama situated above bothridium (pl. 97 : C). A suture between anal and genital plates present
genus 1: **Oribotritia** Jacot, 1925

3 (2) Bothridial squama situated below bothridium (pl. 103 : C). Suture between anal and genital plates absent.

4 (5) Legs monodactyle
genus 4: **Protoribotritia** Jacot, 1938

5 (4) Legs tridactylous.

6 (7) Body posteriorly with a sinus terminalis. Aspis without lateral keel. Genu IV with a solenidium
genus 3: **Maerkelotritia** Hammer, 1967

7 (6) Body posteriorly with a fissura terminalis. Aspis with one lateral keel. Genu IV without solenidium
genus 2: **Mesotritia** Forsslund, 1963

8 (1) Furrow between genital and aggenital plates partly or wholly absent.

9 (10) Genito-aggenital suture evanescent beyond half length of plate, therefore, genital and aggenital plates fused anteriorly
genus 5: **Indotritia** Jacot, 1925

10 (9) Genito-aggenital suture completely absent, therefore, genital and aggenital plates fused in their entire length
genus 6: **Paratritia** Moritz, 1966

Genus 1: ORIBOTRITIA Jacot, 1924

Oribotritia Jacot, 1924 : 83.
Oribotritia Van der Hammen: 1959 : 34.
Oribotritia Märkel: 1964 : 18.
Oribotritia Grandjean: 1967 : 258.
Tritia Berlese, 1883: 6 (1).
Tritia Van der Hammen, 1959 : 34.

Aspis with one lateral keel. Bothridial squama situated above bothridium. Lamellar setae in basal half of aspis, nearly aligned into a transverse row with interlamellar ones. Exobothridial seta developed. 14—15 pairs of notogastral setae present. Ventral plates differentiated: well discernible furrows separating genital plate from aggenital one and anal plate from adanal one. Anogenital suture present, continuing laterally in aggenito-adanal incision. Ventral region terminating in sinus terminalis. 5—10 pairs of genital, 2—3 pairs of aggenital, 1—3 pairs of anal and 3—4 pairs of adanal setae present. Palp 5-jointed. Trochanters of legs III and IV with 3 setae each. Genu of leg IV with developed solenidium φ_1.

Type-species: *Hoplophora decumana* auct., non C. L. Koch, 1836 = *Phthiracarus berlesei* Michael, 1898.

Key to species

1 (8) Two pairs of aggenital setae present.

2 (3) Three pairs of anal setae present. An enormous species, about 3000 μm long.—Body setae (pl. 96 : D) very fine, thin and short (pl. 96 : A). Setae *ro* and *le* of aspis about equal in length, setae *in* considerably longer, reaching bases of hairs *ro*. Sensillus (pl. 96 : B) also very thin, setiform. Notogastral setae shorter than interlamellar ones. Setae of anogenital region minute (pl. 96 : F). Aggenito-adanal incision narrow (pl. 96 : E). Eight pairs of genital setae present. Length of aspis: 1000—1200 μm, length of notogaster: 2160—2320 μm, height of notogaster: 1460—1500 μm. The largest known species of the *Ptyctima*.

Locus typicus: Near Proshiri-sanso, Hidaka, Hokkaido, Japan. Distribution: only the type-locality known

tokukoae Aoki, 1973

3 (2) Only one pair of anal setae present. Small-sized species.

4 (5) Five pairs of genital setae present (pl. 96 : I).—Aspis setae (pl. 96 : H) short, smooth. Sensillus also short, thin and setiform. Notogastral setae straight, rigid but rather short, also smooth (pl. 96 : G). Anal and adanal setae minute. Length of aspis: 420 μm, length of notogaster: 760 μm, height of notogaster: 680 μm.

Locus typicus: Iwakiri, Miyazaki-shi, Miyazaki district, Japan. Distribution: only the holotype known

mollis Aoki, 1959

5 (4) Seven to eight pairs of genital setae present.

6 (7) Eight pairs of genital setae present (pl. 97 : E). Body setae longer, c_1 nearly as long as distance enclosed by c_1—d_1; no essential difference in length between setae c_1 and c_2, or between d_1 and d_2 (pl. 97 : A). Setae straight or only very finely arcuate. Interlamellar seta comparatively long, reaching or approaching half distance between setae *in* and *ro* (pl. 97 : B—C). Length of aspis: 597—760 μm, length of notogaster: 1080—1645 μm, height of notogaster: 900—1205 μm.

Locus typicus: Palermo, Sicily, Italy. Distribution: Europe

berlesei (Michael, 1898)

7 (6) Only seven pairs of genital setae present. Body setae shorter (pl. 97 : F) than in preceding species; c_1 reaching half distance between c_1—d_1, interlamellar seta not reaching half distance between its base and insertion point of rostral seta. Notogastral setae mostly sigmoid or falcate. Great differences in length between setae c_1—c_2 and especially between d_1 and d_2. Length of aspis: 484—580 µm, length of notogaster: 895—1080 µm.

Locus typicus: Strasbourg, France. Distribution: France, Italy

hermanni Grandjean, 1967

8 (1) Three pairs of aggenital and 2 or 3 pairs of anal setae present.

9 (10) Eight of genital setae present (pl. 97 : H). Body setae long, interlamellar seta approaching insertion point of rostral seta (pl. 97 : I—J). Seta c_1 considerably longer than distance between its insertion point and anterior margin of notogaster (pl. 97 : G). Sensillus short, setiform, smooth, evenly tapering. Length of aspis: 555—705 µm, length of notogaster: 1075—1520 µm, height of notogaster: 725—1000 µm.

Locus typicus: "Trärminne Zoological Station", Nylandia, Finland. Distribution: Finland, Sweden

fennica Forsslund et Märkel, 1963

10 (9) Only seven pairs of genital setae present. Body setae shorter; interlamellar ones not reaching half distance between their insertion points and rostral setae.

11 (12) Sensillus wide and smooth (pl. 98 : C) basal half widening then abruptly attenuating. Aspis setae medium long, interlamellar ones longer than lamellar setae. c_1 reaching anterior margin of notogaster (pl. 98 : A). Adanal setae short, shorter than width of adanal plate (pl. 98 : B). Length of aspis: 550—565 µm, length of notogaster: 920—1024 µm, height of notogaster: 700 µm.

Locus typicus: Babadag, Constanta, Roumania. Distribution: Italy, Roumania

storkani Feider et Suciu, 1957

12 (11) Sensillus wide and serrate (pl. 98 : D), basal half not widening. Aspis setae very short (pl. 98 : E), interlamellar ones shorter than lamellar setae. Notogastral setae (pl. 98 : F) also very short, c_1 not reaching anterior margin of notogaster. Adanal setae (pl. 98 : G) comparatively long, ad_1 longer than width of adanal plate. Length of aspis: 460—570 µm, length of notogaster: 791—993 µm, height of notogaster: 607—828 µm.

Locus typicus: Brosteni, region Craiova, Roumania. Distribution: only the type-locality known

serrata Feider et Suciu, 1958

Species interpretabiles

1. berlesei Michael, 1898 : 81 *(Phthiracarus)*.
   decumana Berlese: 1883 : 6, 2 *(Tritia)* non C. L. Koch, 1836.
   decumana Van der Hammen: 1959 : 35.
   berlesei Märkel, 1964 : 19, figs. 2a—g.
   Berlesei Grandjean: 1967 : 261, figs. 3C—E, 4A.
2. fennica Forsslund et Märkel, 1963 : 284, figs. 1A—E, 2A—F.
   fennica Aoki et Ohnishi: 1974 : 156, figs. 22—25.

3. Hermanni Grandjean, 1967 : 262, figs. 2A—B.
4. mollis Aoki, 1959 : 19, figs. 14A—C.
5. serrata Feider et Suciu, 1958 : 31, figs. 1—10.
6. storkani Feider et Suciu, 1957 : 39, figs. 160—165.
   Storkani Grandjean: 1967 : 262, figs. 3A—B, 4B.
7. tokukoae Aoki, 1973 : 241, figs. 1—9.

Genus 2: Mesotritia Forsslund, 1963

Mesotritia Forsslund, 1963 : 284.
Mesotritia Märkel: 1964 : 38.

Aspis with lateral keel. Bothridial squama below bothridium. Exobothridial seta also well developed. Notogaster with 14—15 pairs of setae. Genital and aggenital plates as well as anal and adanal plates separated in their entire length by a furrow. No aggenito-adanal incision. Ventral region usually ending in a long terminal fissure. Ventral plates with 5—11 pairs of genital, 2—4 pairs of aggenital, 1—2 pairs of anal and 3—4 pairs of adanal, setae. Legs tridactylous. Palp 3-jointed. Trochanters of legs III and IV with 2 setae each; solenidium on genu of leg IV present.

Type-species: *Mesotritia testacea* Forsslund, 1963.

Key to species

1 (2) Genital plates with 9—11, adanal plates with 5, pairs of setae (pl. 98 : H). — Aspis (pl. 98 : I) setae very long, as are notogastral ones, all erect, Sensillus (pl. 98 : J) thin, setiform, also long, distally finely ciliate. Length of aspis: 270 µm, length of notogaster: 880 µm.

Locus typicus: Faeroe Islands, Denmark. Distribution: only the type-locality known

faroensis (Sellnick, 1923)

2 (1) Genital plates with 5—8, adanal plates with 3, pairs of setae.

3 (4) Lamellar setae of aspis originating near, while rostral and lamellar ones far removed from, each other (pl. 99 : B); lamellar setae twice longer than interlamellar ones. Lateral carina absent (?). Sensillus short, distally lanceolately attenuating. 14 pairs of notogastral setae present, all thin, fine, comparatively short, reclinate and flagelliform (pl. 99 : A). Length of aspis: 300—360 µm, length of notogaster: 550—600 µm.

Locus typicus: Selap, Nepal. Distribution: only the type-locality known

maerkeli Sheals, 1965

4 (3) Rostral setae of aspis originating near, lamellar and interlamellar ones far removed from, each other.

5 (6) Interlamellar setae emitted far anterior to bothridia, about equal distances from lamellar setae and bothridia. Notogastral setae very fine, hardly perceptible. Sensillus distally fusiform, apex acute. Length of aspis: 330 µm, length of notogaster: 575 µm, height of notogaster: 375 µm.

Locus typicus: Padua, Italy. Distribution: Germany, Italy

nuda (Berlese, 1887)

6 (5) Interlamellar setae nearly aligned with bothridia on an imaginary transverse line, or emitted at least much nearer to bothridia than to lamellar setae. Body setae longer.

7 (8) Anal plates with merely one pair of setae (pl. 99 : C). Terminal fissure comparatively short, with a median pore above its apex (pl. 99 : D). Number of setae varying on genital, aggenital and adanal plates, often asymmetrical setae also occurring. Number of genital setae: 5—8, of aggenitals: 2—4, of adanals: 3—4. Notogastral setae longer (pl. 99 : F). Sensillus (pl. 99 : G) short, fusiform. Length of aspis 305—370 μm, length of notogaster: 535—665 μm, height of notogaster: 385—470 μm.

Locus typicus: Vindeln, Degerfors, Sweden. Distribution: Sweden, Finland

testacea Forsslund, 1963

8 (7) Anal plates with two pairs of setae. Number of genital, aggenital and adanal setae constant. Terminal fissure long, no median pore at its apex (pl. 100 : D).

9 (10) Notogaster with 14 pairs of setae. Basal part of aspis (pl. 100 : B) flatly arcuate in lateral view, anterior part striated. Interlamellar setae short, not even approaching insertion of lamellar setae; these latter originating behind rostral setae. Notogastral setae (pl. 100 : A) short, c_1 originating far from anterior margin of notogaster. Chaetom of anogenital region as in pl. 100 : C. Terminal fissure very long (pl. 100 : D). Length of aspis: 395—480 μm, length of notogaster: 650—890 μm, height of notogaster: 475—675 μm.

Locus typicus: "Donau-Auen", Vienna, Austria. Distribution: Austria, Sweden

piffli Märkel, 1964

10 (9) Notogaster with 15 pairs of setae. Basal part of aspis strongly convex in lateral view (pl. 100 : E), anterior part smooth. Interlamellar setae longer, reaching insertion points of lamellar setae (pl. 100 : F); lamellar and rostral setae nearly aligned on a transverse line. Notogastral setae longer, c_1 emitted on notogastral "collar". Length of aspis: 323—478 μm, length of notogaster: 736—920 μm, height of notogaster: 552—613 μm.

Locus typicus: Harghita, Ciuc, Hungarian Autonomous Region, Roumania. Distribution: Roumania, Soviet Union (SET)

grandjeani (Feider et Suciu, 1957)

Species interpretabiles

1. färöensis Sellnick, 1923 : 41, figs. 8, 19, 31 *(Tritia)*.
2. grandjeani Feider et Suciu, 1957 : 38, figs. 148—158 *(Oribotritia)*.
3. maerkeli Sheals, 1965 : 22, figs. 40—48.
4. nuda Berlese, 1887 : 35, 9 pl. 21. figs. 1—6 *(Tritia)*.
   nuda Sellnick: 1923 [1968 : 40, figs. 6, 17, 29 *(Tritia)*].
   nuda Van der Hammen: 1959 : 35 *(Oribotritia)*.
5. piffli Märkel, 1964 : 39, figs. 9a—f *[Mesotritia (Entomotritia)]*.
6. testacea Forsslund, 1963 : 289.
   testacea Märkel: 1964 : 39, fig. 8.

Genus 3: MAERKELOTRITIA Hammer, 1967

Maerkelotritia Hammer, 1967 : 6.
Maerkelotritia Märkel: 1968 : 726.

Bothridial squama below bothridium. Lateral carina absent. Notogaster with 15 pairs of setae. End of body with terminal sinus. Genital and aggenital plates entirely separated by a suture, same as anal from adanal plates. Aggenito-adanal incision absent or short, occasionally its very short primordium present. 7—10 pairs of genital, 2 pairs of aggenital, 0—1 pair of anal and 4 pairs of adanal, setae present. Palp 3-jointed. Legs tridactylous. Trochanters of legs III and IV with 2—3 setae, solenidium on genu of leg IV only exceptionally absent.

Type-species: *Maerkelotritia alaskensis* Hammer, 1967.

Known only from the Holarctic Region.

Key to species

1 (4) Genital plates with seven pairs of setae, anal setae completely reduced.

2 (3) Well developed solenidia on genu of legs III and IV, in addition to three and two pairs of setae, respectively, on the leg joint. Setae of aspis uniform, all tapering apicad.—Lamellar and interlamellar setae longer than notogastral ones, erect. Sensillus only slightly incrassate (pl. 101 : B), setiform, finely ciliate. Notogastral setae (pl. 101 : A) arcuately proclinate (towards aspis), finely ciliate, comparatively short. Adanal setae (pl. 101 : C) very long, especially ad_3 and ad_4. Length of aspis: 240 µm, length of notogaster: 400 µm (?), height of notogaster: 270 µm.

Locus typicus: "Bergrücken" of Terskei Ala Tau, East Tien-shan, Soviet Union. Distribution: only the type-locality known

krivolutzkii krivolutzkii Märkel, 1968

3 (2) Genu of leg IV without solenidium (pl. 101 : F), present on leg III. Number of setae as in preceding taxon. Lamellar setae apically conical (pl. 101 : D), rostral seta attenuating. In other respects agreeing with nominate subspecies. Length of aspis: 328—353 µm, length of notogaster: 541—705 µm, height of notogaster: 394—542 µm.

Locus typicus: Bajanzogt, Central aimak, Mongolia. Distribution: only the type-locality known

krivolutzkii mongolica Mahunka, 1981

4 (1) Genital plates with 10 pairs of setae anal plates with one pair of setae.

5 (6) Exobothridial seta minute. Aggenital-adanal incision terminally obtuse, situated perpendicularly to ano-adanal plates, at right angles (pl. 101 : H—I). Aspis and notogastral setae long, rigid, *in* considerably longer than distance enclosed by setae *in—le* (fig. 101 : J). Sensillus setiform, smooth, distally slightly arcuate. One pair of anal setae present; ad_1 considerably longer than its mates (pl. 101 : G). Length of aspis: 540—590 µm, length of notogaster: 1040—1130 µm.

Locus typicus: Kare-Numa, Nikko, Japan. Distribution: only the type-locality known

kishidai (Aoki, 1958)

6 (5) Exobothridial seta very long (pl. 100 : H) nearly as long as half length of aspis. Only a primordium of aggenito-adanal incision visible (pl. 100 : G).—Aspis and notogastral setae long. Sensillus (pl. 100 : I) weakly fusiform, slightly ciliate, somewhat arcuate. Seta ad_4 slightly longer than the others. Length of aspis: 420—490 μm, length of notogaster: 800—1000 μm.

Locus typicus: "West" of Pinguicula Lake, Kodiak Island, Alaska, USA. Distribution: USA (Alaska), Soviet Union (FE)

alaskensis Hammer, 1967

Species interpretabiles

1. alaskensis Hammer, 1967 : 8, figs. 1a—f.
2. kishidai Aoki, 1958 : 18, figs. A—D *(Oribotritia)*.
   kishidai Aoki: 1973 : 249.
   sellnicki Walker, 1964 : 49,figs. 20a—e—21 (Aoki: 1973 : 249).
3. krivolutzkii Märkel, 1968 : 725, figs. 1a—e.
4. krivolutskii mongolica Mahunka, 1981 : 300, figs. 10—12 (lapsus calami).

Genus 4: Protoribotritia Jacot, 1938

Protoribotritia Jacot, 1938 : 114.
Protoribotritia Märkel: 1964 : 31.
Protoribotritia Aoki: 1969 : 28.

Aspis without lateral keel. Bothridial squama below bothridium. Exobothridial seta developed. Notogaster with 14 or 15 pairs of setae. Ventral plates differentiated, furrow between genital and aggenital, and anal and adanal, plates well recognizable. Aggenito-adanal incision absent. Notogaster posteriorly with sinus terminalis. Seven pairs of genital, two pairs of aggenital, three pairs of anal and 3—4 pairs of adanal, setae present. Palp 3- or 4-jointed. Legs monodactyle.; trochanters of legs III and IV with two setae each; solenidium on genu of leg IV developed.

Type-species: *Protoribotritia canadaris* Jacot, 1938.

Known only from the Holarctic Region.

Key to species

1 (4) Notogaster with 15 pairs of setae. Palp 3-jointed.

2 (3) Sensillus gradually widening distally, apex obtuse, finely ciliate (pl. 102 : D).—Aspis very convex, strikingly wide in lateral view (pl. 102 : A). Rostral seta at most one-third as long as interlamellar one. Lamellar and interlamellar setae erect, rostral ones proclinate. Notogastral setae directed slightly towards cephalic part of body, finely ciliate. Four long adanal pairs of setae present (pl. 102 : B, E). Length of aspis: 200—210 μm, length of notogaster: 330—380 μm.

Locus typicus: Wilhelmshaven, GFR. Distribution: Germany

aberrans (Märkel et Meyer, 1959)

3 (2) Sensillus apically acute, preapically with two juxtaposed spines, enclosed apex resembling a bud (pl. 102 : F).—Aspis flatter (pl 102 : G—H), no significant difference in length between rostral and lamellar setae. Otherwise highly similar to the preceding nominate subspecies. Adanal setae slightly shorter, or rather ad_1 considerably longer than the others (pl. 102 : I). Length of aspis: 200 µm, length of notogaster: 340 µm.

Locus typicus: "Bergfuss" of Fujiyama, Japan. Distribution: only the type-locality known

ensifer Aoki, 1969

4 (1) Notogaster with merely 14 pairs of setae, ps_4 absent (pl. 103 : A). No great difference in length between aspis setae (pl. 103 : B), lamellar and rostral setae straight, interlamellar seta at most one and a half times longer. Aspis less convex (pl. 103 : C). Sensillus long, only apically slightly lanceolate. Three pairs of adanal setae present (pl. 103 : D). Length of aspis: 180—190 µm, length of notogaster: 290—320 µm, height of notogaster: 220—245 µm.

Locus typicus: Svartberget, Degerfors, Sweden. Distribution: Sweden

oligotricha Märkel, 1963

Species interpretabiles

1. aberrans Märkel et Meyer, 1959 : 339, figs. 7a—d *(Oribotritia)*.
   aberrans Märkel: 1964 : 31, figs. 5a—f.
2. ensifer Aoki, 1969 : 27, figs. 1—5 *(aberrans ensifer)*.
3. oligotricha Märkel, 1963 : 291, figs. 5—6.

Genus 5: Indotritia Jacot, 1929

Indotritia Jacot, 1929: 213.
Indotritia Märkel: 1964 : 24.

Aspis with two lateral keels. Bothridial squama situated above bothridium. Lamellar setae originating very far posteriorly, therefore, nearly forming a transverse row with interlamellar setae. Exobothridial seta reduced, but its insertion point observable. Notogaster with 14 pairs of setae. Genital plate partly separated from aggenital plate, fused anteriorly; also anal and adanal plates separated. Aggenital-adanal incision short. Notogaster ending in a sinus terminalis. Nine pairs of genital, two or three pairs of aggenital, two pairs of anal and two pairs of adanal, setae present. Palp 5-jointed. Trochanters of legs III and IV with 3 setae each, genu of leg IV with developed solenidium δ_1.

Type-species: *Tritia krakatauensis* Sellnick, 1923.

Principally tropical, with only one species occurring in the Palaearctic Region.

— Aspis flat, with a slight concavity between interlamellar and lamellar setae in lateral view (pl. 103 : F). Sensillus (pl. 103 : G) setiform, acute. Interlamellar setae shorter than lamellar ones. Notogastral setae also short (pl. 103 : F), proclinate, finely arcuate. Genital, setae minute, shorter than aggenital ones (pl. 103 : H). Length of aspis: 470 µm, length of notogaster: 675 µm

Locus typicus: Split, Yugoslavia. Distribution: Yugoslavia

consimilis Märkel, 1964

Species interpretabilis

1. consimilis Märkel, 1964:25, figs. 3a—c *(krakatauensis consimilis)*.

Genus 6: PARATRITIA Moritz, 1966

Paratritia Moritz, 1966:374.

Aspis without lateral keel. Bothridial squama situated above bothridium. Notogaster with 15 pairs of setae. Genital and aggenital plates fused, suture between anal and adanal plates well observable. No anogenital furrow present. No ventral triangle formed between plates. Ventral region ending in sinus terminalis. Eight pairs of genital, two pairs of aggenital, two pairs of anal and three pairs of adanal, setae present. Palp 3-jointed. Legs monodactyle. Trochanters of legs III and IV with two setae each. Genu of leg IV without solenidium.

Type-species: *Paratritia baloghi* Moritz, 1966.

Only the European type-species is known.

— Sensillus terminally abruptly widening, lancolate, finely ciliate (pl. 104: B). Aspis setae approximately equal in length, lamellar seta not even approaching basis of rostral one. Notogastral setae also short (pl. 104: A, E), c_1 nearly as long as half distance between c_1—d_1; all setae slightly proclinate. Anal and adanal setae developed (pl. 104: C), sinus terminalis small (pl. 104: D). Length of aspis: 152—179 μm, length of notogaster: 275—339 μm, height of notogaster: 166—212 μm.
Locus typicus: Bad Frankenhausen, southern edge of Kyffhäuser range, GDR. Distribution: GDR, Hungary

baloghi Moritz, 1966

Species interpretabilis

1. baloghi Moritz, 1966: 376, figs. 1—5.

Family 2: EUPHTHIRACARIDAE Jacot, 1930

Euphthiracarini Jacot, 1930: 214.
Euphthiracaridae Van der Hammen: 1959: 35.
Euphthiracaridae Grandjean: 1967: 247.
Tritiinae Jacot, 1923: 180.
Pseudotritiidae Grandjean, 1954: 430.

The four pairs of plates constituting ventral plate fused into two longitudinal pairs, therefore, genital, aggenital, anal and adanal plates no more discrete. Genital and anal plates separated medially by a heavy, often twisted, triangular chitinous thickening. A similar formation occasionally present also in terminal section of anal plate. 4—9 pairs of genital, 2—3 pairs of aggenital, 3—6 pairs of anal and 3—5 pairs of adanal, setae observable. Palp 3-jointed.

A cosmopolitan family, with three Palaearctic genera.

Key to genera

1 (2) Longitudinal suture separating ventral plates with two triangles, one medially one posteriorly
genus 1: **Euphthiracarus** Ewing, 1917
2 (1) Longitudinal suture separating ventral plates with a median triangle only.
3 (4) Genito-aggenital plate with 4—5 genital and 2—3 aggenital setae. Trochanters of legs III and IV with only one hair each
genus 3: **Microtritia** Märkel, 1964
4 (3) Genito-aggenital plate with 8—9 genital and 2 aggenital setae. Trochanters of legs III and IV with two setae each
genus 2: **Rhysotritia** Märkel et Meyer, 1959

Genus 1: Euphthiracarus Ewing, 1917

Euphthiracarus Ewing, 1917 : 125.
Euphthiracarus Märkel: 1958 : 486.
Euphthiracarus Van der Hammen: 1959 : 39.
Euphthiracarus Märkel: 1964 : 63.
Pseudotritia Willmann, 1919 : 552.

Aspis with one or two lateral keels. Bothridial squama situated below bothridium. Exobothridial seta occasionally absent, in general medium long. Notogaster with diverse surface sculpture and 14 pairs of setae; ps_1 originating below lower margin of notogaster. Genital and aggenital plates as well as anal and adanal plates, respectively, fused completely. No anogenital suture present. Plates with two separating ventral triangles, one medially, one posteriorly. Ventral incision ending in terminal fissure. 7—9 pairs of genital, 2 pairs of aggenital, 3—6 pairs of anal and 3—5 pairs of adanal, setae observable. Palp 3-jointed. Legs mono- or tridactylous. Trochanters of legs III and IV with two hairs each. Genu of leg IV with solenidium δ_1 developed.

Type-species: *Phthiracarus flavus* Ewing, 1908.

The species occur in every zoogeographical region. So far 16 species are known, but some need revision. Two subgenera are recognized (*Euphthiracarus* and *Brasilotritia* Märkel, 1964*).

Key to species

1 (6) Legs monodactylous.
2 (5) Exobothridial seta developed. Distance between lamellar setae considerably smaller than that between rostral ones; distance between interlamellar setae twice or even more greater than that between lamellar setae.
3 (4) Lamellar setae originating very near each other, their enclosed distance only one-third of that between interlamellar setae (pl. 104 : H). Body setae comparatively short (pl. 104 : G), c_1 as long as d_1; c_2 shorter than half distance between c_2—d_2.—Interlamellar setae relatively long, longer than distance

* Known from tropical areas only.

between interlamellar and lamellar setae. Chaetom of anogenital region as in pl. 104: I. Length of aspis: 245—255 μm, length of notogaster: 422—498 μm height of notogaster: 223—346 μm.

Locus typicus: Burgdamm bei Bremen, Germany. Distribution: Germany, Austria, Hungary, Poland

monodactylus (Willmann, 1919)

4 (3) Lamellar setae originating more removed from each other, their enclosed distance slightly more than half of that between interlamellar setae (pl. 104: M). Body setae longer (pl. 104: K), notogastral c_1 and c_2 at least twice longer than d_1 or d_2, respectively.—Rostral and interlamellar setae originating at approximately equal distances. Stalk of sensillus (pl. 104: J) slightly shorter that that of the related species. Chaetom of anogenital region depicted in pl. 104: L. Length of aspis: 225—270 μm, length of notogaster: 463—542 μm, height of notogaster: 254—320 μm.

Locus typicus: Elikon Massif, Greece. Distribution: only the type-locality known

mixtus Mahunka, 1979

5 (2) Exobothridial seta absent. Interlamellar setae hardly farther from each other than lamellar (pl. 105: B). Body setae longer (pl. 105: A), c_1 essentially longer than d_1; seta c_2 as long as distance between c_2—d_2. Interlamellar setae very long, approaching insertion points of rostral setae. Sensillus (pl. 105: C) distally considerably widening. Chaetom of anogenital region as in pl. 105: D. Length of aspis: 253 μm, length of notogaster: 508 μm, height of notogaster: 356 μm.

Locus typicus: Valea Rea, Retezat Range, Roumania. Distribution: only the type-locality known

antipai Feider et Calugar, 1969

6 (1) Legs tridactylous.

7 (8) Seven pairs of anal setae present, but six represented only by their insertion points (pl. 105: F).—Three pairs of robust adanal setae similar to an_1. 7—8 pairs of genital setae. Aspis with two lateral keels (pl. 105: E), dorsal keel stronger than ventral one. Bothridial squama comparatively large, conspicuous. Surface of aspis granulose between dorsolateral keels, but not extending anteriorly beyond line of rostral setae and posteriorly beyond that of interlamellar setae (pl. 105: H). Notogastral surface (pl. 105: G) with a fine polygonal sculpture: polygons proper delimited by straight lines. Length of aspis: 430 μm,length of notogaster: 835 μm, height of notogaster: 625 μm.

Locus typicus: Pernegg, Steiermark, Austria. Distribution: Austria

reticulatus alpinus Märkel, 1964

8 (7) Three pairs of anal setae present, but two pairs represented by insertion points only.

9 (10) Five pairs of adanal setae present.—Genito-aggenital plate with eight pairs of genital and two pairs of aggenital setae. Notogaster with foveolate sculpture (pl. 105: K). Sensillus setiform (pl. 105: I), with some hardly recognizable cilia. Aspis and notogastral setae (pl. 105: J, L) ciliate. Interlamellar setae extraordinarily long, reaching insertion points of rostral setae. Length of aspis: 442 μm, length of notogaster: 920 μm, height of notogaster: 644 μm.

Locus typicus: Brosteni, region Craiova, Roumania. Distribution: only the type-locality known

intermedius (Feider et Suciu, 1958)

10 (9) Only three pairs of adanal setae observable.

11 (12) Surface of aspis granulate between lateral keels, granulation extending behind interlamellar setae, up to posterior margin of aspis. Notogaster with polygonal sculpture defined by straight lines.—Berlese illustrated the species with 30 pairs of long, robust, acutely pointed notogastral setae, but the number is probably erroneous.

Locus typicus: Tiarno, Italy. Distribution: Italy, Soviet Union (SET)

reticulatus reticulatus (Berlese, 1913)*

12 (11) Surface of aspis never granulose between lateral keels, but with well discernible foveolae.

13 (14) Lamellar seta essentially longer than rostral one (pl. 106 : B). Upper lateral keel more robustly developed than ventral keel, its edge rough (pl. 106 : B).—Seta c_1 longer than d_1, and considerable differences in length also among the other notogastral setae (pl. 106 : A); every seta ciliate. Chaetom of anogenital region as in pl. 106 : F. Length of aspis: 330 μm, length of notogaster: 610 μm, height of notogaster: 440 μm.

Locus typicus: Norway (without exact locality). Distribution: Europe

cribrarius (Berlese, 1904)

14 (13) Rostral seta essentially longer than lamellar one (pl. 106 : H). No essential difference between upper and lower lateral carinae, upper one without a rough edge (pl. 106 : G).—No great difference in length between notogastral setae, c_1 as long as d_1. Chaetom of anogenital region as in pl. 106 : I. Length of aspis: 260—330 μm, length of notogaster: 475—620 μm.

Locus typicus: Sandakphu, Nepal. Distribution: only the type-locality known

inglisi Sheals, 1965

Species interpretabiles

1. antipai Feider et Calugar, 1969 : 626, figs. 11—23.
2. cribrarius Berlese, 1904 : 23 *(Phthiracarus)*.
   cribrarius Van der Hammen: 1959 : 39.
   cribrarius Märkel: 1964 : 67, figs. 18a—e.
   cribrarius Bernini: 1970 : 391, figs. 1a—f.
3. inglisi Sheals, 1965 : 18, figs. 30—39.
4. intermedia Feider et Suciu, 1958 : 36, figs. 10—23 *(Pseudotritia)*.
5. mixtus Mahunka, 1979 : figs. 38—41.
6. monodactylus Willmann, 1919 : 552, fig. 1 *(Tritia (Pseudotritia))*
   monodactyla Märkel: 1958 : 448, fig. 7 *(Pseudotritia)*.
   monodactyla Van der Hammen: 1959 : 36 *(Pseudotritia)*.
   monodactylus Märkel: 1964 : 69, figs. 19a—d.
7. reticulatus Berlese, 1913 : 102, figs. 8 : 97a, 98 *(Phthiracarus)*.
   reticula Van der Hammen: 1959 : 38.

* In need of urgent redescription!

Genus 2: RHYSOTRITIA Märkel et Meyer, 1959

Rhysotritia Märkel et Mayer, 1959 : 329.
Rhysotritia Märkel: 1964 : 54.

Aspis with 1 or 2 lateral keels. Bothridial squama above bothridium. Exobothridial seta developed, but minute. Notogaster with 14 pairs of setae, ps_1 originating far from ps_2 and ps_3, high above and far from ventral margin of notogaster. Genital and aggenital, as well as anal and adanal, plates wholly fused with each other; no anogenital suture present. Plates medially with a single separating triangle. 5—9 pairs of genital, 2 pairs of aggenital, 3—9 pairs of anal and 3 pairs of adanal, setae observable. Palp 3-jointed, legs mono- or tridactylous. Trochanters of legs III and IV with two setae each; genu of leg IV without solenidium.

Type-species: *Hoplophora ardua* C. L. Koch, 1841.

The genus needs revision, the species are rather difficult to separate from one another.

Key to species

1 (2) Aspis with two lateral keels (pl. 107 : B). Body setae (pl. 107 : A) long, lamellar one essentially longer than distance between lamellar and interlamellar bases; c_1 as long as distance between its insertion point and anterior margin of notogaster. Sensillus (pl. 107 : C) setiform, very finely ciliate. Seta a_1 of anal plate (pl. 107 : D) minute but recognizable near triangle. Leg I bi-, legs II—IV tridactylous. Length of aspis: 290—295 µm, length of notogaster: 560—600 µm, height of notogaster: 410—430 µm.

Locus typicus: Epping Forest, Great Britain. Distribution: Great Britain, Germany, France, Bulgaria, Soviet Union (SET)

duplicata (Grandjean, 1953)

2 (1) Aspis with one lateral keel only.

3 (4) Sensillus short and clavate (pl. 107 : J). Interlamellar setae long, longer than distance between their insertion points and those of lamellar setae. Lamellar and rostral setae about half as long as interlamellar ones, shorter than distance between lamellar and rostral setae. Notogastral setae also short, c_1 considerably shorter than half distance enclosed by c_1—d_1. All notogastral setae smooth. Number of genital setae (pl. 107 : K—L) varying between 5—8 pairs. Legs monodactyle. Length of aspis: 210 µm, length of notogaster: 426 µm, height of notogaster: 308 µm.

Locus typicus: Bouches du Rhone, Aix-en-Provence, France. Distribution: only the type-locality known

clavata sextiana Lions, 1966*

4 (3) Sensillus long, setiform, at most only distally slightly incrassate (pl. 107 : G).

5 (6) Sensillus setiform, only distally ciliate, stalk not thickened. A great difference in length between interlamellar and lamellar setae, lamellar one shorter than

* The nominate subspecies occurs in Peru (see Märkel, 1964 : 61).

distance between lamellar and rostral setae (pl. 107 : G). Notogastral setae also short (pl. 107 : E), c_1 about half as long as distance enclosed by setae c_1—d_1.Leg I bi- legs II—IV tridactylous, but a monodactyle form (subspecies?) is also known. Fissura terminalis long (pl. 107 : F). Length of aspis: 235—300 µm, length of notogaster: 480—620 µm, height of notogaster: 340—375 µm.

Locus typicus: Regensburg, Germany. Distribution: Holarctic Region, Polynesia

ardua ardua (C. L. Koch, 1841)*

6 (5) Sensillus apically heavily clavate (pl. 107 : H), resembling the nominate subspecies, but smaller (pl. 107 : I). Length of aspis: 230—275 µm, length of notogaster: 480—545 µm.

Locus typicus: Madrid, Spain. Distribution: Spain, Greece

ardua penicillata Pérez-Iñigo, 1969

Species interpretabiles

1. ardua C. L. Koch, 1841 : 32, 15 figs. *(Hoplophora)*.
   ardua Märkel: 1958 : 486, fig. 5 *(Pseudotritia)*.
   ardua Van der Hammen: 1959 : 36 *(Pseudotritia)*.
   ardua Pérez-Iñigo: 1968 : 212, figs. 27—28.
   canestrinii Michael, 1898 : 81 *(Phthiracarus)*.
   lentula Berlese, 1887 : fasc. 36 (3) *(Tritia)*.
2. ardua penicillata Pérez-Iñigo, 1969 : 378, figs. 4—5.
   ardua penicillata Pérez-Iñigo, 1968 : 214, fig. 29 (nom. nudum).
3. clavata sextiana Lions, 1966 : 779, figs. 1—4.
4. duplicata Grandjean, 1953 : 157 *(Pseudotritia;* nom. nov. for *ardua* sensu Michael, 1898)
   duplicata Märkel: 1954 : 56, figs. 14a—h.

Species inquirendae

— loricata Rathke, 1799 : 811.**
— sinensis Jacot, 1924 : 178, figs. 20—26 *(Acrotritia)*.
— sinensis triheterodactylis Jacot, 1924 : 181 *(Acrotritia)*.

Genus 3: Microtritia Märkel, 1964

Microtritia Märkel, 1964 : 45.

Aspis with one lateral keel. Bothridial squama situated above bothridium. Exobothridial seta minute or reduced. Notogaster with 14 pairs of setae. Ventral plate ending posteriorly in fissura terminalis or sinus terminalis. Anogenital plates with a single separating triangle, in middle of plates. Four pairs of genital, 2—3 pairs of aggenital, three pairs of anal and three pairs of adanal, setae observable. Palp 3-

* On the basis of literature data, the species has an enormous range. It is probable, however, that the data refer to a number of closely related species still treated under a single specific name.

** See Oudemans 1937 : 2733.

jointed. Legs monodactyle. Trochanters of legs III and IV with merely one seta each; genu of leg IV without solenidium.

Type-species: *Phthiracarus minimus* Berlese, 1904.

The species also occur in the tropics, principally in the Neogaea. Two species are known from the Palaearctic Region.

Key to species

1 (2) Notogaster posteriorly with terminal fissure.—Setae of aspis very short (pl. 108 : C), interlamellar seta shortest of all. Sensillus gradually widening distally, with 3—4 minute spines terminally. Bothridial squama situated behind bothridium. Notogastral setae (pl. 108 : A) short, thin, characteristically recurving. Chaetom of anogenital region as in pl. 108 : B, three pairs of aggenital setae present. Length of aspis: 210 μm, length of notogaster: 360 μm, height of notogaster: 275 μm.

Locus typicus: "Tundra-Landschaften", Altai, Soviet Union. Distribution: only the type-locality known

fissurata Märkel, 1968

2 (1) Notogaster posteriorly with sinus terminalis.

3 (4) Sensillus finely fusiform, gradually tapering apicad, terminally acute (pl. 108 : H).—Interlamellar setae minute, essentially shorter than lamellar ones, but well visible. Bothridial squama situated above bothridium (pl. 108 : G). Anogenital plate (pl. 108 : I) with well developed a_2 and a_3. Length of aspis: 175—215 μm, length of notogaster: 245—345 μm, height of notogaster: 220—285 μm.

Locus typicus: Tingo Maria, Peru. Distribution: Peru, Japan

tropica Märkel, 1964

4 (3) Sensillus gradually widening apicad, terminally obtuse, with minute spines (pl. 108 : D).—Interlamellar setae of aspis as long and as well discernible as lamellar and rostral ones. Bothridial squama situated behind bothridium (pl. 108 : E). Notogastral setae slightly shorter than in preceding species. Anogenital plates (pl. 108 : F) with setae a_2 and a_3 represented by insertion points only. Length of aspis: 180 μm, length of notogaster: 260 μm, height of notogaster: 190 μm.

Locus typicus: "Giardino di Boboli", Florence, Italy. Distribution: Italy, Germany, Hungary

minima (Berlese, 1904)

Species interpretabiles

1. fissurata Märkel, 1968 : 726, figs. 2a—d.
2. minimus Berlese, 1904 : 22 *(Phthiracarus)*.
   minima Van der Hammen: 1959 : 37 *(Pseudotritia)*.
   minima Märkel: 1964 : 46, figs. 10a—e.
   minuta Willmann, 1919 : 552, fig. 2 *(Tritia (Pseudotritia))*.
3. tropica Märkel, 1964 : 48, figs. 11a—e.
   tropica Aoki: 1977 : 91, figs. 10—12.

Species inquirenda

— xilofila Subias, 1977: 86, fig. 41.

Supercohort 5: HOLOSOMATA Grandjean, 1969

Holosomata Grandjean, 1969: 135.*
Holosomata Balogh et Mahunka, 1979: 281.

A collective group nearly defying uniform characterization and probably in need of further subdividing. Body holoid, notogaster holonotic. 15, or if setae f_1 also evolved, 16 pairs of notogastral setae and one pair of exobothridial setae present. Oil gland extant, well discernicle. Diagastria occasionally present. Peranal plate not differentiated.

Composed of three well differentiated superfamilies, with representatives distributed in the whole world.

Key to superfamilies

1 (2) Body macropilic; anal and genital openings large, meeting. A distinct adanal plate present, emitting adanal setae
superfamily 1: **Nothroidea** Berlese, 1896

2 (1) Body brachypilic; adanal plate fused with ventral plate emitting adanal setae.

3 (4) Diagastria well visible on ventral side (fig. 31). Genital and anal plates far from each other. Genital plate with nine pairs of setae aligned in a single longitudinal row
superfamily 2: **Nanhermannioidea** Sellnick, 1928

4 (3) Ventral side uniform, without diagastria. Genital and anal plates large, meeting or approaching each other. Genital plate with nine (6+3) pairs of setae, arranged in two longitudinal rows
superfamily 3: **Hermannioidea** Sellinck, 1928

Superfamily 1: NOTHROIDEA Berlese, 1885

Nothridae Berlese, 1885: 5.
Nothroidea Grandjean: 1954: 431.

Body holoid, notogaster uniform; ventral plate considerably subdivided, of macropilic type; adanal and occasionally also aggenital plate differentiated.

* Grandjean established this superfamilial group in 1969; we have given it a status at supercohort level (1979).

A survey of the genera of the superfamily Nothroidea

Characteristics / Genera	Sensillus	Number of claws	Number of setae: genital	aggenital	anal	adanal	Epimeral setal formula
Nothrus	+	1—3	9	0	2	3	(3—7)—(3—5)—(4—7)—(4—7)
Heminothrus	+	1	9—24	2	2	3	(2—3)—1—3—(3—4)
Camasia	+	1,3	9,14	2	3	3	3—1—(2—3)—3
Trhypochthonius	+	3	6—11	0	1	3	3—1—3—2
Trhypochthoniellus	+,—	3	16—18	0	1	2	3—1—3—2
Malaconothrus	—	1	4—6	0	1	3	3—1—3—3
Trimalaconothrus	—	2,3	4—12	0	1	3	3—1—(2—3)—3
Mucronothrus	+,—	1	18—20	0	2	2	3—1—2—2—2

The superfamily contains five families, four of which occur also in the Palaearctic Region. They are ancient forms and representatives are known already from the Jurassic and the Cretaceous. They inhabit mainly wet habitats.

Key to families

1 (6) Bothridium and sensillus developed, sensillus differing also in form from all other prodorsal setae.

2 (3) Two pairs of aggenital setae present. Epimeral setal formula: (2—3)—(1)—(2—3)—(3—4)

family 2: **Camisiidae** Oudemans, 1900

3 (2) Aggenital setae absent.

4 (5) A heavy epimeral neotrichy observable: epimeres 2 with at least three, epimeres 3 with four, pairs of setae. Possible epimeral setal formulae: (3—7)—(3—5)—(4—7)—(4—7). Heavily chitinized, dark animals. Sensillus mostly setiform

family 1: **Nothridae** Berlese, 1885

5 (4) No epimeral neotrichy present; epimeral setal formula: 3—1—3—2

family 3: **Trhypochthoniidae** Willmann, 1931 (partim)

6 (1) Trichobotrium reduced, sensillus (if present) simple, setiform, not differing from any other prodorsal seta.

7 (8) Epimeres 3 with merely two pairs of setae; 16—20 pairs of genital and two pairs of adanal setae present

family 3: **Trhypochthoniidae** Willmann, 1931 (partim)

8 (7) Epimeres 3 with three pairs of setae; genital plates at most with 12 (4—12) pairs of setae and always with three pairs of adanal setae

family 4: **Malaconothridae** Berlese, 1916

Family 1: NOTHRIDAE Berlese, 1885

Nothridae Berlese, 1885 : 5.

Large-sized, often dark brown animals. Proterosoma large, wide, often comprising one-third of entire body. Bothridium well developed, sensillus always long setiform or rarely bacilliform. Notogaster with 16 pairs of setae, also f_1 developed. A heavy epimeral neotrichy present.

Only two genera of the family are known, but merely one is represented in the Palaearctic Region.

Genus 1: NOTHRUS C. L. Koch, 1836

Nothrus C. L. Koch, 1836 : 2, 17.
Nothrus Sellnick et Forsslund: 1955 : 495.
Nothrus Van der Hammen: 1959 : 62.
Angelia Berlese, 1885 : 17, 6.
Angelia Van der Hammen: 1959 : 62.

Rostrum with a median incision. Notogaster with a rough sculpture of foveolae; notogastral setae rubust, mostly bacilliform or phylliform, terminal ones, especially k_1 (less so pn_1 and pn_2) longer than the others; nine pairs of genital setae present, eight in one longitudinal row, ninth seta slightly laterally. Aggenital hairs absent, three pairs of adanal and two pairs of anal setae present. Epimeral neotrichy varying, framed in the general formula: (3—7)—(3—5)—(4—7)—(4—7). Lengs mono- or tridactylous, seldom 2-clawed.

Type-species: *Nothrus palustris* C. L. Koch, 1839*.

The species occur in the whole world; 13 have been described from the Palaearctic Region.

Key to species

1 (10) Legs monodactyle.

2 (5) At least one pair of slender, narrow, bacilliform setae present among notogastral setae; k_1 essentially longer than pn_1 (pl. 109 : A).

3 (4) Lamellar setae tapering apicad, simple, short, shorter than distance between their insertion points. Five median pairs of notogastral setae: c_1, d_1, d_2, e_1 and f_1 short and setiform, some marginal setae also similar in form (pl. 109 : A); all other setae bacilliform of weakly spatulate; k_1 one and a half times longer than pn_1. Foveolae (pl. 109 : B) in median area of notogaster removed from each other at a distance of at least their own diameter. Epimeral setal formula: 5—3—4—4. Length: 936—940 µm, width: 540—558 µm.

Locus typicus: Zehlanbruches (Königsberg now Kaliningrad), Soviet Union. Distribution: Belgium, Austria, Finland, Sweden, Netherlands, Soviet Union (NET, CET), Canada

pratensis Sellnick, 1929

* Concerning the designation of the type-species, see Van der Hammen, 1959 : 62; we accept his standpoint.

4 (3) Lamellar setae bacilliform or weakly spatulate, as long as distance between their insertion points. Every notogastral seta (pl. 109 : F) weakly but well discernibly spatulate (pl. 109 : F), only terminal k_1 of uniform thickness, not widening, at least twice as long as seta pn_1. Foveolae (pl. 109 : E) in median area of notogaster generally larger, their distances from each other generally shorter than their diameter. Epimeral setal formula: (6—7)—5—6—4. Length: 810 μm width: 396—432 μm.

Locus typicus: "Bois d'Aulnay," Sceaux, France. Distribution: Holarctic Region, Mexico (?)

silvestris Nicolet, 1855

5 (2) Every notogastral seta considerably widened, spatulate; terminal k_1 not longer than pn_1 (pl. 109 : H).

6 (7) Lamellar setae short, considerably shorter than interlamellar ones or those originating on anterior margin of notogaster (pl. 109 : H); essentially shorter than distance between their points of origin. Notogastral setae marginally weakly serrate, their surfaces not squamose but with a pattern resembling venation. Margin of rotund foveolae ornamenting median field of smooth notogaster (pl. 103 : I). Length: 750—790 μm.

Locus typicus: "Valle de la Orotava", Tenerife, Canary Islands, Spain. Distribution: Canary Islands

perezinigoi Mahunka, 1980

7 (6) Lamellar setae long, not shorter than c_1 originating on anterior part of notogaster, about as long as distance between their insertion points.

8 (9) Lamellar setae narrowly spatulate, considerably narrower than interlamellar and notogastral setae (pl. 110 : A). Basal portion of prodorsum with angular, oblong foveolae in interbothridial region.—Lamellar setae finely ciliate, interlamellar and notogastral setae squamose (pl. 110 : B). Notogastral foveolae with smooth, rounded margins. Notogastral seta longer, f_1 approaching insertion point of pn_2. Length: 660 μm, width: 355 μm.

Locus typicus: "Giardino di Boboli", Florence, Italy. Distribution: only the type-locality known

pulchellus (Berlese, 1910)

9 (8) Lamellar setae wide, wider and larger than interlamellar setae (pl. 110 : E). Basal portion of prodorsum with punctiform or minute round foveolae in a disarranged pattern.—Lamellar setae phylliform, their margins smooth, all other setae with very large squamae (pl. 110 : F). Margins of foveolae on notogaster sinuous. Notogastral setae essentially shorter than in preceding species; f_1 by far not approaching insertion point of pn_2. Length: 674—708 μm, width: 333—366 μm.

Locus typicus: environs of Primorsk, Irkutsk, Soviet Union. Distribution: Soviet Union (FE)

parvus Sitnikova, 1975

10 (1) Legs bi- or tridactylous.

11 (12) Legs only bidactylous.—Lamellar setae bacilliform, about as long as distance between their insertion points. Considerable differences in length among notogastral setae (pl. 110 : D), and all slightly incrassate; k_1 about one and a

half times longer than pn_1. Foveolae in median field of notogaster large, removed from each other at distances essentially shorter than their average diameter. Epimeral setal formula: 6—5—5—5. Length: 720—810 μm, width: 378—414 μm.

Locus typicus: exact reference unknown to us; Italy. Distribution: Great Britain, Sweden, Finland, Netherlands, France, Czechoslovakia, Italy, Hungary, Soviet Union (SET, CET, NET)

anauniensis Canestrini et Fanzago, 1877

12 (11) Legs tridactylous.

13 (16) Seta k_1 very long, setiform, acute, considerably longer than all other notogastral setae, reaching half length of body. Seta c_2 originating much nearer to c_3 than to c_1.

14 (15) Seta pn_2 comparatively long, approximately twice longer than f_2 (pl. 110 : H). Posterior margin of notogaster finely concave.—Lamellar setae originating on ends of a heavy chitinous lath. Excepting seta k_1, notogastral setae bacilliform. Genital plate (pl. 110 : G) less chitinized, all genital setae arising along a longitudinal line, excepting a lateral pair basally. Length: 1000—1188 μm, width: 700—756 μm.

Locus typicus: Zweibrücken, Rheinbayern, Germany. Distribution: Europe, Soviet Union (NET, SET, WS, ES, FE)

palustris C. L. Koch, 1839

15 (14) Seta pn_2 (pl. 111 : D) comparatively short, nearly equal in length with f_2 (pl. 111 : A). Posterior margin of notogaster finely convex.—Notogastral setae (pl. 111 : C) slightly widened, phylliform, their surfaces occasionally weakly squamose. Seta k_1 (pl. 111 : E) somewhat shorter than in preceding species. Anterior margin of genital plate chitinized, tuberculate, genital setae not aligned in a row (pl. 111 : B). Length: 1070—1190 μm, width: 660—727 μm.

Locus typicus: Kami-oboro, Akkeshi, Hokkaido, Japan. Distribution: only the type-locality known

asiaticus Aoki et Ohnishi, 1974

16 (13) Seta k_1 similar to the others, shorter, bacilliform or spatulate. Also seta c_2 in nearly median position, originating essentially nearer to c_1 than to c_3.

17 (20) At least notogastral setae k_1, pn_1 and pn_2 bacilliform, not or hardly spatulate, essentially longer than the other notogastral setae.

18 (19) Setae originating in anterior part of notogaster thin, acute, distal half finely ciliate; k_1 and pn_1 equal in length (pl. 111 : H).—Lamellar and interlamellar setae slightly widening, ciliate. Notogastral areolae rotund, separating chitinous laths simple (pl. 111 : I). Genital plate with six setae in one longitudinal row along inner margin of plate, three others more inwards (pl. 111 : J). Length: 860—910 μm, width: 445—455 μm.

Locus typicus: Mugello, Italy. Distribution: only the type-locality known

longipilus (Berlese, 1910)

19 (18) Notogastral setae uniform, bacilliform, smooth, k_1 considerably longer than pn_1 (pl. 111 : G.)—Lamellar and interlamellar setae not widening, smooth. Notogastral areolae slightly angular, separating chitinous laths subdivided by tubercles and incrassations. Epimeral setal formula: 7—5—7—6. Genital

plate (pl. 111 : F)with seven setae originating along inner margin. Length: 954—990 μm, width: 522—558 μm.

Locus typicus was not specified in the description. Distribution: Holarctic Region

borussicus Sellnick, 1929

20 (17) Notogastral setae largely spatulate, k_1, pn_1 and pn_2 nearly all well discernibly widened, not or hardly longer than the other notogastral setae (pl. 112 : A).

21 (22) Sensillus preapically lanceolately widended (pl. 112 : F), seta k_1 apically only slightly spatulate.—Epimeral setal formula: (6—7)—(3—4)—5—(5—7). Length: 1070—1130 μm, width: 620 μm.

Locus typicus: Selap, Nepal. Distribution: only the type-locality known

springsmythi Sheals, 1965

22 (21) Sensillus setiform, preapically never lanceolate. Seta k_1 always well discernibly widely spatulate.

23 (24) Surface and margins of notogastral setae smooth, all setae phylliform. Seta pn_2 essentially shorter than k_1 (pl. 112 : A).—Lamellar setae widened, originating far behind rostral setae. Notogastral foveolae comparatively far removed from one another (pl. 112 : B), hence surface apparently without a polygonal reticulation. Epimeral setal formula: 6—5—4—4. Length: 770—790 μm, width: 396—414 μm.

Locus typicus: Regensburg, Germany. Distribution: Ireland, Greenland, Netherlands, Denmark, Sweden, Germany, Austria, Czechoslovakia, Hungary, Italy, Soviet Union (SET, TC, KZ, FE), Japan

biciliatus C. L. Koch, 1841

24 (23) Surface of notogastral setae squamose, hence their margins not smooth (pl. 112 : E); setae k_1 and pn_2 about equal in length (pl. 112 : D).—Lamellar setae thin, situated near rostral ones. Notogastral foveolae large, distance between them small, hence surface ornamented with a polygonal reticulation. Basal part of prodorsum a similarly densely reticulate surface. Length: 666—708 μm, width: 333—374 μm.

Locus typicus: Sakhalin, Soviet Union. Distribution: only the type-locality known

reticulatus Sitnikova, 1975

Species interpretabiles

1. anauniensis Canestrini et Fanzago, 1876 : 99.
   anauniensis Sellnick et Forsslund: 1955 : 506, figs. 30—32 (*silvestris* var. *anauniensis*).
   anauniensis Van der Hammen: 1959 : 63.
2. asiaticus Aoki et Ohnishi, 1974 : 153, figs. 12—21. (*palustris asiaticus*).
3. biciliatus C. L. Koch, 1841 : 38,2.
   biciliatus Sellnick et Forsslund: 1955 : 502, figs. 33—35.
4. borussicus Sellnick, 1929 : 19.
   borussicus Sellnick et Forsslund: 1955 : 500, figs. 36—38.
5. longipilus Berlese, 1910 : 381 (*Angelia anauniensis* var. *longipila*).
   longipilus Van der Hammen: 1959 : 63.
   longipilus Mahunka: 1980 : 115, figs. 25—27.

6. palustris C. L. Koch, 1839: 29, 13.
palustris Sellnick et Forsslund: 1955: 498, figs. 39—41.
palustris Van der Hammen: 1959: 62.
7. parvus Sitnikova, 1975: 75, figs. 105—106.
8. perezinigoi Mahunka, 1980: 115.
pulchellus Pérez-Iñigo 1972: 194, figs. 4—5, nec Berlese, 1910.
9. pulchellus Berlese, 1910: 381 *(Angelia)*.
pulchella Berlese: 1913: 100, figs. 8: 100a—d *(Angelia)*.
pulchellus Van der Hammen: 1959: 64.
pulchellus Mahunka: 1980: 115, figs. 28—29.
10. pratensis Sellnick, 1929: 19.
pratensis Sellnick, 1929: 347, fig. 9.
pratensis Sellnick et Forsslund: 1955: 503, figs. 42—44.
11. reticulatus Sitnikova, 1975: 75, figs. 119—120.
12. silvestris Nicolet, 1855: 458, pl. 7, fig. 4.
13. springsmythi Sheals, 1965: 26, figs. 49—62.

Family 2: CAMISIIDAE Oudemans, 1900

Camisiinae Oudemans, 1900: 142.
Camisiidae Tuxen: 1952: 392.
Camisiidae Grandjean: 1954: 431.
Camisiidae Van der Hammen: 1959: 65.

Very large, medium sclerotized, often weakly pigmented animals. Cuticle often dirtied by earthen granules, therefore, not readily investigable. Rostrum without median incision. Prodorsum with large apophyses, in general emitting rostral and lamellar setae. Notogaster occasionally with longitudinal laths. Notogaster with 15 pairs of setae, f_2 absent, setae occasionally also arising on apophyses. Epimeres fused, epimeral chaetom generally simple, a basic type of 3—1—3—3 the most frequent, and with small deviations only. Number of genital setae highly varying, between 9—23 pairs; two pairs of aggenital, 2 or 3 pairs of anal and three pairs of adanal, setae present. Legs generally monodactyle, but tridactylous also occur.

Five genera have been described, one of which occurs exclusively in the tropics. Among the four Palaearctic genera we consider the genera *Platynothrus* Berlese, 1916, and *Neonothrus* Forsslund, 1955, as subgenera of the genus *Heminothrus* Berlese, 1913, through there are also protagonists of even further subdivisions.

Key to genera

1 (2) Three pairs of anal setae present. Ventral side with notogastral margin delimiting anogenital region by a straight transverse line. Aggenital setae arising on soft cuticle behind plates
genus 1: **Camisia** Heyden, 1826

2 (1) Only two pairs of anal setae present. Ventral side with notogastral margin forming an attenuating excision corresponding to shape of anal plates. Aggenital setae originating on margin of aggenital plate (pl. 115: J)
genus 2: **Heminothrus** Berlese, 1917

Genus 1: Camisia Heyden, 1826

Camisia Heyden, 1826 : 612.
Camisia Sellnick et Forsslund: 1955 : 478.
Camisia Van der Hammen: 1959 : 65.
Uronothrus Berlese, 1913 : 98.

Large, nearly 1 cm long, weakly pigmented animals, their cuticle usually producing a secretion to which grains of earth and detritus adhere and thus the specimens appear dirty. Trichobothria always present, sensillus short and clavate. Prodorsum and notogaster, especially on its posterior margin, often with chitinous appendages emitting setae. Epimeres separated by soft cuticular areas from each other, no sternal apodeme evolved. Epimeral setal formula: 3—1—(2—3)—3. Nine (in species not occurring in our area exceptionally more) pairs of genital, two pairs of aggenital, three pairs of anal and three pairs of adanal, setae observable; aggenital ones originating on soft body surface between aggenital and genital plates. Legs generally tridactylous, exceptionally monodactyle.

Type-species: *Notaspis segnis* Hermann, 1804.

The species are widely distributed and frequent in the Holarctic Region; about 15 species have been described so far. Earlier authors confused several species, hence the data of previous publications should be interpreted only after careful evaluation.*

Key to species

1 (2) Legs monodactyle.—Notogastral setae weakly yet definitely phylliform (pl. 113 : A). Notogaster greyish yellow, secretion with earthen grains; posteriorly semicircular, convex, without appendages. Prodorsum with small apophyses bearing long lamellar setae. Interlamellar setae short, hardly ciliate. Epimeral setal formula: 3—1—3—3. Length: 864 μm, with: 450 μm.

Locus typicus: Sarek, Sweden. Distribution: Greenland, Sweden, Finland, Germany, Switzerland, Austria, Bulgaria, Soviet Union (NET, CET, FE), Canada

lapponica (Trägårdh, 1910)

2 (1) Legs tridactylous. Notogastral setae simple, ciliate or spiniform.

3 (12) Interlamellar setae long, at least as long as distance between their insertion points; arising often on apophyses (pl. 113 : G).

4 (5) Five pairs of setae originating on margin of notogaster (c_3, d_3, e_2, f_2, pn_3) long, originating on bacilliform chitinous excrescences; end of body with a pair of large branching chitinous appendages bearing setae pn_2 and k_1 (pl. 113 : B).—Setae cited above very long, flagelliform. Lamellar and interlamellar setae also similar and originating on chitinous appendages; only rostral and median notogastral setae shorter, simple. Notogaster usually very squalid, earthen grains fixed also by the setae proper. Epimeral setal formula: 3—1—3—3. Length: 1026 μm, width: 340 μm.

* See Van der Hammen: 1959 : 65.

Locus typicus: Regensburg, Germany. Distribution: Holarctic Region

spinifer (C. L. Koch, 1836)

5 (4) Neither notogastral margin nor end of body with long chitinous appendages.

6 (7) Genital plates (pl. 113: F) with 14 pairs of setae.—Lamellar apophyses of prodorsum short, interlamellar apophyses absent. Ratio of sensillus and prodorsal setae as in pl. 113: E. Notogastral setae short, some arcuate, ciliate or dentate. Body measurements not given in the original description.

Locus typicus: Tsola Tso, Himalaya, Nepal. Distribution: only the type-locality known

polytricha Piffl, 1971

7 (6) Genital plates with merely nine pairs of setae.

8 (9) Interlamellar setae originating on small tubercles instead of chitinous appendages (pl. 113: C), medium long, reaching only to bases of lamellar setae; apophyses of these latter also small, only slightly elevated from prodorsal surface, setae hardly projecting beyond anterior margin of body. Posterior margin of notogaster convex, without chitinous excrescences; setae (h_1, pn_1, pn_2) arising in this region equal in length. Epimeral setal formula: 3—1—3—3. Length: 648—792 μm, width: 360—432 μm.

Locus typicus: Clifton Bridge, Great Britain. Distribution: Ireland, Great Britain, Sweden, Faeroe Islands, Switzerland, Soviet Union (SMA)

invenusta (Michael, 1888)

9 (8) Interlamellar setae originating on chitinous appendages, very long, reaching rostrum. Apophyses of lamellar setae enormous, projecting beyond margin of body. Posterior margin of notogaster concave (pl. 113: D).

10 (11) Notogaster very long, twice longer than wide, with nearly parallel sides, posteriorly with a pair of bifurcating obtuse appendages (pl. 113: D), their inner branches bearing setae k_1; this latter longer than all other notogastral setae. Posterior end of body deeply concave between appendages. Lamellar apophyses with long, flagelliform, heavily ciliate lamellar setae. Epimeral setal formula: 3—1—3—3. Length: 900—1080 μm, width: 440—500 μm.

Locus typicus: without any nearer designation (Regensburg?), Germany. Distribution: Holarctic Region

biurus (C. L. Koch, 1839)

11 (10) Notogaster shorter, at most one and a half times as long as wide, with slightly convex sides, posteriorly with three pairs of very short, tuberculate chitinous appendages bearing setae pn_1, k_1 and pn_3 (pl. 113: G). Posterior end of body slightly concave between k_1. Setae k_1 and pn_3 approximately equal in length. Lamellar apophyses with lamellar setae considerably shorter than in preceding shorter even than apophyses proper. Epimeral setal formula: 3—1—3—3. Length: 900 μm, width: 468 μm.

Locus typicus: Strasbourg, France. Distribution: Europe, Soviet Union (CET, WS)

segnis (Hermann, 1804)

12 (3) Interlamellar setae minute, fine, usually hardly discernible, essentially shorter than half distance between their insertion points; neither chitinous appendages nor tubercles present at their insertion points.

13 (14) Lamellar setae originating on minute chitinous tubercles, short, their tips not reaching rostrum (pl. 114 : B). Posterior margin of notogaster slightly convex, without appendages. Notogaster foveolate. Length: 850 µm, width: 550 µm.
Locus typicus: Klaushavn, West Greenland. Distribution: Greenland, USA (Alaska)

foveolata Hammer, 1955

14 (13) Lamellar setae originating on long chitinous appendages reaching, indeed, considerably projecting — together with lamellar setae—beyond anterior margin of rostrum. Posterior margin of notogaster straight or weakly concave, occasionally with 1—2 pairs of chitinous appendages.

15 (16) Posterior margin of notogaster medially with a flat, trapezoid projection, bearing setae pn_1 on minute appendages of its corners (pl. 114 : A). Posterior margin of trapeze (the line connecting setae pn_1) nearly straight, hardly concave. Epimeral setal formula: 3—1—2—3. Length: 954—1010 µm, width: 468—485 µm.
Locus typicus: Spitsbergen, Norway. Distribution: Greenland, Norway (Jan Mayen Islands, Spitsbergen), Soviet Union (Novaya Zemlya, North Siberia)

borealis (Thorell, 1872)

16 (15) Posterior margin of notogaster medially with two tubiform appendages, their inner margins bearing setae pn_1; line connecting these setae strongly concave.

17 (18) The two tubiform appendages on end of body comparatively large, their terminal width not or hardly smaller than their enclosed distance (pl. 114 : E). Body margin between appendages deeply concave, nearly semicircular. The two appendages far projecting from profile of body margin. Epimeral setal formula: 3—1—2—3. Length: 1080 µm, width: 490 µm.
Locus typicus: presumably Regensburg, Germany. Distribution: Great Britain, Netherlands, Finland, Sweden, Germany, Austria, Switzerland, Hungary, Soviet Union (NET, CET, SET, KZ, WS, ES, FE)

biverrucata (C. L. Koch, 1839)

18 (17) The two tubiform appendages on end of body comparatively small, their enclosed distance at least twice greater than terminal width of appendage (pl. 114 : F); margin between appendages flatly concave; the two appendages hardly projecting from profile of margin. Epimeral setal formula: 3—1—2—3. Length: 873—900 µm, width: 450—460 µm.
Locus typicus: Strasbourg, France. Distribution: Holarctic Region

horrida (Hermann, 1804)

Species interpretabiles

1. biurus C. L. Koch, 1839 : 30, 2 *(Nothrus)*.
   biurus Sellnick et Forsslund: 1955 : 479, fig. 9.
   biurus Van der Hammen: 1959 : 66.
   exuvialis Grandjean, 1939 : 305.
2. biverrucatus C. L. Koch, 1839 : 29, 15, fig. *(Nothrus)*.
   biverrucata Sellnick et Forsslund: 1955 : 482, fig. 10.
   biverrucata Van der Hammen: 1959 : 66.
   horridus Nicolet, 1855: pk. 7, fig. 1 *(Nothrus)*.

3. borealis Thorell, 1872: 697 *(Nothrus)*.
borealis Sellnick et Forsslund: 1955: 434, fig. 11.
4. foveolata Hammer, 1955: 8, fig. 1 *(Camisia)*.
5. horrida Hermann, 1804: 351.
6. invenustus Michael, 1888: 500, figs. 47A: 12, 47: 6 *(Nothrus)*.
invenusta Sellnick et Forsslund: 1955: 488, fig. 13.
7. lapponicus Trägårdh, 1910: 526 *(Nothrus)*.
lapponica Sellnick et Forsslund: 1955: 489, fig. 14.
8. polytricha Piffl, 1971: 23, figs. C1—8.
9. segnis Hermann, 1804: 94, figs. 4: 8,9: X, Y *(Notaspis)*.
segnis Grandjean: 1936: 38, figs. 1—2.
segnis Sellnick et Forsslund: 1955: 491, fig. 15.
segnis Van der Hammen: 1959: 65.
bicarinatus Berlese, 1885: 17: 4 *(Nothrus)*.
10. spinifer C. L. Koch, 1836: 2, 18 *(Nothrus)*.
spinifer Sellnick et Forsslund: 1955: 493, fig. 16.
spinifer Van der Hammen: 1959: 66.

Genus 2: Heminothrus Berlese, 1913

Heminothrus Berlese, 1913: 98.
Heminothrus Sellnick et Forsslund: 1955: 507.
Heminothrus Aoki: 1958: 125.
Heminothrus Van der Hammen: 1959: 69.

Large-sized, mostly flattened, ribbed, heavily pigmented animals. Trichobothrium always present, but sensillus usually setiform or medium long and slightly incrassete. Posterior margins of prodorsum and notogaster often with smaller apophyses bearing setae. Number of genital setae varying between 9—23, all on inner margin of genital plates, delimited by a longitudinal chitinous line; two pairs of aggenital setae on inner margin of genital plates. Legs monodactyle (tridactylous species occur also outside of the Palaearctic Region).

Type-species: *Nothrus Targionii* Berlese, 1885. We have drawn in the former genera *Platynothrus* Berlese, 1913, and *Neonothrus* Forsslund, 1955, and retain them as merely subgeneric taxa.

Key to species

1 (8) Setae pn_1 and k_1 originating on long chitinous appendages (pl. 114: G) (subgenus 1: **Heminothrus** Berlese, 1913)

2 (3) Setae d_1, d_2 and e_1 comparatively long, not much shorter than c_3, d_3 and e_2; apex of d_2 extending beyond insertion point of e_1 (pl. 114: G). Notogaster with two longitudinal chitinous laths.—Number of genital setae uncertain. Length: 700 µm, width: 320 µm.

Locus typicus: "Lago Palu", Sondrio, Italy. Distribution: only the type-locality known.

paolianus Berlese, 1913*

* The species is known only from Berlese's brief original description; it appears to be identifiable on the basis of the strikingly long hairs.

3 (2) Setae d_1, d_2 and e_1 comparatively short, not longer than c_2, d_3 and e_2; apex of d_2 never reaching insertion point of e_1. Notogaster without chitinous ribs (pl. 114: I).

4 (5) Twenty-three pairs of genital setae present (pl. 114: I). Setae d_1, d_2 and e_1 short, widening, marginally ciliate. Distance d_1—d_1 only half as long as distance e_1—e_1. Apices of setae c_3, d_3, e_2 and f_2 not extending beyond insertion points of any subsequently originating setae. Lamellar and interlamellar setae of uniform thickness and ciliation. Sensillus not longer than interlamellar seta. Length: 860—972 μm, width: 505 μm.

Locus typicus: Rua, Padova, Italy. Distribution: Great Britain, Sweden, Finland, Netherlands, Denmark, Germany, Czechoslovakia, Austria, Hungary, Soviet Union (NET, CET, SET, KZ)

targionii (Berlese, 1885)

5 (4) Genital plates with merely nine pairs of setae.

6 (7) Sensillus long, thin, setiform or bacilliform (pl. 115: A).—Smooth interlamellar setae about as long as sensillus, ciliate lamellar ones slightly longer. Notogastral setae in marginal position *(c_3, cp, e_2)* long, thin, occasionally flagellate, each one longer than distance between two neighbouring ones. Prodorsal surface foveolate, notogaster punctate. Length: 675 μm, width: 360 μm.

Locus typicus: Solenhofen, Germany. Distribution: Greenland, Sweden, Norway, Finland, Germany, Soviet Union (NET, WS, ES, FE)

longisetosus Willman, 1925

7 (6) Sensillus distally widened, lanceolate (pl. 115: C—D).—Smooth interlamellar setae of prodorsum considerably longer than sensillus; lamellar setae finely ciliate, arising on very robust chitinous laths. Marginal setae of notogaster (pl. 115: G) slightly incrassate, shorter than in preceding species, finely ciliate or with a narrow marginal velum (pl. 115: E). Configuration and chaetotaxy of anogenital region as in pl. 115: H. Length: 530—55 μm, width: 255 μm.

Locus typicus: Otanomosu-no-taira, Mt. Shiga, Nagano pref., Japan. Distribution: only the type-locality known

minor Aoki, 1969

8 (1) Setae pn_1 and k_1 originating not on a long chitinous appendage each, but based at most on short chitinous pegs, not longer than wide.

9 (10) Notogastral sides nearly parallel; setae partly blade-like (pl. 115: I). Three pairs of setae originating inside longitudinal ribs, d_2 outside (subgenus 2: **Neonothrus** Forsslund, 1955).—Sensillus long, setiform, finely ciliate, lamellar and interlamellar setae similar. Setae on end of body ciliate, setiform, arching towards each another. 12—13 pairs of genital setae present (pl. 115: J). Length: 595—650 μm, width: 245—330 μm.

Locus typicus: not given in the original description, presumably Sweden. Distribution: Sweden, Finland, Soviet Union (NET, CET, SET, FE)

humicola (Forsslund, 1955)

10 (9) Notogaster oval or slightly widening posteriorad. No blade-shaped setae present among notogastral ones (subgenus 3: **Platynothrus** Berlese, 1913).

11 (18) Median zone of notogaster with longitudinal ribs (pl. 116: A).

12 (13) Genital plates with nine pairs of genital setae. Notogastral setae medium long, apically reaching or extending beyond insertion point of next seta, somewhat longer than half length of proterosoma. Outer surface of trochanter of leg III with three setae. Length: 835 m, width: 558 μm.

Locus typicus: no exact locality given, Soviet Union. Distribution: Hungary, Soviet Union **grandjeani** Sitnikova, 1975

13 (12) Genital plates with 11—15 pairs of setae.

14 (15) Eleven pairs of genital setae present (pl. 116: B). Notogastral setae short, d_1 not reaching half distance enclosed by d_1—d_2, and d_2 that of d_2—e_2. Setae pn_1, k_1 and pn_3 short and straight (pl. 116: A). Sensillus short, clavately incrassate. Prodorsal setae also short, interlamellar seta not reaching insertion point of lamellar one. Length: 720—864 μm, width: 400—540 μm.

Locus typicus: Gåskap, Soviet Union. Distribution: northern part of the Holarctic Region: Greenland, Iceland, Sweden, Finland, Soviet Union (Novaya Zemlya, NET, WS, ES, FE), Canada **punctatus** (L. Koch, 1879)

15 (14) 12—15 pairs of genital setae present. Body setae considerably longer, c_1, d_1 and d_2 always extending beyond insertion point of next seta in their longitudinal rows. Setae pn_1, k_1 and pn_3 arcuate.

16 (17) Prodorsal and notogastral setae very long, especially those on end of body flagelliform (pl. 116: C), interlamellar setae projecting beyond anterior margin of prodorsum.—Prodorsal surface with very large foveolae, interstices punctate. Notogaster also with large foveolae. Seta d_2 extending much beyond insertion point of e_2, setae pn_1 essentially longer than their enclosed distance. Setae well discernibly ciliate. 12—14 pairs of genital setae present. Length: 767—820 μm, width: 410—467 μm.

Locus typicus: no exact locality given, Soviet Union. Distribution: Soviet Union (FE), Mongolia **sibiricus** Sitnikova, 1975

17 (16) Body setae medium long, none flagelliform (pl. 116: E), interlamellar ones short, only half as long as distance between their insertion points and anterior margin of prodorsum.—Prodorsal and notogastral surfaces punctate, with only some minor foveolae. Setae d_2 only approaching insertion points of e_2. Setae pn_1 shorter than their enclosed distance. 13—15 pairs of genital setae present. Length: 925—940 μm, width: 540—572 μm.

Locus typicus: Regensburg, Germany. Distribution: Holarctic Region **peltifer** (C. L. Koch, 1839)

18 (11) Notogaster without median longitudinal ribs.

19 (22) Genital plates with 12—13 pairs of setae.

20 (21) Interlamellar seta short, reaching only to insertion point of lamellar one (pl. 117: E); sensillus bacilliform, not or hardly thickening.—Setae in anterior half of notogaster, especially c_2 and d_2, considerably shorter than those originating in posterior half. Length: 954—1000 μm, width: 570—620 μm.

Locus typicus: Norway. Distribution: Eastern Greenland, Iceland, Norway, Sweden, Finland, Germany, Austria, Italy, Hungary, Switzerland, Soviet Union (NET, CET, SET, WS, KZ, ES, FE), Canada **thori** (Berlese, 1904)

21 (20) Interlamellar seta long, projecting far beyond anterior margin of prodorsum (pl. 116: F); sensillus distally heavily incrassate, clavate (pl. 116: G).—Notogastral setae equal in length both anteriorly and posteriorly, neither setae c_2 and d_2 any shorter. Length: 650—690 μm, width: 350—370 μm.

Locus typicus: Sanshiro-Weide, Nagano, Japan. Distribution: Japan, Soviet Union **yamasakii** Aoki, 1958

22 (19) Genital plates with 19—21 pairs genital setae.

23 (26) Interlamellar and notogastral setae very long (pl. 117: A). Interlamellar setae extending beyond insertion points of lamellar ones.

24 (25) Notogaster (pl. 117: A, C) with a characteristic polygonal sculpture, prodorsum with foveolae arranged in very large polygonal fields (pl. 117: B).—Rostral setae arising on prodorsal surface, comparatively long, only slightly shorter than lamellar setae pn_1 of notogaster essentially shorter than pn_2, distance between setae c_1—c_2 about as great as that enclosed by c_2—c_3. Length: 900—1050 μm.

Locus typicus: Granada, Sierra Nevada, Spain. Distribution: only the type-locality known **nevadensis** (Pérez-Iñigo, 1969)

25 (24) Notogater and prodorsum uniformly with a sculpture of foveolae of diverse size and density (pl. 117: G).—Rostral setae arising on prodorsal margins, but removed from each other, short, considerably shorter than lamellar setae. Notogastral setae somewhat longer, but position and proportions of setae resembling those of preceding species. Length: 970 μm.

Locus typicus: Bergamo, Italy. Distribution: only the type-locality known **capillatus** (Berlese, 1914)

26 (23) At least interlamellar setae of all body setae short (pl. 118: A), not even approaching insertion points of lamellar setae. Body surface without large foveolae.

27 (28) All notogastral setae long, thin, reclinate. Distance between setae e_1 not greater than that between setae pn_2, and former also reaching insertion point of latter (pl. 118: A).—Rostrum projecting like an enlarged tubercle, bearing rostral setae close to each other. Prodorsal surface punctate, notogaster with a weak polygonal sculpture still discernible. Distance between setae c_1—c_2 considerably greater than that enclosed by setae c_2—c_3. Length: 1000—1080 μm, width: 680—684 μm.

Locus typicus: exact reference unknown to us; Greenland. Distribution: Greenland, Finland, Czechoslovakia, Soviet Union (WS, KZ) **septentrionalis** (Sellnick, 1944)

28 (27) Notogastral setae considerably shorter (fig. 118: D), some of them, especially posteromarginal ones, in transverse positions. Setae e_1 of notogaster originating at about half distance from that enclosed by setae pn_2, and former not even approaching insertion points of latter.—Rostrum robust, without a projecting swollen part. Prodorsum with small foveolae, notogaster punctate. Sensillus as in pl. 118: C. Length: 949—1010 μm, width: 566—607 μm.

Locus typicus: Numbur, Himalaya, Nepal. Distribution: only the type-locality known **numatai** (Aoki, 1965)

Species interpretabiles

1. capillata Berlese, 1914 : 132, fig. 11 : 25 *(Angelia)*.
   capillatus Aoki: 1965 : 292 *(Platynothrus)*.
   capillatus Mahunka: 1980 : 118, fig. 31.
2. grandjeani Sitnikova, 1975 : 84 *(Platynothrus)*.
3. humicola Forsslund, 1955 : 512, figs. 4—8 *(Neonothrus)* **(comb. nov.)**.
4. longisetosus Willmann, 1925 : 9 *(paolianus* var. *longisetosus)*.
   longisetosus Sellnick et Forsslund: 1955 : 509, figs. 28—29 *(paolianus* Berlese var. *longisetosus)*.
   longisetosus Aoki: 1958 : 121, figs. 1A—M *(paolianus* Berlese var. *longisetosus)*.
5. minor Aoki, 1969 : 121, figs. 7—11.
6. nevadensis Pérez-Iñigo, 1969 : 385, figs. 13—15 *(Platynothrus)*.
   nevadensis Pérez-Iñigo: 1968 : 229 *(Platynothrus)* (nom. nudum).
7. numatai Aoki, 1965 : 290, figs. 5—8 *(Platynothrus)*.
8. paolianus Berlese, 1913 : 99, fig. 8 : 89 *[Nothrus (Heminothrus)]*.
   paolianus Van der Hammen: 1959 : 70.
9. peltifer C. L. Koch, 1839 : 29, 9 *(Nothrus)*.
   peltifer Sellnick et Forsslund: 1955 : 515, figs. 17—18.
   peltifer Van der Hammen: 1959 : 71.
   palliatus C. L. Koch, 1839 : 39, 4 *(Nothrus)*.
   bistriatus Berlese, 1885 : 17, 9 *(Nothrus)*.
10. punctatus L. Koch, 1879 : 114, pl. 3, fig. 25 *(Nothrus)*.
   punctatus Sellnick et Forsslund: 1955 : 517, figs. 19—20 *(Platynothrus)*.
   punctatus Seyd: 1970 : 291*.
   carinata Kramer, 1879 : 81, fig. 3 *(Hermannia)*.
   valentianus Hull, 1916 : 405, fig. 17 : 4.
11. septentrionalis Sellnick, 1944 : 42 *(Platynothrus capillatus* var. *septentrionalis)*.
   septentrionalis Sellnick et Forsslund: 1955 : 519, figs. 21—22 *(Platynothrus capillatus* var. *septentrionalis)*.
   septentrionalis Aoki: 1965 : 292 (Platynothrus).
12. sibiricus Sitnikova, 1975 : 82, fig. 135 *(Platynothrus)*.
13. Targionii Berlese, 1885 : 17, 8 *(Nothrus)*.
   targionii Sellnick et Forsslund: 1955 : 507, figs. 26—27.
   targionii Van der Hammen: 1959 : 69.
14. Thori Berlese, 1904 : 275 *(Angelia)*.
   thori Berlese: 1913 : 99, fig. 7 : 83 *[Nothrus (Heminothrus)]*.
   thori Sellnick et Forsslund: 1955 : 521, figs. 23—25 *(Platynothrus)*.
   thori Van der Hammen: 1959 : 70.
15. yamasakii Aoki, 1958 : 123, figs. 2A—N.

Family 3: Trhypochthoniidae Willmann, 1931

Trhypochthoniinae Willmann, 1931 : 102.
Tryhpochthoniidae Knülle: 1957 : 149.
Trhypochthoniidae Van der Hammen: 1959 : 72.
Tryhpochthoniellidae Knülle, 1957 : 150.

Medium to large-sized, weakly sclerotized, light yellow of brownish yellow, only exceptionally dark animals with a cuticle covered by cerotegumen. Bothridium either developed or wholly reduced. Glandula large, often well visible owing to its darker colour. 15 pairs of notogastral setae. Epimeral setal formula characteristic of all genera: 3—1—(2—3)—2; 5—20 genital setae present, aggenital ones absent. Anal

* With a detailed synonymy.

plates weakly developed, bearing 1 or 2 pairs of anal setae, 2 or 3 pairs of adanal setae observable. Legs mono- or tridactylous.

A comparatively less known heterogeneous group, in need of revision which may result in several changes. Members of the family occur in the whole world; nine genera have been described so far, three of which occur also in the Palaearctic Region.

Key to genera

1 (4) Legs tridactylous. Anal plates with merely one pair of setae.

2 (3) Three pairs of adanal setae present. Trichobothrium always well developed
genus 1: **Trhypochthonius** Berlese, 1904

3 (2) Two pairs of adanal setae present. Trichobothrium developed or wholly reduced (in this latter case, still present in developmental stages and vanished only later, in sexually mature individuals).
genus 2: **Trhypochthoniellus** Willmann, 1928

4 (1) Legs monodactyle. Anal and adanal plates with two pairs of setae each. Trichobothrium reduced
genus 3: **Mucronothrus** Trägårdh, 1931

Genus 1: Trhypochthonius Berlese, 1904*

Trhypochthonius Berlese, 1904 : 27.
Trhypochthonius Van der Hammen: 1956 : 398.
Trhypochthonius Van der Hammen: 1955 : 73.
Trilohmannia Willmann, 1923 : 475.
Tumidalvus Ewing, 1908 : 243.

Prodorsum with well developed trichobothrium, situated always near body margins. Exobothridial setae absent. 6—11 pairs of genital setae present, aggenital setae absent. One pair of anal and three pairs of adanal setae observable. Legs tridactylous. Tibia of leg I with two solenidia, also φ_2 present. Number of setae on genu of legs I—IV : 5 + 1—5 + 1—3 + 1—3 + 0.

Type-species: *Hypochthonius tectorum* Berlese, 1896.

A cosmopolitan genus, with several misinterpreted or confused species. They occur mainly in wet situations, principally in mossy bogs, more seldom in soil mosses and forests.

Key to species

1 (2) Notogastral simple, smooth, setiform, erect, apically pointed. Setae on end of body not always longer than the others. Body elongated, twice longer than its greatest width. End of body considerably darker than its middle section. 6—7 pairs of genital setae present. Length: 530—550 μm, width: 250—275 μm.

* Schweitzer (1956) described three new species as *Trhypochthonius*, however, they are nymphs of higher Oribatids. See Van der Hammen: 1959 : 14.

Locus typicus: "Umgegend Bremens", Germany (several localities without locus typicus). Distribution: Europe, Soviet Union (NET, LET, SET)
cladonicola (Willmann, 1919)

2 (1) Notogastral setae bacilliform, thickened, generally ciliate, distally often widening. Body not twice as long as wide, end of body not strikingly dark. 7—11 pairs of genital setae present.

3 (4) Notogastral setae slightly arcuate, short, subulate. Entire body dark. Length: 525—590 µm, width: 300 µm.
Locus typicus: Büchel, near Wasserburg, Lake Boden, Germany. Distribution: Germany, Czechoslovakia
nigricans Willmann, 1928

4 (3) Notogastral setae rigid, erect, often apically incrassate; centrodorsal setae shorter than posterior and marginal ones. Body light.

5 (6) Posterior end of notogaster widely rounded, without posterior boss (pl. 118 : E). Notogastral setae in general apically incrassate, there more densely ciliate, all notogastral setae ciliate.—Rostral setae shortest, rest of prodorsal setae, lamellar and interlamellar ones without any essential difference in length. Sensillus clavate. Length: 660—775 µm, width: 39—466 µm.
Locus typicus: Florence, Italy (topotypes by Van der Hammen). Distribution: Palaearctic Region
tectorum (Berlese, 1896)

6 (5) Notogastral setae (pl. 118 : J) not incrassate apicad, pointed, ps_2 and ps_3 ciliate (pl. 118 : H), notogaster with posterior boss (pl. 118 : G).—Prodorsal setae considerably longer, especially interlamellar one: projecting beyond rostral apex. Rostral and lamellar setae nearly equal in length. Sensillus (pl. 118 : I) clavate. Plates and chaetotaxy of anogenital region as in pl. 118 : F. Length: 640—690 µm, width: 375—390 µm.
Locus typicus: Mt. Oboshi, Mineson, Japan. Distribution: Japan
japonicus Aoki, 1970

+ americanus Ewing
silvestris Jacot

Species interpretabiles

1. cladonicola Willmann, 1919: 553, fig. 4 *(Camisia)*.
   cladonicola Van der Hammen: 1952: 23, fig. 4d.
   cladonicola Van der Hammen: 1956: 403.
2. japonicus Aoki, 1970: 404, figs. 18—29.
3. nigricans Willmann, 1928: 4, figs. 4—5.
4. tectorum Berlese, 1896: 78, 8 *(Hypochthonius)*.
   tectorum Van der Hammen: 1956: 402.
   tectorum Van der Hammen: 1959: 73.

Species inquirendae

— arnoldii Bashkirova, 1958: 206, figs. 3: a—v.
— congregator Grandjean, 1940: 66, figs. 2D—E *(tectorum* var. *congregator)*.
  congregator Van der Hammen: 1956: 402.
— monodactylus Mihelcic, 1947: 45, fig. 1
— sclerotricha Willmann, 1923: 475, pl. 7, figs. 5—7 (*cladinicola* var.).

Genus 2: Trhypochthoniellus Willmann, 1928

Trhypochthoniellus Willmann, 1928 : 3 *[(Trhypochthonius (Trhypochthoniellus)].*
Trhypochthoniellus Grandjean: 1939 : 306.
Trhypochthoniellus Van der Hammen: 1959 : 73.

Light yellow to yellowish brown animals. Trichobothrium developed or wholly absent. Rostrum anteriorly rounded or pointed. Exobothridial seta present. 16—18 pairs of genital setae observable, aggenital ones absent; one pair of anal and two pairs of adanal setae present. Epimeral setal formula: 3—1—3—2. Tibia of leg I with one solenidium only, φ_2 absent. Setal formula of genu of legs I—IV : 3 + 1—3 + 1—2 + 1—2 + 0. Legs tridactylous.

Type-species: *Trhypochthonius (Trhypochthoniellus) setosus* Willmann, 1928.

A Holarctic genus; the status of some species is questionable, the species of the genera *Trhypochthonius* and *Trhypochthoniellus* may easily be confused. About 10 species are known; all inhabit very wet biotopes, mainly *Sphagnum* bogs.

Key to species

1 (2) Trichobothrium absent. Rostral seta short, interlamellar and lamellar ones very long, lamellar seta extending by half its own length beyond rostrum (pl. 119 : A). A short and simple seta originating in position of trichobothrium. Notogastral setae medium long, thin. Length: 500—541 μm, width: 270—300 μm.

Locus typicus: Riedschachen, Schussenried. Württemberg, Germany. Distribution: Europe, Soviet Union (NET), North America

setosus Willmann, 1928

2 (1) Trichobothrium well developed, a regular sensillus present.

3 (4) Interlamellar setae minute, considerably shorter than sensillus.—Sensillus distally aciculate. Body setae long. Length: 500 μm, width: 280 μm.

Locus typicus: "Greenhouse", Florence, Italy. Distribution: the single known locality of occurrence speaks for an accidental introduction on some ornamental, probably tropical, plant

longisetus (Berlese, 1904)

4 (3) Interlamellar setae very long, essentially longer than sensillus; this latter distally smooth

5 (6) Sensillus long, stalk sigmoid, distally fusiform, considerably longer than seta c_1. Posterior part of body with three pairs of setae much longer than all other notogastral ones. Anterior part of notogaster with some impressed lines. Body chestnut brown. Length: 550 μm, width: 310 μm.

Locus typicus: "Cascine", Florence, Italy. Distribution: Italy, Poland

badius (Berlese, 1905)

6 (5) Sensillus comparatively short, shorter than c_1. Notogaster posteriorly without three pairs of setae much longer than the rest.

7 (8) Notogastral setae short; c_1 apically not reaching insertion point of d_1 (pl. 119 : B). Length: 530 μm, width: 320 μm.

Locus typicus: Neuendamm bei Osterholz, Germany. Distribution: Europe, North and South America, New Zealand

excavatus (Willmann, 1919)

8 (7) Notogastral setae longer, apex of c_1 extending beyond insertion point of d_1. Length: 555 μm, width: 300 μmm.

Locus typicus: Wiederholz, Basel, Switzerland. Distribution: Switzerland

trichosus (Schweizer, 1922)

Species interpretabiles

1. badius Berlese, 1905 : 237 *(Trhypochthonius)*.
   badius Berlese: 1913 : 155, pl. 2, fig. 12 *(Trhypochthonius)*.
   badius Van der Hammen: 1959 : 74.
2. excavata Willmann, 1919 : 553, fig. 3 *(Camisia)*.
   excavatus Grandjean: 1939 : 306.
   excavatus Van der Hammen: 1952 : 24.
3. longisetus Berlese, 1904 : 27, pl. 2, fig. 44 *(Trhypochthonius)*.
4. setosus Willmann, 1928 : 3, fig. 3 *[Trhypochthonius (Trhypochthoniellus)]*.
   setosus Grandjean: 1939 : 306.
5. trichosus Schweizer, 1922: 75, fig. 32 *(Tripochthonius)*.

Genus 3: Mucronothrus Trägårdh, 1931

Mucronothrus Trägårdh, 1931 : 31.
Mucronothrus Knülle; 1957 : 167.

Large-sized, slender animals. Rostrum with a characteristic nasiform projection, bearing rostral setae near each other. Bothridium almost completely reduced, represented by a simple bristle. Exobothridial setae developed. Epimeral setal formula: 3—1—2—2. 18—20 pairs of genital setae present, aggenital ones absent, two pairs of anal and two pairs of adanal setae observable. Legs with a single strong claw.

Type-species: *Mucronothrus rostratus* Trägårdh, 1931 = *M. nasalis* (Willmann, 1929).

Probably a consmopolitan genus, its single known species has been recorded from very many localities.

— Prodorsum with a transverse lath between lamellar setae. Lamellar and interlamellar setae long and fine. Notogaster nearly parallel-sided, its setae with great differences in length (pl. 119 : D): c_3, e_2, h_1, h_2 and ps_2 long, all others, especially c_1, c_2 and d_1, very short. Length: 630—800 μm, width: 341 μm.

Locus typicus: Herdla Island, Bergen, Norway. Distribution: Palaearctic Region, Argentina, Bolivia, Peru, New Zealand

nasalis (Willmann, 1929)

Species interpretabiles

1. nasalis Willmann, 1929 : 4, figs. 1—3 *(Malaconothrus (?))*.
   nasalis Hammer: 1965 : 3, fig. 3.
   rostratus Trägårdh, 1931 : 31, figs. 60—62.

Family 4: Malaconothridae Berlese, 1916

Malaconothridae Berlese, 1916: 175.
Malaconothridae Knülle: 1957: 154.
Malaconothridae Van der Hammen: 1959: 75.

Medium to small-sized animals; in general weakly pigmented, yellowish white, dirty white or fuscous. Body cylindrical or oval. Trichobothrium reduced, interlamellar and well developed exobothridial setae originating near each other. Notogaster with 15 pairs of setae. Epimeral setal formula: 3—1—(2—3)—3. Number of setae on genital plates varying (4—12), aggenital setae absent, one pair of anal and three pairs of adanal setae present. Legs mono- or tridactylous.

The species are distributed all over the world. Three genera are so far known; two occur in the Palaearctic Region.

Key to genera

1 (2) Tarsus short, bearing one claw on every leg. Body cylindrical, notogaster largely parallel-sided
genus 1: **Malaconothrus** Berlese, 1901

2 (1) Tarsus elongate, bearing three (exceptionally two) claws on every leg. Notogaster in general widening posteriorad
genus 2: **Trimalaconothrus** Berlese, 1916

Genus 1: Malaconothrus Berlese, 1904

Malaconothrus Berlese, 1904: 24.
Malaconothrus Knülle: 1957: 155.
Malaconothrus Van der Hammen: 1959: 75.

Small-sized, yellowisch to greyish animals. Body cylindrical, anteriorly and posteriorly obtuse, sides parallel. Epimeral setal formula: 3—1—3—3. Genital and ano-adanal plates wide. Genital plate with 4—6 pairs of setae, anal plate bearing one pair (exceptionally absent), adanal plate with three pairs of setae. Legs short, thick, especially tarsi stout.

Fastigial seta *(ft")* on ends of tarsi 1—3 modified into a spine. Solenidium on tarsus of palp clavate. Every leg monodactyle.

Type-species: *Lohmannia (Malaconothrus) egregia* Berlese, 1904.

Species belonging in the genus are known from every continent and several have been described also from the Palaearctic Region; some of these are, however, doubtful* and the group thus needs further investigations and a thorough revision. The known species live in very wet habitats.

* See Knülle 1957: 157.

1 (2) Setae d_1 considerably nearer to each other than c_1 or e_1 (pl. 119 : E, G). Body margin between legs I and II terminally projecting, acute-angled (pl. 119 : F).—Setae c_1 and d_1 equal in length, essentially longer than c_2 and especially c_3. Sculpture of prodorsal and notogastral cerotegumen as in pl. 119 : H. Lateral lines of body considerably convergent posteriorad. Length: 375—440 μm.

Locus typicus: Rockanje, Voorne's Duin, Netherlands. Distribution: Netherlands

processus Van der Hammen, 1952

2 (1) Setae d_1 at, or nearly at, same distance from each other as setae c_1 and e_1.

3 (6) Notogastral setae long, e_1 nearly as long as distance between e_1 nearly as long as distance between e_1 and h_1, and as long as, or longer than, that between setae c_1 and e_1.

4 (5) Dorsal setae long, interlamellar ones longer than distance between their insertion points (pl. 120 : A), d_1 apically extending beyond insertion of e_1.—Projection of body margin (pl. 120 : B) between legs I and II obtuse-angled. Configuration and chaetom of ventral side as in pl. 120 : D Length: 415—429 μm, width: 188 μm.

Locus typicus: De Lutte Bavel, Netherlands. Distribution: Europe, Soviet Union (NET, CAT), USA

gracilis Van der Hammen, 1952*

5 (4) Dorsal setae shorter, interlamellar ones considerably shorter than distance between them (pl. 120 : F), d_1 reaching at most to insertion point of e_1.—Projection of body margin between legs I and II obtuse, but not rounded. Length: 410—450 μm width: 210—230 μm.

Locus typicus: Yellow Knife, Canada. Distribution: Europe, North and South America, Soviet Union (CET)

mollisetosus Hammer, 1952

6 (3) Notogastral setae comparatively short, e_1 at most as long as distance enclosed by e_1—h_1, and shorter than distance of e_1—e_2.

7 (8) Setae d_2 of notogaster in a marginal position, more or less aligned with setae cp—e_2 (pl. 120 : E).—Notogaster with a sinuous rib and posteriorly two obtuse tubercles. Projection of body margin between legs I and II obtuse-angled. Length: 420 μm.

Locus typicus: Ram, Agaba, Jordan. Distribution: only the type-locality known

ramensis Hammer, 1966

8 (7) Notogaster with setae d_2 originating far from body margin, at equal distances from e_1 and e_2.

9 (10) Distance between setae c_1 and $_2$ on notogaster agreeing with that between c_2 and c_3, setae c_1, d_1 and e_1 short, equal in length (pl. 121 : A).—Prodorsal setae comparatively longer than notogastral ones, especially lamellar and interlamellar pairs. Projection of body margin between legs I and II very long,

* See Knülle 1957 : 156.

acute-angled. Seta ps_1 recurving to notogastral surface. Configuration and chaetom of ventral side as in pl 121 : C. Length: 500—530 µm, width: 245—280 µm.

Locus typicus: Mt. Soma, near Lake Haruna, Gumma, Japan. Distribution: Japan

japonicus Aoki, 1966

10 (9) Distance between setae c_1 and c_2 on notogaster considerably greater than that between c_2 and c_3. Setae c_1, d_1 and e_1 of divers lengths.

11 (12) Setae d_1 long, longer than distance enclosed by c_1—d_1 (pl. 121 : E), also all other notogastral setae comparatively long, hence e_1 approaching insertion point of e_2.—Interlamellar setae of prodorsum originating considerably nearer to each other than distance between rostral and lamellar setae. Projection of body margin between legs I and II rectangular. Chaetom of anogenital region as in pl. 121 : D. Length 405—450 µm, width: 160—210 µm.

Locus typicus: "Stazione di Entomologia Agraria", Florence, Italy. Distribution: Italy

egregius Berlese, 1904

12 (11) Seta d_1 short, essentially shorter than distance enclosed by c_1—d_1; also all other notogastral setae shorter.

13 (14) Lamellar setae originating about same distance from each as from rostral setae (pl. 121 : H). Marginal setae of notogaster obliquely reclinate. — Lamellar setae longer than interlamellar ones. c_1 and c_2 of notogaster considerably shorter than c_3. Prodorsum with denser (pl. 121 : F), notogaster with sparser (pl. 121 : G)), foveolae. Length: 323—378 µm, width: 186—200 µm.

Locus typicus: Mt. Shiga, Otanomosu-taira, Nagano, Japan. Distribution: only the type-locality known

pygmaeus Aoki, 1969

14 (13) Lamellar setae originating much nearer to each other than from rostral setae. Marginal setae of notogaster exclinate to proclinate (pl. 122 : A).—Lamellar setae shorter than interlamellar ones. Sculpture of prodorsum and notogaster as in pl. 122 : A. Length: 330—405 µm.

Locus typicus: "pool between Spier and Wijster", Spier, Netherlands. Distribution: West and Central Europe, Soviet Union (NET)

punctulatus Van der Hammen, 1952

Species interpretabiles

1. egregia Berlese, 1904: 24, pl. 2, fig. 38 *[(Lohomannia (Malaconothrus)]* (lapsus calami).
   egregius Van der Hammen: 1959 : 76.
   egregius Pérez-Iñigo: 1969 : 230, figs. 43—44.
   pulcher Mihelcic, 1957 : 46, fig. 2 (Pérez-Iñigo: 1969 : 231).
2. gracilis Van der Hammen, 1952 : 27, figs. 2c, d, g, k, l.
   gracilis Knülle: 1957 : 156, figs. 5—8, 28—29.
3. japonicus Aoki, 1966: 169, figs. 1—6.
4. mollisetosus Hammer, 1952 : 22, fig. 19.
   mollisetosus Knülle: 1957 : 158.
5. processus Van der Hammen, 1952 : 25, figs. 2a, b, e, h, i.

6. punctulatus Van der Hammen, 1952: 27, figs. 2f, j, m, n.
7. pygmaeus Aoki, 1969: 123, figs. 12—14.
8. ramensis Hammer, 1966: 272, figs. 1—2.
ramensis Hammer, 1975: 189, fig. 1.

Species inquirendae

— globiger Trägårdh, 1910: 537, figs. 327—333a.
globiger Knülle: 1957: 158.
— monodactylus Michael, 1888: 528, pl. 45, fig. 10—14 *(Nothrus)*.
monodactylus Knülle: 1957: 158.

Genus 2: Trimalaconothrus Berlese, 1916

Trimalaconothrus Berlese, 1916: 336 *[(Malaconothrus (Trimalaconothrus)]*.
Trimalaconothrus Knülle: 1957: 159.
Trimalaconothrus Van der Hammen: 1959: 76.
Tyrphonothrus Knülle, 1957: 160 *[(Trimalaconothrus (Tyrphonothrus)]*.

Medium-sized, yellowish brown to fuscous animals. Body widening posteriorad. Epimeral setal formula: 3—1—(2—3)—3. Genital plates with 4—12 pairs of setae, anal plates with one pair, adanal plates with three pairs, of setae. Tarsi longer, thinner, fastigial hair *(Fb)* thin, setiform. Legs tridactylous. Solenidium on tarsus of palp setiform.

Type-species: *Malaconothrus (Trimalaconothrus) indusiatus* Berlese, 1916.

Representatives of the family occur in every zoogeographical regions, many are known also from the Palaearctic Region. There are several incompletely described species, in need of revision or clarification.

Key to species

1 (10) Genital plates with 7—12 pairs of genital setae.

2 (3) Setae d_1 originating at equal distances from c_1 and e_1. Notogaster widening posteriorad (pl. 122: D), its sides nearly straight, posterior margin slightly concave medially; its surface ornamented with evenly spaced minute foveolae. Epimeral setal formula: 3—1—2—3. Length: 417—570 μm, width: 240—300 μm.

Locus typicus: Sager Meer, Oldenburg, Germany. Distribution: only the type-locality known

foveolatus Willmann, 1931

3 (2) Setae d_1 originating considerably nearer to c_1 than to e_1.

4 (5) Notogaster with a double sculpture: finely punctate between foveolar interspaces (pl. 122: B). Distance between foveolae hardly greater than diameter of foveolae. Margins of notogaster convely arcuate, body widest between setae e_2 and f_2. Chaetom of ventral side as in pl. 122: C. Length: 539—613 μm, width: 294—343 μm.

Locus typicus: Neulinum, Kr. Kulm, Danzig (near Gdansk, Poland). Distribution: Holarctic Region, ? Chile

novus (Sellnick, 1921)

5 (4) Notogaster merely finely punctate, without an additional foveolate pattern.

6 (7) Rostral and lamellar setae reclinate.—Interlamellar setae relatively short, hardly longer than half length of prodorsum and not reaching bases of lamellar setae (pl. 122: E). Integument and cerotegument equally finely punctate. Notogastral setae very fine and thin, some very long; setae ps_1 not shorter than h_1. Measurements unknown.

Locus typicus: Tsola Tso, Himalaya, Nepal. Distribution: Nepal

altissimus Piffl, 1971

7 (6) Lamellar and rostral setae proclinate.

8 (9) Epimeres 3 emitting only two pairs of setae. Notogaster with seta h_2 more than thrice longer than ps_1 (pl. 123: A). Prodorsal setae, especially interlamellar ones, extending far beyond anterior margin of prodorsum. Extremely great differences in length between setae originating in anterior third and posterior of notogaster. Length: 564—686 µm, width: 319—392 µm.

Locus typicus: Epping Forest, England. Distribution: England, Germany

glaber (Michael, 1888)

9 (8) Epimeres 3 with three pairs of setae (pl. 123: D). Notogastral seta h_2 at most one and a half times to twice longer than ps_1 (pl. 123: B).—Prodorsal setae considerably shorter, even longest interlamellar setae not reaching rostral apex. Differences in length also smaller between notogastral setae. Length: 635—672 µm, width: 345—373 µm.

Locus typicus: "Dainichi-ana Cave", Shimokawazawa, Shizuoka, Japan. Distribution: only the type-locality known

nipponicus Yamamoto et Aoki, 1971

10 (1) Genital plates with merely 4—6 pairs of setae.

11 (18) Six pairs of genital setae present.

12 (13) Notogastral setae basally abruptly incrassate, heavily ciliate (pl. 123: E), distally abruptly acuate, smooth.—Rostral seta smooth, exobothridial one phylliform. Six pairs of genital setae equal in length (pl. 123: F), smooth, adanal setae ciliate. Length: 510—560 µm, width: 260—290 µm.

Locus typicus: Hatakuku, Hakone, Kanagawa, Japan. Distribution: only the type-locality known

barbatus Yamamoto, 1977

13 (12) Notogastral setae evenly tapering apicad.

14 (15) A small and slender species: body approximately two and a half times longer than wide, notogastral sides nearly parallel. (pl. 124: A).—Four pairs of notogastral setae (e_2, h_1, h_2, ps_2) flagelliform and essentially longer than the others, seta e_2 about four times longer than f_2. Notogaster with a thick, sinous chitinous lath on both sides in superior view. Anal setae very long (pl. 124: B). Length: 392 µm, width: 172 µm.

Locus typicus: Osenberge, between Oldenburg and Sandkrug, GFR. Distribution: GFR

sculptus Knülle, 1957

15 (14) Larger, wider species. Body at most twice longer than wide, margins of notogaster rather divergent posteriorad.

16 (17) Setae c_2, c_3 and cp approximately equal in length (pl. 124: D).—Every body seta heavily ciliate. All prodorsal setae long, lamellar and rostral ones originating near rostrum. Notogaster with longitudinal and in posterior third also with transverse ribs. No great difference in length between setae, ps_2 slightly shorter than h_2. Ventral side as in pl. 124: E. Anal setae minute. Length: 563—588 μm, width: 270—588 μm.

Locus typicus: Brugnes, Barcelona, Spain. Distribution: Spain, Soviet Union (CET)

saxosus Knülle, 1957

17 (16) Setae c_2 and c_3 only half as long as cp (pl. 124: C). Body setae only partly or very finely ciliate. Prodorsal setae longer, through interlamellar one longer, not reaching beyond rostral apex. Notogaster without discernible ribs. Essential differences in length between notogastral setae: ps_2 considerably longer than h_2. Length: 640—650 μm, width: 370—375 μm.

Locus typicus: "Bosco della Sila", Cosenza, Italy. Distribution: only the type-locality known

indusiatus Berlese, 1916

18 (11) Genital plates with merely 4 or 5 pairs of setae.

19 (20) Notogaster (pl. 125: A) with a polygonal ornamentation only. Four pairs of genital setae present (pl. 125: B).—Body setae wholly smooth. Notogastral setae c_1 and c_3 equal in length, c_2 shorter, h_1 essentially shorter than ps_1 (actually shortest dorsal hair). Length: 480—483 μm, width: 235—240 μm.

Locus typicus: Hatajuku, Hakone, Kanagawa, Japan. Distribution: only the type-locality known

reticulatus Yamamoto, 1977

20 (19) Notogaster without polygonal sculpture. Genital plates with five pairs of setae present.

21 (24) Seta d_1 long, nearly as long as distance between d_1 and cp. Anal setae reduced.

22 (23) Last (fifth) pair of genital setae originating anterior to half length of genital plate (pl. 125: F). Seta h_1 short, hardly extending beyond notogastral margin.—Every body seta straight, relatively short. Lamellar setae originating far from interlamellar ones (pl. 125: E). Length: 350—445 μm, width: 175—220 μm.

Locus typicus: Hatajuku, Hakone, Kanahawa, Japan. Distribution: only the type-locality known

hakonensis Yamamoto, 1977

23 (22) Last (fifth) pair of genital setae originating considerably beyond half length of genital plate. Seta h_1 long, extending far beyond posterior margin of notogaster (pl. 125: C).—Notogastral setae in general short, d_1 about as long as distance enclosed by c_1—d_1. Notogaster narrow, elliptical. Genital setae extremely long, g_3 reaching insertion point of g_1 (pl. 125: D). Genital and anal plates approximately equal in length. A medium-sized species. Length: 482 μm, width: 228 μm.

Locus typicus: Land's End, Cornwall, Great Britain. Distribution: Europe, Soviet Union (CET)

tardus (Michael, 1888)

24 (21) Seta d_1 short, shorter than half distance between d_1 and *cp*.

25 (26) Notogastral sides parallel, body narrow (pl. 126 : A). Distance between setae c_1 great, the setae originating considerably farther from each other than distance enclosed by c_1—c_2. Small animals. Length: 327—343 μm, width: 140—147 μm.

Locus typicus: exact reference unknown to us; Germany. Distribution: only the type-locality known **vietsi** Willmann, 1925

26 (25) Lateral sides of body convex, wide species above 500 μm in length.

27 (28) Notogastral setae without any great differences in length (pl. 126 : B), e_2 at most twice longer than e_1; all setae finely ciliate. Length: 547—592 μm, width: 285—307 μm.

Locus typicus: Maladsko, Strandza Planina, Bulgaria. Distribution: only the type-locality known **buresi** Kunst, 1959

28 (27) Notogastral setae with great diversity in length, e_2 twice to four times longer than e_1; setae smooth (pl. 126 : D).—Length: 625 μm, width: 330—470 μm.

Locus typicus: Rockanje, Voornes Anin, Netherlands. Distribution: only the type-locality known **grandis** Van der Hammen, 1952

Species interpretabiles

1. altissimus Piffl, 1971 : 26, figs. T1—T15.
2. barbatus Yamamoto, 1977 : 38, figs. 4—5 : 7—11.
4. buresi Kunst, 1959 : 54, figs. 2a—b.
4. foveolatus Willmann, 1931 : 106, fig. 45.
   foveolatus Knülle: 1957 : 162, fig. 33.
5. glaber Michael, 1888 : 524 *(Nothrus)*.
   glaber Knülle: 1957 : 161, fig. 32.
6. grandis Van der Hammen, 1952 : 28, figs. 3a—b.
7. hakonensis Yamamoto, 1977 : 40, figs. 6 : 12—14.
8. indusiatus Berlese, 1916 : 336 *[(Malaconothrus (Trimalaconothrus)]*.
   indusiatus Van der Hammen: 1959 : 76.
   indusiatus Mahunka: 1980 : 118, fig. 32.
9. nipponicus Yamamoto et Aoki, 1971 : 579, figs. 1—9.
10. novus Sellnick, 1921 : 75, figs. 3—4 *(Malaconothrus)*.
    novus Van der Hammen: 1952 : 28.
    novus Knülle: 1957 : 160, figs. 1—4, 9—13, 18, 24—25, 30—31.
    monodactylus Oudemans, 1925 : 409 *(Camisia)* (see Van der Hammen: 1952 : 28).
11. reticulatus Yamamoto, 1977 : 36, figs. 2—3 : 2—6.
12. saxosus Knülle, 1957 : 164, figs. 36—37.
13. sculptus Knülle, 1957 : 163, figs. 34—35.
14. tardus Michael, 1888 : 526, p. 47, fig. 14 *(Nothrus)*.
    tardus Knülle: 1957 : 165, figs. 40—41.
    tardus Van der Hammen: 1959 : 77.
15. vietsi Willmann, 1925 : 8, figs. 3a—c *[(Malaconothrus (Trimalaconothrus)]*.
    vietsi Knülle: 1957 : 165, figs. 38—39.

Species inquirendae

— angulatus Willmann, 1931 : 107, figs. 46—46a.
— optatus Berlese, 1908 : 12 *(Malaconothrus)*.
  optatus Berlese: 1910 : 223, pl. 19, fig. 46 *(Malaconothrus)*.
  optatus Van der Hammen: 1959 : 77.

— intermedius Cooremann, 1941 : 6, fig. 1.
— sphagnicola Trägårdh, 1910 : 540, figs. 334—341 *(Malaconothrus)*.
sphagnicola Knülle: 1957 : 166.

Superfamily 2: Nanhermannioidea Sellnick, 1928

Nanhermanniidae Sellnick, 1928 : 17.
Nanhermannidae Woolley et Higgins: 1958 : 913.
Nanhermannioidea Balogh: 1972 : 144.

Elongate, nearly elliptical animals; body heavily sclerotized, with rough and always well discernible sculpture of areolae. Posterior margin of prodorsum usually with teeth or appendages directed towards end of body; notogaster often displaying tubercles and excrescences. 15 pairs of notogastral setae present, setae f_1 absent. Epimeral setal formula varying, characteristic of genera. Diagastria recognizable. Nine pairs of genital, two pairs of aggenital, two pairs of anal and three pairs of adanal, setae observable. Tarsi monodactyle.

A single family known.

Family 1: Nanhermanniidae Sellnick, 1928

Nanhermannidae Sellnick, 1928 : 17.

Characterizable by the features given for the superfamily.

Four genera of the family are known so far, and every one of them also occur in the Palaearctic Region.

Key to genera

1 (4) Notogastral setae simple, setiform. Epimeral setal formula: 3—1—3—3.

2 (3) Notogastral profile even, smooth, without tubercles or excrescences. Notogastral setae long, their apices extending beyond insertion points of next pair posteriorad
genus 1: **Nanhermannia** Berlese, 1913

3 (2) Notogastral profile uneven owing to prescence of tubercles and identations. Notogastral setae setae short, considerably shorter than distance between any two neighbouring pairs of setae arranged in longitudinal rows
genus 2: **Cyrthermannia** Balogh, 1958

4 (1) Notogastral setae T-shaped branching, or widended phylliform. Epimeral setal formula: 3—1—3—4 or 3—1—5—5.

5 (6) Notogastral setae T-shaped, branches distally flagellate. Posterior elevation considerably wider than long. Epimeral setal formula: 3—1—3—4.
genus 3: **Masthermannia** Berlese, 1913

6 (5) Notogastral setae simple phylliform. Elevation at end of body extremely long, about as long as wide. Epimeral setal formula: 3—1—5—5.
genus 4: **Cosmohermannia** Aoki et Yoshida, 1970

Genus 1: NANHERMANNIA Berlese, 1913

Nanhermannia Berlese, 1913: 100.
Nanhermannia Van der Hammen: 1959: 78.
Nanhermannia Strenzke: 1953: 69.
Nanhermannia Woolley et Higgins: 1958: 915.

Notogastral profile even, without tubercles or excrescences in superior view. Prodorsum medially usually with a definite area of sculpture differring from the neighbouring surface. Prodorsal condyles large. Notogastral setae long, setiform or acicular, any one of setae arranged in longitudinal rows projecting far beyond insertion point of next one posteriorad. Epimeral setal formula: 3—1—3—3. Nine pairs of genital, two pairs of aggenital, two pairs of anal and three pairs of adanal, setae observable.

Type-species: *Nothrus nanus* Nicolet, 1855.

A very large genus in need of revision; representatives occur in every zoogeographical region; 11 species are known from the Palaearctic Region.

Key to species

1 (4) Prodorsal condyles considerably narrowing posteriorad, approximately triangular, usually with one rarely with two, apices.

2 (3) Central area of prodorsum clearly defined, widening towards rostrum (pl. 127: A).—Lamellar setae originating near rostrum and reaching anterior margin of body. Prodorsal surface finely punctate laterally; condiles (pl. 127: B) abruptly tapering, their bases unconnected by a transverse band. No considerable differences in size among areolae on notogastral surface (pl. 127: C). Length: 575 μm, width: 250 μm.

Locus typicus: Bois de la Breche, France. Distribution: Palaearctic Region **nana** (Nicolet, 1855)*

3 (2) Central area of prodorsum ill defined, borders obsolete, area not widening towards rostrum (pl. 127: E). Lamellar setae originating far from rostrum, by far not approaching anterior margin of body. Prodorsum laterally with a sculpture of rough, large foveolae. Prodorsal appendages wider, gradually tapering apicad, often with obtuse or double apices basally connected by a transverse band. Notogaster with also very large foveolae (pl. 127: D), generally larger than preceding species. Length: 535 μm, width: 216 μm.

Locus typicus: Tiarno (Trentino), Italy. Distribution: Europe, Soviet Union (NET), North America

elegantula Berlese, 1913

4 (1) Prodorsal condyles either wide and quadratic or narrow, lath- or crista-like, in both cases with several teeth directed towards notogaster (pl. 127: H).

5 (14) Prodorsal condyles lath-like, narrow, hardly or not projecting above notogaster (pl. 127: H).

* Incorrectly known for a long time; e.g. confused with *elegantula* Berlese, 1913; most Hungarian data given for *elegantula*, for instance, refer to this species.

6 (7) Prodorsal condyles fused with contiguous transverse laths (pl. 128 : G). Central field not sharply defined, with a double sculpture of dots and foveolae, lateral margins of prodorsum without any sculpture. Sensillus comparatively short, slightly widened (clavate), essentially shorter than interlamellar setae. Foveolae of notogastral surface small, irregular, often angular. Length: 508—541 μm, width: 241—250 μm.

Locus typicus: Gorno Altaysk (?), Soviet Union. Distribution: Soviet Union (WS, FE) **gorodkovi** Sitnikova, 1975

7 (6) Prodorsal condyles separate, if connected by a weak transverse band, a well discernible hollow still discernible between the two condyles.

8 (11) Prodorsal surface finely punctate or irregularly rugulose, never with round foveolae.

9 (10) Lamellar setae originating far from rostrum, not reaching anterior margin of body (pl. 127 : I). Prodorsal condyles (pl. 127 : H) with merely 3—4 wide, flatly rounded teeth (pl. 127 : H).—Notogastral setae comparatively short, c_1 extending only by its 1/5 length beyond insertion of d_1. Prodorsum laterally rugulose, notogaster with comparatively minute foveolae (pl. 127 : G). Length: 496—610 μm, width: 225—270 μm.

Locus typicus: Külbäcksliden, Västerbotten, Degerfors, Sweden. Distribution: Sweden, Soviet Union **sellnicki** Forsslund, 1958

10 (9) Lamellar setae originating near rostrum, projecting beyond anterior margin of body, like also rostral setae (pl. 128 : A). Prodorsal condyles with 6—7 small, round spots. Prodorsal surface punctate only in central field, laterally smooth. Notogastral setae considerably longer than in preceding species, c_1 extending by about 1/2 its length beyond insertion point of d_1. Length: 566—608 μm, width: 274—308 μm.

Locus typicus: Samarkand (?), Soviet Union. Distribution: Soviet Union (SMA, FE) **laevis** Sitnikova, 1975

11 (8) Prodorsal surface with a double sculpture: finely punctate surfaces displaying also large foveolae (pl. 128 : B). Prodorsal condyles with better differentiated teeth.

12 (13) Central field of prodorsum extremely large, reaching lateral margins (pl. 128 : C). Lamellar appendages with 7—8 approximately similar teeth. Notogaster with slightly angular foveolae of similar size. Genu and tibia of legs I and II with large, bifurcating setae (pl. 128 : B). Length: 640—660 μm, width: 330—380 μm.

Locus typicus: Logny, France. Distribution: France, GFR, Italy, Soviet Union (NET, SET) **comitalis** Berlese, 1916

13 (12) Central field of prodorsum essentially smaller, its anterior margin far removed from prodorsal margin. Lateral setae short, far removed from rostrum, not even approaching its margin. Prodorsal condyles with 4—5 teeth of divers size. Notogaster ornamented with round foveolae of divers size. Apical seta of genu legs I and II not bifurcate. Length: 575 μm, width: 250 μm.

Locus typicus: Lake City, Florida, U.S.A. Distribution: Holarctic Region **coronata** Berlese, 1913

14 (5) Prodorsal condyles wide, projecting far above anterior margin of notogaster.

15 (16) Central field of prodorsum finely punctate, foveolae present laterally only; central field terminating far from anterior margin of prodorsum (pl. 128 : D), sides of its anterior part parallel. Sensillus thin, setiform. Posterior margin of prodorsal condyle with 3—4 teeth aligned on a transverse line. Notogastral foveolae of divers sizes, far removed from each other. Length: 540 µm, width: 224 µm.

Locus typicus: "Shore of Lake Krummer", Holstein, GFR. Distribution: Germany, Soviet Union (NET, KZ)

pectinata Strenzke, 1953

16 (15) Both central field and lateral areas of prodorsum with foveolae. Central field considerably widening towards rostrum (pl. 128 : E).

17 (18) Prodorsal condyles medially connected by a wide, robust band; foveolae in central field considerably smaller than those in lateral areas (pl. 128 : E). Sensillus thick, setiform, longer than interlamellar setae. Notogastral setae gradually tapering apicad. Length: 594 µm, width: 271 µm.

Locus typicus: St. Tyrvää, Finland. Distribution: Finland

forsslundi Karppinen, 1958

18 (17) Prodorsal condyles medially inconnected; foveolae of central field not smaller but more densely arranged than on lateral areas (pl. 128 : F). Sensillus not longer than interlamellar setae, weakly fusiform. Notogastral setae tapering only in their apical third. Length: 535—620 µm, width: 238—297 µm.

Locus typicus: Ceska Trebora, East Bohemia, Czechoslovakia. Distribution: Czechoslovakia, Soviet Union (JET, CET)

komareki Kunst, 1956

Species interpretabiles

1. comitalis Berlese, 1916 : 335.
   comitalis Woolley et Higgins: 1958 : 917, fig. 1.
   ussurica Golosova, 1969 : 1340, figs. 5—6.
2. coronata Berlese, 1913 : 100, pl. 7, fig. 85.
3. elegantula Berlese, 1913 : 100, pl. 7, fig. 84.
   elegantula: Van der hammen: 1959 : 80.
   areolata Strenzke, 1953 : 69, fig. 1.
4. forsslundi, Karppinen, 1958 : 193, figs. 1—2.
5. gorodkovi Sitnikova, 1975 : 99, fig. 176.
6. komareki Kunst, 1956 : 267, figs. 1—3.
7. laevis Sitnikova, 1975 : 99, fig. 177.
8. nana Nicolet, 1855 : 458, pl. 7, fig. 5 *(Nothrus)*.
   nana Van der Hammen: 1959 : 78.
9. pectinata Strenzke, 1953 : 70, fig. 2.
10. sellnicki Forsslund, 1958 : 75, figs. 1—3.

Genus 2: Cyrthermannia Balogh, 1958

Cyrthermannia Balogh, 1958 : 3.
Cyrthermannia Aoki: 1965 : 153.

Notogaster ellipsoid, profile in superior view not evenly arcuate, with tubercles and impressions. Prodorsal and notogastral setae simple, comparatively short, surfaces always with a rough sculpture. Epimeral setal formula: 3—1—3—3. Nine pairs of genital, two pairs of aggenitals (far removed from each other), two pairs of anal and three pairs of adanal, setae present.

Type-species: *Cyrthermannia tuberculata* Balogh, 1958.

The majority of species are known from the Holotropic area, and only two have hitherto been found in the Palaearctic Region.

Key to species

1 (2) Prodorsal condyles wide and thick, their posterior margins with 3—4 rounded teeth (pl. 129: F), condyles medially connected by a transverse lath.—Lamellar setae originating far from rostrum, on surface of prodorsum. Sensillus bacilliform. Foveolae of notogastral surface irregularly shaped, often angular. (pl. 129: E). Lateral and posterior tubercles of notogaster weak, body margin merely sinuous (in superior view). Notogastral setae very short, simple. Length: 465—480 μm, width: 162—169 μm.

Locus typicus: Kunitachi, near Tokyo, Japan. Distribution: Japan, Korea **parallela** (Aoki, 1961)

2 (1) Prodorsal condyles divided into two sharp teeth each, without a medially connecting transverse lath (pl. 129: G).—Lamellar setae originating near rostral ones, in vicinity of anterior margin of body. Sensillus slightly clavate. Foveolae of notogaster of divers sizes, all round (pl. 129: G), notogastral setae longer and weakly widended, their majority weakly T-shaped. Notogastral profile highly uneven, especially two posterior tubercles projecting from its outline. Length: 590 μm, width: 338 μm.

Locus typicus: Alexandria, Egypt. Distribution: only the holotype known **ezzati** Bayoumi et Mahunka, 1977

Species interpretabiles

1. ezzati Bayomi et Mahunka, 1977: 46, fig. 1.
2. parallela Aoki, 1961: 66, figs. 3A—B *(Nanhermannia)*.

Genus 3: MASTHERMANNIA Berlese, 1913

Masthermannia Berlese, 1913: 100.
Masthermannia Van der Hammen: 1959: 81.
Masthermannia Grandjean: 1964: 555.
Posthermannia Grandjean, 1954: 298.

Lamellar, interlamellar, notogastral and adanal setae T-shaped, stalks short, branches extremely long, flagellate. 15 pairs of notogastral setae present. Epimeral setal formula: 3—1—3—4. Setae *3c, 4b* and *4c* incrassate, sausage-shaped, all other setae short and simple. Nine pairs of thin genital, two pairs of minute aggenital (far removed from each other), two pairs of anal and three pairs of adanal, setae observable.

Type-species: *Angelia mammillaris* Berlese, 1904.

The species occur principally in the Holotropic area and only one represents it in the Palaearctic Region.

— Prodorsal surface finely punctate; laterally a robust lath each (pl. 130 : A, E). Sensillus long, hardly discernible incrassate, setiform, finely ciliate. Prodorsal condyles weak, but connected with a transverse lath. Notogastral with robust tubercles (pl. 130 : C), hence profile very uneven. Setae with very long branches, all originating from tubercles (pl. 130 : B). Notogaster with rotund foveolae. Length: 400—485 μm, width: 200 μm.

Locus typicus: Cascine, Florence, Italy. Distribution: France, Italy, Yugoslavia, Austria

mammillaris (Berlese, 1904)

Species interpretabilis

1. mammillaris Berlese, 1904 : 275 *(Angelia)*.
mammillaris Berlese: 1913 : 100, pl. 8, fig. 92.
mammillaris Van der Hammen: 1959 : 81.
nematophora Grandjean, 1954 : 298, figs. 1—3 *(Posthermannia)*.

Genus 4: Cosmohermannia Aoki et Yoshida, 1970

Cosmohermannia Aoki et Yoshida, 1970 : 1.

Body elliptical, considerably lengthened posteriorly, lateral margins sinous. Excepting rostral, epimeral, genital and aggenital setae, major part of body setae widended, characteristically phylliform or modified. 15 pairs of notogastral setae present, h_3 originating ventrally. Epimeral setal formula: 3—1—(4—5)—(4—5), setae *1a, 2a* and *3a* short, obtuse, guttiform, all others thin, setiform. Nine pairs of genital, two pairs of aggenital (near each other), two pairs of anal and three pairs of adanal, setae observable.

Type-species: *Cosmohermannia frondosus* Aoki et Yoshida, 1970.

Only two species are known, one of them inhabiting the Palaearctic Region.

— Lamellar setae originating considerably nearer to rostrum than to interlamellar ones. Prodorsal condyles small, marginally sinuous. Sensillus short, slightly widening. Notogastral foveolae of irregular shapes (pl. 129 : C)) but approximately equal in length. No essential differences in length among notogastral setae (pl. 129 : A). Ventral setae as in pl. 129 : D. Length: 963—1120 μm, width: 410—448 μm.

Locus typicus: Hiraichi, Yaku-shima, Japan. Distribution: only the type-locality known

frondosa Aoki et Yoshida, 1970

Species interpretabilis

1. frondosus Aoki et Yoshida, 1970 : 1, figs. 1—7.
frondosa Aoki : 1977 : 231.

Superfamily 3: HERMANNIOIDEA Sellnick, 1928

Hermanniidae Sellnick, 1928 : 18.
Hermannioidea Balogh, 1972 : 144 *(Hermannoidea)*.

Large-sized, dark, heavily pigmented species. Prodorsum wide, notogaster strikingly convex, both with a heavy sculpture. Notogaster emitting 16 (rarely 15) pairs of setae. Epimeral setal formula: 3—1—3—4, but occasionally neotrichy may occur. Genital and anal openings large, often touching. Nine pairs of genital, two (exceptionally more) pairs of aggenital, two pairs of anal and three pairs of adanal, setae observable. Fissura *iad* absent. Legs mono- or tridactylous.

Containing a single family.

Family 1: HERMANNIIDAE Sellnick, 1928

Hermanniidae Sellnick, 1928 : 18.

Characterizable by the features of the superfamily.

Three genera have hitherto been recognized, but the genus *Phyllhermannia* Berlese, 1916—occurring anyway in Far East localities—can be separated or defined with reservations only (in view of certain transitions); pending future investigations we do not consider it a distinct genus.

Genus 1: HERMANNIA Nicolet, 1855

Hermannia Nicolet, 1855 : 418.
Hermannia Sellnick: 1923 : 3.
Hermannia Bäumler: 1958 : 258.
Hermannia Van der Hammen: 1959 : 82.
Phyllhermannia Berlese, 1916 : 65.

Characterizable by the features of the superfamily.

Type-species *Nothrus gibbus* C. L. Koch, 1840.

Representatives of the genus are known from all parts of the world. In the Palaearctic Region, 11 species (including those relegated to the genus *Phyllhermannia*) have hitherto been recorded.

Key to species

1 (4) Epimeral region with neotrichy: epimeres 3 and 4 with 5—8 setae; more than two pairs of aggenital setae present.

2 (3) Three pairs of weakly spatulate aggenital setae present. Anterior part of notogaster with a closed polygonal sculpture, the single units with sinuous borders; sculpture disintegrating more posteriorad, replaced by longitudinal sinous lines (pl. 131 : A). Sensillus gradually incrassate, not distinctly capitate. Interlamellar setae long, half as long as distance between their insertion points. Phylliform notogastral setae long, c_1 approaching insertion point of d_1.

Epimeral setal formula: 3—1—5—6. Length: 837—956 μm, width: 441—550 μm.

Locus typicus: Sikotan, Kuril Islands, Soviet Union. Distribution: Soviet Union (FE)

clara Sitnikova, 1975

3 (2) Five pairs of thin, pointed aggenital setae present. Notogastral surface (pl. 131 : G) punctate and with an irregular and sparse polygonal reticulation boders of single units straight. Sensillus abruptly capitate. Interlamellar setae short, only one-fourth as long as distance between their insertion points. Notogastral setae short, c_1 only half as long as distance enclosed by *c—d.* Epimeral setal formula: 3—1—5—6. Length: 1193—1211 μm, width: 691—700 μm.

Locus typicus: Norilsk, Soviet Union. Distribution: Soviet Union (WS)

gigantea Sitnikova, 1975

4 (1) Epimeral region without neotrichy, epimeral setal formula: 3—1—3—4. Only two pairs of aggenital setae present.

5 (12) Sensillus long, at least as long as half distance between bothridia, thin, bacilliform, at most hardly perceptibly spatulate distally (pl. 131 : D).

6 (7) Extraordinarily large animals (longer than 1150 μm). Notogaster finely punctate, also some darker spots visible in even distribution. Notogastral setae simple, bacilliform or very narrowly spatulate (pl. 131 : B). Length: 1150—1365 μm, width: 660—750 μm.

Locus typicus: Regensburg, Germany. Distribution: Central Europe, Soviet Union (FE), Japan

convexa (C. L. Koch, 1840)

7 (6) Smaller animals (at most 900 μm long). Notogastral setae perceptibly widened, spatulate or phylliform.

8 (9) Notogastral setae broadly phylliform, widest at middle height (pl. 131 : D), comparatively short, setae in longitudinal sequence essentially shorter than distance between any two neighbouring pairs. Notogastral surface with a sculpture of large foveolae. Genital setae (pl. 131 : C) long, thin, at most anterior one of outer row slightly thicker than the others. Femur of leg I (pl. 131 : F) dorsally with two short, obtuse setae. Length: 572 μm, width: 296—310 μm.

Locus typicus: Sonai, Irimote Islands, Japan. Distribution: only the type-locality known

pulchra (Aoki, 1973)

9 (8) Notogastral setae narrower, or only basally slightly incrassate, or only distally and hence spatulate.

10 (11) Notogastral setae (pl. 133 : C) thickest basally, apically acute, relatively long, those aligned in longitudinal sequence reaching insertion point of next one posteriorly (pl. 133 : A). Notogastral surface finely punctate. Genital plate with three short spiniform setae in outer row (pl.133 : E). Femur of leg I (pl. 133 : D) dorsally with two short and one long obtuse setae. Length: 750—830 μm, width: 410—490 μm.

Locus typicus: Shiraishi, Mitagawa-mura, Miyazaki, Japan. Distribution: Japan

kanoi (Aoki, 1959)

11 (10) Notogastral setae slightly widened distally, spatulate. Notogastral surface with a tuberculate sculpture (pl. 132: A—B). Every genital seta thin, simple (pl. 132: C). Femur of leg I dorsally with three pairs of obtuse but long setae. Length: 790—900 µm, width: 525—550 µm.

Locus typicus: Regensburg, Germany. Distribution: Palaearctic Region

gibba (C. L. Koch, 1840)

12 (5) Sensillus short, one-fourth or exceptionally one-third as long as distance enclosed by bothridia, heabily incrassate, spatulate or clavate.

13 (14) Entire notogastral surface ornamented with large foveolae; setae widely phylliform (pl. 132: D, G). Sensillus very short (pl. 132: F), abruptly and clavately incrassate. Lamellar and interlamellar setae (pl. 132: E) originating near each other. Some epimeral setae (pl. 132: J) thickened, genital and anal setae thin and simple, adanal ones again incrassate (pl. 132: H—I). Length: 740—780 µm, width: 351—400 µm.

Locus typicus: Mt. Fuji, Fuji-Fuketsu, Yamanashi, Japan. Distribution: only the type-locality known

areolata (Aoki, 1970)

14 (13) Notogastral sculpture different, setae spatulate of bacilliform. Setae spatulate of bacilliform. Setae of ventral side also simple.

15 (16) Notogastral surface with a definite closed polygonal, sculpture (pl. 131: H) composed of robust laths, well discernible at least in centre of notogaster. Notogastral setae short, bacilliform. Sensillus short, distally clavate. Interlamellar setae short, half as long as c_1. Length: 770—880 µm, width: 420—525 µm.

Locus typicus: exact reference unknown to us; Spitsbergen, Norway. Distribution: North Europe, Soviet Union (NET)

reticulata Thorell, 1871

16 (15) Notogastral surface, especially medially, without any closed polygonal sculpture composed of robust laths.

17 (18) Sensillus abruptly incrassate, globular. Notogastral surface punctate, with sparse tubercles displaying a pattern only sketchily resembling a polygonal sculpture (pl. 133: F). Interlamellar setae considerably shorter than c_1. Notogastral setae distally weakly widened. Length: 900—1055 µm, width: 555—635 µm.

Locus typicus: Cap Grebeni, Novaya Zemlya, Soviet Union. Distribution: North Europe, Soviet Union (NET)

scabra (L. Koch, 1879)

18 (17) Sensillus gradually spatulate distally (pl. 133: H), notogastral sculpture different.

19 (20) Interlamellar setae long, their enclosed distance twice greater than length of setae (pl. 133: G). Notogastral surface marginally with a polygonal sculpture composed of small fields, medially disintegrating into longitudinal lines. Length: 790—900 µm, width: 430—500 µm.

Locus typicus: Kushiro City, Hokkaido, Japan. Distribution: only the type-locality known

hokkaidensis Aoki et Ohnishi, 1974

20 (19) Interlamellar setae short, their enclosed distance four times greater than length of setae (pl. 133 : K). Notogastral surface punctate, with sparsely distributed foveolae. Length: 875—990 μm, width: 530—620 μm.

Locus typicus: Kuhaarden, Germany. Distribution: Finland, Denmark, Germany

subglabra Berlese, 1910

Species interpretabiles

1. areolata Aoki, 1970 : 71, figs. 1—11 *(Phyllhermannia)* **comb. nov.**
2. clara Sitnikova, 1975 : 104 figs. 183.
   clara Sitnikova: 1977 : 164, figs. 4—6 (non sp. nov.!).
3. convexus C. L. Koch, 1839 : 29, 1, fig. *(Nothrus)*.
   convexa Sellnick: 1923 (1968 : 30).
4. gibbus C. L. Koch, 1840 : 29, 4, fig. *(Nothrus)*.
   gibba Sellnick: 1923 (1968 : 30).
   gibba Van der Hammen: 1959 : 83.
   gibba Bäumler: 1970 : 258, figs. 1—18.
   piceus Berlese, 1885 : 9 *(Angelia)* (Van der Hammen: 1959 : 83).
5. gigantea Sitnikova, 1975 : 104, 179*.
   gigantea Sitnikova: 1977: 165; figs. 7—9 (non sp. nov.!).
6. hokkaidensis Aoki et Ohnishi, 1974 : 150, figs. 8—11.
7. kanoi Aoki, 1959: 4, figs. 3A—E.
8. pulchra Aoki, 1973 : 86, figs. 1—5.
9. reticulata Thorell, 1871 : 696
   reticulata Sellnick, 1923 : (1968 : 31).
10. scaber L. Koch, 1879: 113 *(Nothrus)*.
    scaber Sellnick: 1923 (1968 : 31).
11. subglabra Berlese, 1910 : 380.
    subglabra Forsslund: 1957 : 592, fig. 4.
    pulchella Willmann, 1952 : 172.

Species inquirendae

— doliaris C. L. Koch, 1839 : 29, 5—6 *(Nothrus)*.
— grandis Berlese, 1910 : 380.
  grandis Van der Hammen: 1959 : 83.

LITERATURE

Aoki, J. (1958): Zwei Heminothrus-Arten aus Japan (Acarina: Oribatei). — *Annotnes zool. jap.*, 31 : 121—125.

Aoki, J. (1958): Einige Phthiracariden aus Utsukushigahra, Mitteljapan (Acarina: Oribatei). — *Annotnes zool. jap.*, 31: 171—175.

Aoki, J. (1958): Eine Neue Art von der Gattung Oribotritia Jacot (Acarina: Oribatei). — *Acta arachn., Tokyo*, 16: 18—20.

Aoki, J. (1959): Die Moosmilben (Oribitaei) aus Südjapan. — *Bull. biogeogr. Soc. Japan*, 21: 1—22.

Aoki, J. (1960): Eine dreikrallige Gattung der Familie Perlohmannidae (Acarina: Oribatei). — *Jap. J. Zool.*, 12: 507—511.

Aoki, J. (1961): Beschreibungen von neuen Oribatiden Japans. — *Jap. J. app. Zool.*, 5: 64—69.

Aoki, J. (1963): Einige Neue Oribatiden aus dem Kaiserlichen Palastgarten Japans. — *Annotnes zool. jap.*, 36: 218—224.

Aoki, J. (1965): Notes on the species of the Genus Epilohmannia from the Hawaiian Islands (Acarina: Oribatei). — *Pacif. Insects*, 7 (2): 309—315.

Aoki, J. (1965): Oribatiden (Acarina) Thailands. I. — *Nature and Life in Southeast Asia*, 4: 129—193.

Aoki, J. (1965): Oribatid mites (Acarina: Oribatei) from Himalaya with descriptions of several new species. — *J. Coll. Arts and Sciences, Chiba*, 4 (3): 289—302.

Aoki, J. (1966): Studies on the Oribatid Mites of Japan. II. Trichthonius simplex spec. nov. — *Bull. natn. Sci. Mus., Tokyo*, 9 (1): 1—7.

Aoki, J. (1966): A New Species of Soil Mite, Malaconothrus japonicus, from Central Japan (Cryptostigmata; Malaconothridae). — *Annotnes. zool. jap.*, 39 (3): 169—172.

Aoki, J. (1969): Taxonomic Investigations on Free-living Mites in the Subalpine Forest on Shiga Heights IBP Area III. Cryptostigmata. — *Bull. natn. Sci. Mus., Tokyo*, 12 (1): 117—141.

Aoki, J. (1969): Eine neue Unterart der Bodenmilben aus Fujisan (Fudschijama) (Acari: Euphthiracaridae). — *Acta arachn., Tokyo*, 22 (2): 27—30.

Aoki, J. (1970): The Oribatid Mites of the Islands of Tsushima. — *Bull. natn. Sci. Mus., Tokyo*, 13 (3): 395—442.

Aoki, J. (1970): A Peculiar New Species of the Genus Phyllhermannia Collected at Mt. Fuji (Acari: Hermanniidae). — *Bull. natn. Sci. Mus., Tokyo*, 13 (1): 71—75.

Aoki, J. (1971): A New Species of Oribatid Mite, Thamacarus moribei, from the West Coast of the Red Sea, with Record of Two Other Species of the Family Lohmanniidae. — *Jap. J. Zool.*, 16 (3): 127—129.

Aoki, J. (1973): Oribatid Mites from Iriomote jima, the Southernmost Island of Japan (I). — *Mem. Nat. Sci. Mus., Tokyo*, 6: 85—101.

Aoki, J. (1973): Oribatid Mites from Mt. Poroshiri in Hokkaido, North Japan. — *Annotnes, zool. jap.*, 46 (4): 241—252.

Aoki, J. (1975): Taxonomic Notes on Some Little Known Oribatid Mites of Japan. — *Bull. natn. Sci. Mus., Ser. A (Zool.), Tokyo*, 1 (1): 57—65.

Aoki, J. (1975): Two Species of the Primitive Oribatid Genus Gehypochthonius from Japan. — *Annotnes. zool. jap.*, 48 (1): 55—59.

Aoki, J. (1977): Three species of the genus Eohypochthonius from Japan. — *Acarologia*, 19 (1): 117—122.

Aoki, J. (1977): Studies on Oribatei (Acarina) from the South Pacific. III. Cosmohermannia monstruosa, n. sp. from New Guinea. — *Pacif. Insects*, 17 (2—3): 229—232.

Aoki, J. (1977): New and Interesting Species of Oribatid Mites from Kakeroma Island, Southwest Japan. — *Acta arachn., Tokyo*, 27: 85—93.

Aoki, J. and Ohnishi, J. (1974): New Species and Record of Oribatid Mites from Hokkaido, North Japan. — *Bull. natn. Sci. Mus., Tokyo*, 17 (2): 149—156.

Aoki, J. and Yoshida, K. (1970): A New Oribatid Mite, Cosmohermannia frondosus, gen. n. et sp. n. from Yaku-shima Island. — *Bull. biogeogr. Soc., Japan,* 26 (1): 1—4.

Baker, E. W. and Wharton, G. W. (1952): *An introduction to Acarology.* — MacMillan Co., New York, 465 pp.

Balogh, J. (1943): Magyarország páncélosatkái (Conspectus Oribateorum Hungariae). — *Mat. Természettud. Közlem.,* 39, 5: 1—202.

Balogh, J. (1958): Oribatides nouvelles de l'Afrique tropicale. — *Rev. Zool. Bot. Afr.,* 58: 1—34.

Balogh, J. (1959): On the preparation and observation of Oribatids. — *Acta Zool. Hung.,* 5: 241—253.

Balogh, J. (1961): An outline of the family Lohmanniidae Berl., 1916 (Acari: Oribatei). — *Acta Zool. Hung.,* 7: 19—44.

Balogh, J. (1961): Identification keys of world Oribatid (Acari) families and genera. — *Acta Zool. Hung.,* 7: 243—344.

Balogh, J. (1961): Identification keys of world Oribatid (Acari) families and genera. — *Acta Zool. Hung.,* 7: 243—344.

Balogh, J. (1962): An outline of the family Microzetidae Grandjean, 1936 (Acari: Oribatei). — *Opusc. Zool. Budapest,* 4: 35—58.

Balogh, J. (1965): A synopsis of the world Oribatid (Acari) genera. — *Acta Zool. Hung.,* 11: 5—99.

Balogh, J. (1972): *The Oribatid genera of the World.* — Akadémiai Kiadó, Budapest, 188 pp., 71 pls.

Balogh J. and Mahunka, S. (1966): New Oribatids from Australian Soils. — *Folia ent. hung.,* 19: 553—568.

Balogh, J. and Mahunka, S. (1968): The Scientific Results of the Hungarian Soil Zoological Expeditions to South America 5. Acari: Data to the Oribatid Fauna of the Environment of Cordoba, Argentina. — *Opusc. Zool. Budapest,* 8: 317—340.

Balogh, J. and S. Mahunka (1979): New taxa in the system of the Oribatida (Acari). — *Ann. Hist.-nat. Mus. Nat. Hung.,* 71: 279—290.

Balogh, J. és S. Mahunka (1980): Atkák XV — Acari XV. — *Magyarország Állatvilága (Fauna Hungariae),* 18, 19: 1—177.

Bashkirova E. Ya.—Башкирова, Е. Я. (1958): Фауна клещей-орибатид целинной степи юго-востока европейской части СССР. — *Зоол. Ж.* 37: 193—209.

Bayoumi, B. M. and S. Mahunka (1976): Contributions to the knowledge of the genus Epilohmannia Berlese, 1916 (Acari: Oribatida). — *Folia ent. hung.,* 29: 5—21.

Bayoumi, B. M. and Mahunka, S. (1977): Cyrthermannia ezzati sp. n. and Further Data to the Knowledge of Egyptian Oribatid Fauna. — *Opusc. Zool. Budapest,* 14 (1—2): 45—49.

Bäumler, W. (1970): Zur Morphologie, Biologie und Ökologie von Hermannia gibba (C. L. Koch) (Acarina: Oribatei) unter Berücksichtigung einiger Begleitarten. — *Zeitschr. f. angew. Ent.,* 66: 257—277.

Beck, L. (1962): Beiträge zur Kenntnis der neotropischen Oribatidenfauna I. Eohypochthonius und Cosmochthonius (Arach., Acari). — *Senckenberg. biol.,* 43: 227—236.

Berlese, A. (1896): *Acari Myriopoda et Scorpiones hucusque in Italia reperta. Ordo Cryptostigmata (Oribatidae). (Cryptostigmata II).* Portici.

Berlese, A. (1904): Acari nuovi. Manipulus III. — *Redia,* 2: 10—32

Berlese, A. (1904): Acari nuovi. Manipulus I^{us}. — *Redia,* 1: 235—252.

Berlese, A. (1904): Acari nuovi. Manipulus II^{us}. — *Redia,* 1: 258—280.

Berlese, A. (1905): Acari nuovi. Materiali pel "Manipulus V". *Redia,* 2: 231—238.

Berlese, A. (1908): Elenco di generi e specie noube di Acari. — *Redia,* 5: 1—15.

Berlese, A. (1910): Acari nuovi. Manipulus V. — *Redia,* 6: 199—234.

Berlese, A. (1910): Brevi diagnosi di generi e specie nuovi di Acari. — *Redia,* 6: 346—388.

Berlese, A. (1913): Acari nuovi. Manipoli VII—VIII. — *Redia,* 9: 77—111.

Berlese, A. (1914): Acari nuovi. Manipulus IX. — *Redia,* 10: 113—150.

Berlese, A. (1916): Centuria secondo di Acari nuovi. — *Redia,* 12: 125—177.

Berlese, A. (1916): Centuria terza di Acari nuovi. — *Redia,* 12: 289—338.

Berlese, A. (1923): Centuria sesta di Acari nuovi. — *Redia,* 15: 237—262.

Bernini, F. (1971): Notulae Oribatologicae. III. Ridescrizione di Steganacarus brevipilus (Berlese) e note sui generi Tropacatus e Steganacarus (Acarida, Oribatei). — *Redia,* 52: 549—567.

Bernini, F. (1973): Notulae Oribatologicae VII. Gli Oribatei (Acarida) dell'isolotto di Basiluzzo (Isole Eolie). — *Lav. Soc. It. Biogeogr. n. s.* 3: 355—480.

Bulanova-Zachvatkina, E. M. — Буланова-Захваткина, Е. М. (1960): Новые представители примитивных панцирных клещей надсемейства Perlohmannoidae Grandjean, 1958 (Acariformes, Oribatei). — *Зоол. ж.*, 39: 1835—1848.
Bulanova-Zachvatkina, E. M. — Буланова-Захваткина, Е. М. (1967): Панцирные Клещи — Орибатиды. — *Высшая Школа, Москва,* 254 pp.
Bulanova-Zachvatkina, E. M. — Буланова-Захваткина, Е. М. (1978): Новые Орибатидные клещи (Oribatei, Mixonomata) из термитников в средней Азии. — *Энтомол. обозр.*, 57: 918—922.
Călugăr, M. et Vasiliu, N. (1976): Nouvelles Espéces d'Oribates (Acari: Oribatei). — *Rev. Roum. Biol. Anim.*, 21: 13—17.
Canestrini, G. (1884): Prospetto dell'Acarofauna italiana. Parte I. Oribatini e Gamasini. — *Att. R. Ist. Ven. Sci. Lett. Art.* 6,2: 1397—1443.
Canestrini, G. e Fanzago, F. (1876): Nuovi Acari italiani. — *Att. Soc. Ven.-Trent. Sci. Nat. Pad.*, 3: 99—111.
Chinone, S. (1974): Further contribution to the knowledge of the family Brachychthoniidae from Japan. — *Bull. Biogeogr. Soc. Japan*, 30: 1—28.
Chinone, S. (1978): Additional report on the soil mites of the family Brachychthoniidae from Japan. — *Bull. Biogeogr. Soc. Japan*, 33: 9—32.
Chinone, S. and Aoki, J. (1972): Soil mites of the family Brachychthoniidae from Japan. — *Bull. Nat. Sci. Mus.*, 15: 217—251.
Cooremann, J. (1947): Association d'Acariens vivant dans les nids de Laurus ridibundus ridibundus Linné, a Kalmthout. — *Bull. Mus. Hist, nat, Belg, Bruxelles,* 23 (30): 1—14.
Csiszár, J. (1961): Neue Oribatiden (Acari). — *Fol. ent. hung.*, 14: 447—450.
Evans, G. (1952): British mites of the genus Brachychthonius Berl., 1910. — *Ann, Mag. Nat. Hist.* 12,5: 227—239.
Evans, G. O (1952): Terrestrial Acari new to Britain. — II. — *Ann. Mag. Nat. Hist.*, 12, 5: 660—675.
Ewing, H. E. (1908): Two new species of the genus Phthiracarus. — *Ent. News,* 19; 449—451.
Ewing, H. E. (1908): A new genus and species of Oribatidae. — *Ent. News.*, 19: 243—245.
Ewing, H. E. (1909): New American Oribatoidea. — *Journ. N. Y. Ent. Soc.*, 17: 116—136.
Ewing, H. E. (1917): A synopsis of the genera of beetle Mites with special reference to the North American fauna. — *Ann. ent. Soc. Am.*, 10: 117—132.
Feider, Z. et Calugar, M. (1968): Nouveaux Oribatides pour la science et pour la faune de Roumanie. — *Trav. Mus. Hist, nat. Gr. Antipa,* 8: 621—640.
Feider, Z. et Calugar, M. (1969): Chetotaxia picioarelor la specile genurilor Phthiracarus si Hoplophthiracarus (Oribates), din Romania. — *Anal. stiint. Univ. Al. I. Cuza, Sect. II.*, 15: 75—86.
Feider, Z. Suciu, I. (1957): Contributie la cunoastera Oribatelor (Acari) din R. P. R. Familia Phthiracaridae Perty, 1841. — *Studii Cerc. stiint. Iasi,* 8: 23—48.
Feider, Z. et Suciu, I. (1958): O noua contributie la cunoasterea Oribatidelor (Acari) din R. P. R. — *Studii Cerc. biol.*, 10: 31—44.
Feider, Z. si Calugar, M. (1970): Acarieni din Familia Phthiracaridae de la portile de fier. — *Comunic. zool.:* 17—32.
Forsslund, K. H. (1942): Schwedische Oribatei (Acari). I. — *Ark. Zool.* 34A, 10: 1—11.
Forsslund, K. H. (1957): Notizen über Oribatei (Acari). I. — *Ark. Zool.* 2,10: 583—593.
Forsslund, K. H. (1957): Schwedische Oribatei (Acari). III. — *Entomol. Ts.*, 77: (1956) 210—218.
Forsslund, K. H. (1958): Notizen über Oribatei (Acari). II. — *Entomol. Ts.*, 79: 75—86.
Forsslund, K. H. (1964): Liochthonius muscorum n. sp. und L. lapponicus (Träg.) (Acari, Oribatei). — *Entomol. Ts.* 85: 236—239.
Forsslund, K. H. und Märkel, K. (1963): Drei neue Arten der Fam. Euphthiracaridae (Acari, Oribatei). — *Entomol. Ts.*, 84: 284—296.
Fujikawa, T. and Aoki, J. (1969): Notes on Two Species of the Genus Perlohmannia Berlese (Acari, Perlohmanniidae) Taxonomic Notes on Oribatid Mites of Hokkaido. II. — *Annotnes. zool. jap.*, 42 (4): 220—225.
Ghilarov M. S. — Гиляров, М. С. и Криволуцкий, Д. А. (1975): *Определитель обитающих в почве клещей (Sarcoptiformes).* — Наука, Москва, 500 пп.
Golosova L. D. — Голосова, Л. Д. (1969): Фауна Орибатид (Acariformes, Oribatei) пойменных лесов южной части приморского края. — *Зоол. журн.*, 48: 1337—1341.

Gordeeva, E. V. — Гордеева, Е. В. (1978): Новый род панцирных клещей из восточного Крыма. — *Зоол. журн.*, 57: 1099—1101.
Gordeeva E. V. — Гордеева, Е. В. (1980): Панцирные клещи семейства Cosmochthoniidae (Oribatei). — Зоол. Журн., 59: 838—850.
Grandjean, F. (1928): Sur un Oribatidé pourvu d'yeux. — *Bull. Soc. zool. France,* 53: 235—242.
Grandjean, F. (1931): Observations sur les Oribates (1re série). — *Bull. Mus. nat. Hist. natur.* (2), 3: 131—144.
Grandjean, F. (1931): Observations sur les Oribates (2e série). — *Bull. Mus. nat. Hist. natur.* (2), 3: 651—665.
Grandjean, F. (1932): Au sujet des Palaeacariformes Trägardh. — *Bull. Mus. nat. Hist. natur.* (2) 41: 411—426.
Grandjean, F. (1932): La famille des Protoplophoridae (Acariens). — *Bull. Soc. zool. France,* 57: 10—36.
Grandjean, F. (1933): Structure de la région ventrale chez quelques Ptyctima (Oribates). — *Bull. Mus. nat. Hist. natur.* (2), 5: 309—315.
Grandjean, F. (1933): Oribates de l'Afrique du Nord (1re série). — *Bull. Soc. Hist. natur. Afr. Nord,* 24: 308—323.
Grandjean, F. (1933): Observations sur les Oribates (4e série). — *Bull. Mus. nat. Hist. natur.* (2), 5: 215—222.
Grandjean, F. (1933): Observations sur les Oribates (5e série). — *Bull. Mus. nat. natur.* (2) 5: 461—468.
Grandjean, F. (1933): Etude sur le développement des Oribates. — *Bull. Soc. zool. France,* 58: 30—61.
Grandjean, F. (1934): Oribates de l'Afrique du Nord (2me série). — *Bull. Soc. Hist. natur. Afr. Nord,* 25: 235—252.
Grandjean, F. (1934): Observations sur les Oribates (6e série). — *Bull. Mus. nat. Hist. natur.* (2), 6: 353—360.
Grandjean, F. (1934): Observations sur les Oribates (Arach. Acar.) (7e série). — *Bull. Mus. nat. Hist. natur.* (2), 6: 423—431.
Grandjean, F. (1934): La notation des poils gastronotiques et des poils dorsaux du prodosoma chez les Oribates (Acariens) — *Bull. Soc. zool. France,* 59: 12—44.
Grandjean, F. (1934): Les organes respiratoires secondaires des Oribates (Acariens). — *Ann. Soc. ent. France,* 103: 109—146.
Grandjean, F. (1934): Phthiracarus anonymum n. sp. — *Rev. franc. Ent.,* 1: 51—58.
Grandjean, F. (1935): Les poils et les organes sensitifs portés par les pattes et le palpe chez les Oribates. — *Bull. Soc. zool. France,* 60: 6—39.
Grandjean, F. (1936): Les Oribates de Jean Frédéric Hermann et de son pére (Arachn. Acar.). — *Ann. Soc. ent. France,* 105: 27—110.
Grandjean, F. (1939): Observations sur les Oribates (12e série). — *Bull. Mus. nat. Hist. natur.* (2), 11: 300—307.
Grandjean, F. (1939): Observations sur les Oribates (13e série). — *Bull. Mus. nat. Hist. natur.* (2), 12: 62—69.
Grandjean, F. (1941): Observations sur les Acariens (6e série). — *Bull. Mus. nat. Hist. natur.* (2), 13 : 532— *France.* 66: 33—50.
Grandjean, F. (1941): Observations sur les Acariens (6[e]série). — *Bull. Mus. nat. Hist. natur.* (2), 13: 532—539.
Grandjean, F. (1946): Les poils et les organes sensitifs portés par les pattes et le palpe chez les Oribates. Troisiéme partie. — *Bull. Soc. zool. France,* 71: 10—29.
Grandjean, F. (1947): Les Enarthronota (Acariens). Premiére série. — Ann. Sci. natur. Zool. (11), 8: 213—248.
Grandjean, F. (1947): L'origine de la pince mandibulaire chez les Acariens actinochitineux. — *Arch. Sci. phys. natur. Genéve* (5), 29: 305—355.
Grandjean, F. (1948): Les Enarthronota (Acariens) (2e série). — *Ann. Sci. natur. Zool.* (11), 10: 29—58.
Grandjean, F. (1949): Formules anales, gastronotiques, génitales et aggénitales du développement numérique des poils chez les Oribates. — *Bull. Soc. zool. France,* 74: 201—225.
Grandjean, F. (1950): Observations sur les Oribates (20e série). — *Bull. Mus. nat. Hist. natur.,* 22: 73—80.
Grandjean, F. (1950): Étude sur les Lohmanniidae (Oribates, Acariens). — *Arch. Zool. exp. gén.,* 87: 95—162.
Grandjean, F. (1950): Les Enarthronota (Acariens) (3e série). — *Ann. Sci. natur., Zool.* (11), 12: 85—107.
Grandjean, F. (1952): Au sujet de l'ectosquelette du podosoma chez les Oribates supérieurs et de sa terminologie. — *Bull. Soc. zool. France,* 77: 13—36.

Grandjean, F. (1952): Observations sur les Palaeacarcides (2e série) — *Bull. Mus. nat. Hist. natur.* (2), 24: 460—467.
Grandjean, F. (1953): Observations sur les Oribates (25e série). — *Bull. Mus. nat. Hist. natur.* (2), 25: 155—162.
Grandjean, F. (1954): Observations sur les Oribates (28e série). — *Bull. Mus. nat. Hist. natur.* (2), 26: 204—211.
Grandjean, F. (1954): Essai de classification des Oribates (Acariens). — *Bull. Soc. zool. France,* 78: 421—446.
Grandjean, F. 1954): Les Enarthronota (Acariens) (4e série). — *Ann. Sci. natur. Zool.* (11), 16. 311—335.
Grandjean, F. (1954): Étude sur les Palaeacaroides (Acariens, Oribates). — *Mém. Mus. nat. Hist. natur. (n. s.), sér. A, Zool.,* 7: 179—274.
Grandjean, F. (1954): Posthermannia nematophora n. g., n. sp. (Acarien, Oribate). — *Rev. franc. Ent.,* 21: 298—311.
Grandjean, F. (1956): Caracteres chitineux de l'ovipositeur, en structure normale, chez les Oribates (Acariens). — *Arch. Zool, exp. gén., 93, Notes et Revue,* 2: 96—106.
Grandjean, F. (1956): Observations sur les Oribates (35e série). *Bull. Mus. nat. Hist. natur* (2), 28: 282—289.
Grandjean, F. (1957): L'infracapitulum et la manducation chez les Oribates et d'autres Acariens. — *Ann. Sci. Natur., Zool.* (11), 19: 233—281.
Grandjean, F. (1958): Perlohmannia dissimilis (Hewitt) (Acarien, Oribate). — *Mém. Mus. nat. Hist. natur. (n. s.), sér. A, Zool.,* 16: 57—119.
Grandjean, F. (1961): Perlohmannia coiffaiti n. sp. (Oribate). — *Acarologia,* 3: 604—619.
Grandjean, F. (1962): Nouvelles observations sur les Oribates (2e serie). — *Acarologia,* 4: 396—422.
Grandjean, F. (1963): Sur deux espéces de Brachychthoniidae et leur développement (Oribates). — *Acarologia,* 5: 122—151.
Grandjean, F. (1964): Nouvelles observations sur les Oribates (3e série). — *Acarologia,* 6: 170—198.
Grandjean, F. (1964): La solénidiotaxia des Oribates. — *Acarologia,* 6: 529—556.
Grandjean, F. (1965): Nouvelles observations sur les Oribates (4e série). — *Acarologia,* 7: 91—112.
Grandjean, F. (1965): Complément á mon travail de 1953 sur la classification des Oribates. — *Acarologia,* 7: 713—734.
Grandjean, F. (1966): Collohmannia gigantea Selln. (Oribate). Premiére partie. — *Acarologia,* 8: 328—357.
Grandjean, F. (1967): Nouvelles observations sur les Oribates (5e série). — *Acarologia,* 9: 242—272.
Grandjean, F. (1969): Considérations sur le classement des Oribates. Leur division en 6 groupes majeurs. — *Acarologia,* 11: 127—153.
Grandjean, F. (1970): States. Actinopiline. Rappel de ma classification des Acariens en 3 groupes majeurs. Terminologie en some. — *Acarologia,* 11: 796—827.
Grishina, L. G. — Гришина, Л. Г. (1968): Новый вид панцирного клеща Perlohmannia altaica sp. n. с Алтая. — *Зоол. журн.,* 47: 297—298.
Grishina, L. G. — Гришина, Л. Г. (1971): Новый вид панцирного клеща рода Atopochthonius (Oribatei, Atopochthoniidae)с Алтая. — *В сб. Новые и малоизвест. виды фауны Сибири. Новосибирск, Наука:* 71—73.
Haller, G. (1884): Beschreibung einiger neuen Milben. — *Arch. Naturg.,* 50: 217—236.
Hammen, L. van der (1950): On two interesting Oribatei from the Sint Pietersberg (Southern Limburg, Netherlands). — *Mededelingen van de Commissie inzake wetenschappelijk onderzoek van de St. Pietersberg, no. 11. Naturhist Maandbl.,* 39: 129—132.
Hammen, L. van der (1952): The Oribatei (Acari) of the Netherlands. — *Zool. Verh.,* no. 17.
Hammen, L. van der (1956): Notes on the Oribatei (Acari) of Durch New Guinea. IV. Description of Trhypochthonius montanus nov. spec., and preliminary revision of the genus Trhypochthonius. — *Proc. Kon. Ned. Ak. Wet., ser. C.,* 59: 398—404.
Hammen, L. van der (1959): Berlese's primitive Oribatid mites. — *Zool. Verh., Leiden,* 40: 1—93.
Hammen, L. van der (1963): The Oribatid family Phthiracaridae I. Introduction and Redescription of Hoplophthiracarus pavidus (Berlese). — *Acarologia,* 5: 306—317.
Hammen, L. van der (1963): The Oribatid family Phthiracaridae. II. Redescription of Phthiracarus laevigatus (C. L. Koch). — *Acarologia,* 5: 704—715.
Hammen, L. van der (1963): The Oribatid family Phthiracaridae III. Redescription of Phthiracarus nitens (Nicolet). — *Acarologia,* 6: 400—411.

Hammen, L. van der (1968): Introduction generale a la classification, la terminologie morfologique, l'ontogenese et l'evolution des Acariens. — *Acarologia,* 10: 401—412.
Hammer, M. (1952): Investigations on the Microfauna of Northern Canada. Part 1, Oribatidae. — Acta Arctica, 4: 1—108.
Hammer, M. (1955): Alaskan Oribatids. — *Acta Artica,* 7: 1—36.
Hammer, M. (1958): Investigations on the Oribatid fauna of the Andes Mountains. I. The Argentina and Bolivia. — *Biol. Skr. Dan. Vid. Selsk.,* 10: 1—129 + 34 Pl.
Hammer, M. (1965): Are low temperatures a Species-preserving Factor? Illustrated by the Oribatid Mite Mucronothrus Masalis (Willm.) — *Acta Univ. lund.,* 2 (2): 3—10.
Hammer, M. (1966): A few Oribatid mites from Ram, Jordan. — *Zool. Anz.,* 177: 272—276.
Hammer, M. (1967): Some Oribatids from Kodiak Island near Alaska. — *Acta Arctica,* 14: 1—25.
Hammer, M. (1975): On some oribatids from Central Sahara (Acari: Oribatidae). — *Steenstrupia,* 3: 187—196.
Harding, D. J. L. (1976): A new species of Phthiracarus (Acari: Cryptostigmata) from Great Britain. — *Acarologia,* 18: 163—169.
Hartmann, A. G. (1949): A new species of Nanhermannia with notes on the genus —*Proc. Ent. Soc. Wash.,* 51: 169—171.
Hermann, J. F. (1804): Mémoire Aptérologique. — Strassburg.
Hewitt, C. G. (1908): On a new phytaphagous mite, Lohmannia insignis Berl. var. dissimilis n. var., with notes on other species of economic importance. — *Mem. Proc. Manch. Lit. Phil. Soc. (Manch. Mem.),* 52, 5: 1—10.
Hull, J. E. (1914): British Oribatidae. Notes on new and critical species. — *Naturalist, Hull* 1914: 215—288.
Hull, J. E. (1916): Terrestrial Acari of the Tyne province. I. Oribatidae. — *Trans. nat. Hist. Soc. Northumb.,* 4: 381—410.
Jacot, A. P. (1923): Revision of the Ginglymosoma (Acarina), and the creation of two new genera. — *China J. Sci. and Art,* 1: 161—162.
Jacot, A. P. (1924): Oribatoidea sinensis III. — *Journ. N. Chim. Branch Roy. As. Soc.,* 55: 78—83.
Jacot, A. P. (1924a): Oribatid mites: Euphthiracarus depresculus sp. n. and Euphthiracarus flavus (Ewing). — *Trans. Am. Micr. Soc.,* 43: 90—96.
Jacot, A. P. (1929): New Oribatoid mites. — *Psyche,* 35: 213—215.
Jacot, A. P. (1930): Oribatid mites of the subfamily Phthiracarinae of the Northeastern United States. — *Proc. Bost. Soc. Nat. Hist.,* 39: 209—261.
Jacot, A. P. (1933): Phthiracarid mites of Florida. — *J. Elisha Mitchell scient. Soc.,* 48: 232—267.
Jacot, A. P. (1936): Les Phthiracaridae de Carl Ludwig Koch. — *Revue suisse Zool.,* 43: 161—187.
Jacot, A. P. (1936): Some primitive moss-mites of North Carolina. — *J. Elisha Mitchell scient. Soc.* 52: 20—26.
Jacot, A. P. (1936): New moss-mites, chiefly Midwestern. — *Amer. Midl. Nat.,* 17: 346—553.
Jacot, A. P. (1936): More primitive moss-mites of North Carolina. — *J. Elisha Mitchell scient. Soc.,* 52: 247—253.
Jacot, A. P. (1937): New moss-mites, chiefly Midwestern II. — *Amer. Midl. Nat.,* 18: 237—250.
Jacot, A. P. (1937): Journal of North-American moss-mites. — *Journ. N.Y. Ent. Soc.,* 45: 353—375.
Jacot, A. P. (1938): New moss-mites, chiefly Midwestern III. — *Amer. Midl. Nat.,* 19: 647—657.
Jacot, A. P. (1938): More box-mites of the Northeastern United States.—*Journ. N. Y. Ent. Soc.,* 46: 109—145.
Jacot, A. P. (1938): More primitive moss-mites of North Carolina III. — *J. Elisha Mitchell scient. Soc.,* 54: 127—137.
Karppinen, E. (1958): Mitteilungen über einige für Finnland neue Oribatiden (Acar.). — *Ann. Ent. Fenn.,* 24: 192—196.
Knülle, W. (1957): Morphologische und Entwicklungsgeschichtliche Untersuchungen zum phylogenetischen System der Acari: Acariformes Zachv. I. Oribatei: Malaconothridae.—*Mitt. Zool. Mus. Berl.,* 33: 97—213.
Koch, C. L. (1836—1841): *Deutschlands Crustaceen, Myriapoden und Arachniden.* — Regensburg.
Koch, L. (1879): Acariden aus Sibirien und Novaja Semlja. — *Kongl., Sv. Vet. Akad. Handl.,* 16 (5): 1—136.
Krivolutsky, D. A. — Криволуцкий Д. А. (1964): Sellnickochthonius g. n. новый род панцирных клещей из семейства Brachychthoniidae Balogh, 1943 (Acariformes, Oribatei). — *Зоол. ж.,* 43 в. 6: 395—396.

Krivolutsky, D. A. — Криволуцкий, Д. А. (1965): Новые виды панцирных клещей (Acariformes, Oribatei) из таежной зоны СССР. — *Энтомол. обозр.*, т. 44, в. 3: 705—708.

Krivolutsky, D. A. — Криволуцкий, Д. А. (1965): О таксономии голарктических видов рода Hypochthonius C. L. Koch, 1835 (Acariformes, Oribatei). — *Зоол. ж.*, 44 (6): 934—937.

Krivolutsky, D. A.—Криволуцкий, Д. А. (1966): Некоторые материалы по панцирным клещам (Oribatei, Acariformes) из окрестностей Г. Благовещенска. — *Бюлл. М. О-ва Исп. Природы. отд. Биол.*, 71: 125—129.

Krivolutsky, D. A. — Криволуцкий, Д. А. (1971): Новые виды Oribatei из восточной Киргизии. — *Зоол. ж.*, 50 (6): 939—942.

Krivolutsky, D. A. (1971): Some new Oribatoid mites (Acariformes) from Central Asia. — *Vest. Cesk. spol. zool.*, 35 (1): 35—39.

Kunst, M. (1956): Nanhermannia Komáreki n. sp., eine neue Oribatiden Art aus Böhmen (Acarina: Oribatei). — *Act. Soc. Zool. Boh.*, 20: 267—272.

Kunst, M. (1959): Bulgarische Oribatiden (Acarina) III. — *Act. Univ. Carol., Biol.*, 1: 51—74.

Lange, A. B. — Ланге, А. Б. (1954): Морфология клеща Zachvatkinella belbiformis gen. n. et sp. n. — нового представителя группы Palaeacariformes (Acariformes). —*Зоол. ж.*, 33 (5): 1042—1052.

Lange, A. B. — Ланге, А. Б. (1972): Новые виды клещей палеакарид (Palaeacaridae). — *Вестник МГУ, серия биол.*, 4: 103—106.

Lange, A. B. — Ланге, А. Б. (1972): Новые виды клещей рода Zachvatkinella (Palaeacaridae, Acariformes) с Дальнего Востока, Байкала и Кавказа. — *Зоол. ж.*, 51 (12): 1889—1892.

Lange, A. B. — Лапге, А. Б. (1974): Gilarovella demetrii Lange, gen. nov. et sp. nov. — новый представитель палеакароидных клещей (Palaeacaroidea, Ctenacaridae) из Туркмении. — *Вестпик Моск. ун-та.*, И: 18—20.

Lions, J.-C. (1966): Contribution a l'étude du Genre Rhysotritia (Acarien, Oribate): Rhysotritia clavata sextiana n. subsp. — *Vie et Milieu*, 17 (2): 779—794.

Lions, J.-C. (1968): Oribates (Acariens) de la Sainte Baume (1re série). Steganacarus travei n. sp. — *Acarologia*, 10 (3): 500—515.

Macfarlane, D. and Sheals, J. G. (1965): A new Phthiracarid mite from Israel. — *Acarologia.* 7 (3): 577—583.

Mahunka, S. (1966): A study of Oribatids collected by prof. Dr. F. di Castri on the Mt. Spitz (Recoaro, Italy). — *Atti Ist. Ven. Sci. Lett. Art.*, 124: 369—386.

Mahunka, S. (1969): Beiträge zur Kenntnis der Milbenfauna Ungarns I. — *Folia ent. hung.*, 22: 21—30.

Mahunka, S. (1973): Zwei neue Lohmanniiden-Arten aus Korea (Acari, Oribatida). — *Folia ent. hung.*, 26: 49—56.

Mahunka, S. (1974): Neue und interessante Milben aus dem Genfer Museum XII. Beitrag zur Kenntnis der Oribatiden-Fauna Griechlands (Acari). — *Revue suisse Zool.*, 81: 569—590.

Mahunka, S. (1977): Neue und interessante Milben aus dem Genfer Museum XXXI. A remarkable sample of archaic soil mites from Kenya (Acari: Oribatida). — *Revue suisse Zool.*, 84: 463—479.

Mahunka, S. (1977): Neue und interessante Milben aus den Genfer Museum XXVIII. Zwei neue Oribatiden-Arten (Acari) aus Israel. — *Acarologia*, 19: 132—135.

Mahunka, S. (1977): Neue und interessante Milben aus dem Genfer Museum XXXIII. Recent data on the Oribatid fauna of Greece (Acari: Oribatida). — *Revue suisse Zool.*, 84: 541—556.

Mahunka, S. (1979): Complementary data to the knowledge of some Oribatid species (Acari). — *Folia ent. hung.*, 32: 139—152.

Mahunka, S. (1979): Neue und interessante Milben aus dem Genfer Museum XLI. Vierter Beitrag zur Kenntnis der Oribatiden-Fauna Griechenlands (Acari: Oribatida). — *Revue suisse Zool.*, 86: 541—571.

Mahunka, S. (1980): Data to the knowledge of mites preserved in the "Berlese Collection" (Acari: Tarsonemina, Oribatida) II. — *Folia ent. hung.*, 41 (33): 105—121.

Mahunka, S. (1980): Oribatids from Tunesian soils (Acari: Oribatida) I. — *Folia ent. hung.*, 41 (33): 123—134.

Mahunka, S. (1982): Neue und interessante Milben aus dem Genfer Museum XXXIX. Fifth Contribution to the Oribatid Fauna of Greece (Acari: Oribatida). — *Revue suisse Zool.*, 89: 497—515.

Märkel, K. (1958): Über die Hornmilben (Oribatei) in der Rohhumusauflage älterer Fichtenbestände des Osterzgebirges. — *Arch. f. Forstwesen*, 7: 459—501.

Märkel, K. (1964): Die Euphthiracaridae Jacot, 1930, und ihre Gattungen (Acari, Oribatei). — *Zool. Verh. Leiden*, 67: 1—78.

Märkel, K. (1968): Zwei Euphthiracariden-Arten aus dem Tienschan und dem Altai-Gebirge (Acari, Oribatei). — *Acarologia*, 10: 725—732.
Märkel, K. und Mayer, I. (1959): Zur Systematik der deutschen Euphthiracarini (Acari, Oribatei). — *Zool. Anz.*, 163: 327—342.
Michael, A. D. (1885): New British Oribatidae. — *Journ. Roy. Micr.*, 2, 5: 385—397.
Michael, A. D. (1888): British Oribatidae. — *Ray Soc.*, 2, 65.
Michael, A. D. (1898): Oribatidae. — *Das Tierreich, Acarina*, 3: 1—91.
Mihelcic, F. (1956): Oribatiden Südeuropas III. — *Zool. Anz.*, 156: 9—29.
Mihelcic, F. (1957): Oribatiden Südeuropas VII. — *Zool. Anz.*, 159: 41—68.
Moritz, M. (1966): Paratritia baloghi n. g., n. sp., eine neue Gattung und Art der Familia Euphthiracaridae (Acari: Oribatei) aus Deutschland. — *Acarologia*, 8 (2): 374—381.
Moritz, M. (1974): Beiträge zur Kenntnis der Oribatiden (Acari) Europas VI. Neoctenacarus hastilis nov. gen., nov. spec., eine neue Ctenacaride aus der DDR (Palaeosomata, Ctenacaroidea). — *Abh. Ber. NaturkMus.-ForschStelle, Görlitz*, 48 (4): 1—15.
Moritz, M. (1976): Revision der europäischen Gattungen und Arten der Familie Brachychthoniidae (Acari, Oribatei). Teil 1. Allgemeiner Teil: Brachychthoniidae Thor, 1934. Spezieller Teil: Liochthonius v. d. Hammen, 1959, Verachthonius nov. gen. und Paraliochthonius nov. gen. — *Mitt. Zool. Mus., Berlin*, 52: 27—136.
Moritz, M. (1976): Revision der europäischen Gattungen und Arten der Familie Brachychthoniidae (Acari, Oribatei). Teil. 2. Mixochthonius Niedbala, 1972. Neobrachychthonius nov. gen., Synchthonius v. Hammen, 1952, Poecilochthonius Balogh, 1943, Brachychthonius Berlese, 1910, Brachychochthonius Jacot, 1938. — *Mitt. Zool. Mus., Berlin*, 52: 227—319.
Nicolet, H. (1855): Histoire naturelle des Acariens qui se trouvent aux environs de Paris. — *Arch. Mus. Hist. Nat.*, 7: 381—482.
Niedbala, W. (1971): Liochthonius occultus sp. nov. (Acari, Oribatei, Brachychthoniidae) a new species of moss mite from Poland. — *Bull. Acad. pol. Sci. Cl. II. Sér. Sci. biol.*, 19 (10): 677—679.
Niedbala, W. (1972): Mechowce (Acari, Oribatei) nowe dla fauny Polski. II. — *Fragm. faun.*, 18 (3): 33—44.
Niedbala, W. (1972): Mechowce (Acari, Oribatei) nowe dla fauny Polski oraz uwagi o kilku innych. III. — *Fragm. faun.*, 17 (10): 183—196.
Niedbala, W. (1972): Studies on the family Brachychthoniidae (Acari, Oribatei) I. Revision of the genera and description of one new species. — *Bull. Acad. pol. Sci. Cl. II. Sci. biol.*, 20 (9): 661—669.
Niedbala, W. (1973): Studies on the family Brachychthoniidae (Acari, Oribatei) II. Notes concerning the evolution of genera. — *Bull. Acad. pol. Sci. Cl. II. Sér. Sci. biol.*, 21 (1): 59—73.
Niedbala, W. (1974): Studia nad rodzina Brachychthoniidae (Acari, Oribatei), III. — *Annls. zool., Warsz.*, 31 (6): 459—524.
Niedbala, W. (1976): Brachychthoniidae Polski (Acari, Oribatei). — *Monografie fauny Polski*, 6: 144 pp.
Niedbala, W. (1977): An attempt of distinguishing subgeneric taxa in the genus Liochthonius Hammen (Acari, Oribatei, Brachychthoniidae). — *Annls zool., Warsz.*, 33 (23): 477—494.
Norton, R. A., Metz, L. J. and Sharma, G. D. (1978): Observations on Epilohmannoides Jacot, 1936 (Acarina: Oribatei), with the description of a new species. — *J. Georgia Entomol. Soc.*, 13 (2): 134—148.
Ohkubo, N. (1979): A new species of the genus Epilohmannoides (Acarina, Oribatida) from Japan. — *Annotnes, zool. jap.*, 52 (4): 261—265.
Oudemans, A. C. (1900): New List of Dutch Acari, 1st Part. — *Tijdschr. Ent.*, 43: 150—171.
Oudemans, A. G. (1901): Remarks on the denomination of the genera and higher groups in "Das Tierreich, Oribatidae". — *Tijdschr. Ent.*, 43 (1900): 140—149.
Oudemans, A. C. (1915): Overzicht der tot 1898 beschreven Phthiracaridae. — *Ent. Ber.*, 4: 212—220.
Oudemans, A. C. (1917): Acarologische Aanteekiningen LXII. — *Ent. Ber.*, 4: 341—348.
Oudemans, A. C. (1917): Notizen über Acari, 25. Reihe (Trombididae, Oribatidae, Phthiracaridae). — *Arch. Naturg.*, 82: 1—84.
Oudemans, A. C. (1923): Studie over de sedert 1877 ontworpen systemen der Acari; nieuwe classificatie; phylogenetische beschouwingen. — *Tijdschr. Ent.*, 66: 49—85.
Oudemans, A. C. (1925): Notizen über Acari, 27. Reihe. — *Arch. f. Naturg.*, 91 (8).
Parry, B. W. (1978): A new species of Steganacarus (Acari, Cryptostigmata) from Israel. —*Bull. Br. Mus. nat. Hist. (Zool.)*, 33 (4): 279—285.
Parry, B. W. (1979): A new species of Phthiracarus (Acari, Cryptostigmata) from Austria. — *Bull. Br. Mus. nat. Hist. (Zool.)*, 37 (1): 73—79.

Parry, B. W. (1979): A revision of the British species of the genus Phthiracarus Perty, 1841 (Cryptostigmata: Euptyctima). — *Bull. Br. Mus. nat. Hist. (Zool.)*, 35: 323—363.
Pérez-Iñigo, C. (1967): Les Lohmanniidae d'Espagne (Acari, Oribatei). — *EOS*, 43: 157—170.
Pérez-Iñigo, C. (1968): Acaros oribatidos de suelos de Espana peninsular e islas Baleares (1.a parte). — *Graellsia*, 24: 143—238.
Pérez-Iñigo, C. (1969): Bioespeleologia de la cueva de Ojo Guarena. Ácaros Oribátidos. — *Bol. R. Soc. Española, Hist. Nat. (Biol.)*, 67: 143—160.
Pérez-Iñigo, C. (1969): Nuevos Oribátidos de Suelos Españoles (Acari. Oribatei). — *EOS*, 44: 377—403.
Pérez-Iñigo, C. (1972): Ácaros oribatidos de la isla de Tenerife. — *Bol. R. Soc. Española, Hist. Nat. (Biol.)*, 70: 185—206.
Pérez-Iñigo, C. (1974): Ácaros oribátidos de suelos de España peninsular e islas Baleares (Acari, Oribatei). Parte V. — *EOS*, 48: 367—475.
Perty, M. (1839): Isis, Oken, 1839. — 11 (12): column 847.
Piffl, E. (1963): Heptacarus notoneotrichus eine neue Hornmilbe aus Agypten (Oribatei — Lohmanniidae). — *Anz. Akad. Wiss. Wien*, 2: 24—30.
Piffl, E. (1966): Heptacarus supertrichus nov. spec., eine neue Lohmanniidenart (Oribatei — Acari) aus dem Irak. — *Anz. Akad. Wiss. Wien*, 9: 162—169.
Piffl, E. (1971): Neue Oribatiden (Acari) aus dem Himalaya. — *Khumbu Himal.*, 4 (1): 23—54.
Rafalski, J. (1966): Materialy do znajomości fauny mechowców (Acari, Oribatei) Polski. I. — *Fragm. faun.*, 12 (21): 347—372.
Rajski, A. (1967): Autecological-zoogeographical analysis of moss mites (Acari, Oribatei) on the basis of fauna in the Pozneń environs. — *Polskie Pismo ent.*, 37 (1): 69—166.
Reeves, R. M. and Marshall, V. G. (1971): Redescription and chaetotaxy of Brachychthonius lydiae adults and nymphs (Acarina: Oribatei). — *Ann. ent. Soc. Am.*, 64 (2): 317—325.
Ryabinin, N. A. and Krivolutzky, D. A. — Рябинин, Н. А. и Криволуцкий, Д. А. (1977): Новый род и виды панцирных клещей из семейства Heterochthoniidae. —*Зоол. ж.*, 56 (II): 1729—1731.
Schuster, R. (1957): Wiederfund und Beschreibung von Steganacarus phyllophorus (Berl.); (Oribatei, Acari). — *Zool. Anz.*, 158: 97—102.
Schuster, R. (1960): Die europäischen Arten der Gattung Perlohmannia Berlese (Acari, Oribatei). — *Zool. Anz.*, 164 (5—6): 185—195.
Schuster, R. (1960): Über die Ökologie und Verbreitung von Bodenmilben (Oribatei) am Alpen-Ostrand, insbesondere in der Steiermark. — *Mitt. naturw. Ver. Steierm.*, 90: 132—149.
Schuster, R. (1960): Über die Morphologie und Artengliederung der Gattung Epilohmannia Berlese, 1917; (Ac., Oribatei). — *Zool. Anz.*, 165 (5—6): 197—213.
Schuster, R. (1965): Über die Morphologie und Verbreitung einiger in Mitteleuropa seltener Milben (Acari-Oribatei). — Mitt. naturw. Ver. Steierm., 95: 211—228.
Schweizer, J. (1922): Beitrag zur Kenntnis der terrestrischen Milbenfauna der Schweiz. — *Vehr. naturf. Ges. Basel*, 33: 23—112.
Schweizer, J. (1948): Die Landmilben aus der Umgebung des schweizerischen Nationalparks. — *Ergebn. wiss. Unters. schweiz. Nationalparks*, 2 (20): 1—28.
Schweizer, J. (1956): Die Landmilben des schweizerischen Nationalparks, Teil 3: Sarcoptiformes Reuter, 1909. — *Ergebn. wiss. Unters. schweiz. Nationalparks*, 5 (34): 215—377.
Sellnick, M. (1920): Neue und seltene Oribatiden aus Deutschland. — *Schr. Phys.-ökon. Ges. Königsberg*, 61, 62: 35—42.
Sellnick, M. (1922): Milben der Sammlung des Deutschen Entomologischen Instituts. I. Oribatidae — *Ent. Mitt.*, 11: 18—20.
Sellnick, M. (1923): Die Gattung Hermannia Nicolet. — Acari — Blätter für Milbenkunde (Lötzen), 2: 3—6. Nachdruck in: *Abh. Ber. Naturkundemus. Görlitz*, 43: 29—31 (1968).
Sellnick, M. (1923): Die mir bekannten Arten der Gattung Tritia Berl. — Acari — Blätter für Milbenkunde (Lötzen), 3: 7—22. Nachdruck in: *Abh. Ber. Naturkundemus. Görlitz*, 43: 32—47 (1968).
Sellnick, M. (1928): Formenkreis: Hornmilben, Oribatei. — *Tierw. Mitteleur.*, 3, 9: 1—42.
Sellnick, M. (1929): Die Oribatiden (Hornmilben) des Zehlaubruches. — *Schr. phys.-ökon. Ges. Königsberg*, 66: 324—351.
Sellnick, M. (1931): Acari. Zoologische Forschungsreise nach den Jonischen Inseln und dem Peloponnes von M. Beier. XVI. Teil. — *Sber. Akad. Wiss. Wien. Abt. I.* 140: 693—776.
Sellnick, M. (1932): Oribatiden aus dem Karst. — *Zool. Jb. (Syst.)*, 63: 701—714.

Sellnick, M. (1944): Acarina. In Hammer: Studies on the Oribatids and and Collemboles of Greenland. — *Meddr. Grønland*, 141: 40—48.
Sellnick, M. (1960): Formenkreis: Hornmilben, Oribazei (Nachtrag). — *Tierw. Mitteleur.*, 3 (Ergänzung): 45—134.
Sellnick, M. et Forsslund, K. H. (1955): Die camisiidae Schwedens (Acar. Oribat.). — *Ark. Zool.* 2, 8: 473—530.
Seyd, E. L. (1970): Note on Heminothrus valentianus, together with a further record from Britain and synonymy of Platynothrus punctatus (Acari: Oribatei.). — *J. Zool., Lond.*, 160: 291—296.
Sheals, J. G. (1965): Primitive cryptostigmatid mite from Rhododendron forests in the Nepal Himalaya. — *Bull. British Mus (Nat. Hist.) London*, 13: 1—35.
Sitnikova, L. G. — Ситникова, Л. Г. (1977): Новые виды панцирных клещей семейств Ameronothridae и Hermanniidae (Oribatei) Фауны Советского Союза. — *Паразит. сб.*, 27: 160—168.
Strenzke, K. (1951): Die norddeutsche Arten der Brachychthonius und Brachychochthonius (Acarina, Oribatei). — *Dt. zool. Z.*, 1: 234—249.
Strenzke, K. (1953): Zwei neue Arten der Oribatiden-Gattung Nanhermannia. — *Zool. Anz.*, 150: 69—75.
Strenzke, K. (1962): Der erste Fund der Oribatide Palaecarus hystricinus in Deutschland (Acarina). — *Zool. Anz.*, 168 (7—10): 310—311.
Strenzke, K. (1963): Entwicklung und Verwandtschaftsbeziehungen der Oribatidengattung Gehypochthonius (Arach., Acari). — *Senck. biol.*, 44: 231—255.
Subias, L.-S. (1977): Taxonomia y Ecologia de los Oribatidos Saxicolas y Arboricolas de la Sierra del Guadarrama (Acarida, Oribatida). — *Publ. Dept. Zool. Fac. Biol. Univ. Compl., Madrid*, 24: 1—379.
Subias, L.-S. y Pérez-Iñigo, C. (1978): Descripción de Bursoplophora iberica nov. g., nov. sp. (Acari, Oribatei) y algunas consideraciones sobre la familia Protoplophoridae. — *EOS*, 52: 387—398.
Štorkán, J. (1925): Přispěvky ku známosteb o českych Oribatidech. Acarina. — *Spisy prirodov Gak. Karl. Univ.*, 42: 1—40.
Tarman, K. (1962): The Oribatids fauna of Macedonia and Montenegro. — *Izdan. Publ. Zavod. ribar. NRM, Skopje*, 3 (2): 138—154.
Thor, S. (1930): Beiträge zur Kenntnis der Invertebraten-Fauna von Svalbard. — *Skr. Svalbard. Ishavet Oslo*, 27: 1—156.
Thor, S. (1934): Neue Beiträge zur Kenntnis der Invertebraten-Fauna von Svalbard. — *Zool. Anz.*, 107: 114—139.
Thorell, T. (1872): Om Arachnider fran Spitsbergen och Beeren-Eiland. — *Öfvers. K. Vetensk. Akad. Föhr.*, 28: 683—702 (1871).
Trägårdh, I. (1910): Acariden aus dem Sarekgebirge. — *Naturw. Unters. Sarekgebirges in Schw. — Lappland, Stockholm*, 4: 375—586.
Trägårdh, I. (1931): Terrestrial Acarina. — *Zoology of the Faroes*, 49: 1—69.
Trägårdh, I. (1932): Palaeacariformes, a new suborder of Acari. — *Ark. Zool.* 24B, 2: 1—6.
Tuxen, S. L. (1952): Die Jugendstadien der norischen Camisiiden (Acar. Orib.). — *Saertryk ent. Medd.*, 26: 392—403.
Walker, N. A. (1964): Euphthiracaroidea of California Sequoia litter; with a reclassification of the families and genera of the world. — *Fort. Hays Stud. N. S., sci. Ser.*, 3: 1—154.
Wallwork, J. A. (1960): Some Oribatei from Ghana I. Sampling localities II. Some members of the Enarthronota Grandj. — *Acarologia*, 2: 368—388.
Walzl, M. G. (1974): Mesoplophora graeca nov. spec. — *Acarologia*, 15: 534—539 (1973).
Weis-Fogh, T. (1948): Ecological investigations on mites and collemboles in the soil. — *Natura Jutlandica Aarhus*, 1: 135—270.
Willmann, C. (1919): Diagnosen einiger neuen Oribatiden aus der Umgegend Bremens. — *Abh. Naturw. Ver. Bremen*, 24: 552—554.
Willmann, C. (1923): Oribatiden aus Quellmoosen. — *Arch. Hydrobiol*, 14: 470—477.
Willmann, C. (1925): Neue und seltene Oribatiden. — *Jahresber. Ent. Bremen*, 13: 7—11.
Willmann, C. (1928): Die Oribatidenfauna nordwestdeutscher und einiger süddeutscher Moore. — *Abh. naturw. Ver. Bremen*, 27: 143—176.
Willmann, C. (1928): Neue Oribatiden I. — *Zool. Anz.*, 76: 1—5.
Willmann, C. (1928): Die Oribatidenfauna nordwestdeutscher und einiger süddeutscher Moore. — *Abh. Naturw. Ver. Bremen*, 27: 143—176.

Willmann, C. (1930): Neue und bemerkenswerte Oribatiden der Sammlung Oudemans. — *Abh. naturw. Ver. Bremen*, 28: 1—12.
Willmann, C. (1931): Moosmilben oder Oribatiden (Oribatei). — *Tierw. Deutschl.*, 22: 79—200.
Willmann, C. (1932): Die Oribatiden des Dummersdorfer Ufers (Untertrave). — *Das Linke Untertraveufer (Dummersdorfer Ufer)*, pp. 122—443.
Willmann, C. (1936): Neue Acari aus schlesischen Wiesenböden. — *Zool. Anz.*, 113: 273—290.
Willmann, C. (1939): Die Moorfauna des Glatzer Schneeberges, 3: Die Milben der Schneebergmoore. — *Beitr. Biol. Glatzer Schneeb.*, 5: 427—458.
Willmann, C. (1939): Die Arthropodenfauna von Madeira nach den Ergebnissen der Reise von Prof. dr. O. Lundblad, Juli-Aug. 1935, XIV: Terrestrische Acari. — *Ark. Zool.*, 31 (A): 1—42.
Willmann, C. (1942): Acari aus nordwestdeutschen Mooren. — *Abh. naturw. Ver. Bremen*, 32: 163—183.
Willmann, C. (1943): Terrestrische Milben aus Schwedisch-Lappland. — *Arch. Hydrobiol.*, 40: 208—239.
Willmann, C. (1951): Die hochalpine Milbenfauna der mittleren Hohen Tauern insbesondere des Grossglockner-Gebietes (Acari). — *Bonn. Zool. Beitr.*, 2: 141—176.
Willmann, C. (1952): Die Milbenfauna der Nordseeinsel Wangerooge. — *Veröff. Inst. Meeresforsch. Bremerhaven*, 1: 139—186.
Willmann, C. (1953): Neue Milben aus den österlichen Alpen. — *Sitz.-Ber. öst. Akad. Wiss., Abt.* 1, 162: 449—519.
Willmann, C. (1954): Mährische Acari, hauptsächlich aus dem Gebiet des Mährischen Karstes. — *Ceskoslov. parasitol.*, 1: 213—272.
Willmann, C. (1956): Milben aus m Naturschutzgebiet auf dem Spieglitzer (Glatzer) Schneeberg. — *Ceskoslov. parasit.*, 3: 211—273.
Woolley, T. A. and Higgins, H. G. (1958): A Revision of the Family Nanhermaniidae (Acari: Oribatei). — *Proc. Tenth Int. Congress Ent.*, 1: 913—923 (1956).
Yamamoto, Y. (1977): Oribatid fauna of the experimental forest of Tamagawa University in Hakone, Central Japan I. Oribatei Inferiores. — *Bull. biogeogr. Soc., Japan*, 32: 33—42.
Yamamoto, Y. and Aoki, J. (1971): The fauna of the lava caves around Mt. Fuji-san VII. Malaconothridae (Acari, Cryptostigmata). — *Bull. natn. Sci. Mus., Tokyo*, 14 (4): 579—583.
Zachvatkin A. A. — Захваткин, А. А. (1945): К морфологии Beklemishevia galeodula n. g. et n. sp. нового представителя группы Palaeacariformes (Acarina). — *Бюлл. М. О-Ва исп. природы, отд. Биол.*, 50: 60—71.
Zakhvatkin, A. A. (1945): On the discovery of Palaeacariformes (Acarina) in the USSR. — *Comt. Rend. (Dokl.) de l'Acad. Sci. URSS*, 47: 673—676.

PLATES

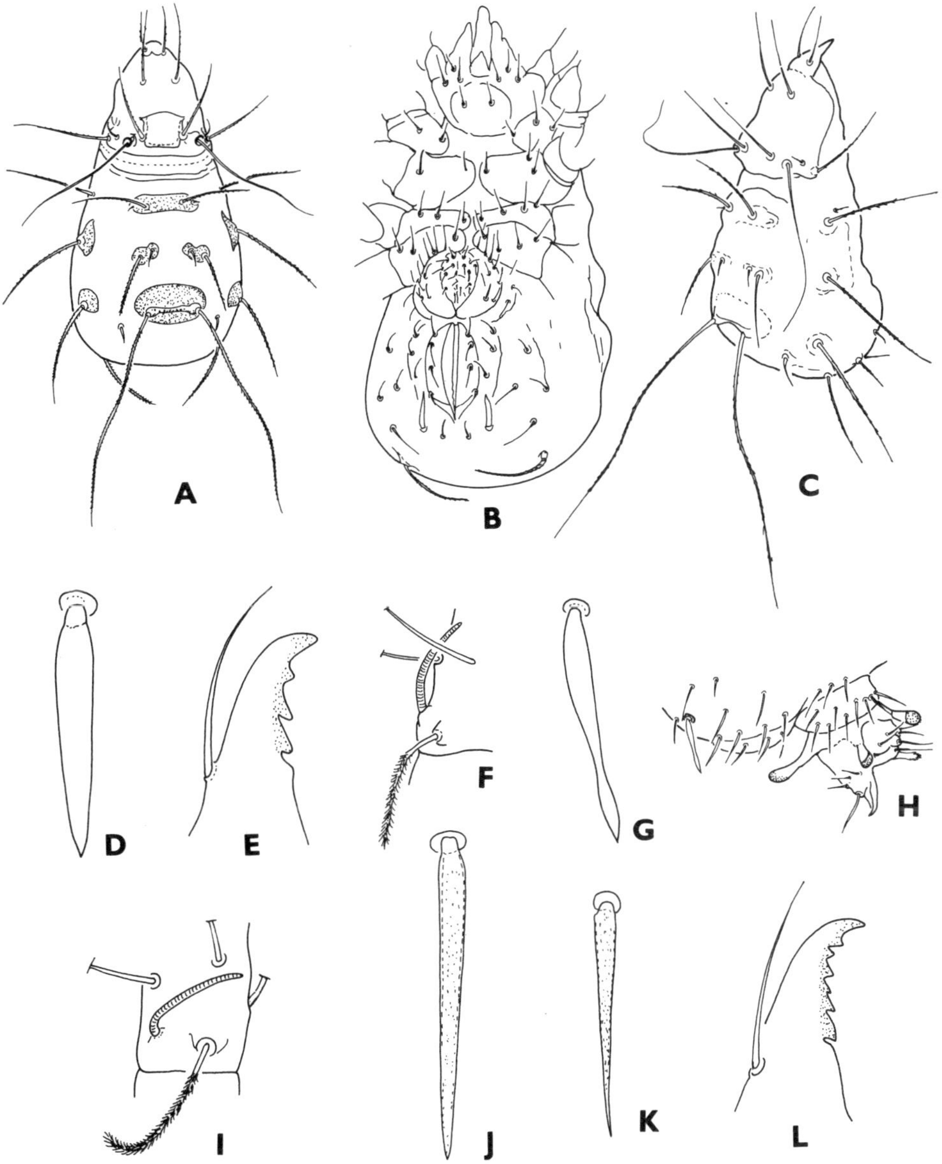

Plate 1. A—E: *Zachvatkinella belbiformis* Lange, 1954 — A: dorsal side, B: ventral side, C: lateral side, D: seta ps_2, E: chelicera; F—H: *Zachvatkinella doryura* Lange, 1972 — F: solenidium ω_p of leg I, G: seta ps_2, H: anogenital region from lateral side; I—J: *Zachvatkinella baicalica* Lange, 1972 — I: solenidium ω_p of leg I, J: seta ps_2; K—L: *Zachvatkinella caucasica* Lange, 1972 — K: seta ps_2, L: chelicera (after Lange)

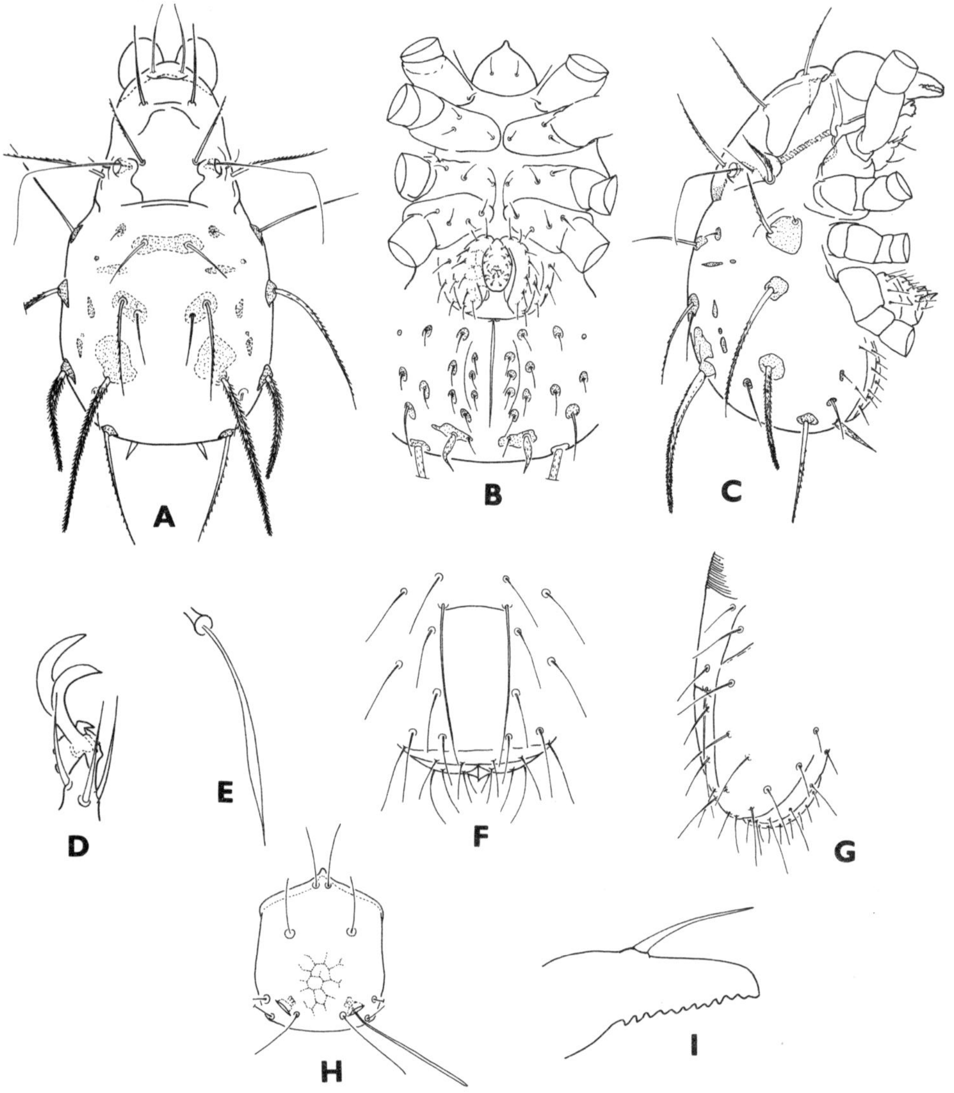

Plate 2. A—D: *Acaronychus tragardhi* Grandjean, 1932 — A: dorsal side, B: ventral side, C: lateral side, D: end of tarsus I; E: *Acaronychus grandjeani* Lange, 1975 — sensillus; F—H: *Adelphacarus sellnicki* Grandjean, 1952 — F: distal end of notogaster from dorsal side, G: distal end of notogaster from lateral side, H: prodorsum; I: *Amuracarus voskresenskii* Lange, 1975 — chelicera (after Grandjean: A—H and Lange: I)

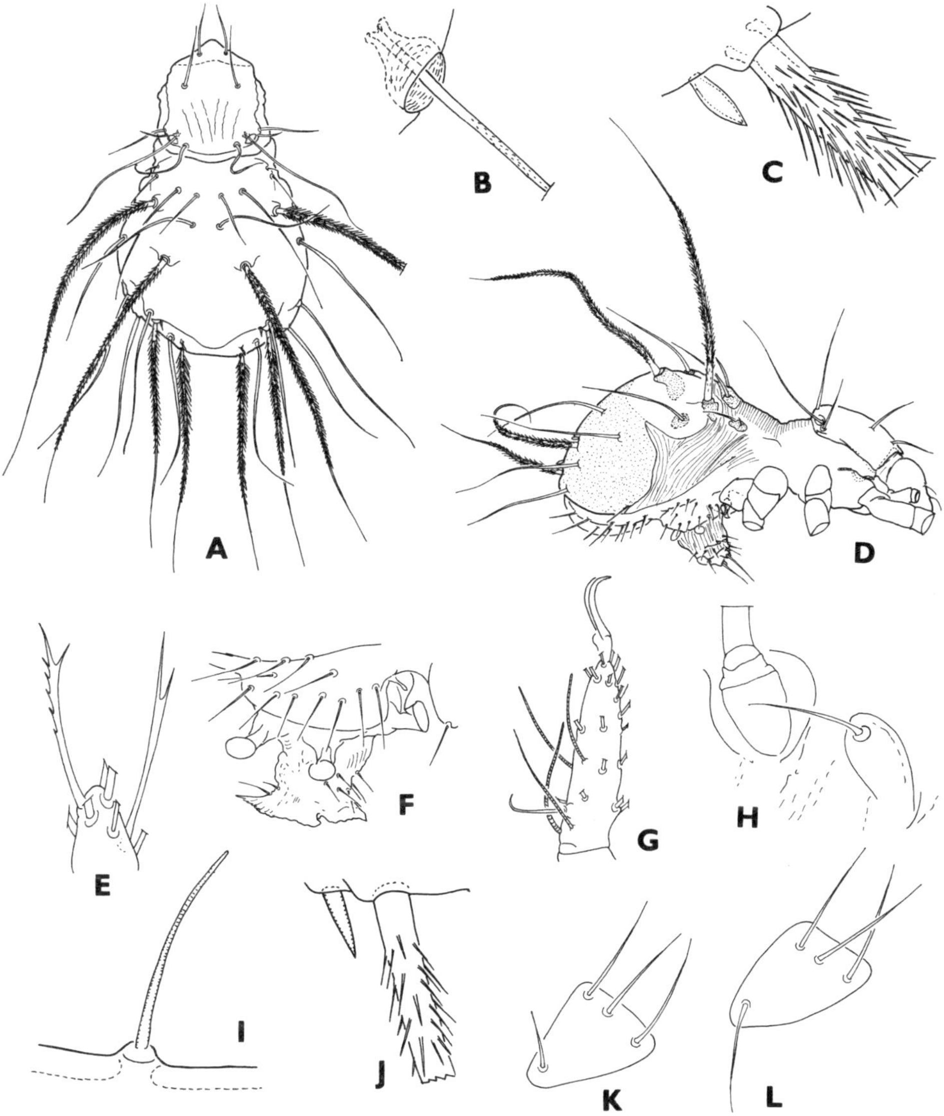

Plate 3. A—D: *Palaeacarus hystricinus* Trägårdh, 1932 — A: dorsal side, B: bothridium, C: setae h_1 and h_2, D: lateral side; E—H: *Palaeacarus caucasicus* Lange, 1972 — E: distal end of palp, F: genital region of male, G: tarsus of leg I, H: setae d_2 and *cp*; I—J: *Palaeacarus kamensikii* Zachvatkin, 1945 — I: famulus (ε) of leg I, J: setae h_1 and h_2; K: *Palaeacarus orientalis* Bulanova-Zachvatkina, 1967—coxa of leg I; L: *Palaeacarus lapshovi* Zachvatkin, 1945 — coxa of leg I (after Aoki: A, Grandjean: B—D, I—J, Lange: E—H and Bulanova-Zachvatkina: K—L)

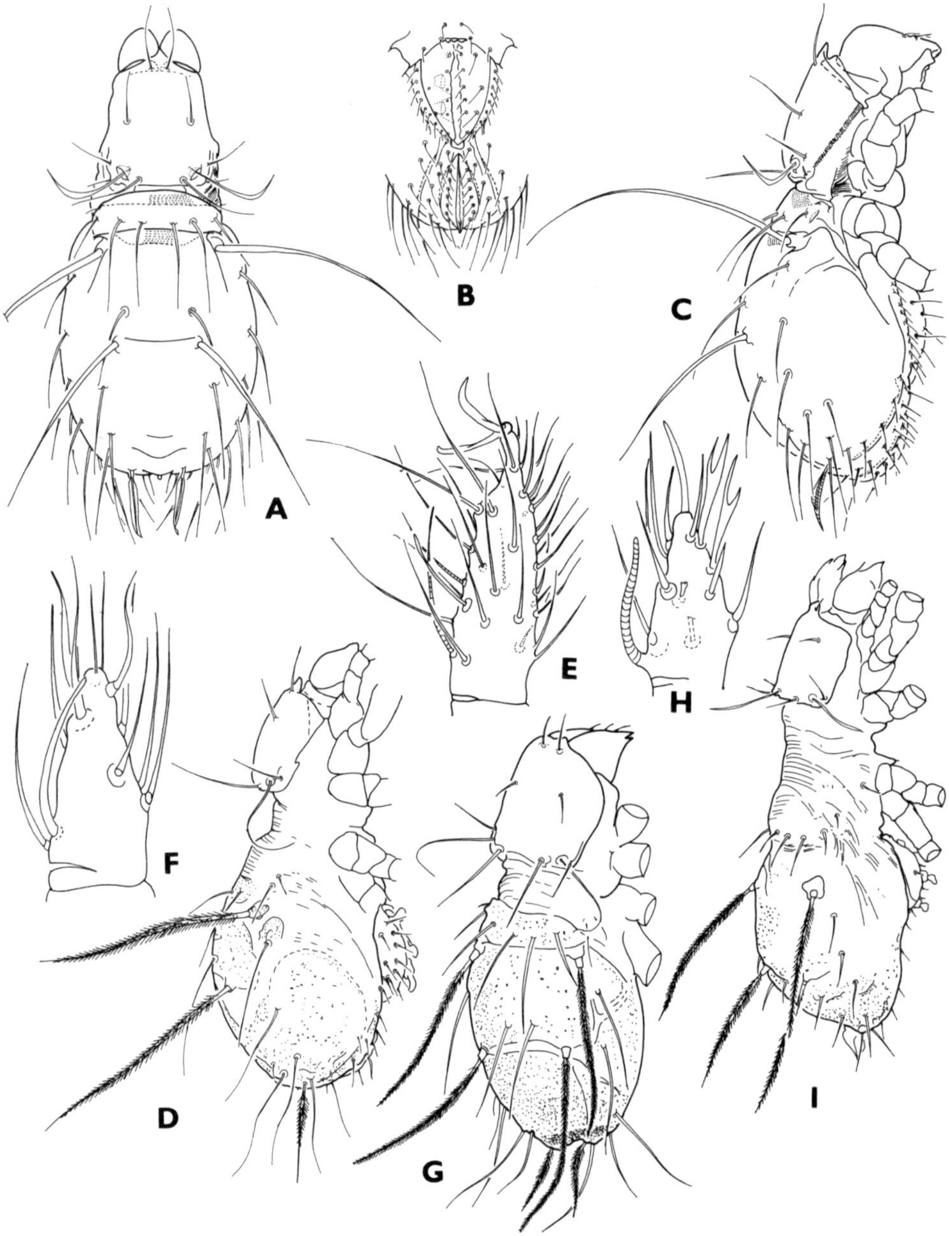

Plate 4. A—C: *Ctenacarus araneola* (Grandjean, 1932) — A: dorsal side, B: anogenital region, C: lateral side; D—F: *Palaeacaroides pacificus* Lange, 1972 — D: lateral side, E: tarsus of leg I, F: distal end of palp; G: *Beklemishevia galeodula* Zachvatkin, 1945 — dorsal side; H—I: *Gilarovella demetrii* Lange, 1974 — H: distal end of palp, I: dorsal side (after Grandjean: A—C and Lange: D—I)

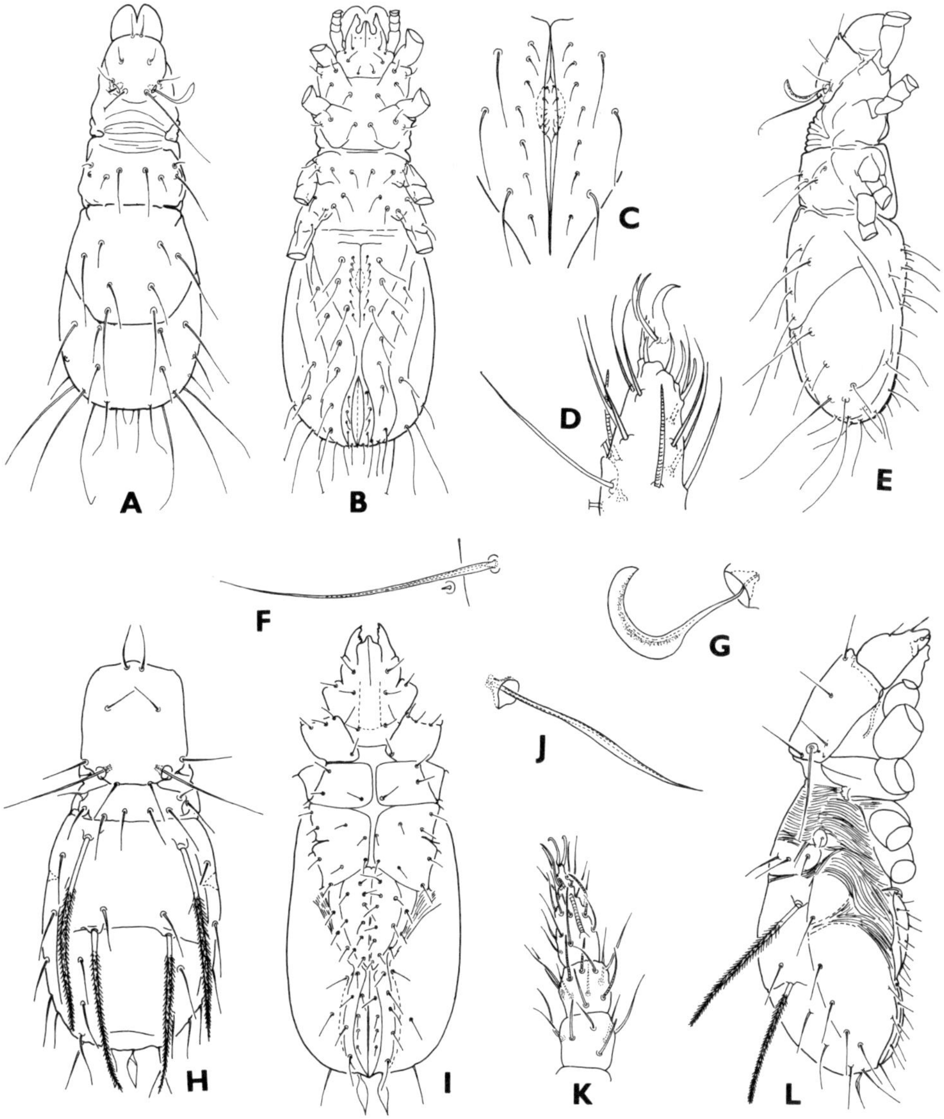

Plate 5. A—G: *Aphelacarus acarinus* (Berlese, 1910) — A: dorsal side, B: ventral side, C: genital region, D: distal end of tarsus I, E: lateral side, F: setae e_1 and f_1, G: sensillus; H—L: *Neoctenacarus hastilis* Moritz, 1974 — H: dorsal side, I: ventral side, J: sensillus, K: leg I, L: lateral side (after Grandjean: A—G and Moritz: H—L)

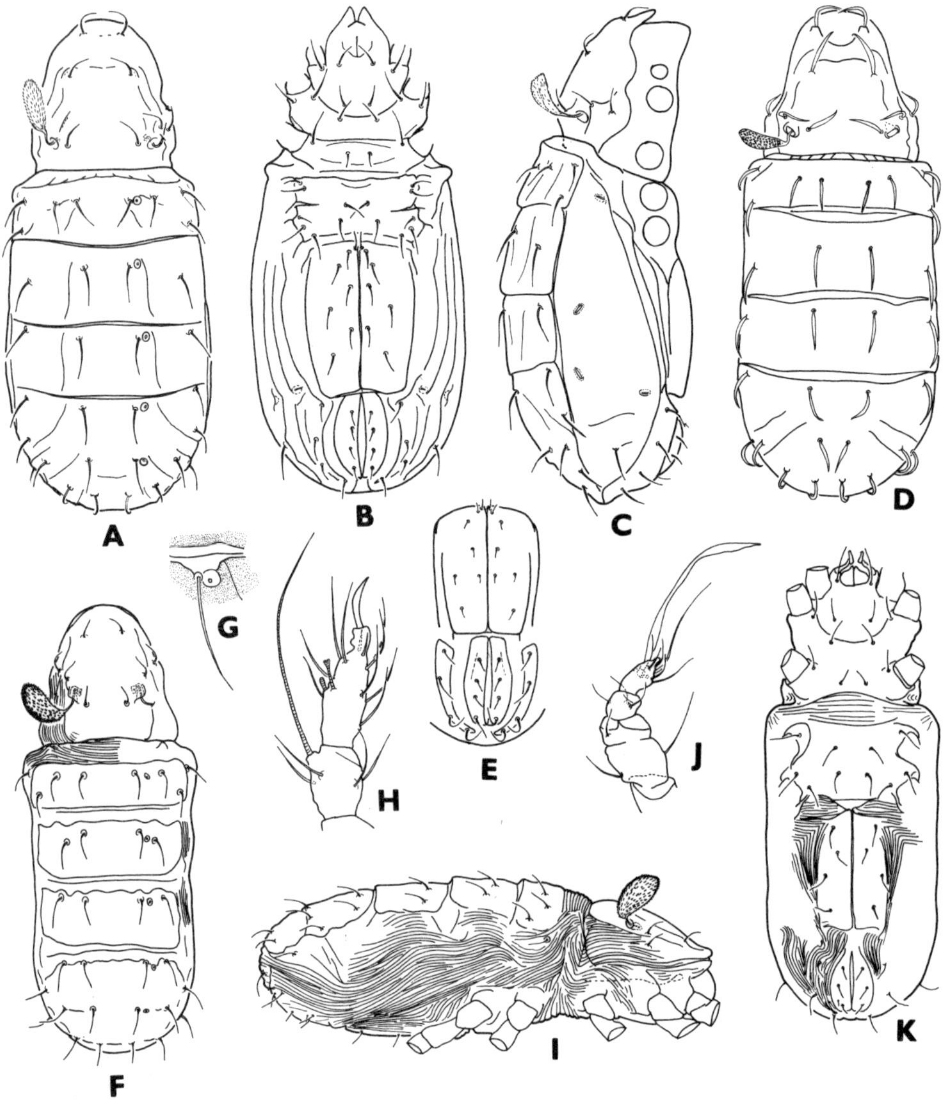

Plate 6. A—C: *Haplochthonius simplex* Willmann, 1930 — A: dorsal side, B: ventral side, C: lateral side; D—E: *Haplochthonius sanctaeluciae* Bernini, 1973 — D: dorsal side, E: anogenital region; F—K: *Amnemochthonius taeniophorus* Grandjean, 1948 — F: dorsal side, G: seta e_1, H: tarsus of leg I, I: lateral side, J: palp, K: ventral side (after Grandjean: A—C, F—K and Moritz: D—E)

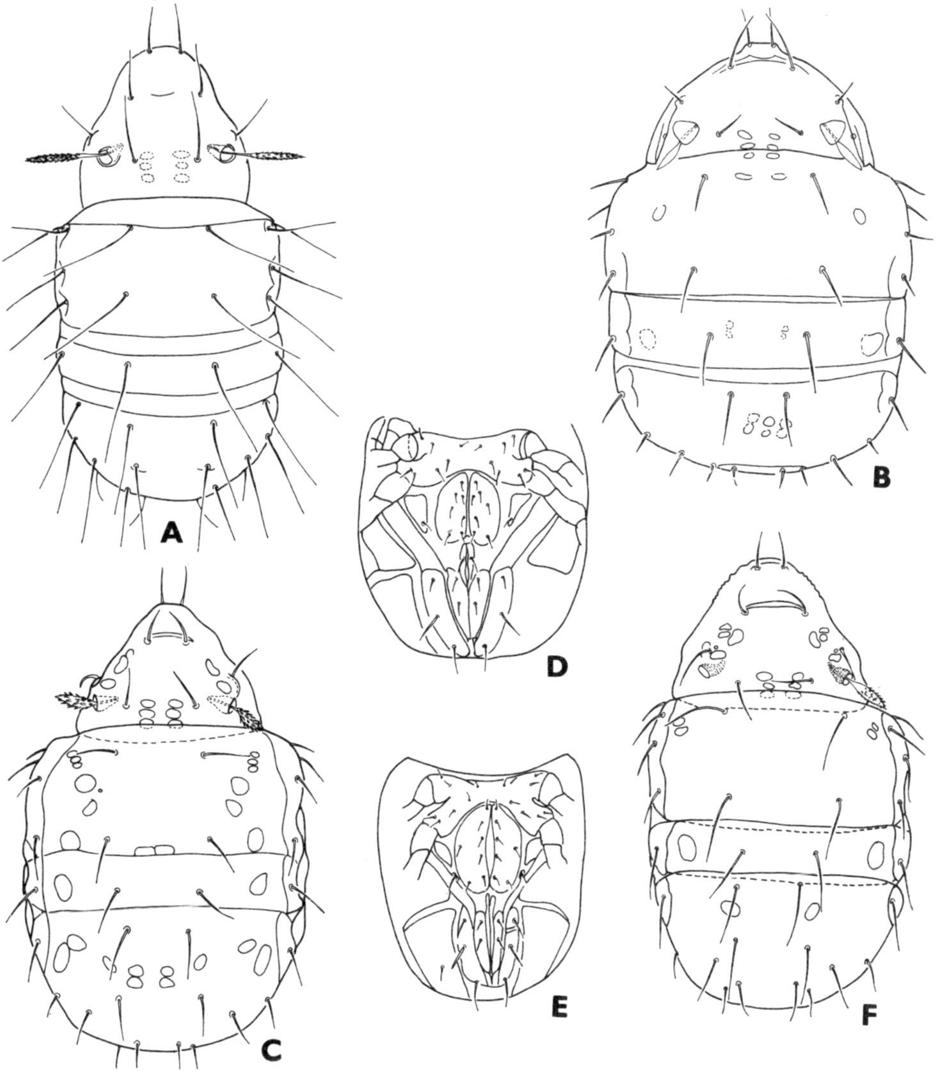

Plate 7. A: *Eobrachychthonius longisetosus* Csiszár, 1961 — dorsal side; B: *Eobrachychthonius mooseri* (Van der Hammen, 1950) — dorsal side; C—D: *Eobrachychthonius latior* (Berlese, 1910) — C: dorsal side, D: ventral side; E—F: *Eobrachychthonius borealis* Forsslund, 1942 — E: ventral side, F: dorsal side (after Csiszár: A, Van der Hammen: B and Niedbala: C—F)

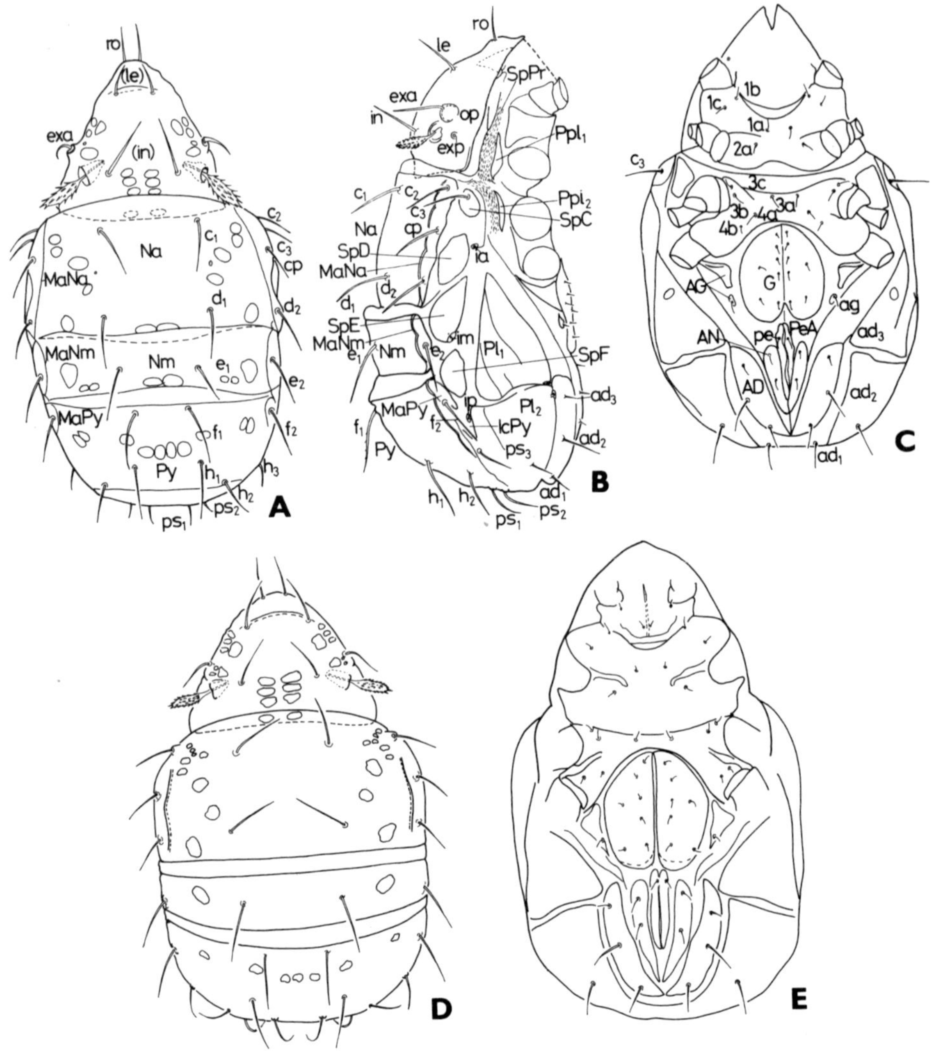

Plate 8. A—C: *Eobrachychthonius oudemansi* Van der Hammen, 1952 — A: dorsal side, B: lateral side, C: ventral side; D—E: *Eobrachychthonius similis* Mahunka, 1979 — D: dorsal side, E: ventral side (after Niedbala: A, C and Moritz: B; D—E original)

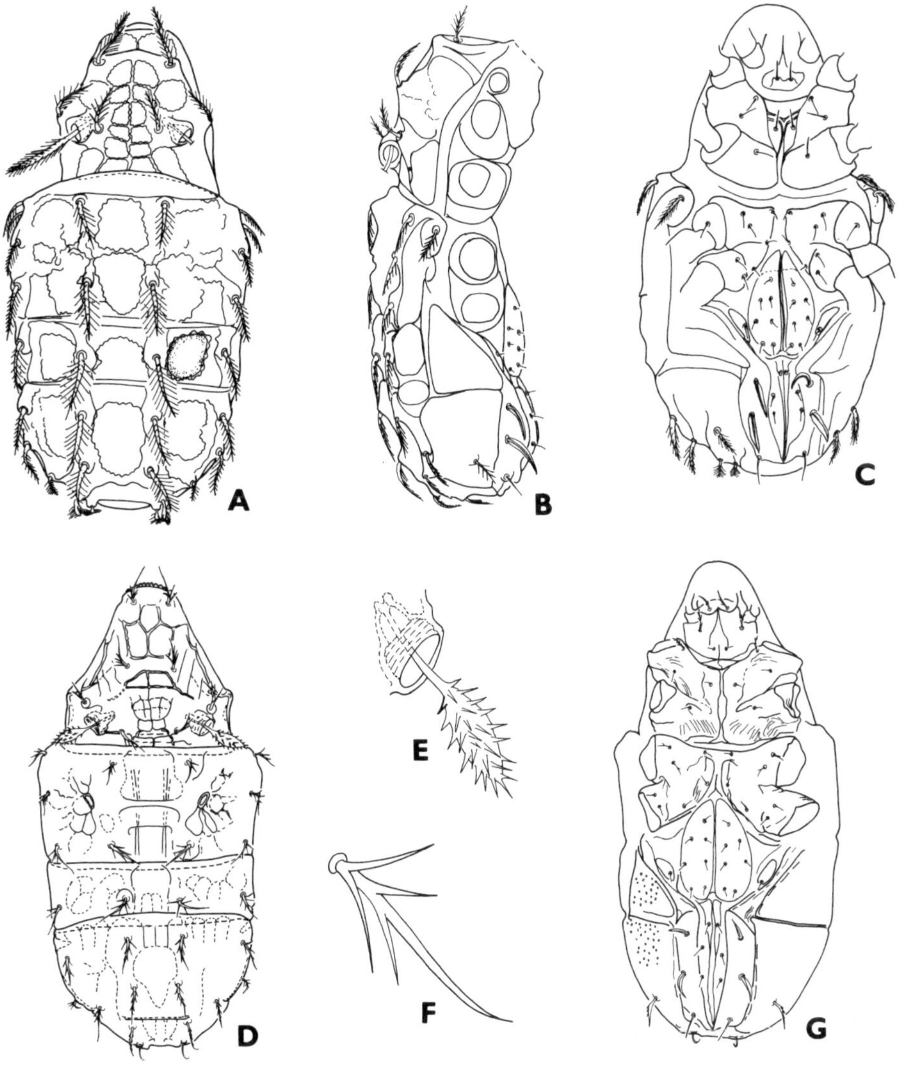

Plate 9. A—C: *Brachychochthonius lydiae* Jacot, 1938 — A: dorsal side, B: lateral side, C: ventral side; D—G: *Brachychochthonius japonicus* (Chinone, 1974) — D: dorsal side, E: sensillus, F: seta d_1, G: ventral side (after Reeves and Marshall: A—C and Chinone: D—G)

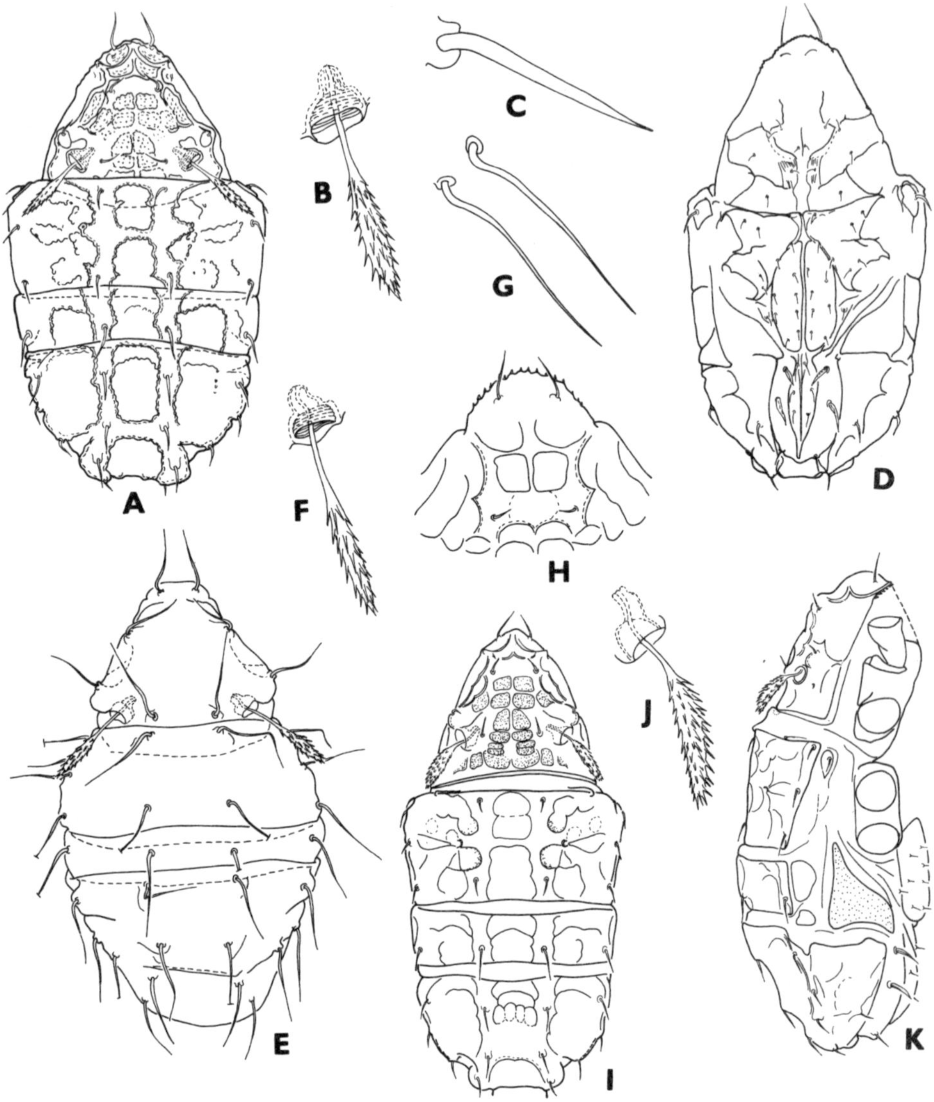

Plate 10. A—D: *Brachychochthonius gracilis* (Chinone, 1974) — A: dorsal side, B: sensillus, C: seta e_1, D: ventral side; E—G: *Brachychochthonius planus* (Chinone, 1974) — E: dorsal side, F: sensillus, G: setae *in* and e_1; H—K: *Brachychochthonius immaculatus* Forsslund, 1942 — H: rostral part of prodorsum, I: dorsal side, J: sensillus, K: lateral side (after Chinone: A—G and Moritz: H—K)

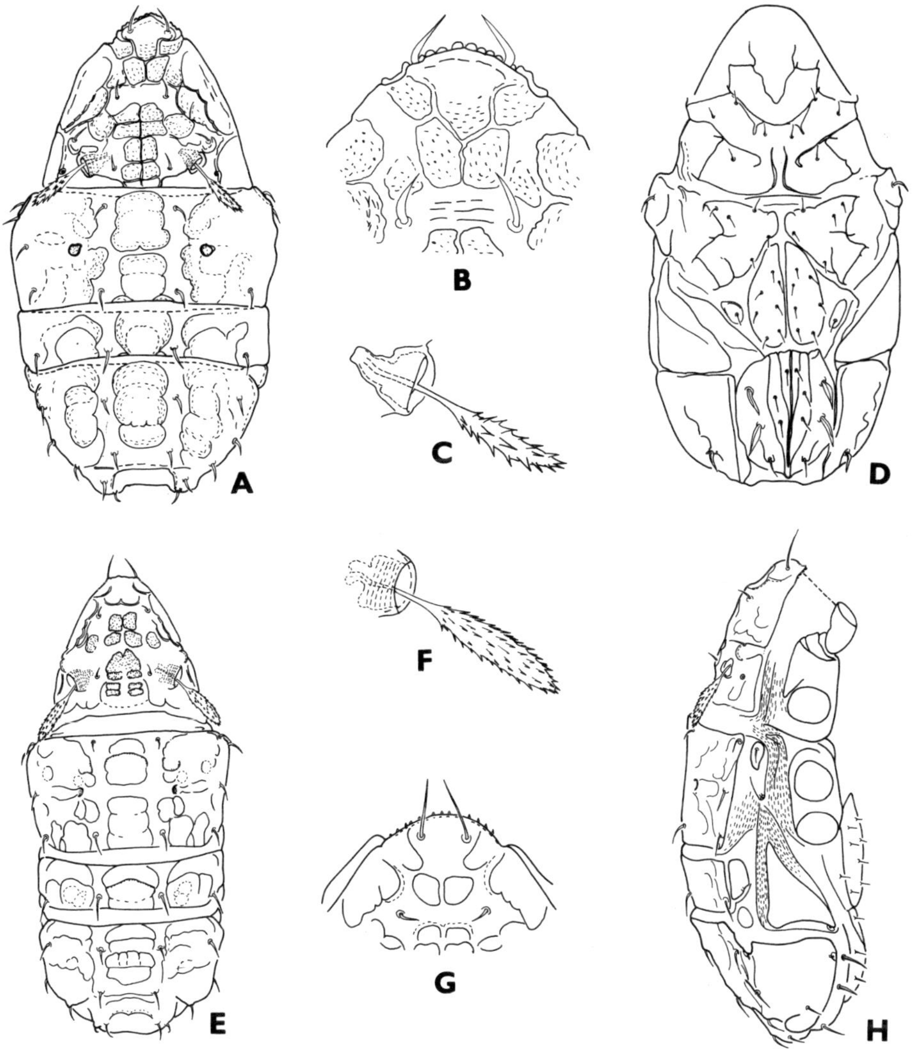

Plate 11. A—D: *Brachychochthonius miyauchii* Chinone, 1978 — A: dorsal side, B: rostral part of prodorsum, C: sensillus, D: ventral side; E—H *Brachychochthonius formosus* (Cooremann, 1947) — E: dorsal side, F: sensillus, G: rostral part of prodorsum, H: lateral side (after Chinone: A—D and Moritz: E—H)

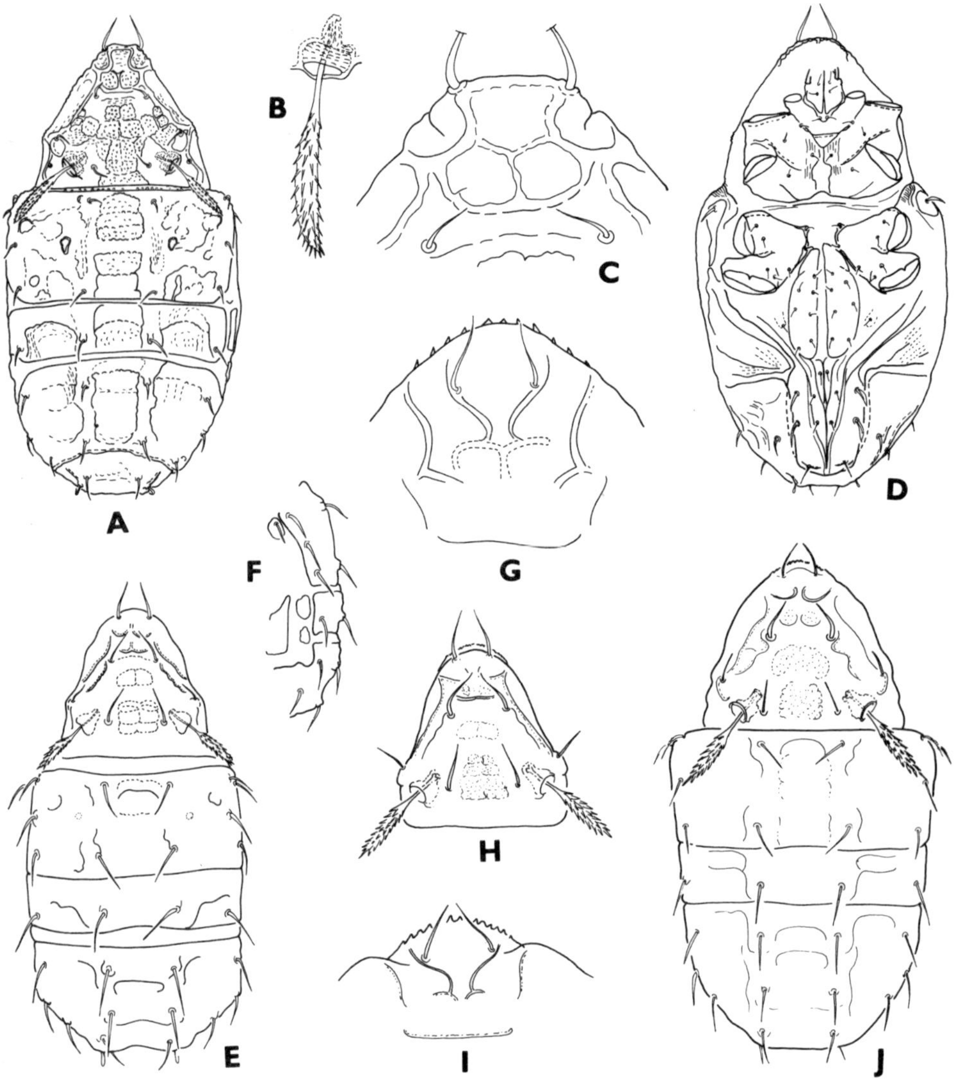

Plate 12. A—D: *Brachychochthonius aokii* (Chinone, 1974): A: dorsal side, B: sensillus, C: rostral part of prodorsum, D: ventral side; E—H: *Brachychochthonius guanophilus* Mahunka, 1979 — E: dorsal side, F: suprapleural plates, G: rostral part of prodorsum, H: prodorsum; I—J: *Brachychochthonius maroccanus* Mahunka, 1980 — I: rostral part of prodorsum, J: dorsal side (after Chinone: A—D; E—J original)

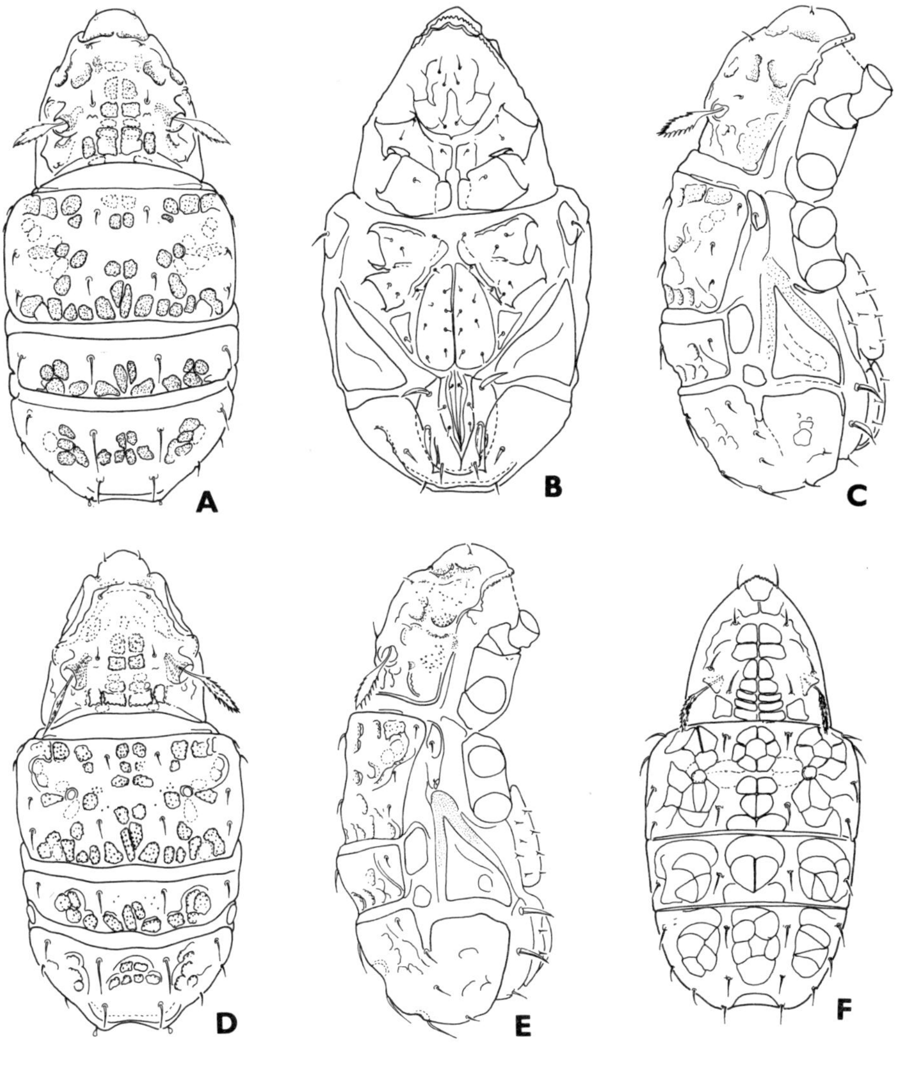

Plate 13. A—C: *Brachychochthonius hungaricus* (Balogh, 1943) — A: dorsal side, B: ventral side, C: lateral side; D—E: *Brachychochthonius rostratus* (Jacot, 1936) — D: dorsal side, E: lateral side; F: *Brachychochthonius borealis* D. Krivolutsky, 1965 — dorsal side (after Moritz: A—E and D. Krivolutsky: F)

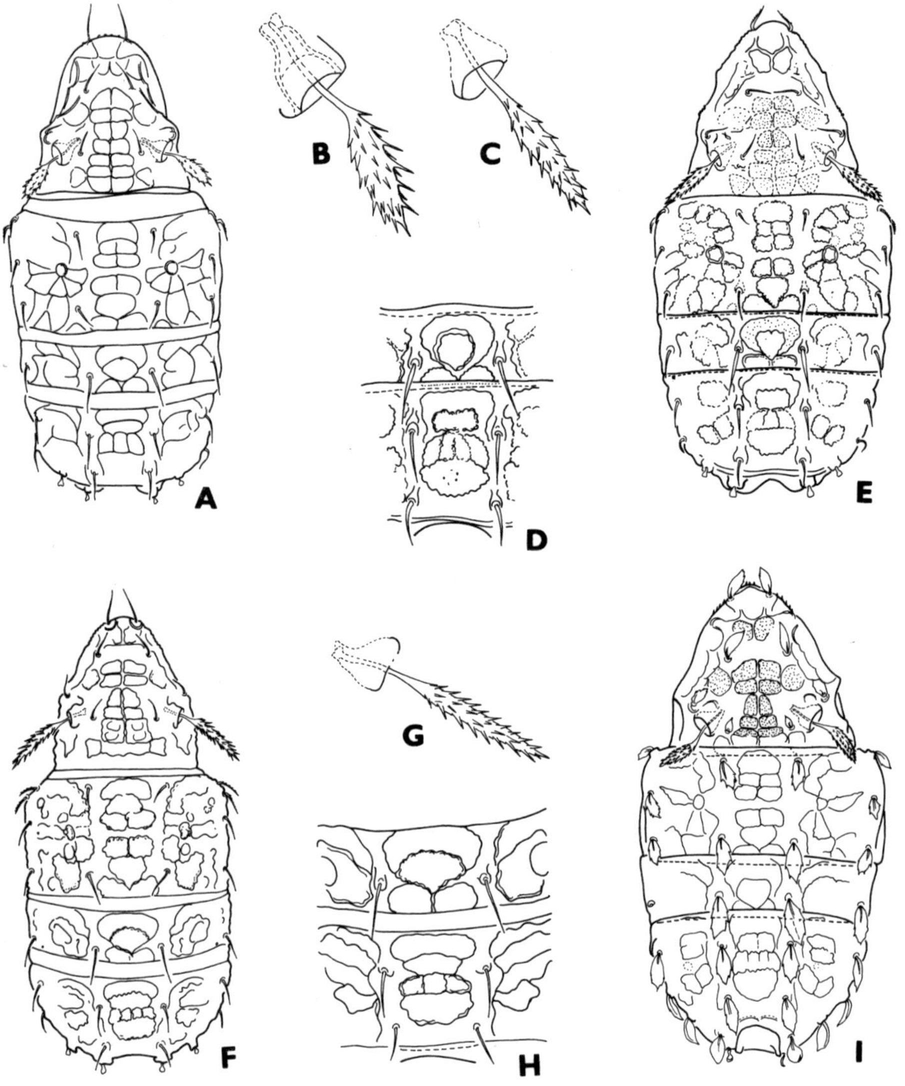

Plate 14. A—B: *Brachychochthonius cricoides* Weis-Fogh, 1948 — A: dorsal side, B: sensillus; C—E: *Brachychochthonius oesziae* Balogh et Mahunka, 1979 — C: sensillus, D: sculpture of notogaster, E: dorsal side; F—H: *Brachychochthonius subcricoides* Balogh et Mahunka, 1979 — F: dorsal side, G: sensillus, H: sculpture of notogaster; I: *Brachychochthonius phyllophorus* Moritz, 1976 — dorsal side (after Moritz: A—B, I; C—H original)

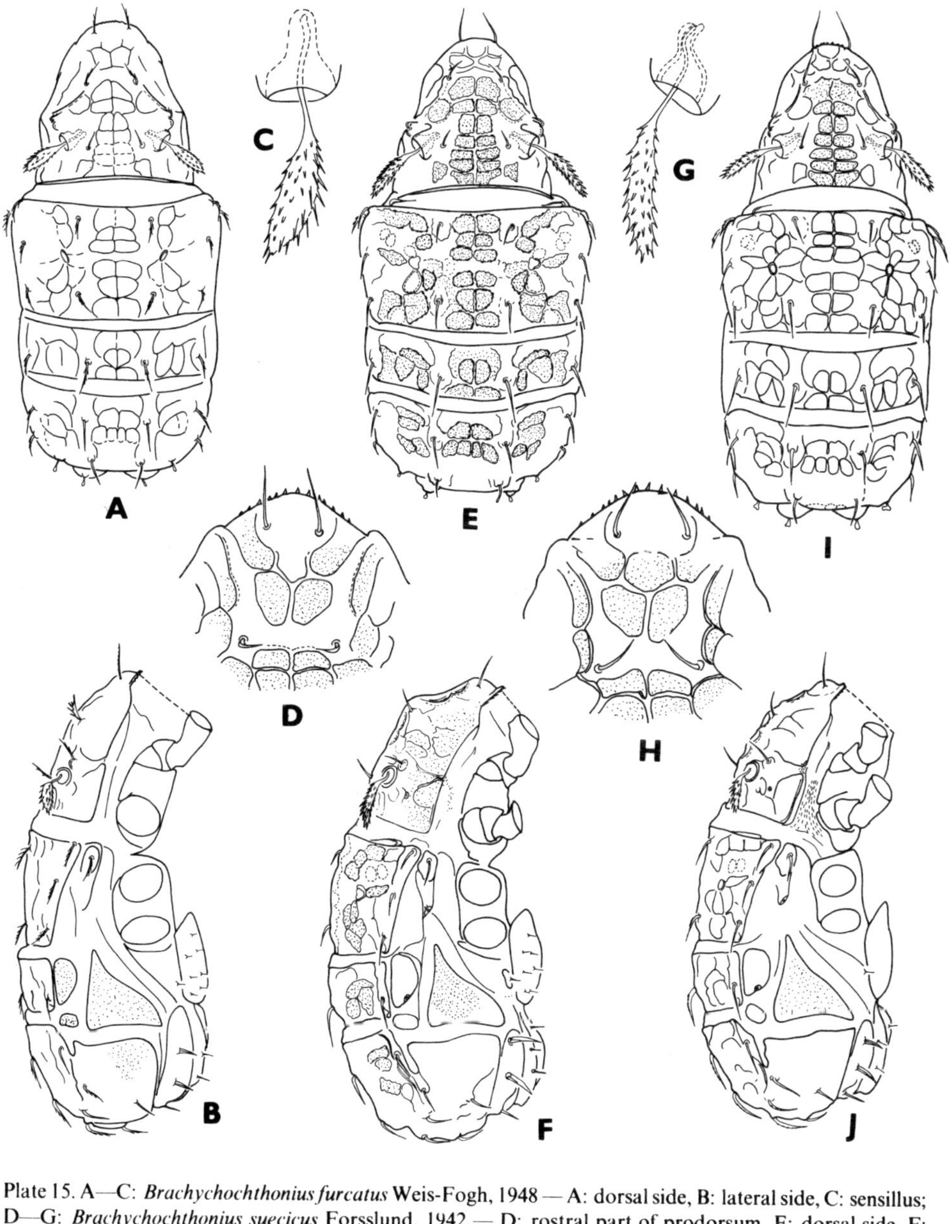

Plate 15. A—C: *Brachychochthonius furcatus* Weis-Fogh, 1948 — A: dorsal side, B: lateral side, C: sensillus; D—G: *Brachychochthonius suecicus* Forsslund, 1942 — D: rostral part of prodorsum, E: dorsal side, F: lateral side, G: sensillus; H—J: *Brachychochthonius jacoti* (Evans, 1952): H: rostral part of prodorsum, I: dorsal side, J: lateral side (after Moritz)

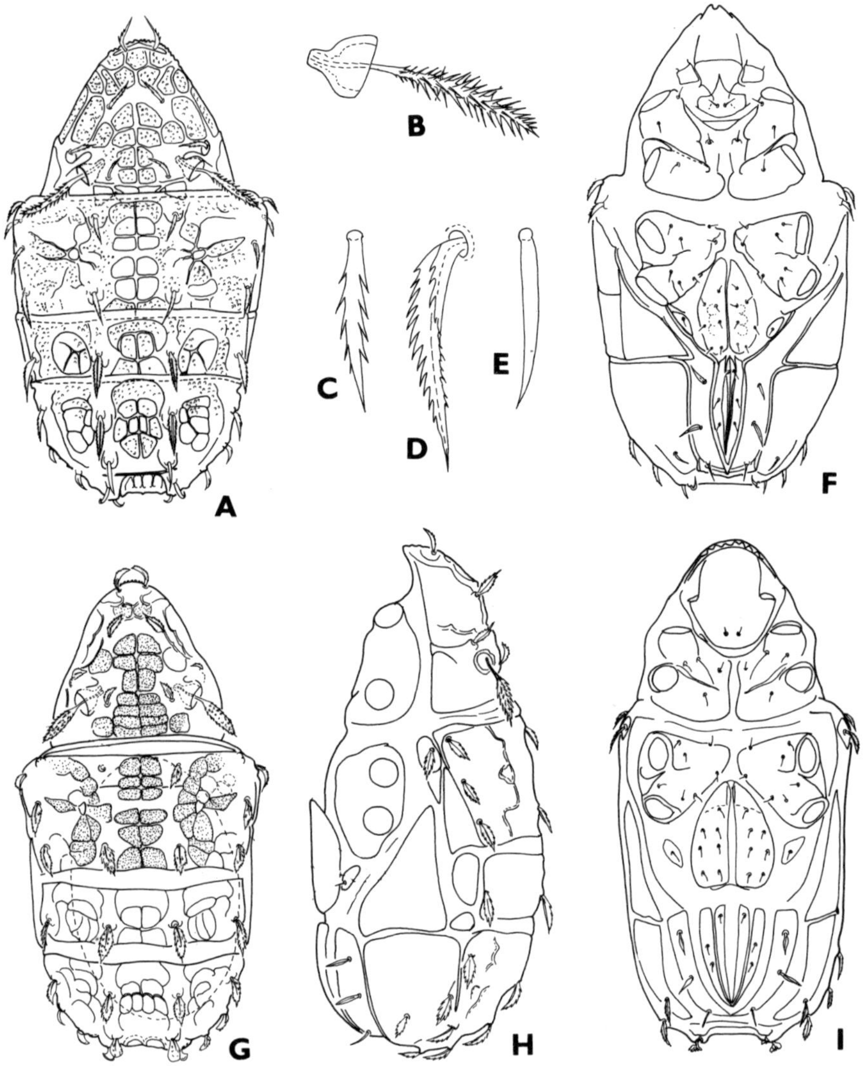

Plate 16. A—F: *Brachychochthonius jugatus* Jacot, 1938 — A: dorsal side, B: sensillus, C: seta c_1, D: seta e_2, E: seta h_1, F: ventral side; G—I: *Brachychochthonius meridionalis* Bernini, 1973 — G: dorsal side, H: lateral side, I: ventral side (after Chinone and Aoki: A—F and Bernini: G—I)

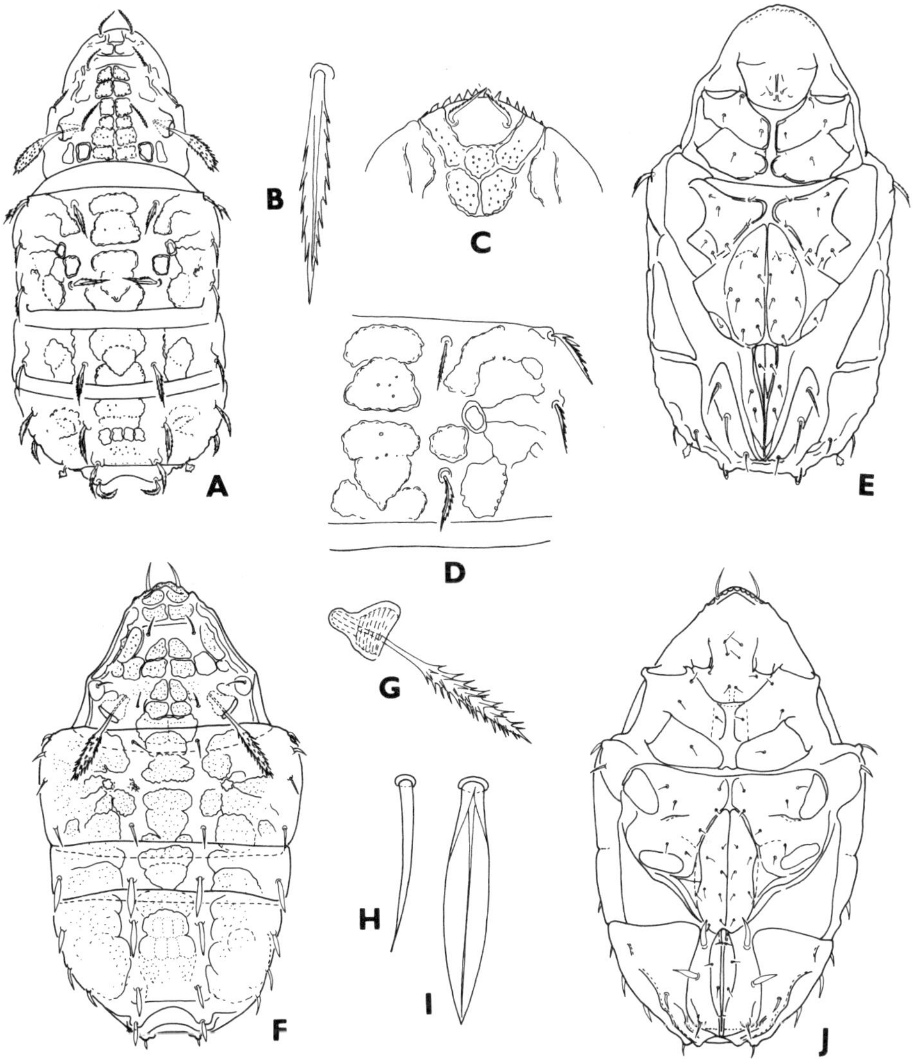

Plate 17. A—E: *Brachychochthonius hauserorum* Mahunka, 1979 — A: dorsal side, B: seta e_1, C: rostral part of prodorsum, D: sculpture of notogaster, E: ventral side; F—J: Brachychochthonius *elsosneadensis* (Hammer, 1958): F: dorsal side, G: sensillus, H: seta d_1, I: seta f_1, J: ventral side (A—E original; after Chinone and Aoki: F—J)

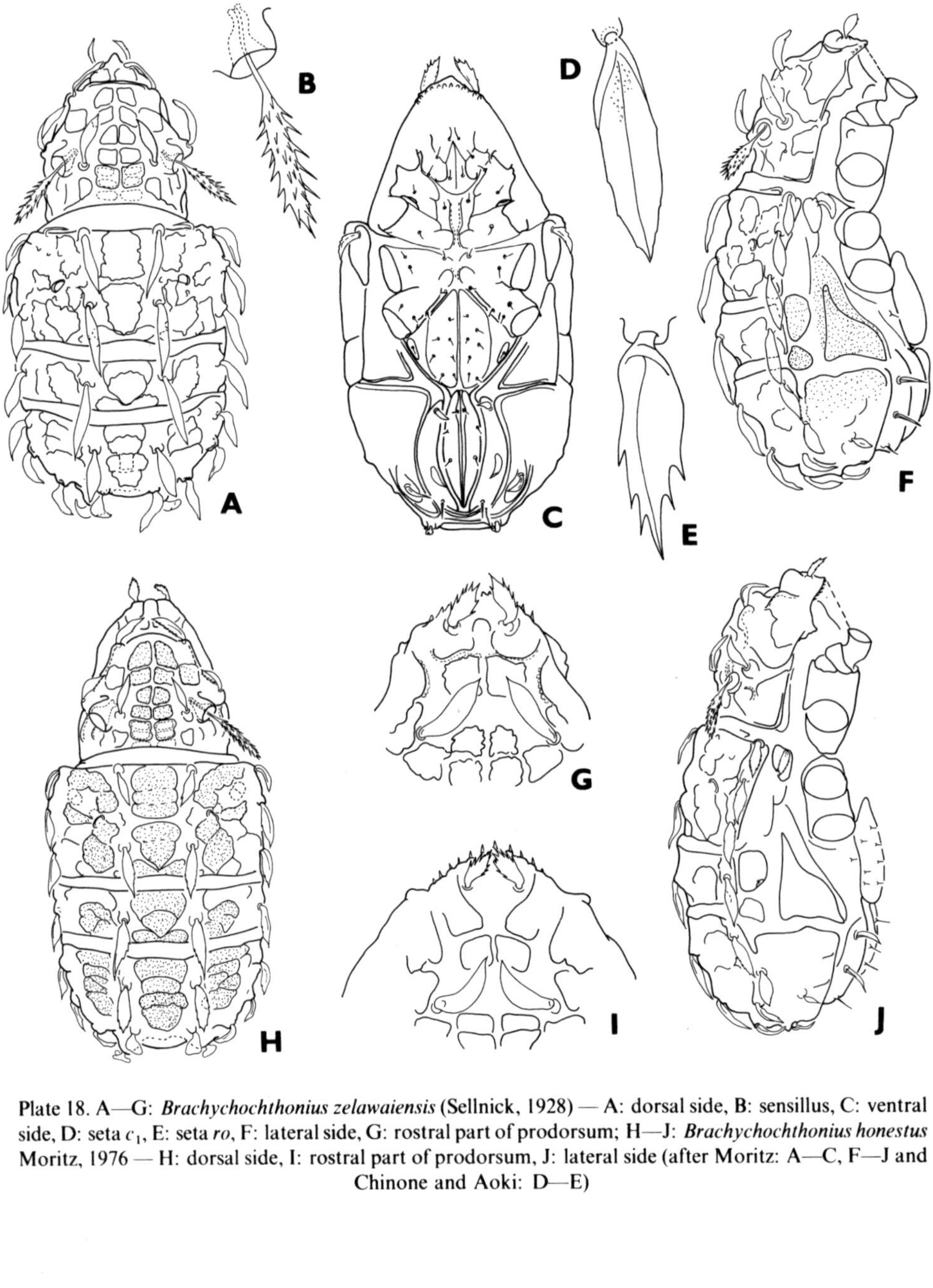

Plate 18. A—G: *Brachychochthonius zelawaiensis* (Sellnick, 1928) — A: dorsal side, B: sensillus, C: ventral side, D: seta c_1, E: seta *ro*, F: lateral side, G: rostral part of prodorsum; H—J: *Brachychochthonius honestus* Moritz, 1976 — H: dorsal side, I: rostral part of prodorsum, J: lateral side (after Moritz: A—C, F—J and Chinone and Aoki: D—E)

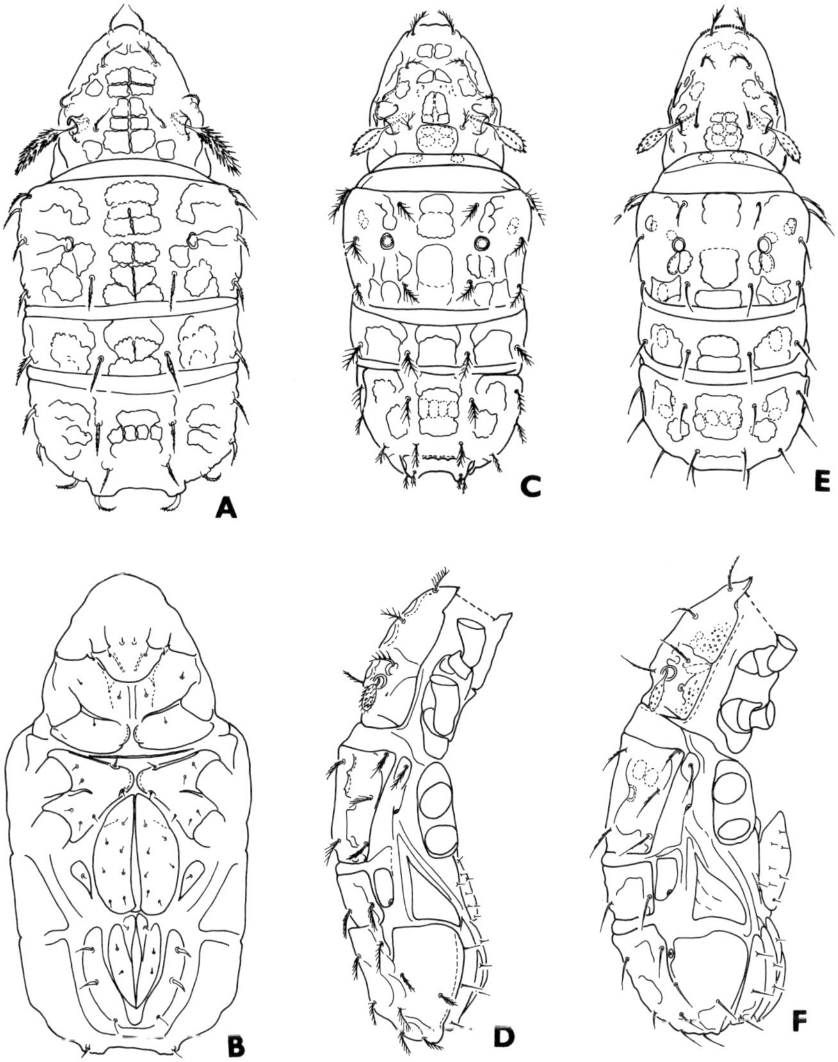

Plate 19. A—B: *Brachychochthonius variabilis* Mahunka, 1982 — A: dorsal side, B: ventral side; C—D: *Brachychthonius bimaculatus* Willmann, 1936 — C: dorsal side, D: lateral side; E—F: *Brachychthonius hirtus* Moritz, 1976 — E: dorsal side, F: lateral side (A—B original; after Moritz: C—F)

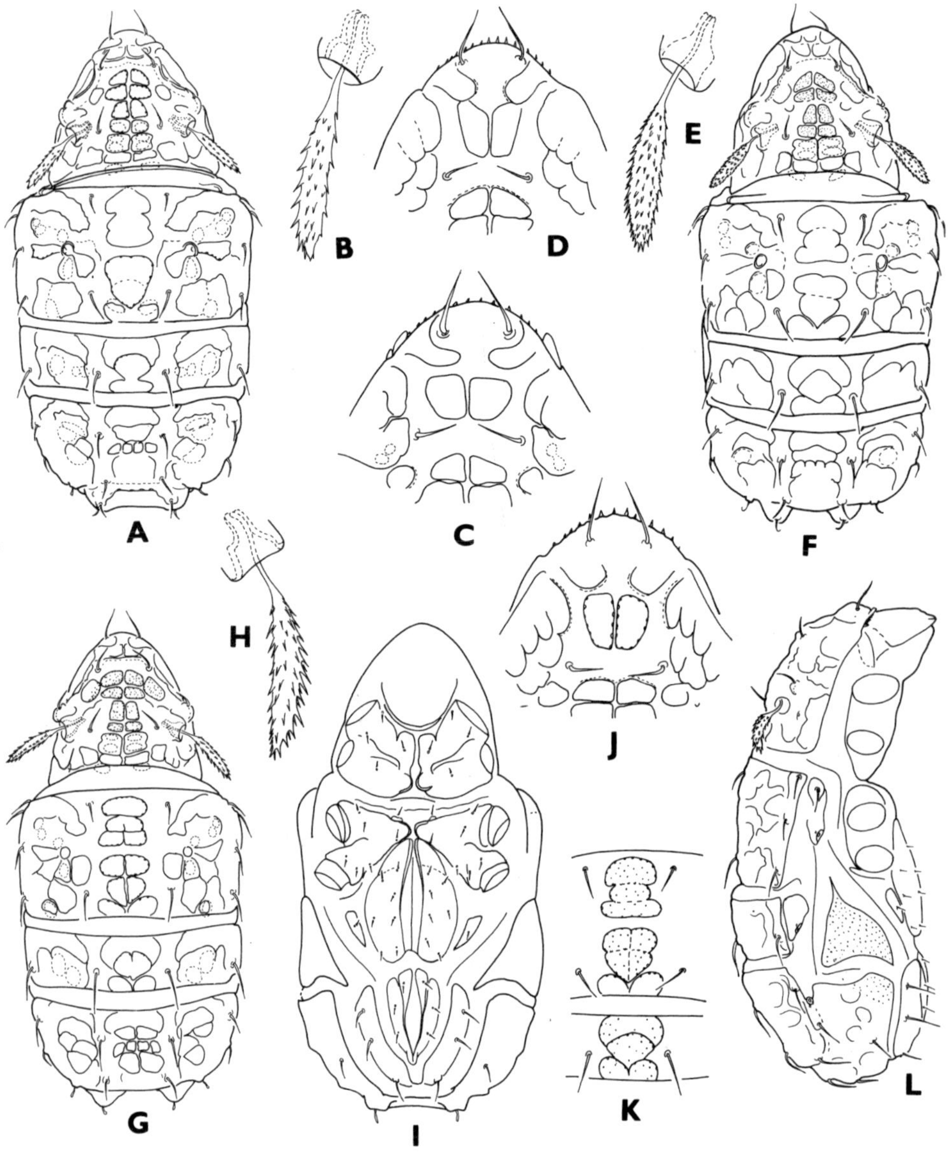

Plate 20. A—C: *Brachychthonius berlesei* Willmann, 1928 — A: dorsal side, B: sensillus, C: rostral part of prodorsum; D—F: *Brachychthonius pius* Moritz, 1976 — D: rostral part of prodorsum, E: sensillus, F: dorsal side; G—L: *Brachychthonius impressus* Moritz, 1976 — G: dorsal side, H: sensillus, I: ventral side, J: rostral part of prodorsum, K: sculpture of notogaster, L: lateral side (after Moritz)

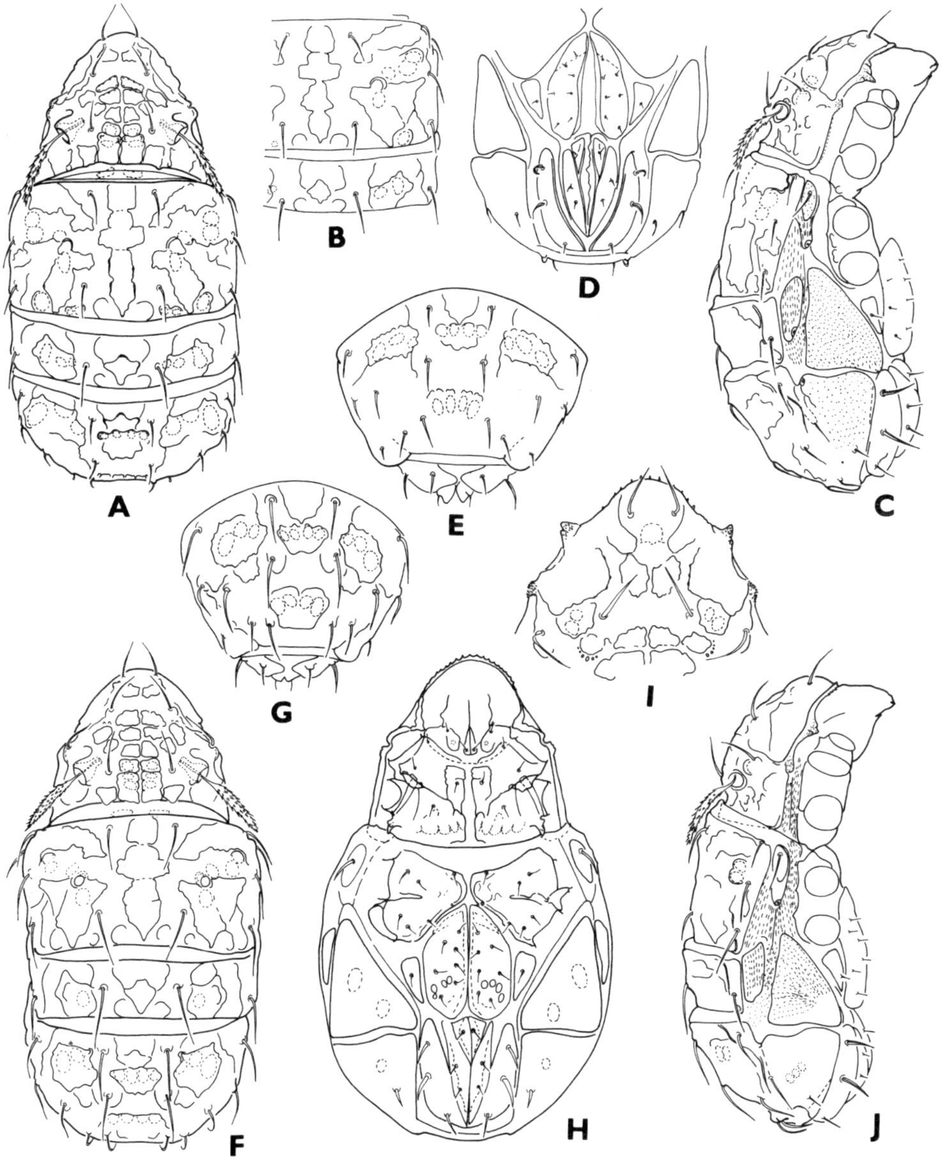

Plate 21. A—E: *Poecilochthonius spiciger* (Berlese, 1910) — A: dorsal side B: sculpture of notogaster, C: pygidium, D: anogenital region, E: lateral side; F—J: *Poecilochthonius italicus* (Berlese, 1910) — F: dorsal side, G: pygidium, H: ventral side, I: rostral part of prodorsum, J: lateral side (after Moritz)

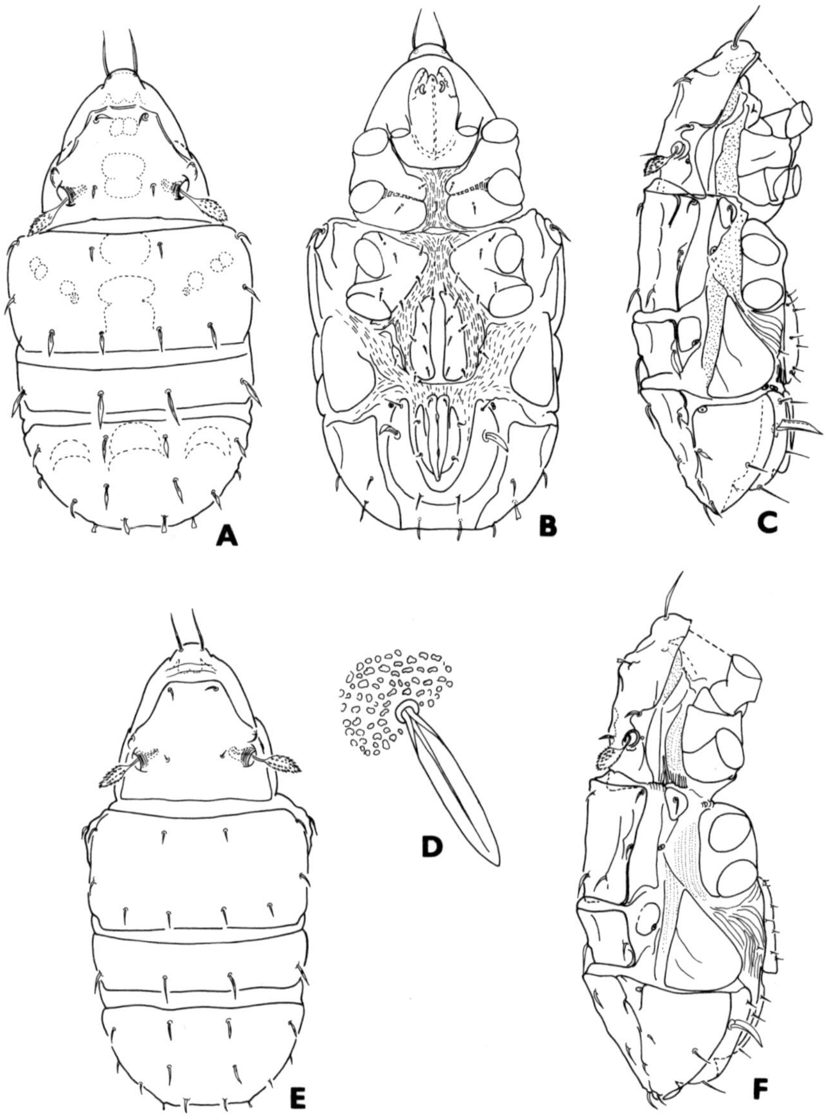

Plate 22. A—D: *Neobrachychthonius magnus* Moritz, 1976 — A: dorsal side, B: ventral side, C: lateral side D: seta e_1 and epidermis; E—F: *Neobrachychthonius marginatus* (Forsslund, 1942) — E: dorsal side, F: lateral side (after Moritz: A—C; E—F and Chinone: D)

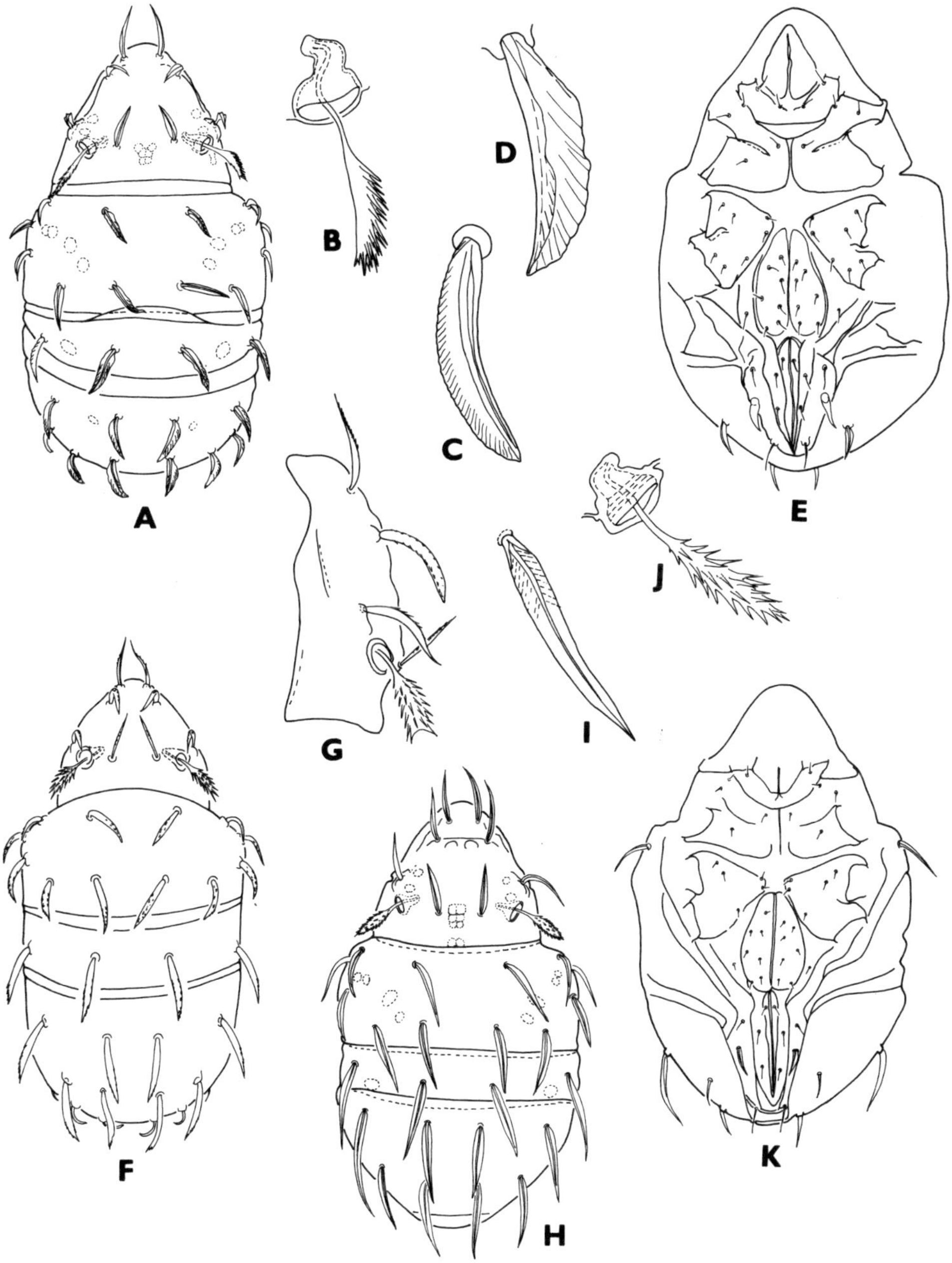

Plate 23. A—E: *Liochthonius penicillus* Chinone, 1978 — A: dorsal side, B: sensillus, C: seta d_2, D: seta f_1, E: ventral side; F—G: *Liochthonius phitosi* Mahunka, 1982 — F: dorsal side, G: prodorsum from lateral side, H—K: *Liochthonius intermedius* Chinone et Aoki, 1972 — H: dorsal side, I: seta d_1, J: sensillus, K: ventral side (after Chinone: A—E and Chinone and Aoki and Aoki: H—K; F—G original)

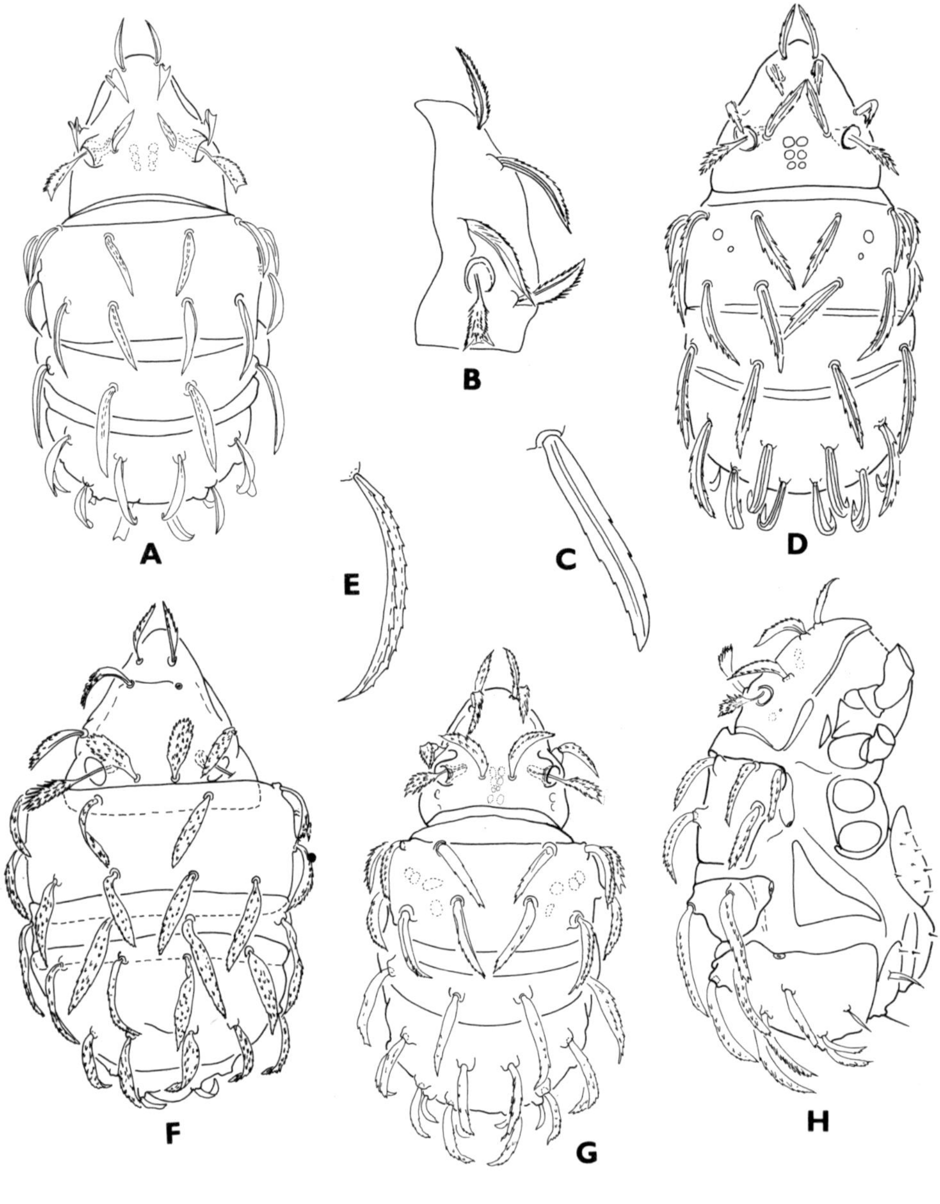

Plate 24. A: *Liochthonius evansi* (Forsslund, 1958) — dorsal side; B—D: *Liochthonius forsslundi* Mahunka, 1969 — B: lateral part of prodorsum, C: seta e_1, D: dorsal side; E—F: *Liochthonius gisini* (Schweizer, 1948) — E: seta e_1, F: dorsal side; G—H: *Liochthonius horridus* (Sellnick, 1928) — G: dorsal side, H: lateral side (after Moritz: A, G—H; B—F original)

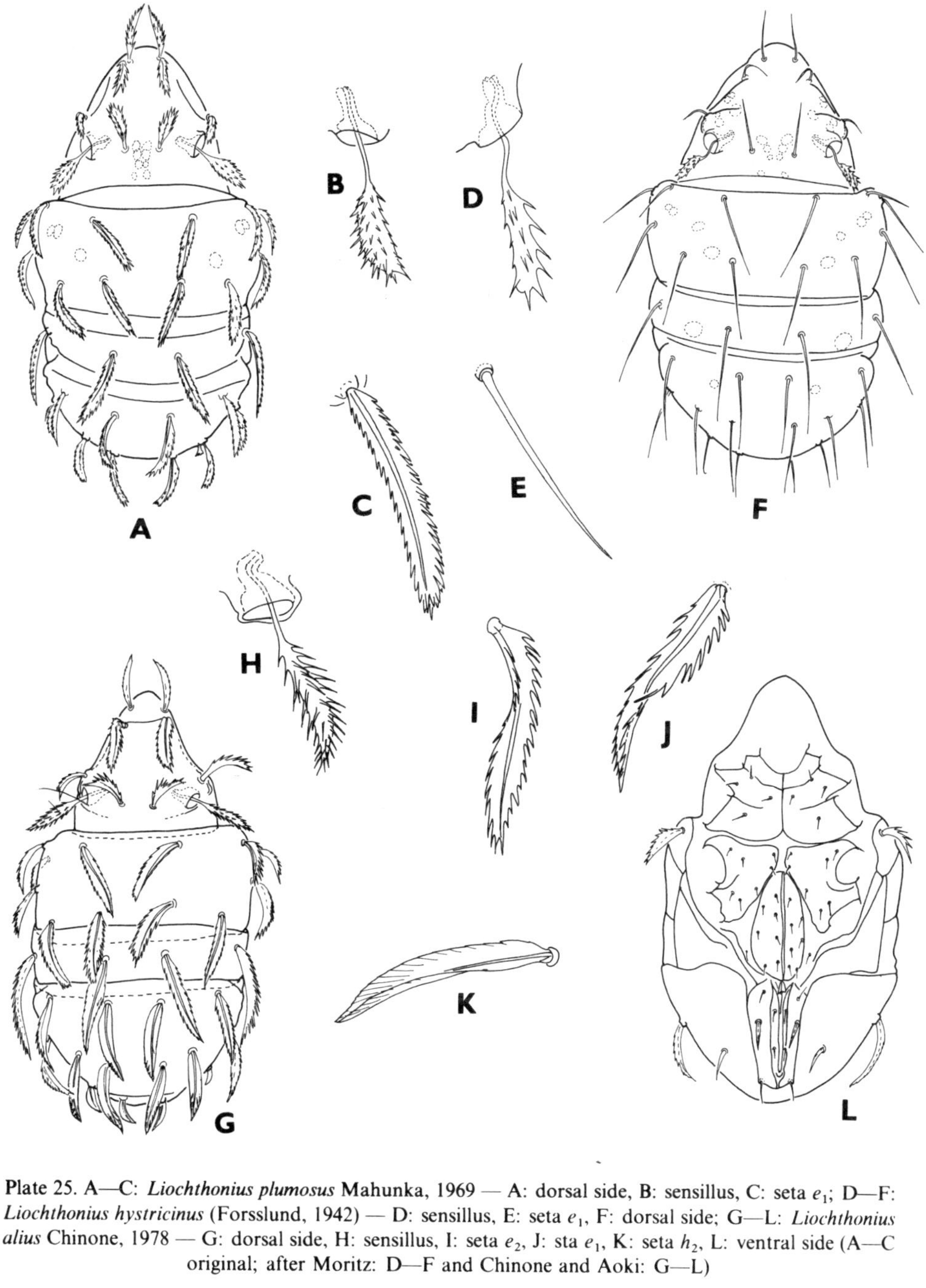

Plate 25. A—C: *Liochthonius plumosus* Mahunka, 1969 — A: dorsal side, B: sensillus, C: seta e_1; D—F: *Liochthonius hystricinus* (Forsslund, 1942) — D: sensillus, E: seta e_1, F: dorsal side; G—L: *Liochthonius alius* Chinone, 1978 — G: dorsal side, H: sensillus, I: seta e_2, J: sta e_1, K: seta h_2, L: ventral side (A—C original; after Moritz: D—F and Chinone and Aoki: G—L)

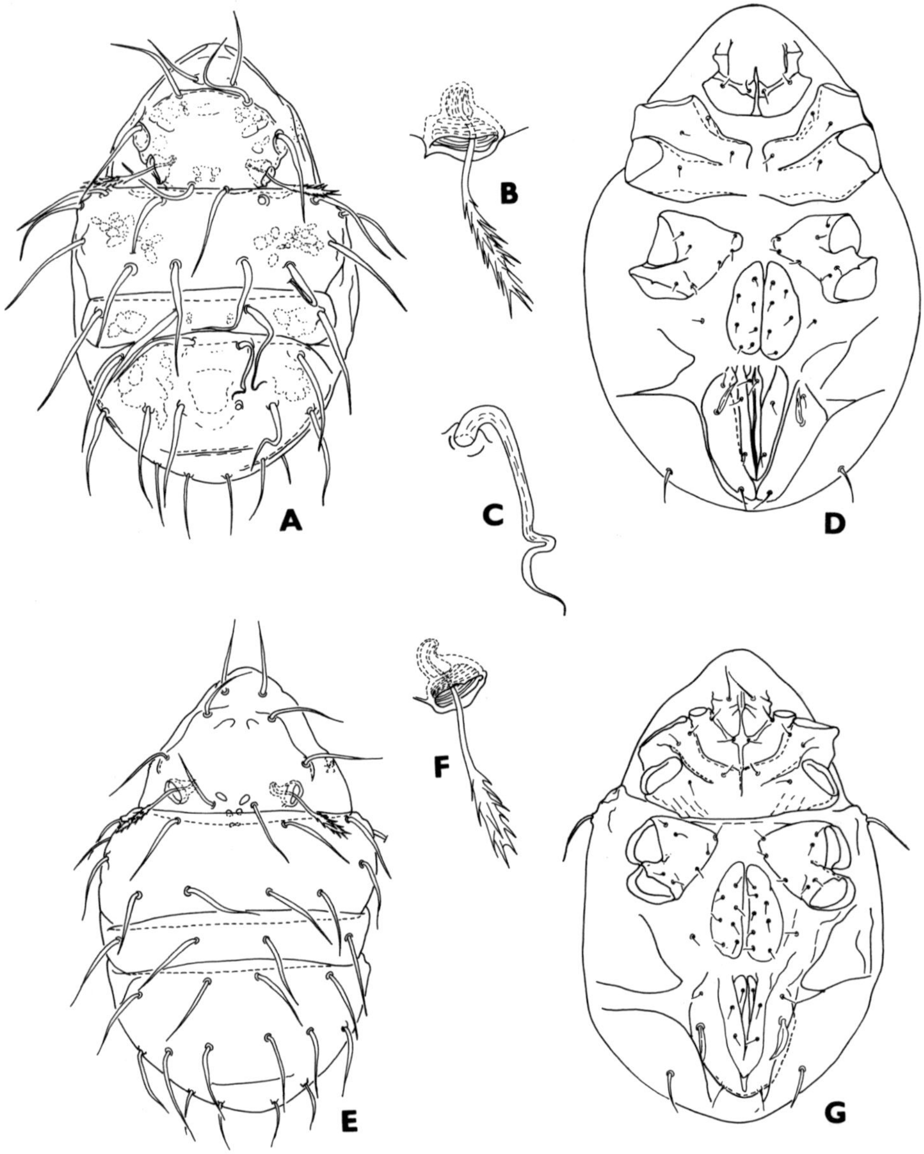

Plate 26. A—D: *Liochthonius lentus* Chinone, 1974 — A: dorsal side, B: sensillus, C: seta f_1, D: ventral side; E—G: *Liochthonius moritzi* nom. nov. — E: dorsal side, F: sensillus, G: ventral side (after Chinone)

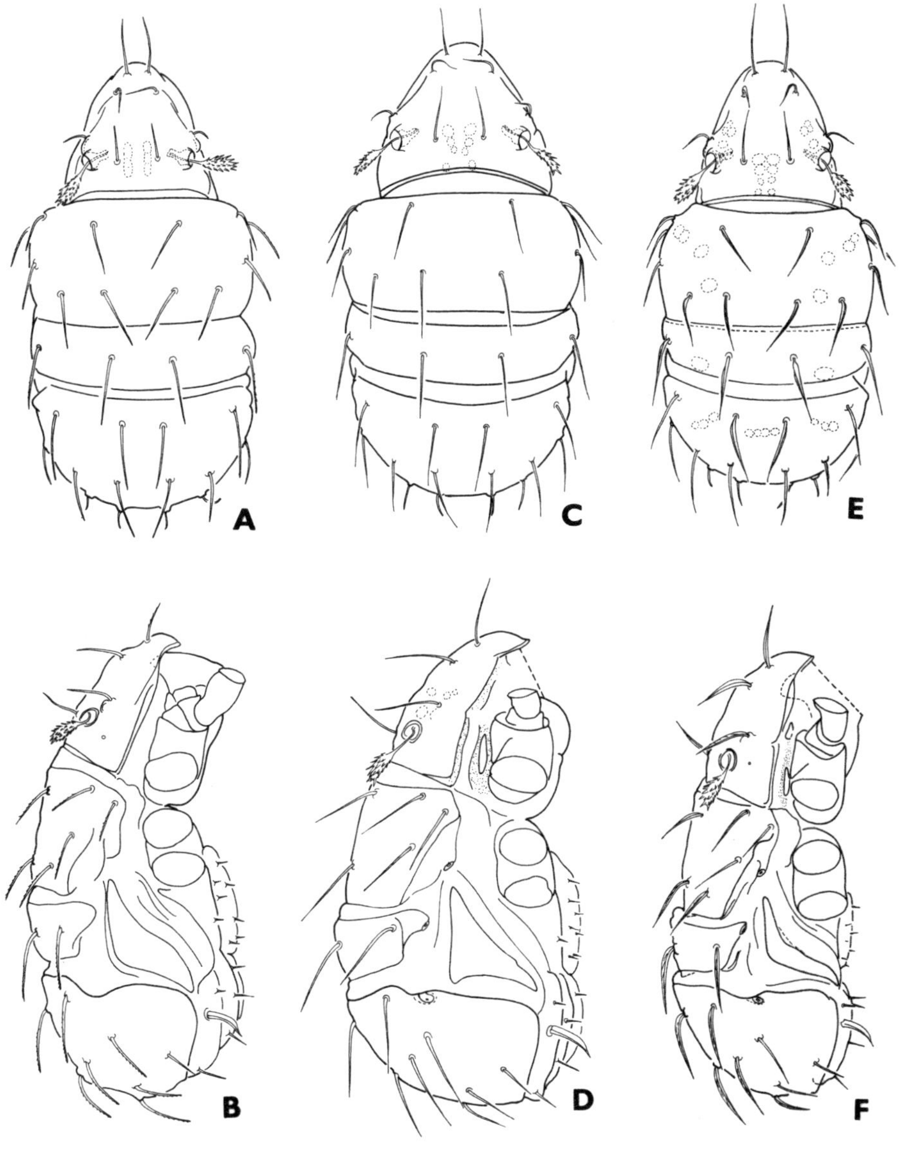

Plate 27. A—B: *Liochthonius neglectus* Moritz, 1976 — A: dorsal side, B: lateral side; C—D: *Liochthonius tuxeni* (Forsslund, 1957) — C: dorsal side, D: lateral side; E—F: *Liochthonius brevis* (Michael, 1888) — E: dorsal side, F: lateral side (after Moritz)

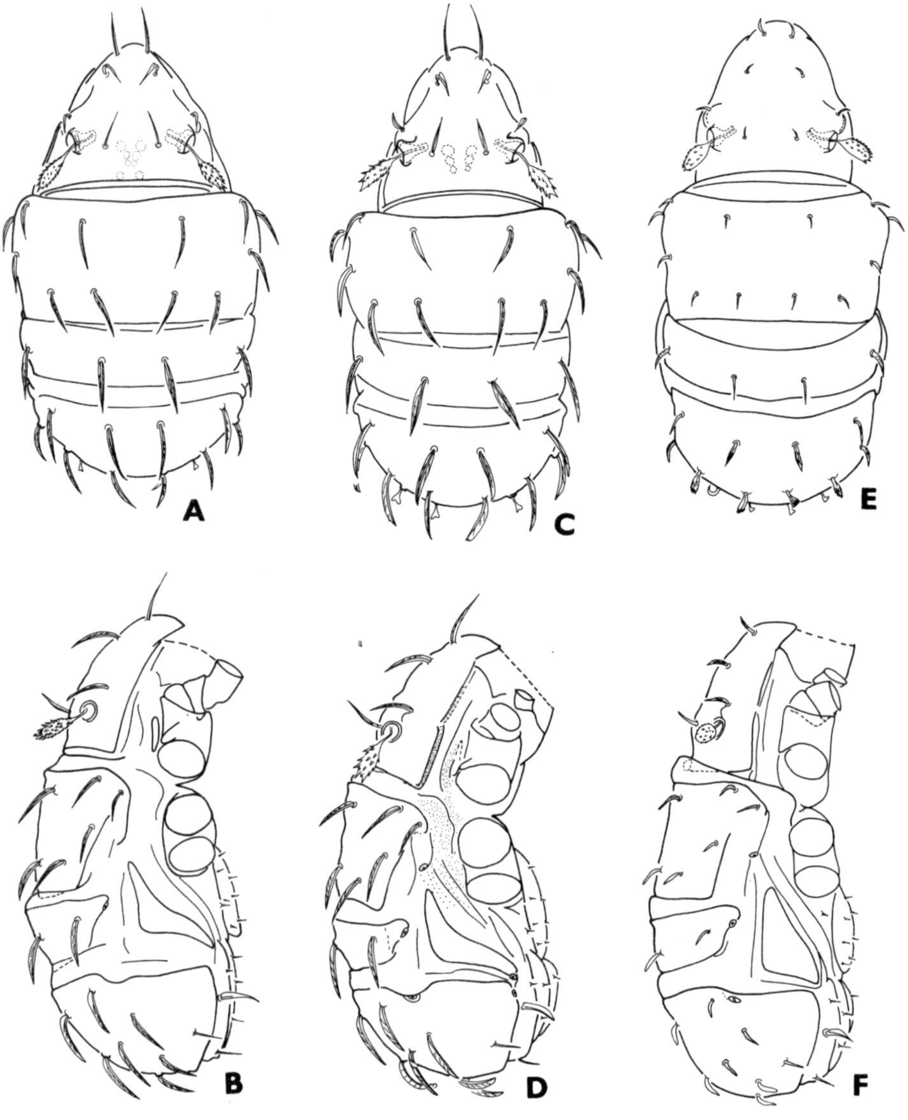

Plate 28. A—B: *Lichthonius dilutus* Moritz, 1976 — A: dorsal side, B: lateral side; C—D: *Liochthonius leptaleus* Moritz, 1976 — C: dorsal side, D: lateral side; E—F: *Liochthonius perelegans* Moritz, 1976 — E: dorsal side, F: lateral side (after Moritz)

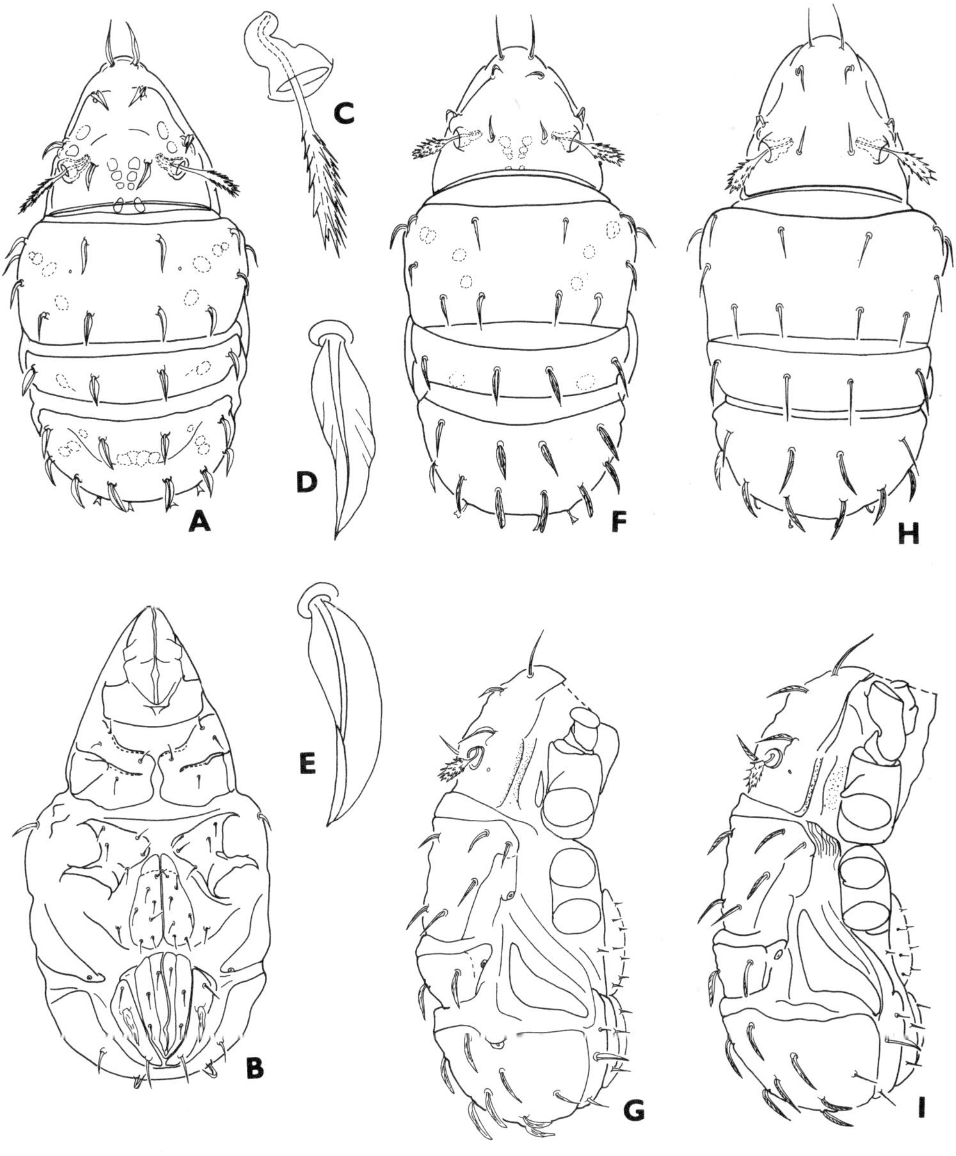

Plate 29. A—E: *Liochthonius asper* Chinone, 1978 — A: dorsal side, B: ventral side, C: sensillus, D: seta f_1, E: seta h_2; F—G: *Liochthonius propinquus* Niedbala, 1972 — F: dorsal side, G: lateral side; H—I: *Liochthonius simplex* (Forsslund, 1942) — H: dorsal side, I: lateral side (after Chinone: A—E and Moritz: F—I)

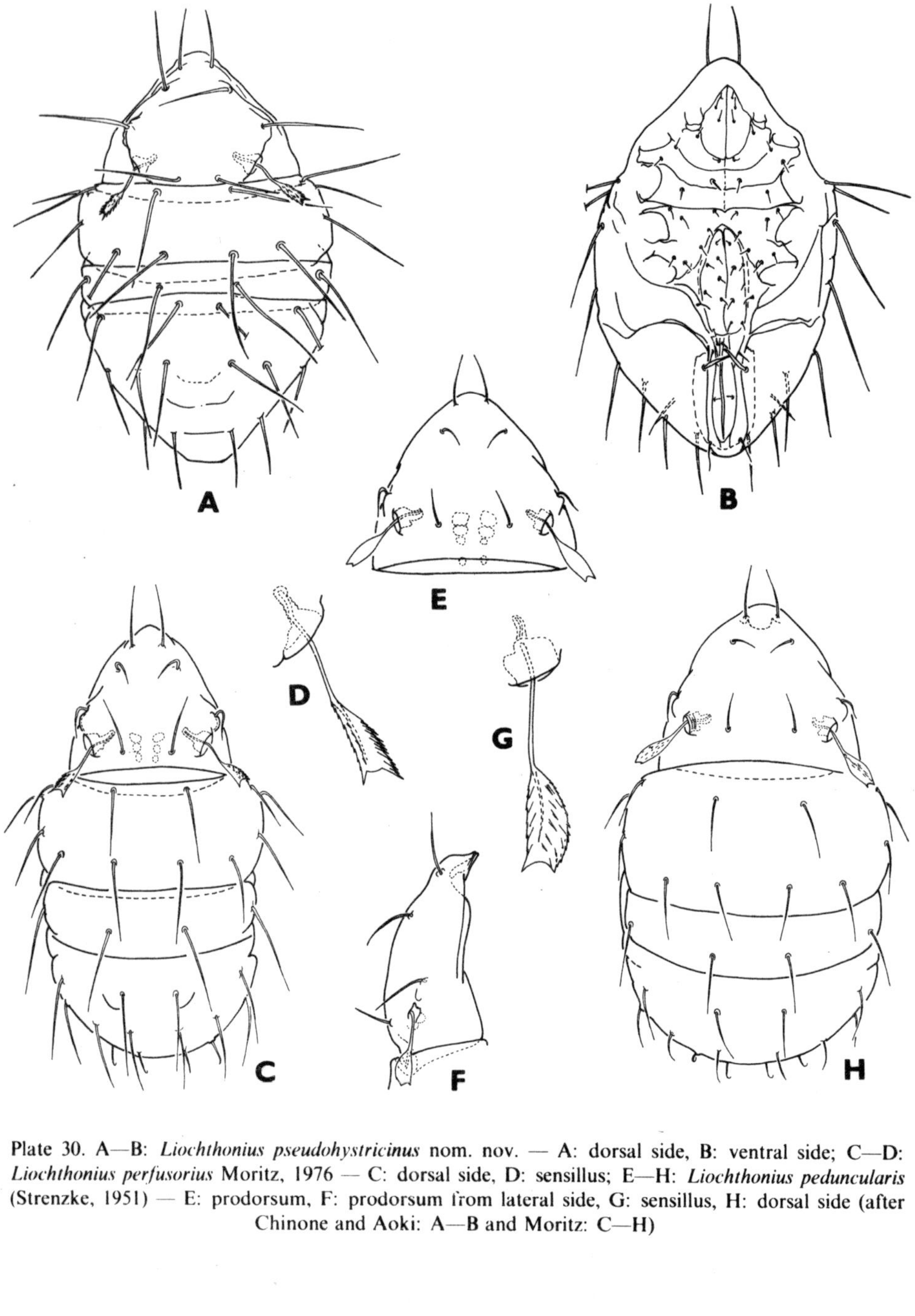

Plate 30. A—B: *Liochthonius pseudohystricinus* nom. nov. — A: dorsal side, B: ventral side; C—D: *Liochthonius perfusorius* Moritz, 1976 — C: dorsal side, D: sensillus; E—H: *Liochthonius peduncularis* (Strenzke, 1951) — E: prodorsum, F: prodorsum from lateral side, G: sensillus, H: dorsal side (after Chinone and Aoki: A—B and Moritz: C—H)

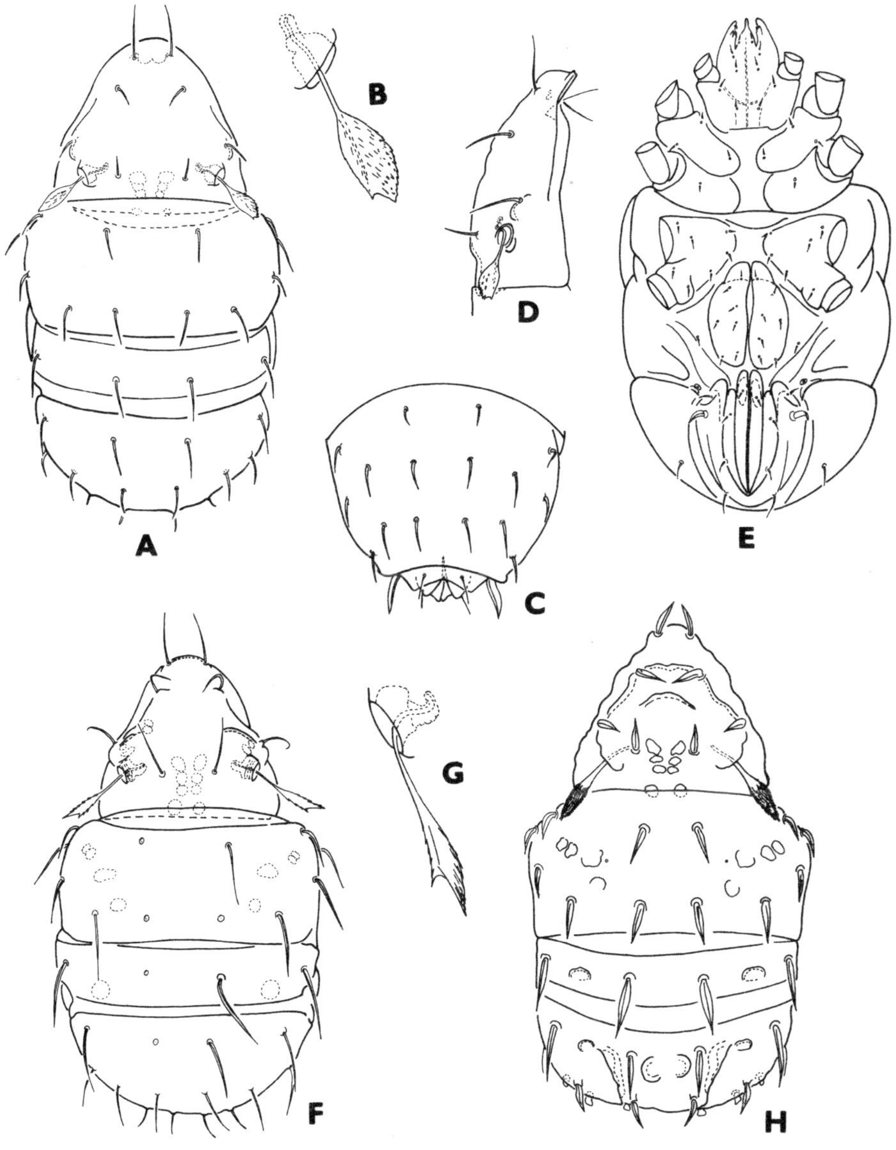

Plate 31. A—E: *Liochthonius alpestris* (Forsslund, 1958) — A: dorsal side, B: sensillus, C: pygidium, D: prodorsum from lateral side, E: ventral side; F—G: *Liochthonius furcillatus* (Willmann, 1942) — F: dorsal side, G: sensillus; H: *Liochthonius kirghisicus* D. Krivolutsky, 1971 — dorsal side (after Moritz: A—G; H original)

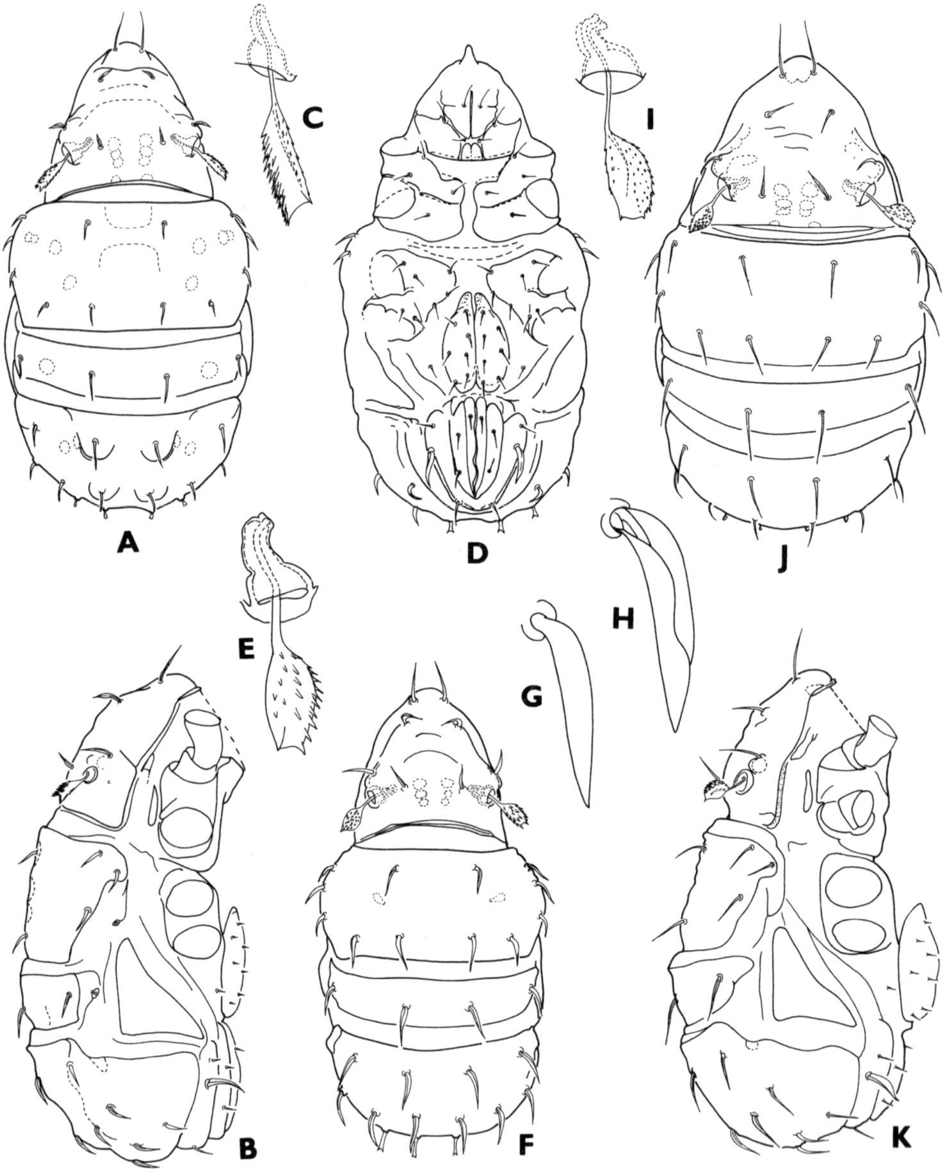

Plate 32. A—D: *Liochthonius strenzkei* Forsslund, 1963 — A: dorsal side, B: lateral side, C: sensillus, D: ventral side; E—H: *Liochthonius ohnishii* Chinone, 1978 — E: sensillus, F: dorsal side, G: seta *c* H: seta f_1; I—K: *Liochthonius clavatus* (Forsslund, 1942) — I: sensillus, J: dorsal side, K: lateral side (after Moritz: A—D, I—K and Chinone: E—H)

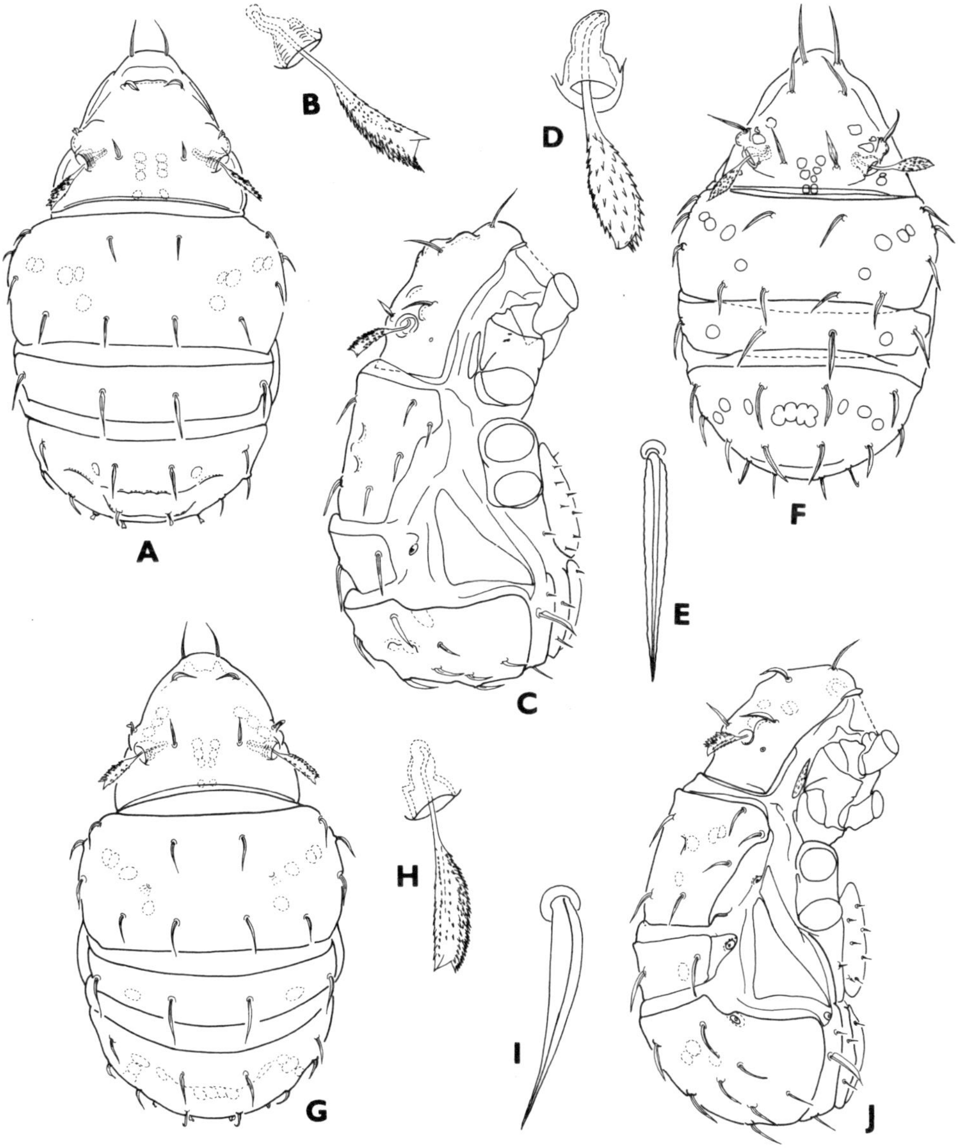

Plate 33. A—C: *Liochthonius sellnicki* (Thor, 1930) — A: dorsal side, B: sensillus, C: lateral side; D—F: *Liochthonius galba* Chinone, 1978 — D: sensillus, E: seta e_1, F: dorsal side; G—J: *Liochthonius muscorum* Forsslund, 1964 — G: dorsal side, H: sensillus, I: seta e_1, J: lateral side (after Moritz: A—C, G—H, J and Chinone: D—F, I)

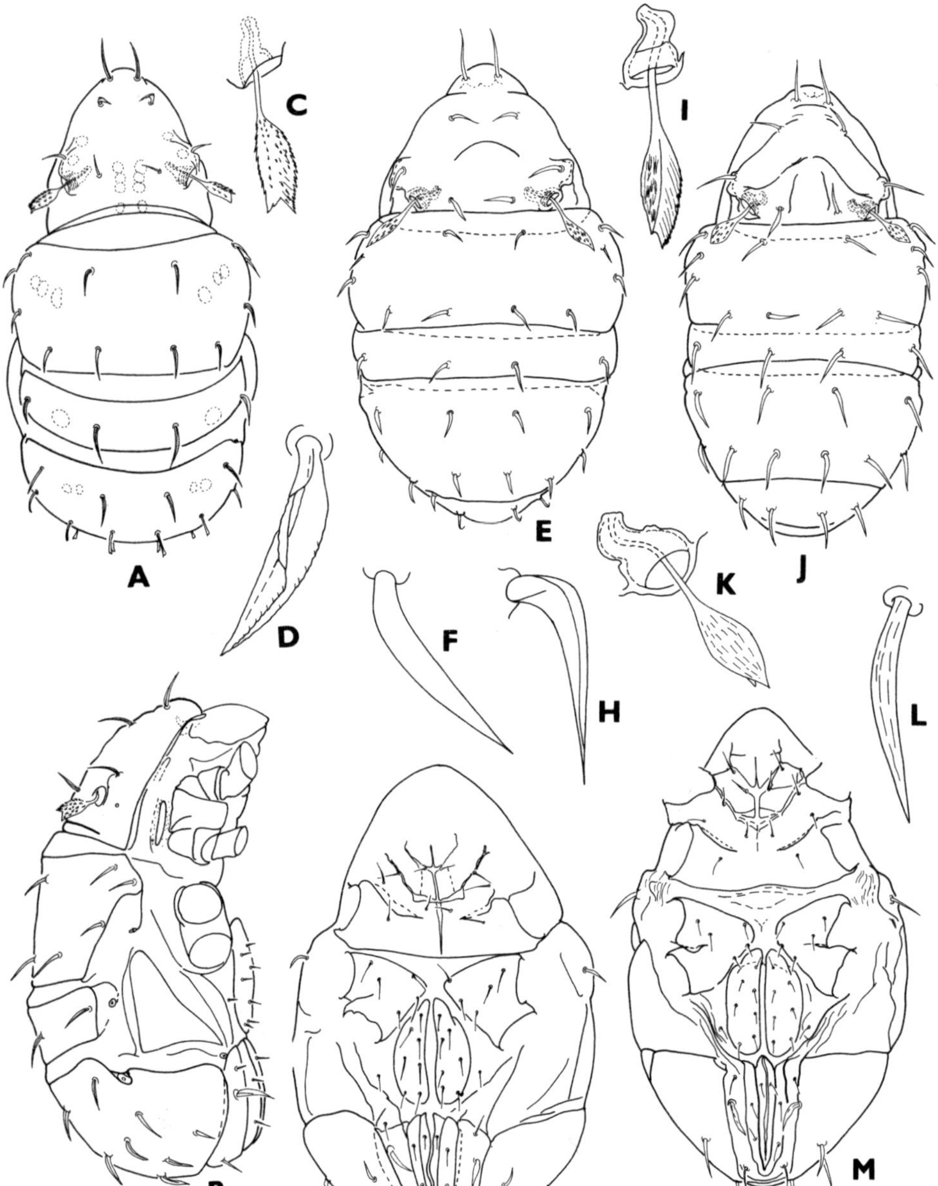

Plate 34. A—D: *Liochthonius lapponicus* (Trägårdh, 1910) — A: dorsal side, B: lateral side, C: sensillus, D: seta f_1; E—I: *Liochthonius pusillus* Chinone, 1978 — E: dorsal side, F: seta *ex*, G: ventral side, H: seta f_1, I: sensillus; J—M: *Liochthonius crassus* Chinone, 1978 — J: dorsal side, K: sensillus, L: seta e_1, M: ventral side (after Moritz: A—C and Chinone: D—M)

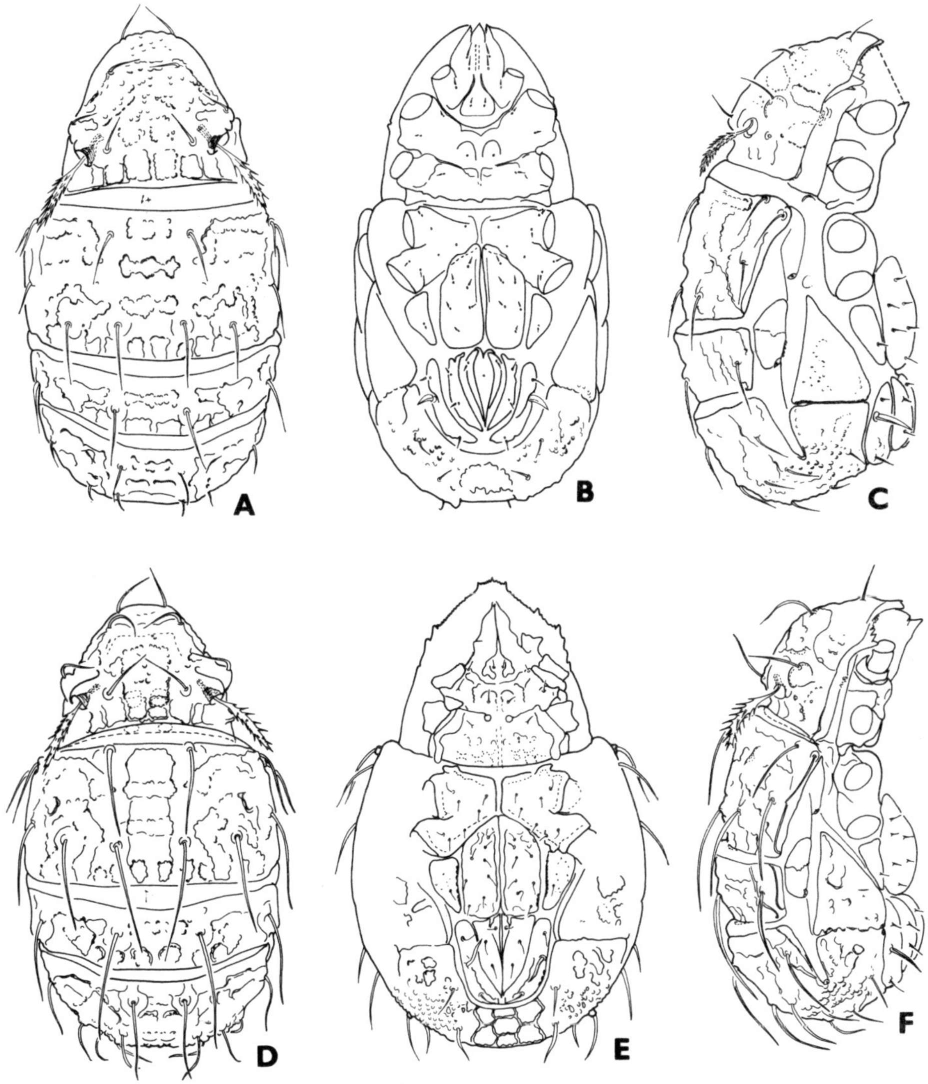

Plate 35. A—C: *Synchthonius crenulatus* (Jacot, 1938) — A: dorsal side, B: ventral side, C: lateral side; D—F: *Synchthonius elegans* Forsslund, 1956 — D: dorsal side, E: ventral side, F: lateral side (after Moritz)

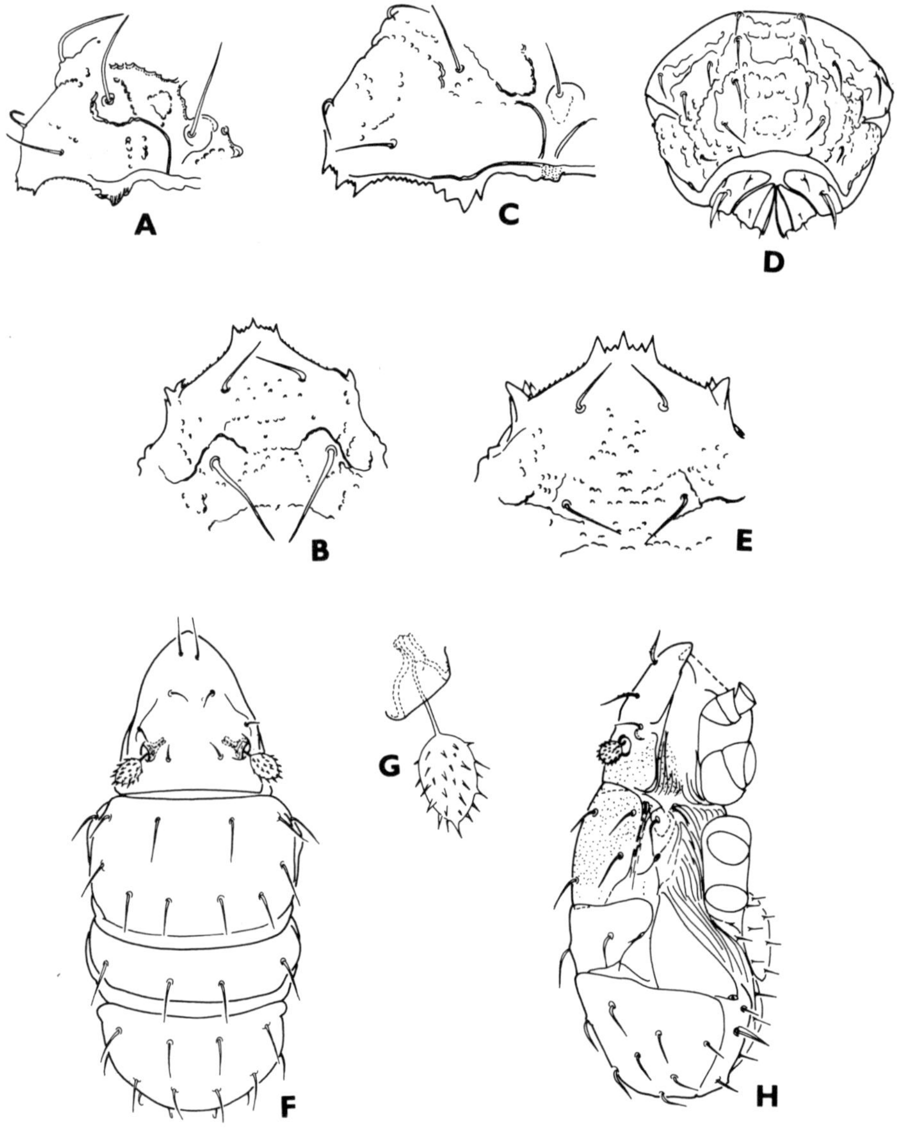

Plate 36. A—B: *Synchthonius elegans* Forsslund, 1956 — A: prodorsum from lateral side, B: rostral part of prodorsum; C—E: *Synchthonius crenulatus* (Jacot, 1938) — C: prodorsum from lateral side, D: pygidium, E: rostral part of prodorsum; F—H: *Paraliochthonius piluliferus* (Forsslund, 1942) — F: dorsal side, G: sensillus, H: lateral side (after Moritz)

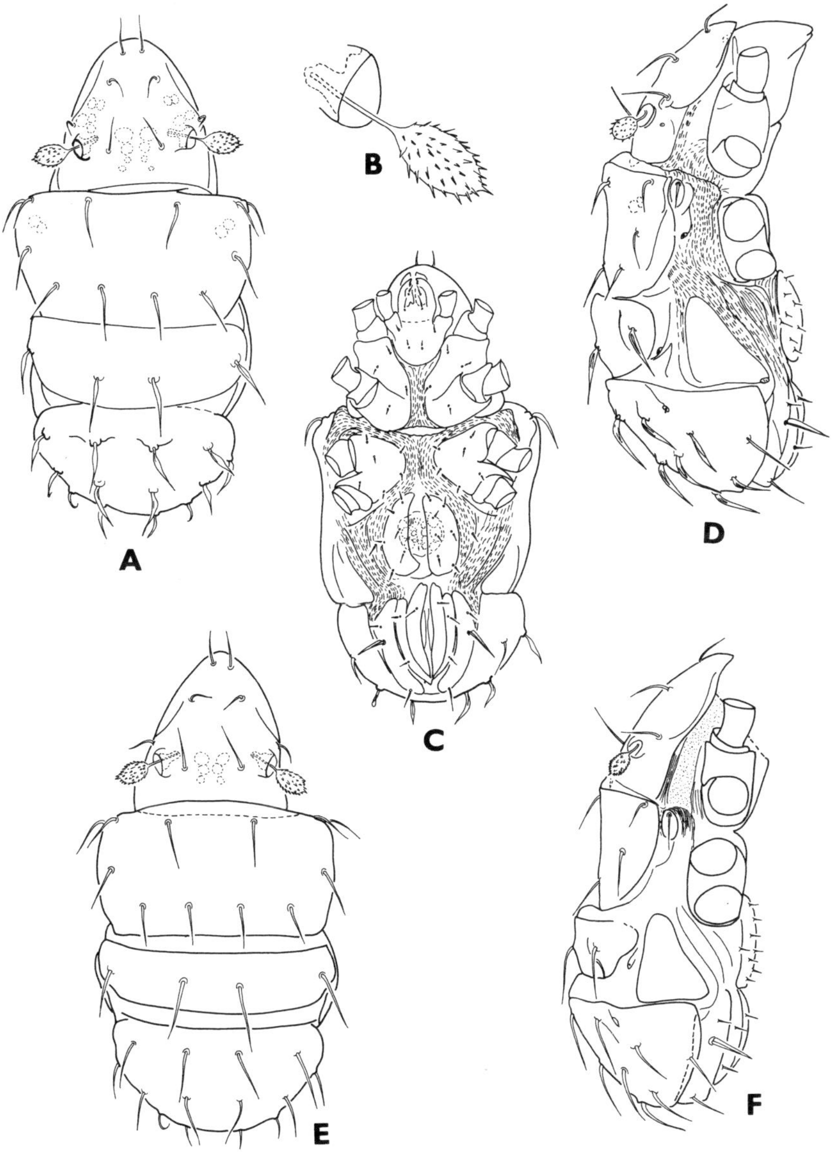

Plate 37. A—D: *Paraliochthonius globuliferus* (Strenzke, 1951) — A: dorsal side, B: sensillus, C: ventral side, D: lateral side; E—F: *Paraliochthonius occultus* (Niedbala, 1971) — E: dorsal side, F: lateral side (after Moritz)

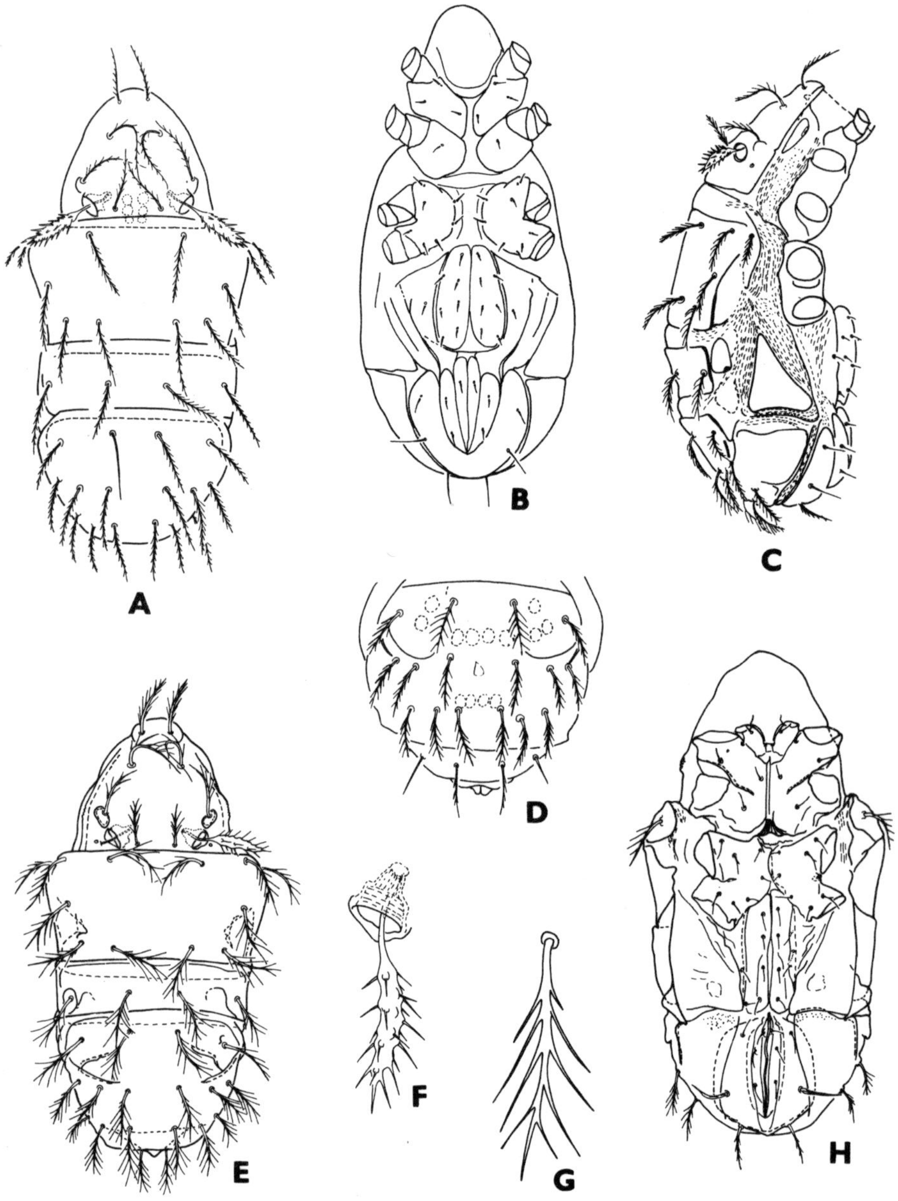

Plate 38. A—D: *Mixochthonius pilososetosus* (Forsslund, 1942) — A: dorsal side, B: ventral side, C: lateral side, D: pygidium; E—H: *Mixochthonius concavus* (Chinone, 1974) — E: dorsal side, F: sensillus, G: seta e_1, H: ventral side (after Moritz: A—D and Chinone: E—H)

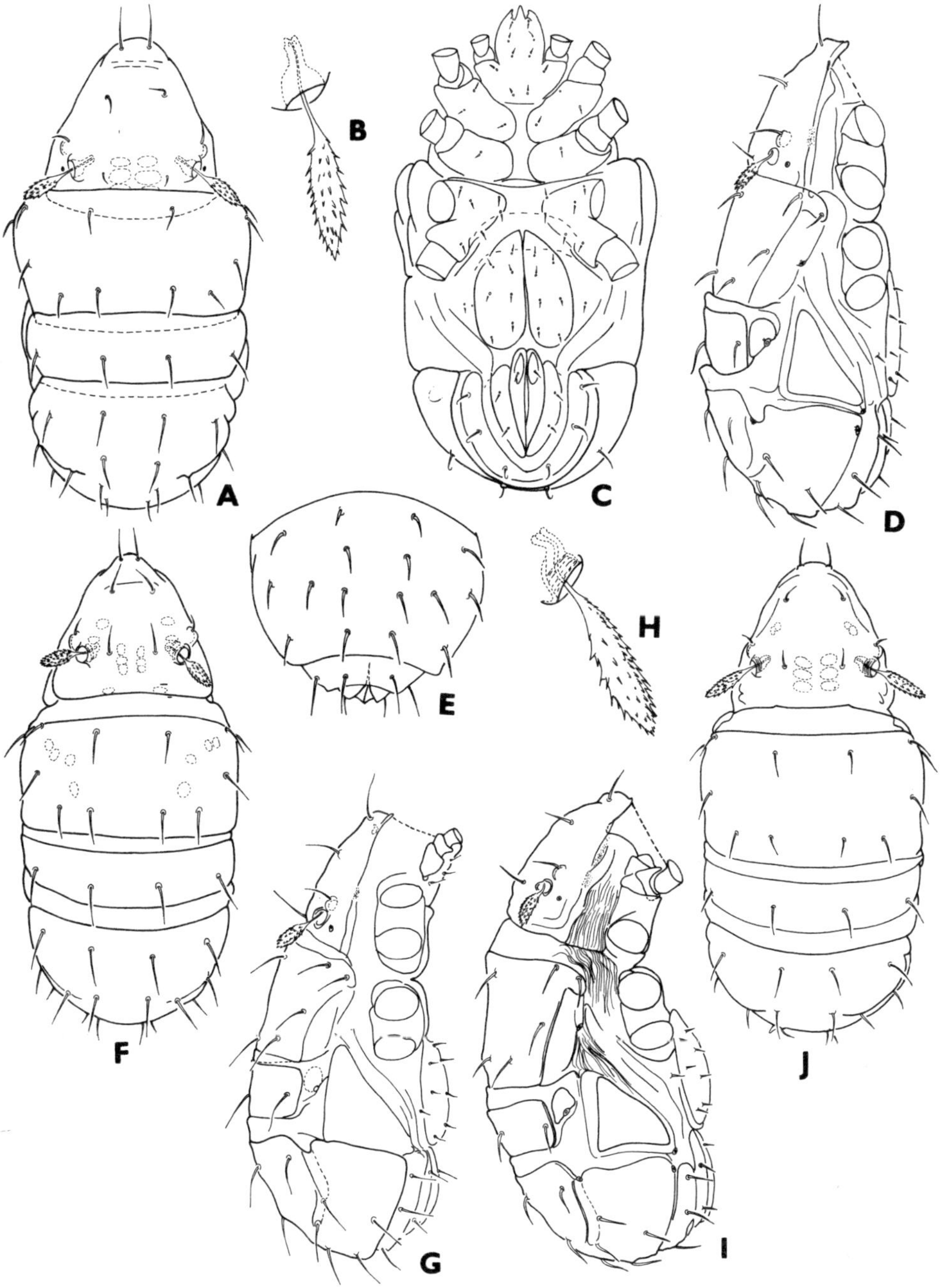

Plate 39. A—E: *Verachthonius laticeps* (Strenzke, 1951) — A: dorsal side, B: sensillus, C: ventral, side, D: lateral side, E: pygidium; F—G: *Verachthonius congruus* Moritz, 1976 — F: dorsal side, G: lateral side; H—J: *Verachthonius diversus* Moritz, 1976 — H: sensillus, I: lateral side, J: dorsal side (after Moritz)

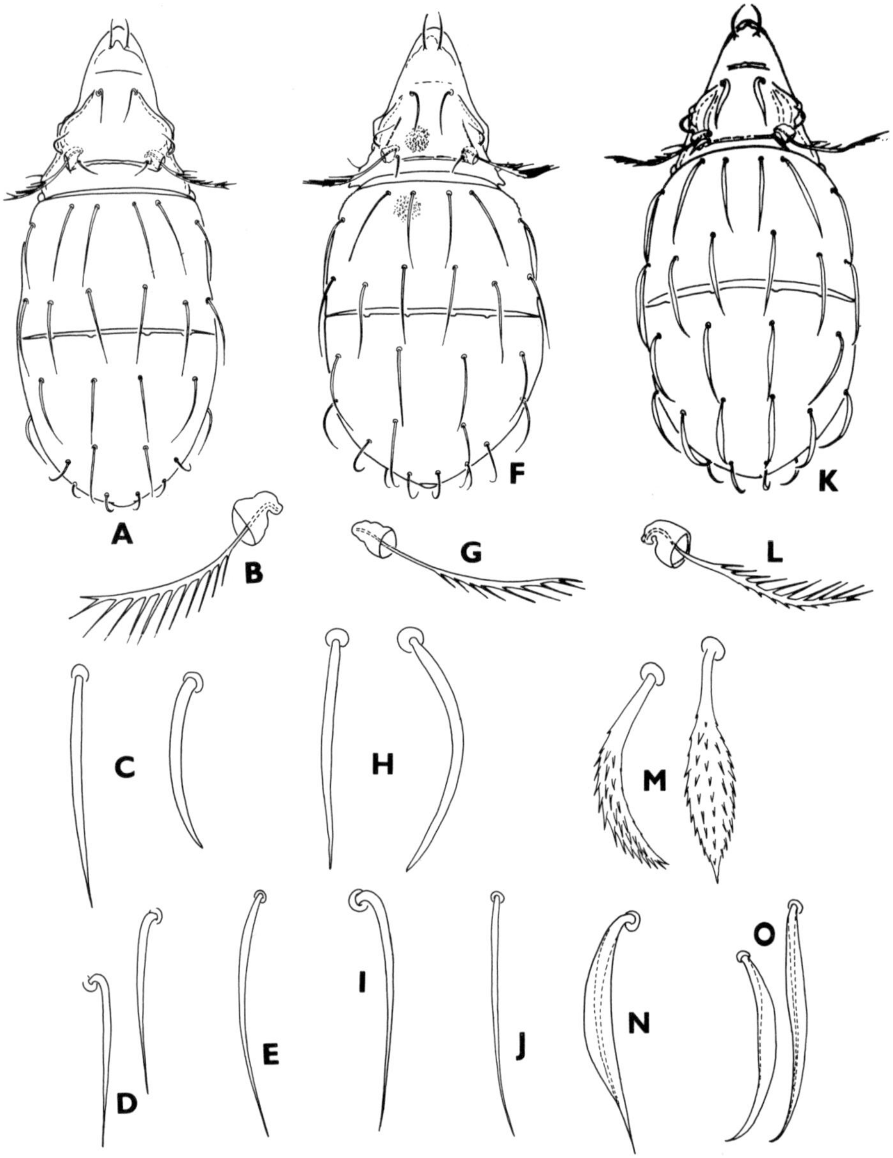

Plate 40. A—E: *Eohypochthonius parvus* Aoki, 1977 — A: dorsal side, B: sensillus, C: seta *in*, D: setae *le*, E: seta *ro;* F—J: *Eohypochthonius magnus* Aoki, 1977 — F: dorsal side, G: sensillus, H: setae *in*, I: seta *le*, J: seta *ro;* K—O: *Eohypochthonius crassisetiger* Aoki, 1959 — K: dorsal side, L: sensillus, M: setae *in*, N: seta *le*, O: setae *ro* (after Aoki)

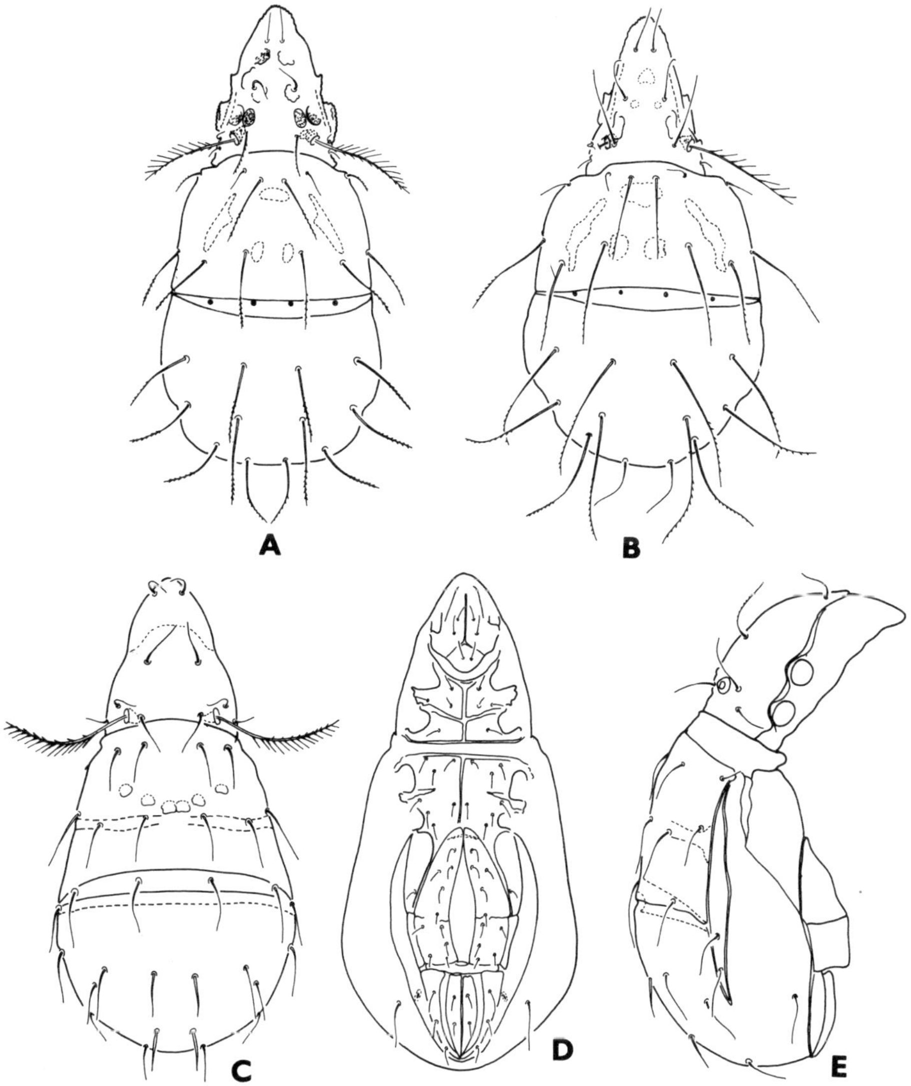

Plate 41. A: *Hypochthonius luteus* Oudemans, 1917 — dorsal side; B: *Hypochthonius rufulus* C. L. Koch, 1836 — dorsal side; C—E: *Hypochthoniella minutissima* (Berlese, 1904) — C: dorsal side, D: ventral side, E: lateral side (original)

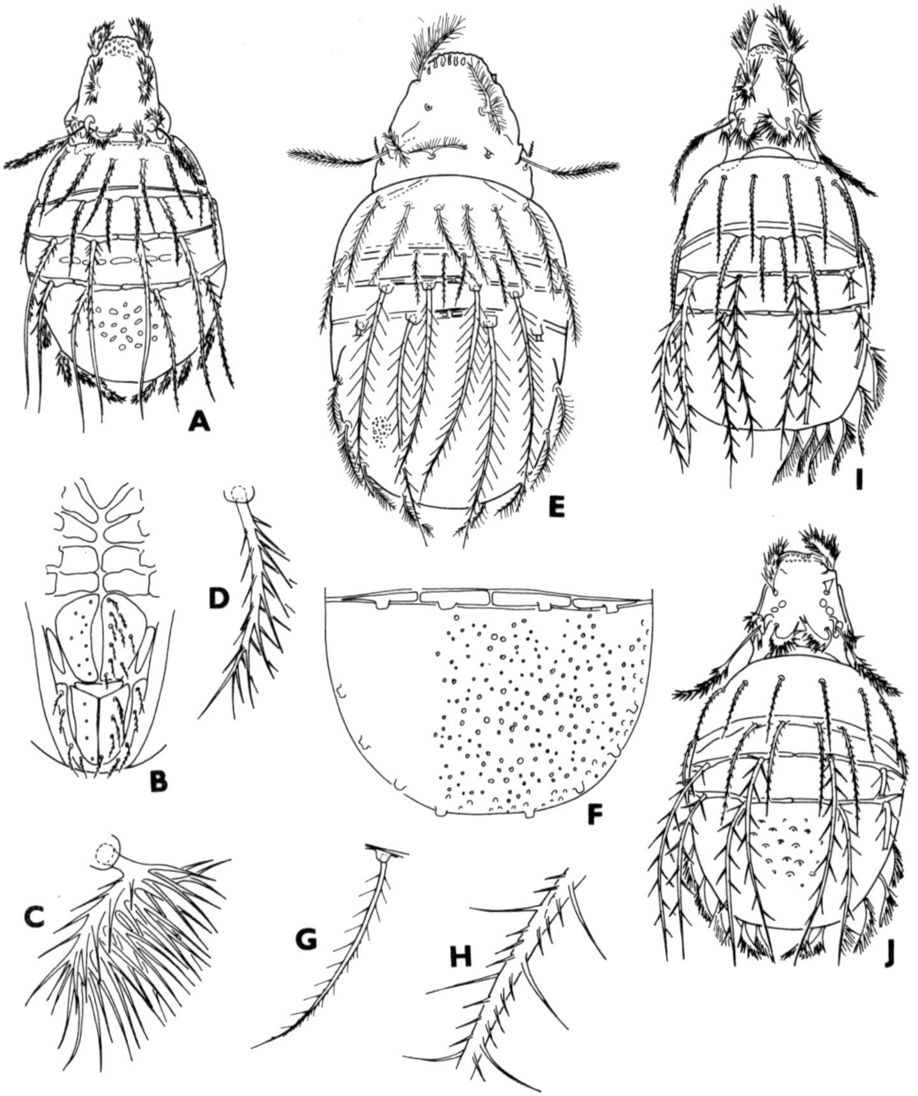

Plate 42. A—B: *Cosmochthonius ponticus* Gordeeva, 1980 — A: dorsal side, B: anogenital region; C—H: *Cosmochthonius plumatus* Berlese, 1910 — C: seta *ex*, D: seta h_2, E: dorsal side, F: sculpture of notogaster, G—H: seta f_1 and detail; I: *Cosmochthonius asiaticus* Gordeeva, 1980 — dorsal side; J: *Cosmochthonius tenuisetus* Gordeeva, 1980 — dorsal side (after Gordeeva)

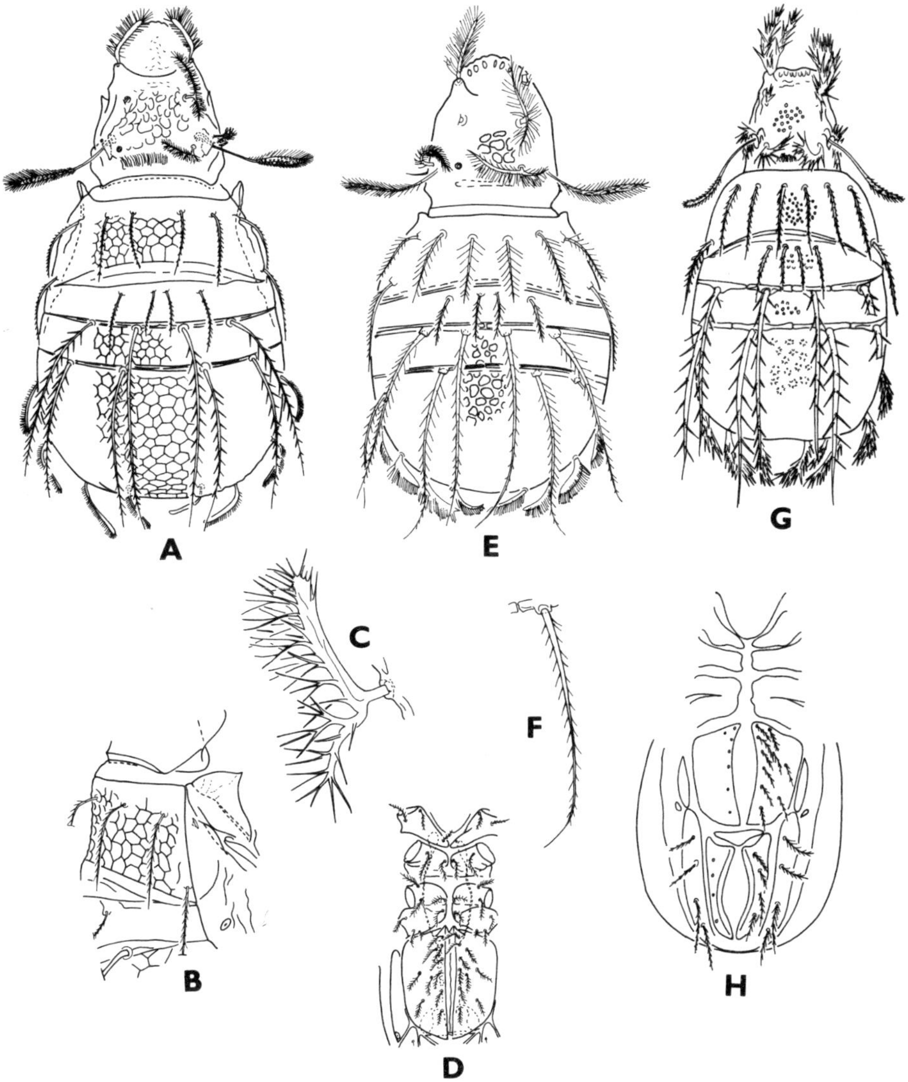

Plate 43. A—D: *Cosmochthonius reticulatus* Grandjean, 1947 — A: dorsal side, B: anterior part of notogaster, C: seta *le*, D: genital region; E—F: *Cosmochthonius lanatus* (Michael, 1885) — E: dorsal side, F: setae f_1; G—H: *Cosmochthonius ugamaensis* Gordeeva, 1980 — G: dorsal side, H: anogenital region (after Grandjean: A—F and Gordeeva: G—H)

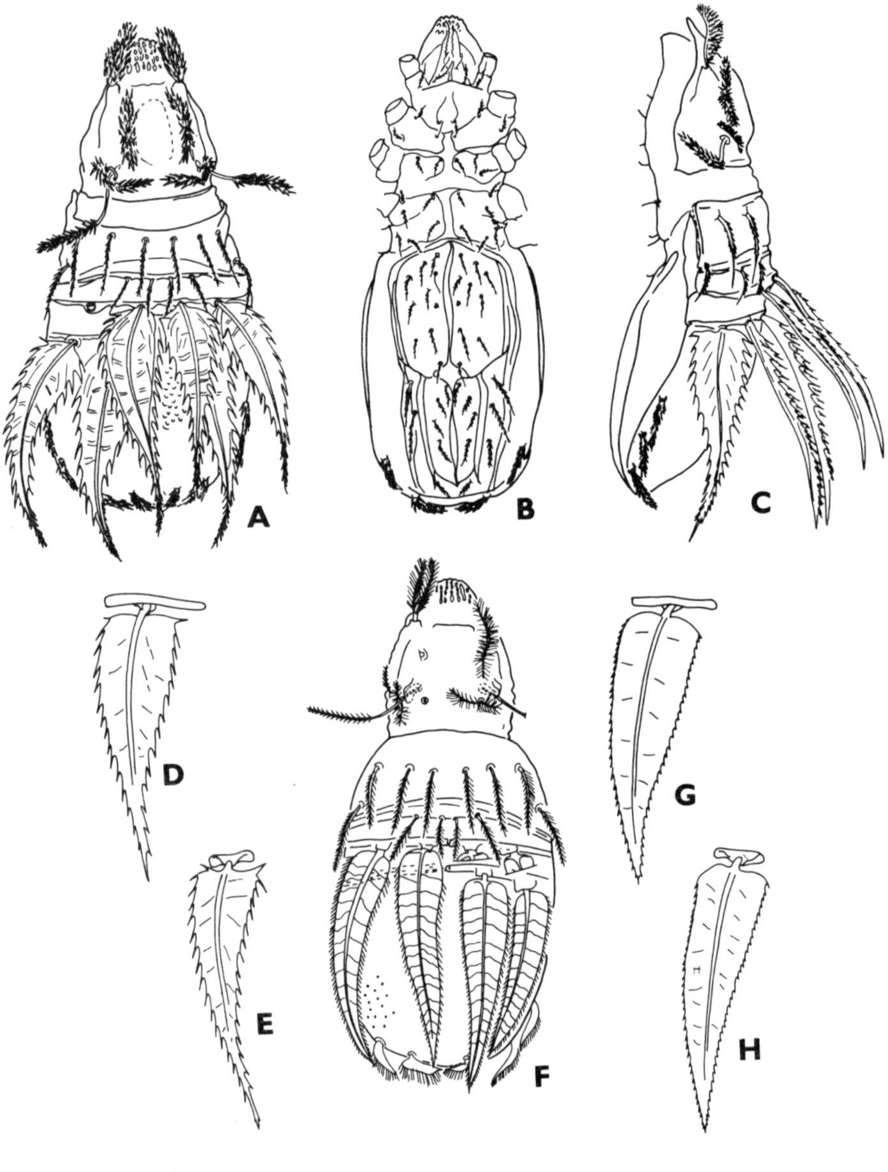

Plate 44. A—E: *Phyllozetes tauricus* Gordeeva, 1978—A: dorsal side, B: ventral side, C: lateral side, D: seta f_1, E: seta e_1; F—H: *Phyllozetes emmae* (Berlese, 1910) — F: dorsal side, G: seta f_1, H: seta e_1 (after Gordeeva: A—E, G—H; F original)

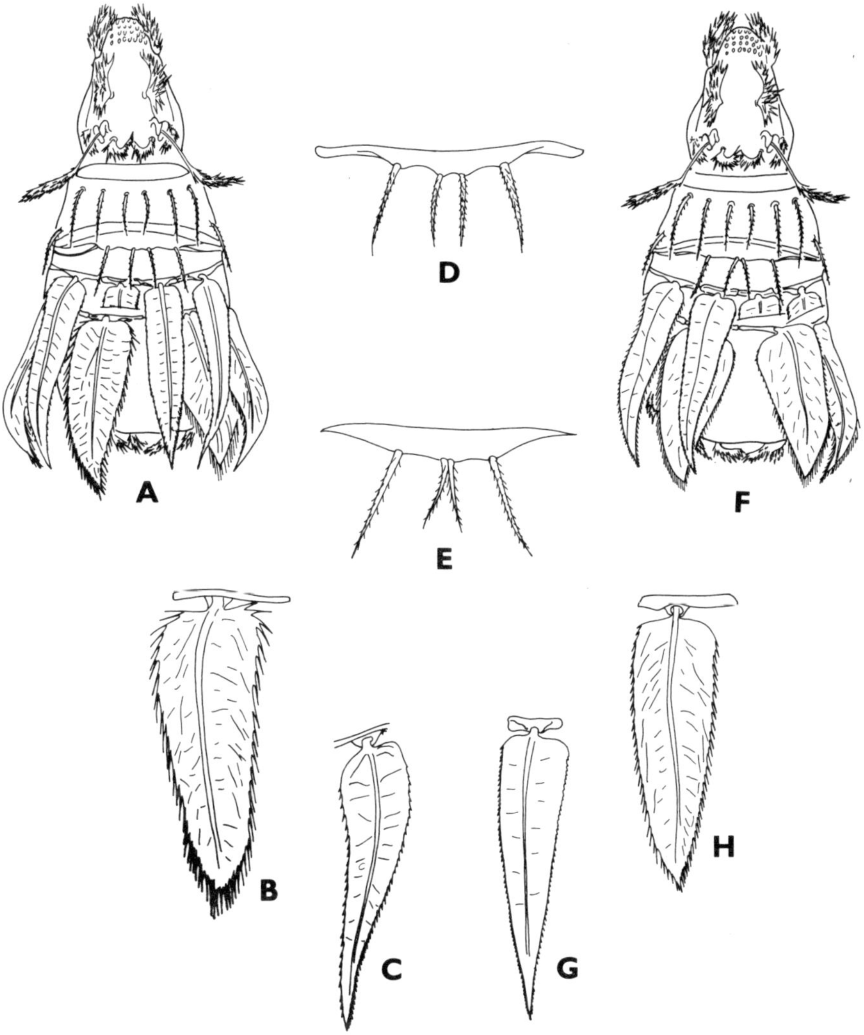

Plate 45. A—D: *Phyllozetes latifolius* Gordeeva, 1980 — A: dorsal side B: seta f_1, C: seta e_1, D: notogastral segment; E—H: *Phyllozetes osithchnjukovi* Gorodeeva, 1980 — E: notogastral segment, F: dorsal side, G: seta e_1, H: seta f_1 (after Gordeeva)

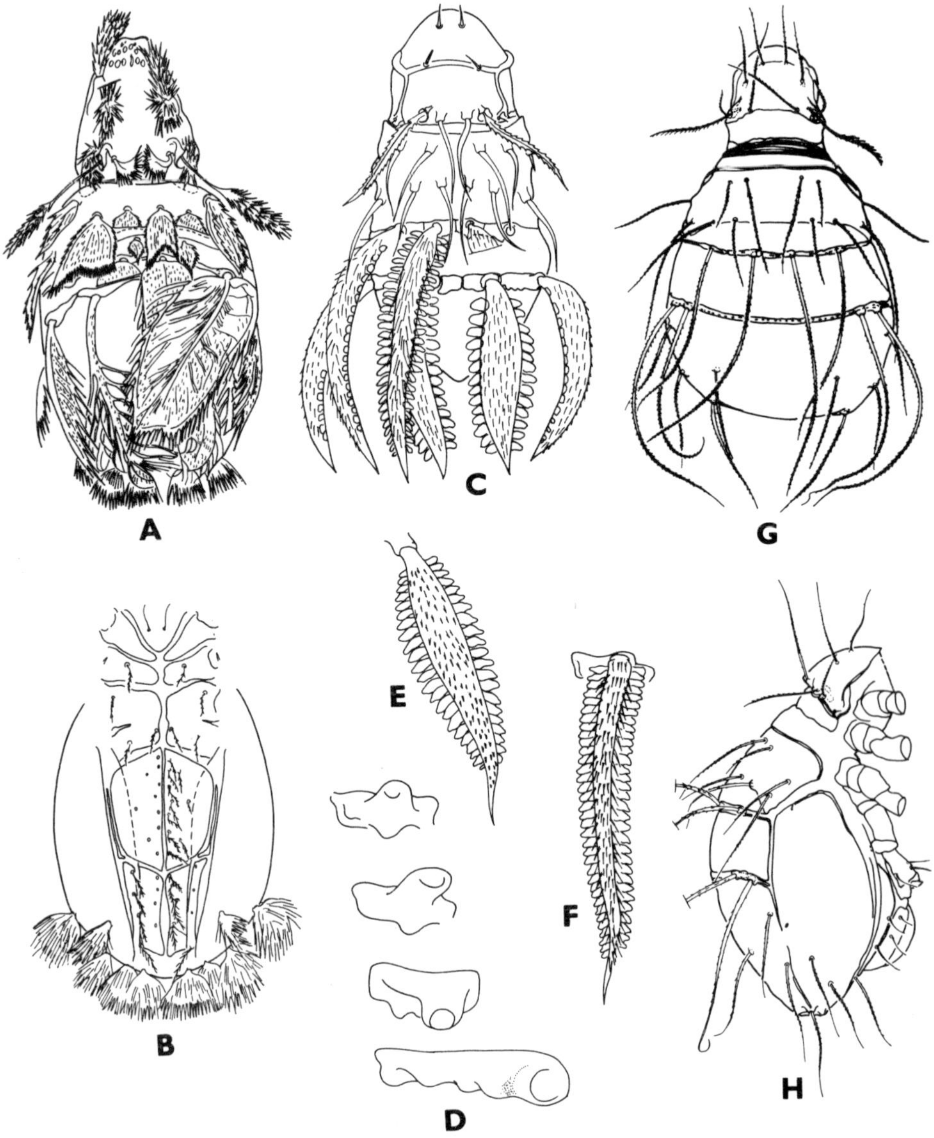

Plate 46. A—B: *Krivolutskiella pubescens* Gordeeva, 1980 — A: dorsal side, B: anogenital region; C—F: *Gozmanyina golosovae* (Gordeeva, 1980) — C: dorsal side, D: intercalary scelerites, E: seta f_1; G—H: *Nipponiella simplex* (Aoki, 1966) — G: dorsal side, H: lateral side (after Gordeeva: A—F and Aoki: G—H)

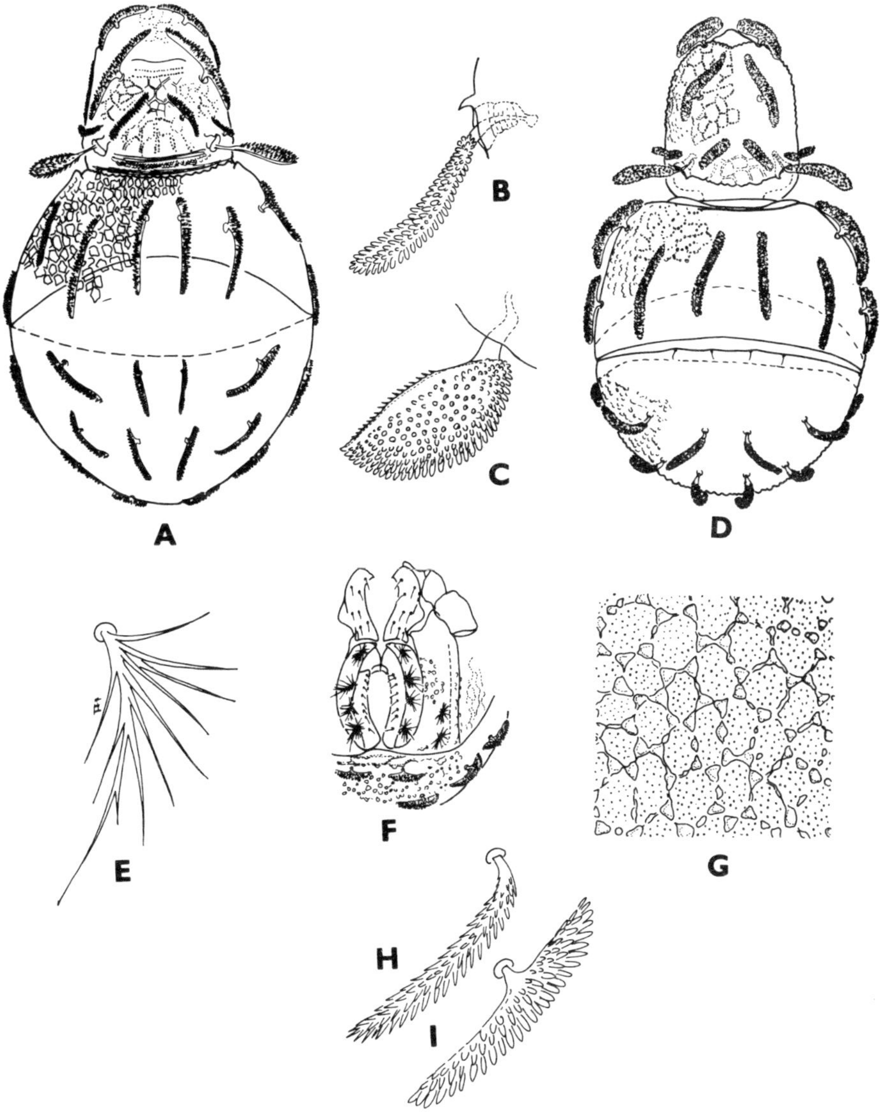

Plate 47. A: *Sphaerochthonius splendidus* (Berlese, 1904) — dorsal side; B—I: *Sphaerochthonius suzukii* Aoki, 1977 — B—C: sensilli, D: dorsal side, E: setae ps_2, F: anogenital region, G: cerotegument, H: seta e_1, I: seta c_1 (A original; after Aoki: B—I)

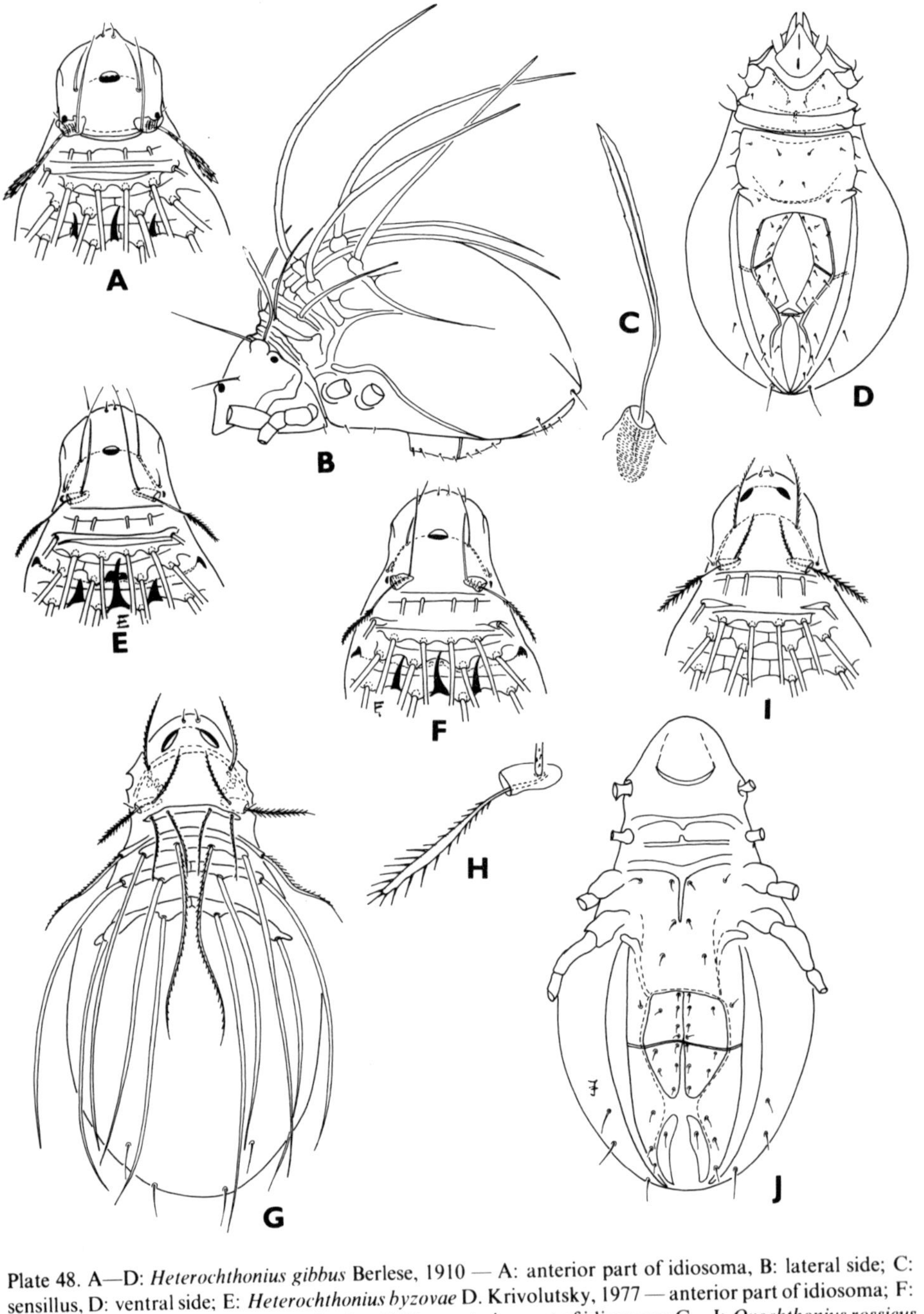

Plate 48. A—D: *Heterochthonius gibbus* Berlese, 1910 — A: anterior part of idiosoma, B: lateral side; C: sensillus, D: ventral side; E: *Heterochthonius byzovae* D. Krivolutsky, 1977 — anterior part of idiosoma; F: *Heterochthonius caucasicus* D. Krivolutsky, 1977 — anterior part of idiosoma; G—J: *Ovochthonius rossicus* Ryabinin, 1977 — G: dorsal side, H: sensillus, I: anterior part of idiosoma, J: ventral side (after D. Krivolutsky and Ryabinin: A, D—J, Grandjean: B and Mahunka: C)

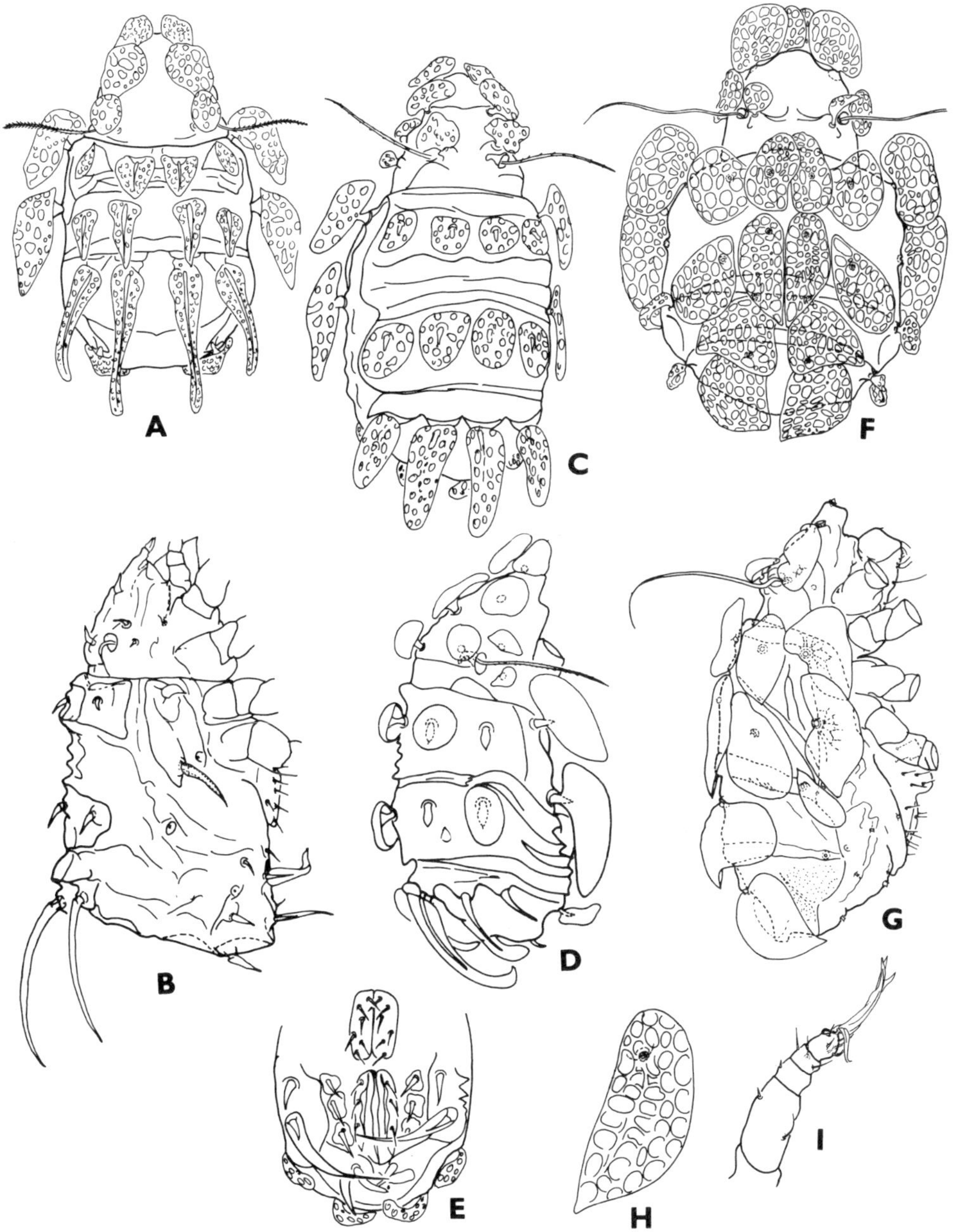

Plate 49. A—B: *Atopochthonius artiodactylus* Grandjean, 1948 — A: dorsal side, B: lateral side; C—E; *Atopochthonius maimaensis* Grishina, 1971 — C: dorsal side, D: lateral side, E: anogenital region; F—I: *Pterochthonius angelus* (Berlese, 1910) — F: dorsal side, G: lateral side, H: seta d_1, I: palp (after Grandjean: A—B, G—I, Grishina: C—E and Aoki: F)

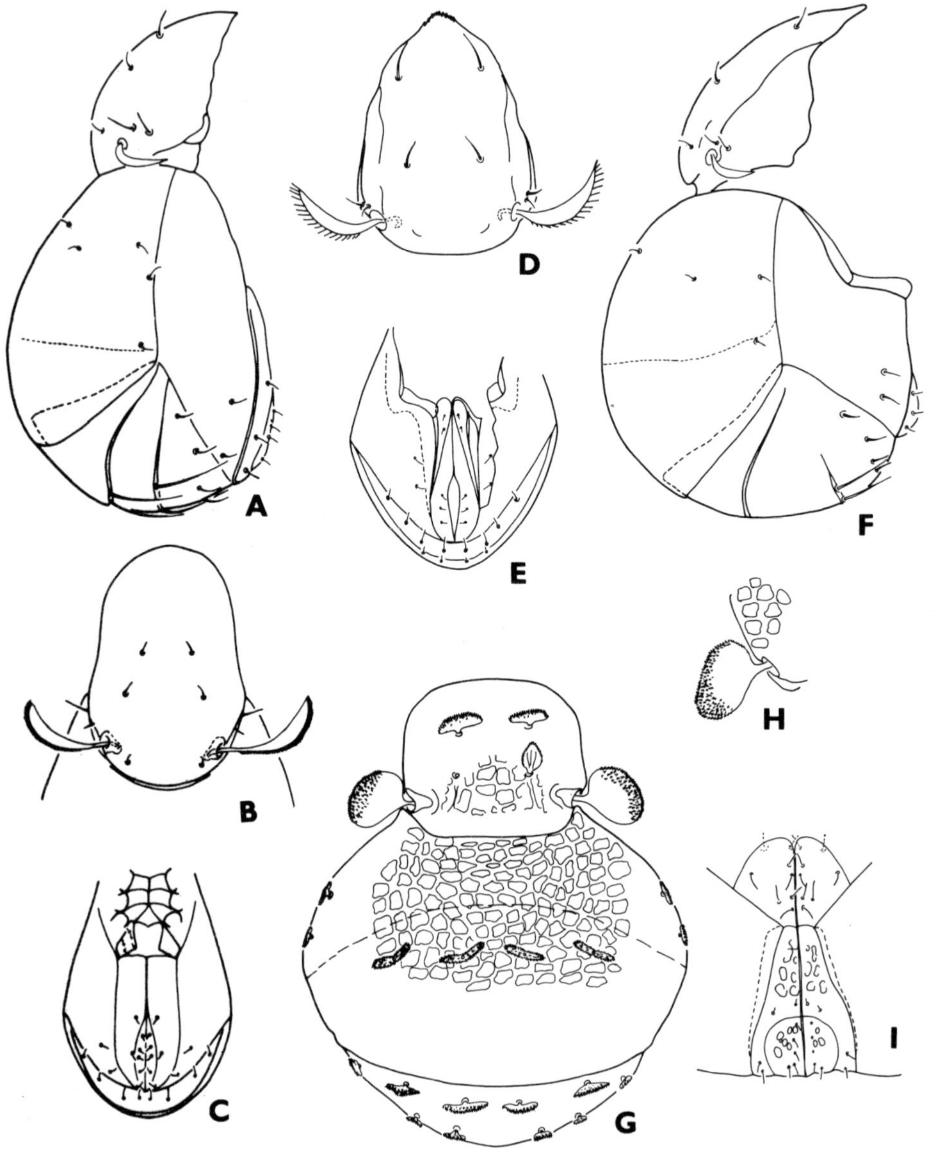

Plate 50. A—C: *Protoplophora palpalis* Berlese, 1910 — A: lateral side, B: prodorsum, C: anogenital region; D—F: *Grandjeanoplophora mauritanica* (Grandjean, 1932) — D: prodorsum, E: anogenital region, F: lateral side; G—I: *Hauseroplophora phitosi* Mahunka, 1977 — G: dorsal side, H: sensillus, I: anogenital region (after Grandjean: A—F; G—I original)

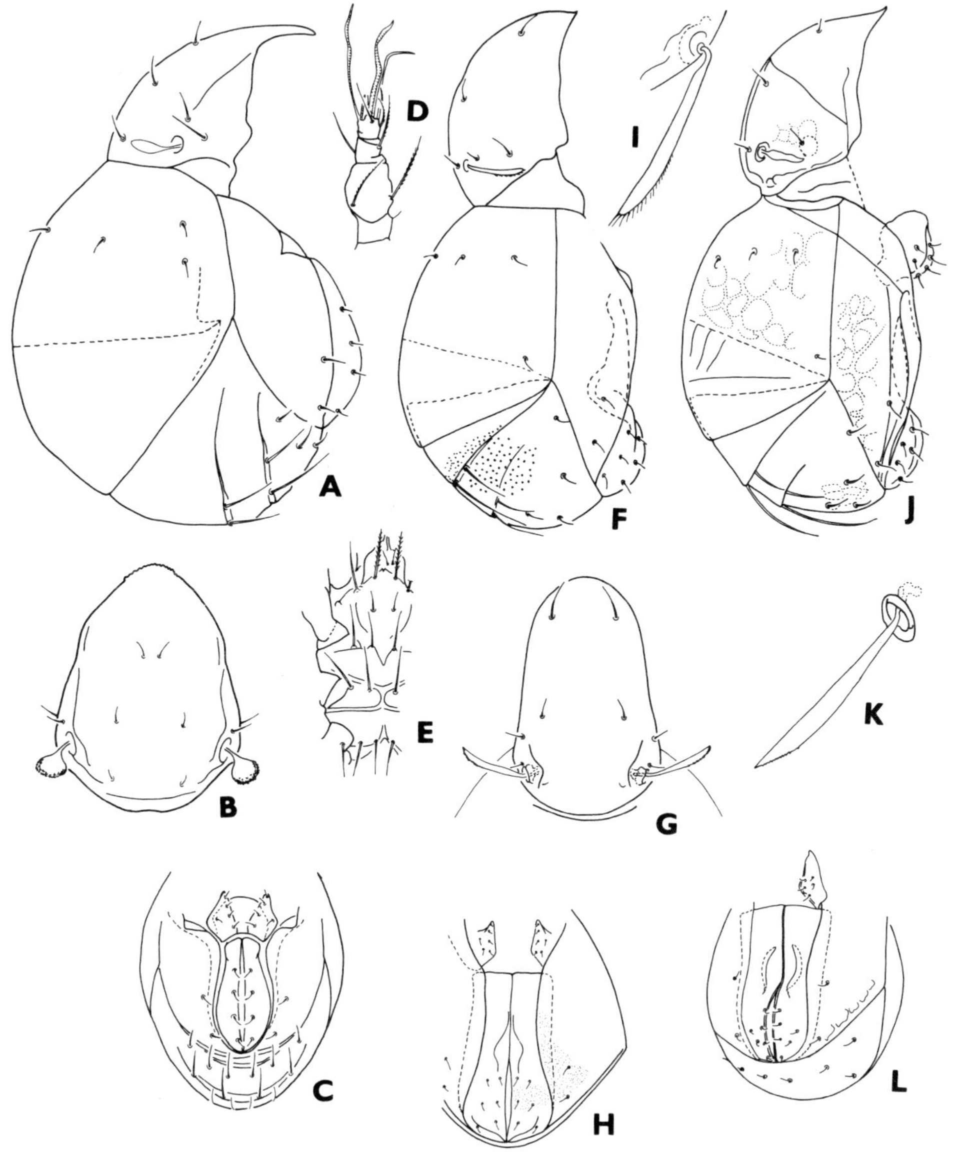

Plate 51. A—E: *Cryptoplophora abscondita* Grandjean, 1932 — A: lateral side, B: prodorsum, C: anogenital region, D: palp, E: epimeral region; F—I: *Bursoplophora bivaginata* (Grandjean, 1932) — *F:* lateral side, G: prodorsum, H: anogenital region, I: sensillus; J—L: *Bursoplophora iberica* Subias et Pérez-Iñigo, 1978 — J: lateral side, K: sensillus, L: anogenital region (after Grandjean: A—I and Subias and Pérez-Iñigo: J—L)

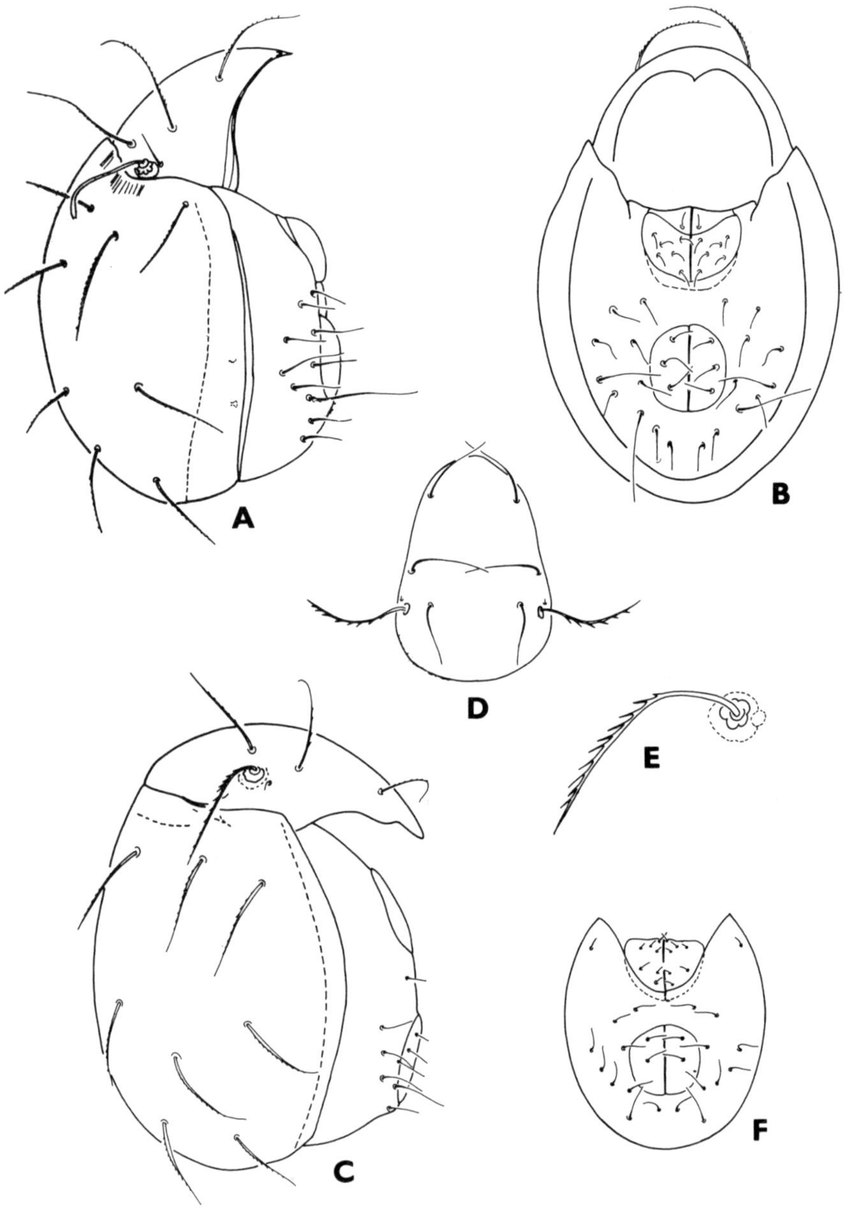

Plate 52. A—B: *Mesoplophora pulchra* Sellnick, 1928 — A: lateral side, B: ventral side; C—F: *Mesoplophora pectinata* Mahunka, 1979 — C: lateral side, D: prodorsum, E: sensillus, F: anogenital region (after Grandjean: A—B; C—F original)

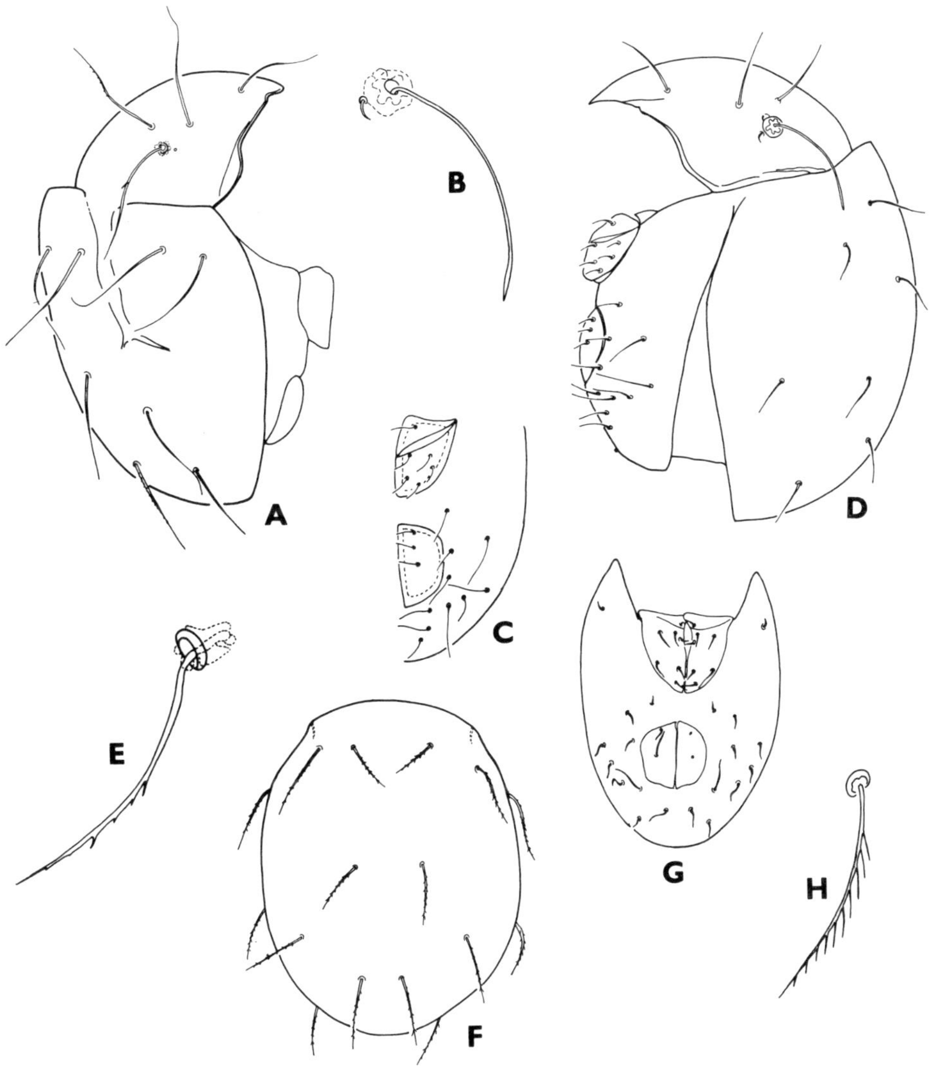

Plate 53. A: *Mesoplophora michaeliana* Berlese, 1904 — lateral side; B—D: *Mesoplophora japonica* Aoki, 1970 — B: sensillus, C: anogenital region, D: lateral side; E—F: *Mesoplophora graeca* Walzl, 1974 — E: sensillus, F: notogaster; G: *Mesoplophora caucasica* D. Krivolutsky, 1975 — sensillus (A original; after Aoki: B—D, Walzl: E—F and D. Krivolutsky: G)

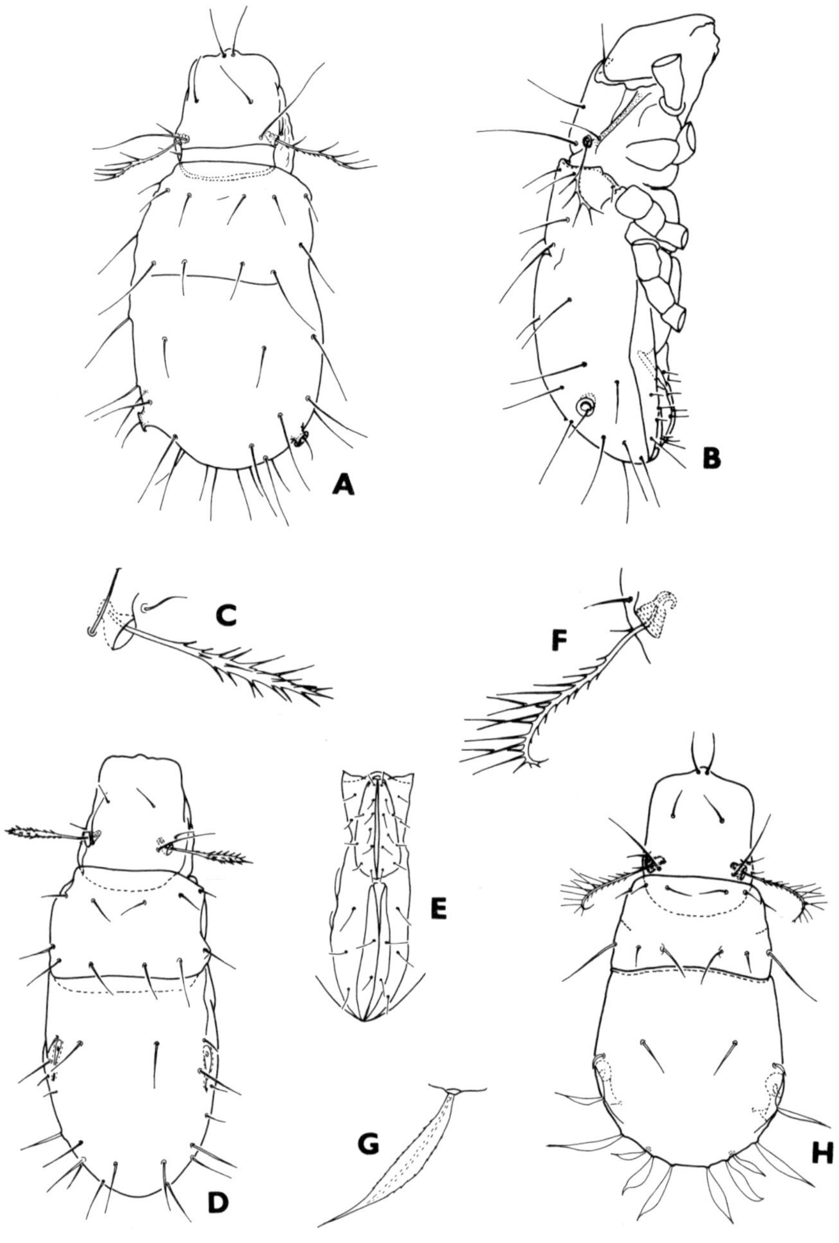

Plate 54. A—B: *Parhypochthonius aphidinus* Berlese, 1904 — A: dorsal side, B: lateral side; C—E: *Geohypochthonius rhadamanthus* Jacot, 1936 — C: sensillus, D: dorsal side, E: anogenital region; F—H: *Geohypochthonius frondifer* Aoki, 1975 — F: sensillus, G: seta h_2, H: dorsal side (after Aoki)

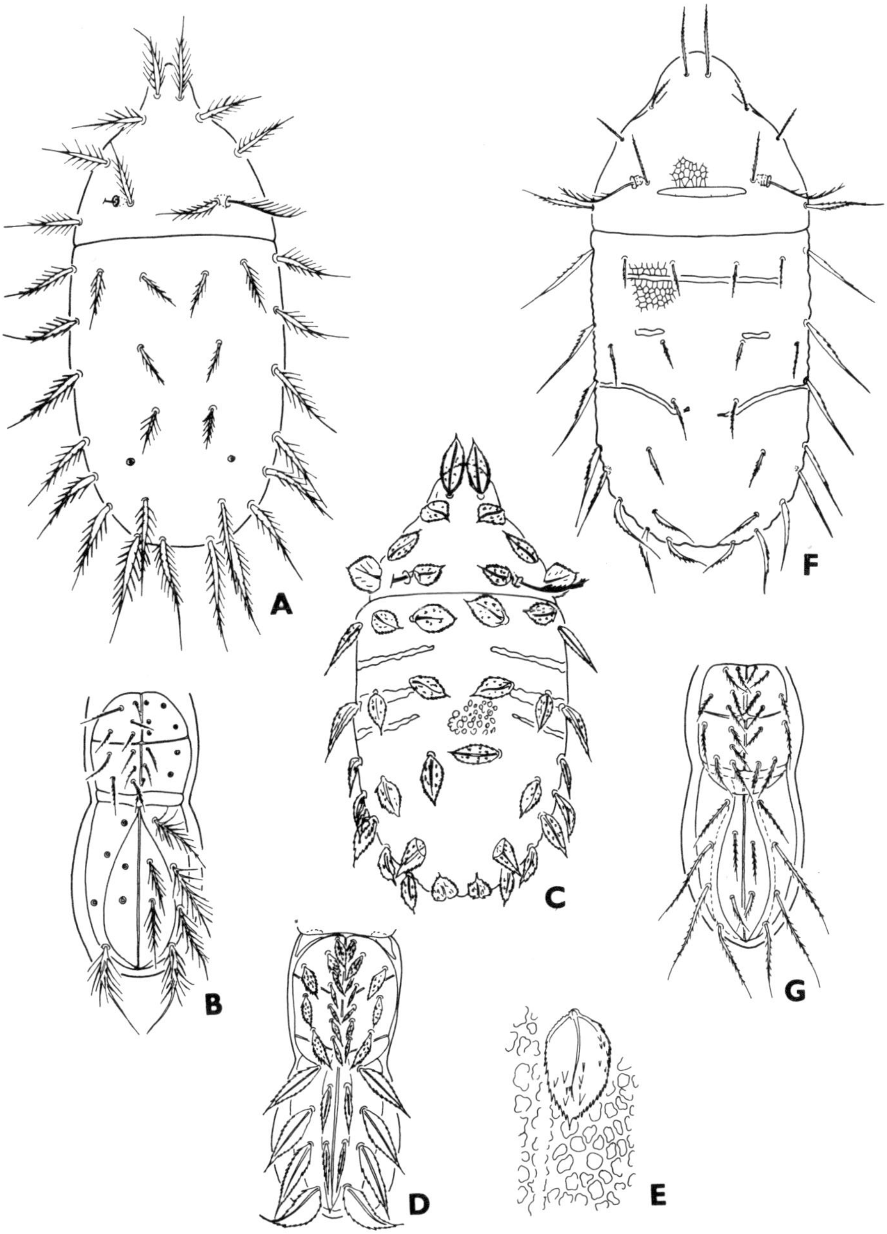

Plate 55. A—B: *Lohmannia regalis* Berlese, 1923 — A: dorsal side, B: anigenital region; C—E: *Lohmannia hungarorum* Mahunka, 1980 — C: dorsal side, D: anogenital region, E: seta d_1 and sculpture of notogaster; F—G: *Lohmannia reticulata* Sellnick, 1931 — F: dorsal side, G: anogenital region (original)

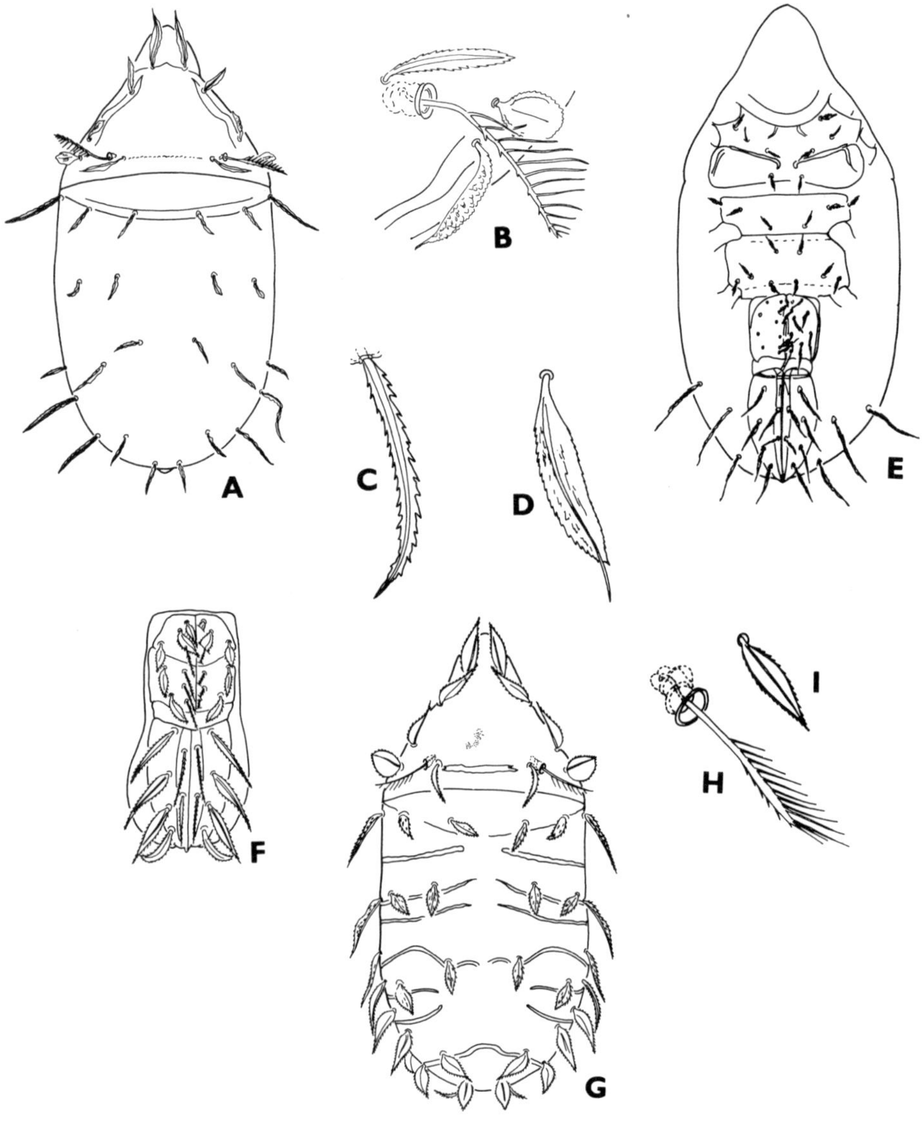

Plate 56. A—E: *Lohmannia turcmenica* Bulanova-Zachvatkina, 1960 — A: dorsal side, B: bothridial region, C: seta of notogaster, D: seta *ro*, E: ventral side; F—I: *Lohmannia loebli* Mahunka, 1977 — F: anogenital region, G: dorsal side, H: sensillus, I: seta c_1 (after Pérez-Iñigo: A—E; F—I original)

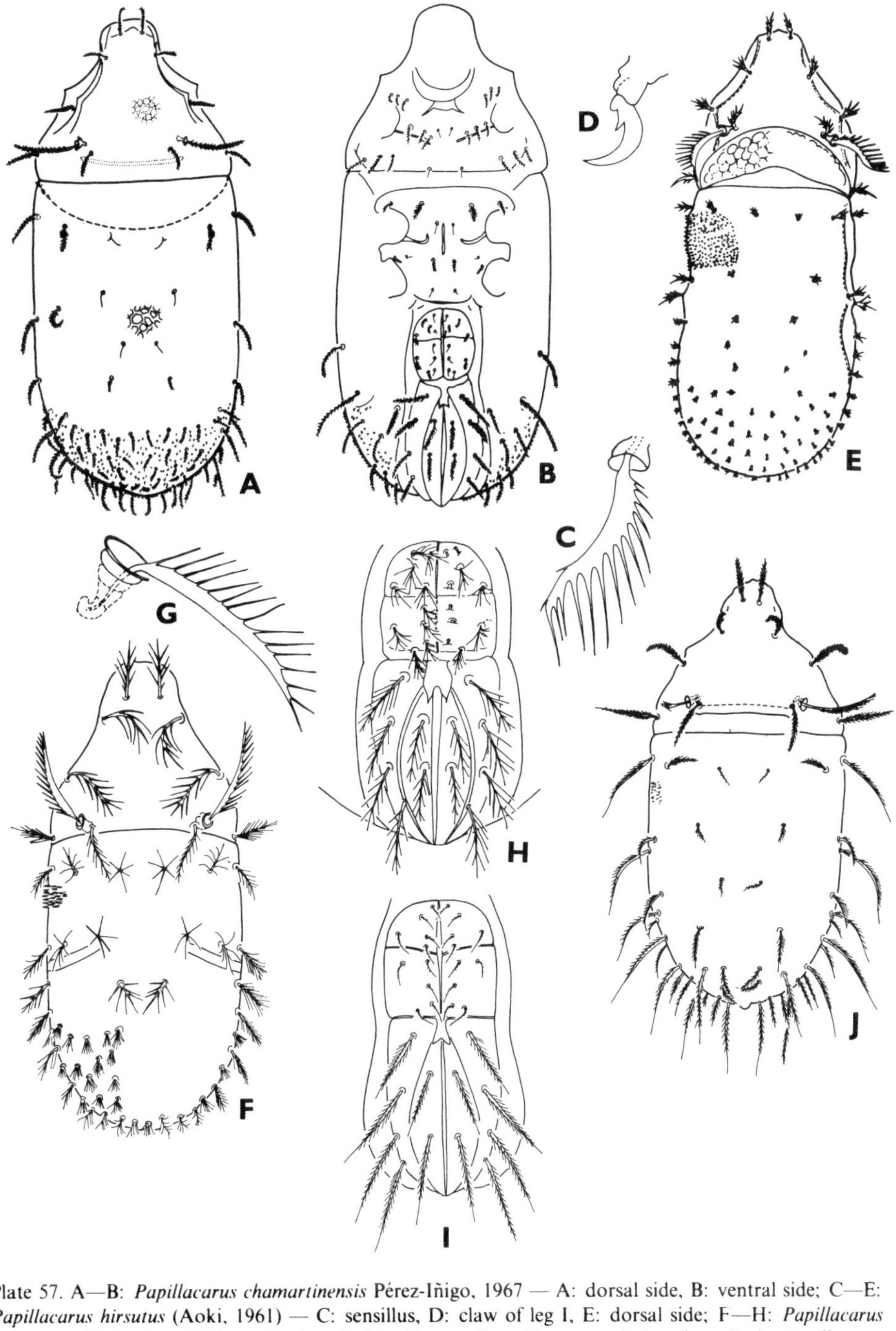

Plate 57. A—B: *Papillacarus chamartinensis* Pérez-Iñigo, 1967 — A: dorsal side, B: ventral side; C—E: *Papillacarus hirsutus* (Aoki, 1961) — C: sensillus, D: claw of leg I, E: dorsal side; F—H: *Papillacarus koreanus* (Mahunka, 1973) — F: dorsal side, G: sensillus, H: anogenital region; I—J: *Papillacarus pseudoaciculatus* Mahunka, 1980 — I: anogenital region, J: dorsal side (after Pérez-Iñigo: A—B and Aoki: C—E; F—J original)

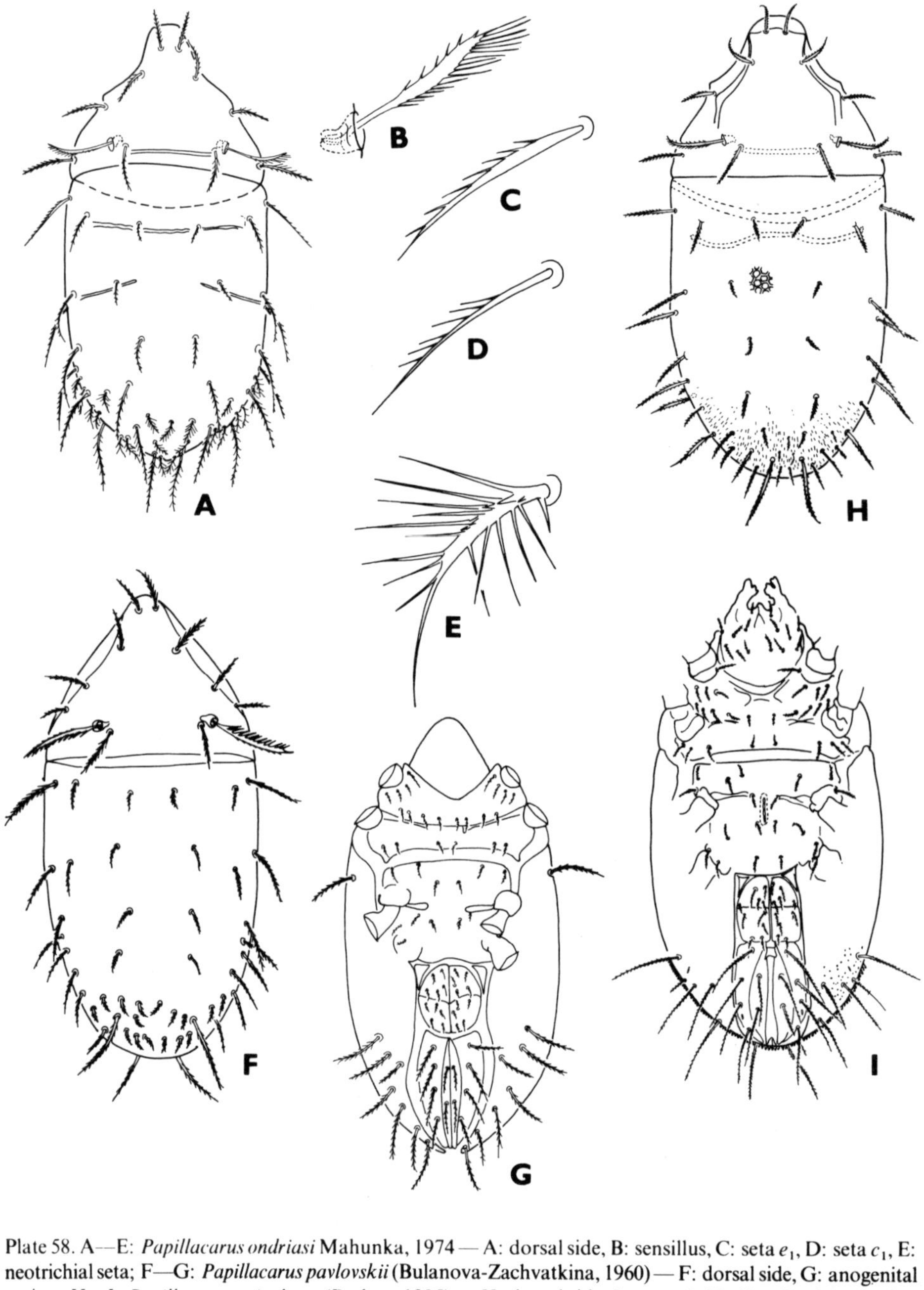

Plate 58. A—E: *Papillacarus ondriasi* Mahunka, 1974 — A: dorsal side, B: sensillus, C: seta e_1, D: seta c_1, E: neotrichial seta; F—G: *Papillacarus pavlovskii* (Bulanova-Zachvatkina, 1960) — F: dorsal side, G: anogenital region; H—I: *Papillacarus aciculatus* (Berlese, 1905) — H: dorsal side, I: ventral side (A—E original; after Bulanova-Zachvatkina: F—G and Kunst: H—I)

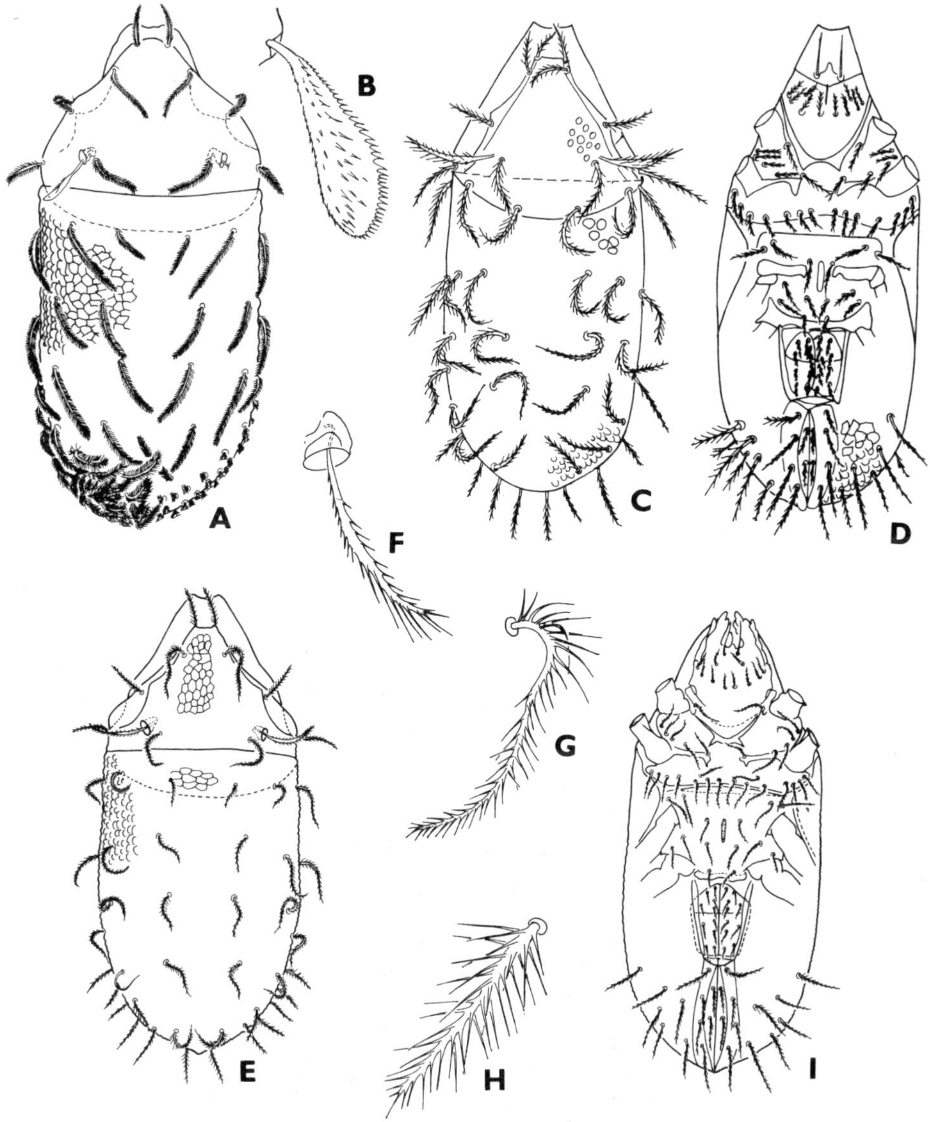

Plate 59. A—B: *Thamnacarus deserticola* (Grandjean, 1934) — A: dorsal side, B: sensillus; C—D: *Thamnacarus smirnovi* Bulanova-Zachvatkina, 1978 — C: dorsal side, D: ventral side; E—I: *Thamnacarus elongatus* (D. Krivolitsky, 1971) — E: dorsal side, F: sensillus, G: seta *le*, H: setae of notogaster, I: ventral side (after Grandjean: A—B, Bulanova-Zachvatkina: C—D and Aoki: E—I)

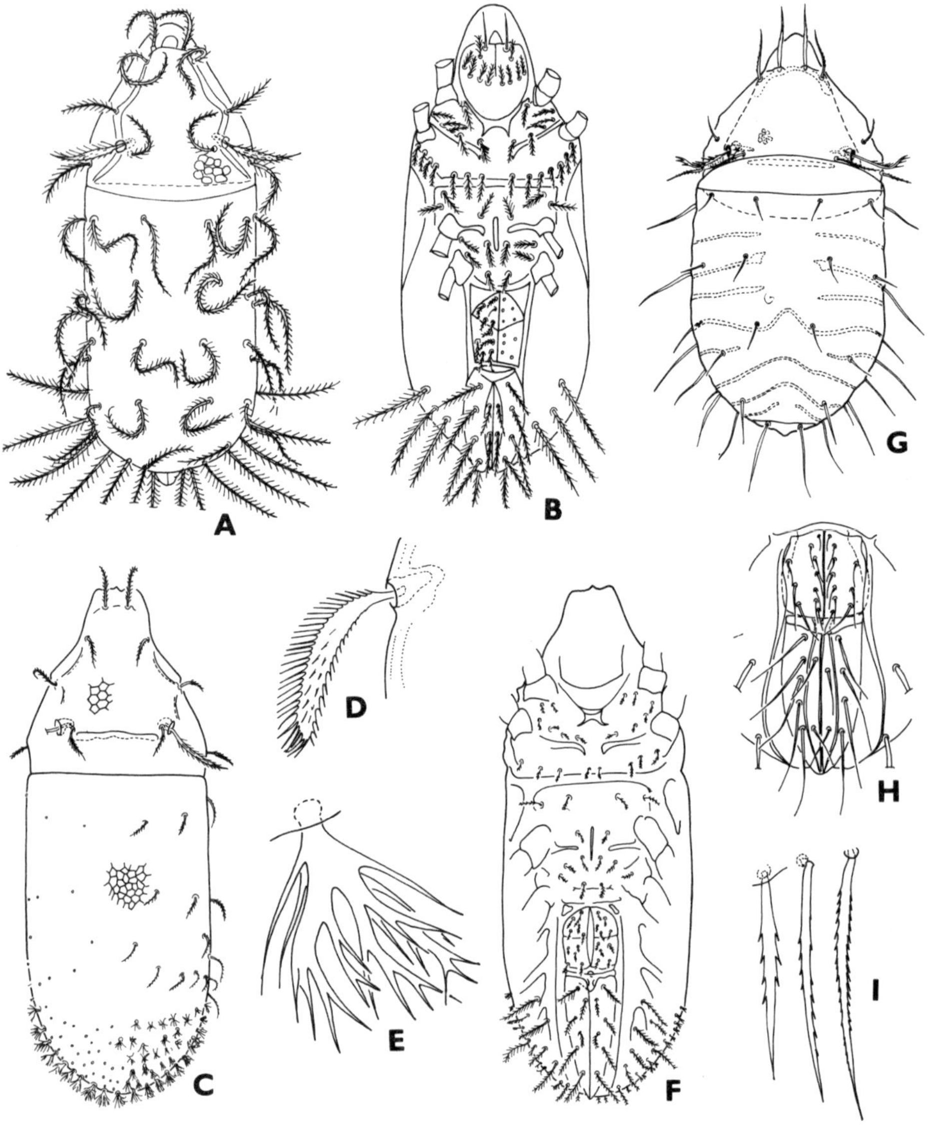

Plate 60. A—B: *Thamnacarus longisetosus* Bulanova-Zachvatkina, 1978 — A: dorsal side, B: ventral side; C—F: *Cryptacarus promecus* Grandjean, 1950 — C: dorsal side, D: sensillus, E: notogastral seta, F: ventral side; G—I: *Mixacarus exilis* Aoki, 1970 — G: dorsal side, H: anogenital region, I: seta *le* (after Bulanova-Zachvatkina: A—B, Grandjean: C—F and Aoki: G—J)

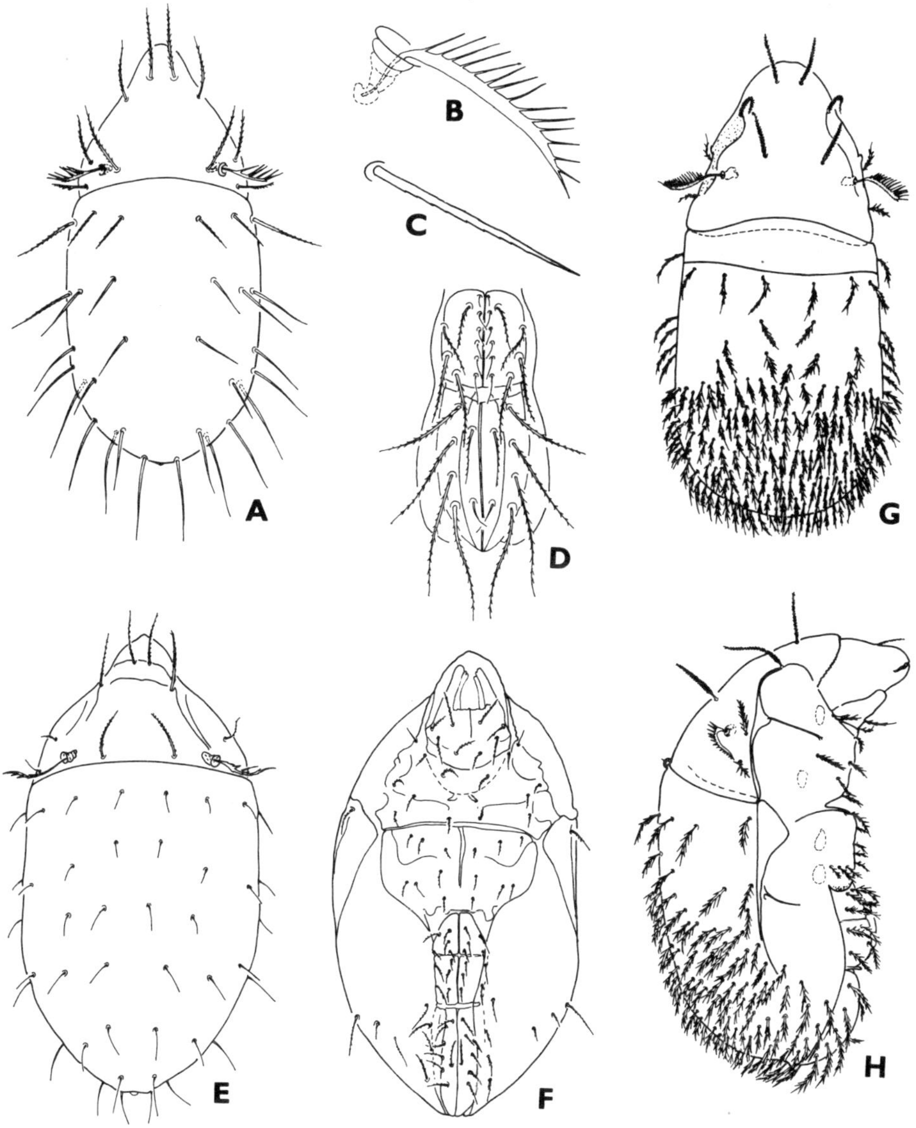

Plate 61. A—D: *Mixacarus vanhonggui* Mahunka, 1973 — A: dorsal side, B: sensillus, C: seta e_1, D: anogenital region; E—F: *Heptracarus notoneotrichus* Piffl, 1963 — E: dorsal side, F: ventral side; G—H: *Heptacarus supertrichus* Piffl, 1967 — G: dorsal side, H: ventral side (A—D original; after Piffl: E—H)

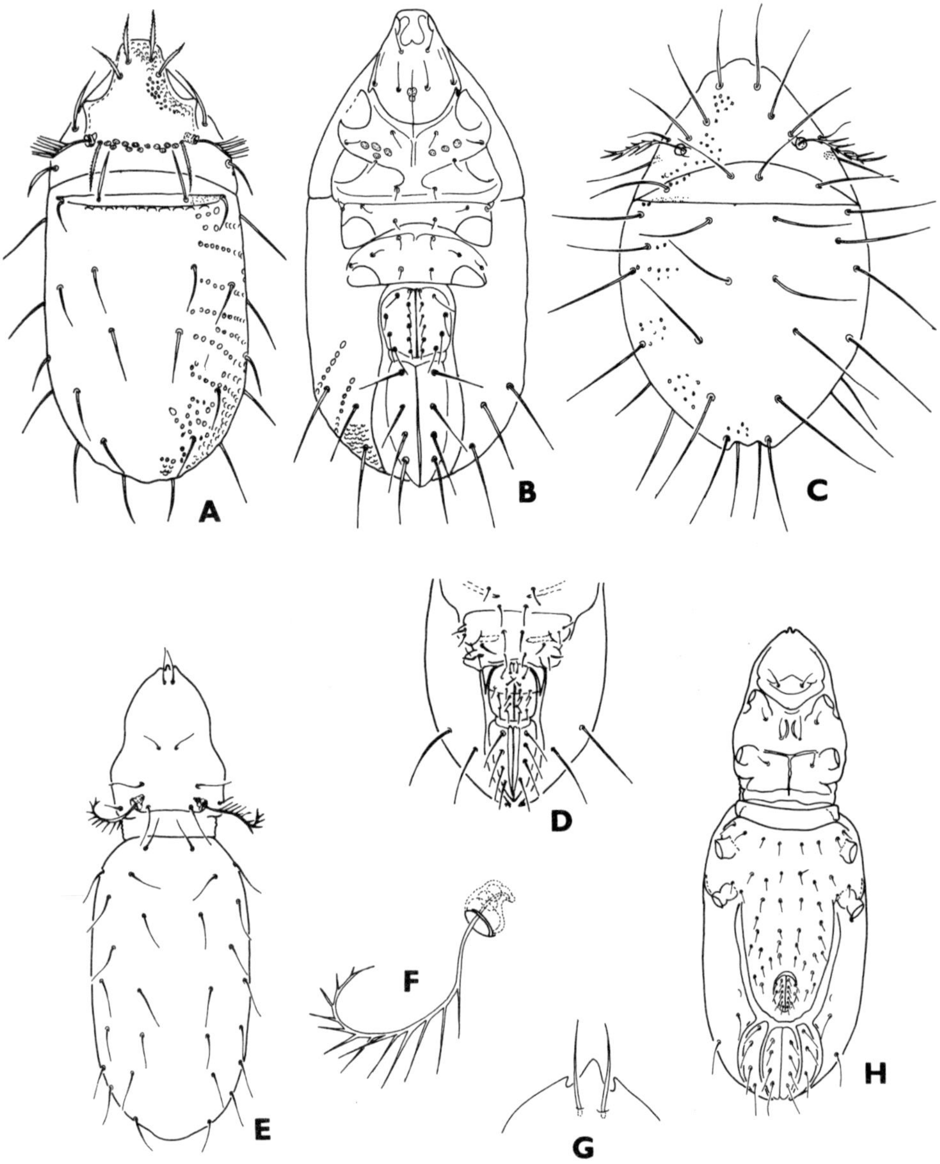

Plate 62. A—B: *Javacarus kuehnelti* Balogh, 1961 — A: dorsal side, B: ventral side; C—D: *Meristolohmannia chinensis* (Bulanova-Zachvatkina, 1960) — C: dorsal side, D: anogenital region; E—H: *Eulohmannia ribagai* Berlese, 1910 — E: dorsal side, F: sensillus, G: rostrum, H: ventral side (after Balogh: A—B, Bulanova-Zachvatkina: C—D and Aoki: E—H)

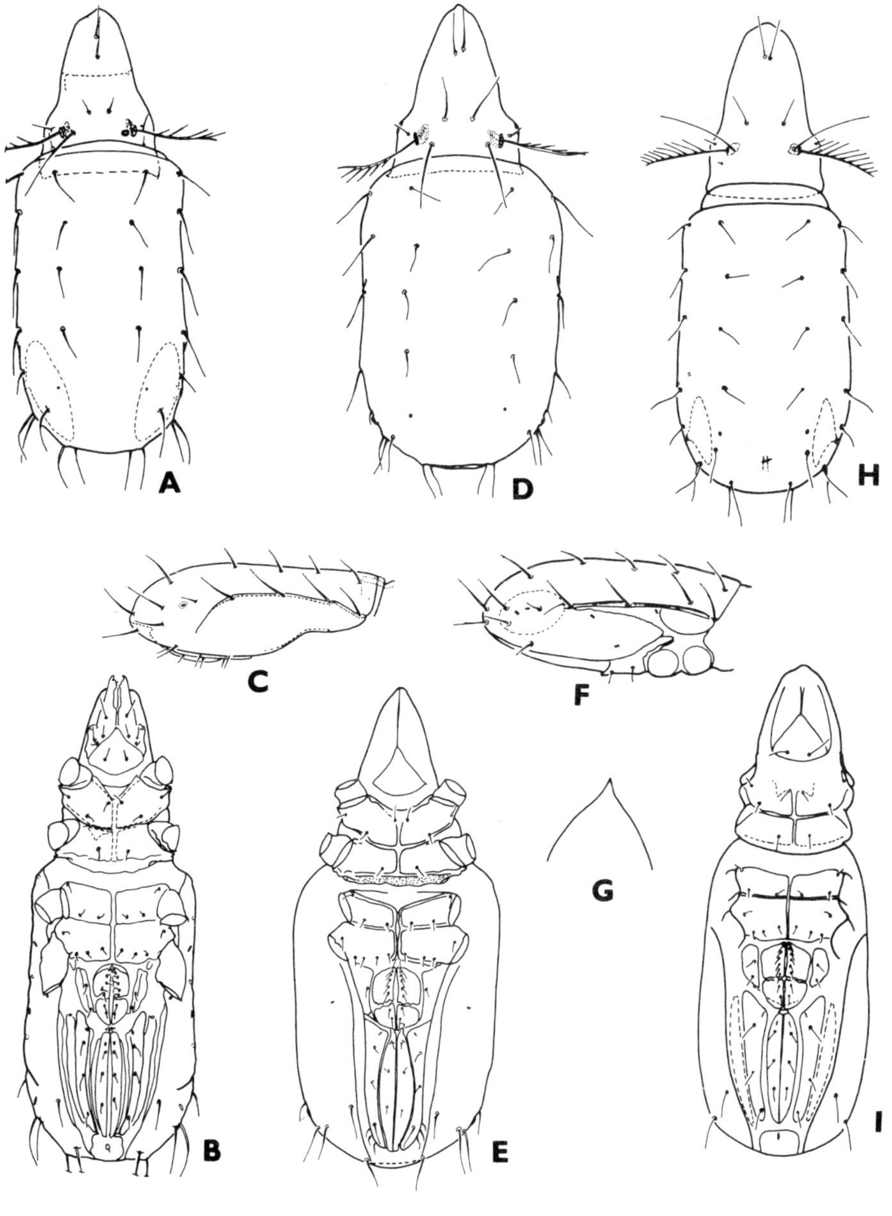

Plate 63. A—C: *Perlohmannia dissimilis* (Hewitt, 1908) — A: dorsal side, B: ventral side, C: anterior part of notogaster from lateral side; D—F: *Perlohmannia coiffaiti* Grandjean, 1961 — D: dorsal side, E: ventral side, F: anterior part of notogaster form lateral side; G: *Perlohmannia insignis* (Berlese, 1904) — rostrum; H—I: *Perlohmannia altaica* Grishina, 1968 — H: dorsal side, I: ventral side (after Grandjean: A—F and Schuster: G; H—I original)

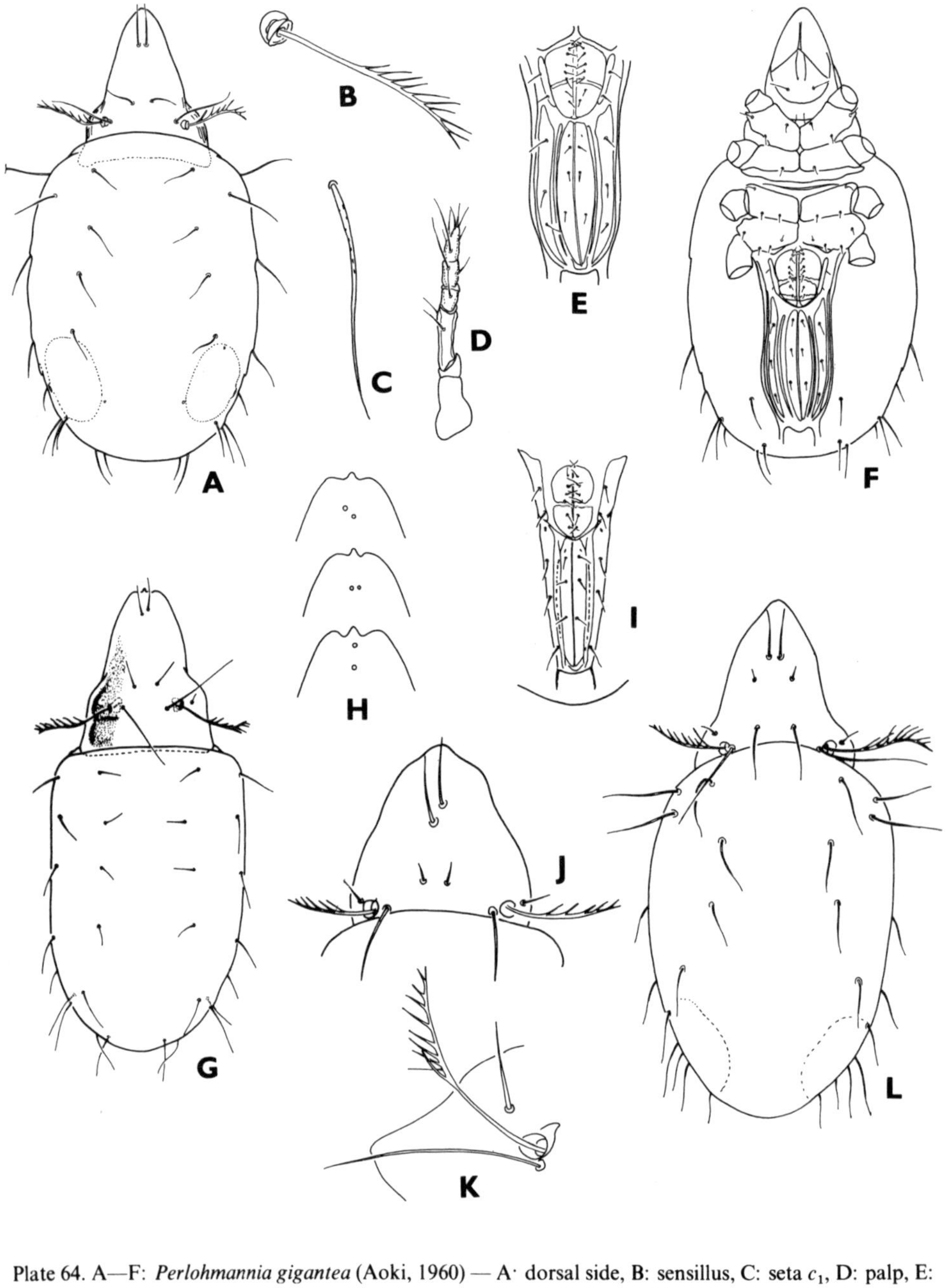

Plate 64. A—F: *Perlohmannia gigantea* (Aoki, 1960) — A· dorsal side, B: sensillus, C: seta c_1, D: palp, E: anogenital region, F: ventral side; G—I: *Perlohmannia nasuta* Schuster, 1960 — G: dorsal side, H: variation of rostrum, I: anogenital region; J—L: *Perlohmannia zachvatkini* (Bulanova-Zachvatkina, 1960) — J: prodorsum, K: trichobothrium, L: dorsal side (after Aoki: A—F, Schuster: G—I and Bulanova-Zachvatkina: J—L)

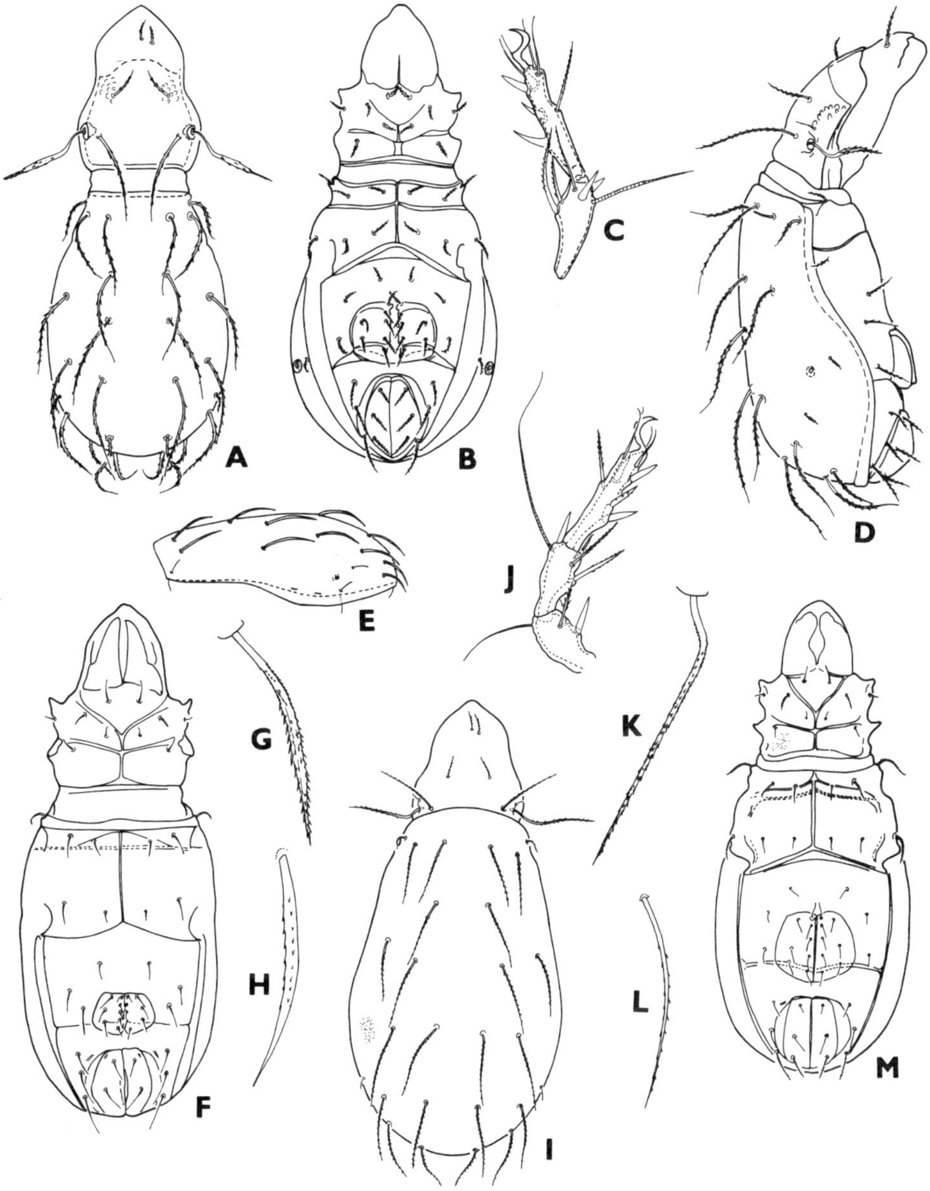

Plate 65. A—D: *Epilohmannia schusteri* Calugar et Vasiliu, 1976 — A dorsal side, B: ventral side, C: leg IV, D: lateral side, E—H: *Epilohmannia* gigantea Berlese, 1916 — E: notogaster from lateral side, F: ventral side, G: sensillus, H: seta c_1; I—M: *Epilohmannia imreorum* Bayoumi et Mahunka, 1976 — I: dorsal side, J; leg IV, K: sensillus, L: seta c_1, M: ventral side (after Calugar and Vasiliu: A—D and Schuster: E; F—M original)

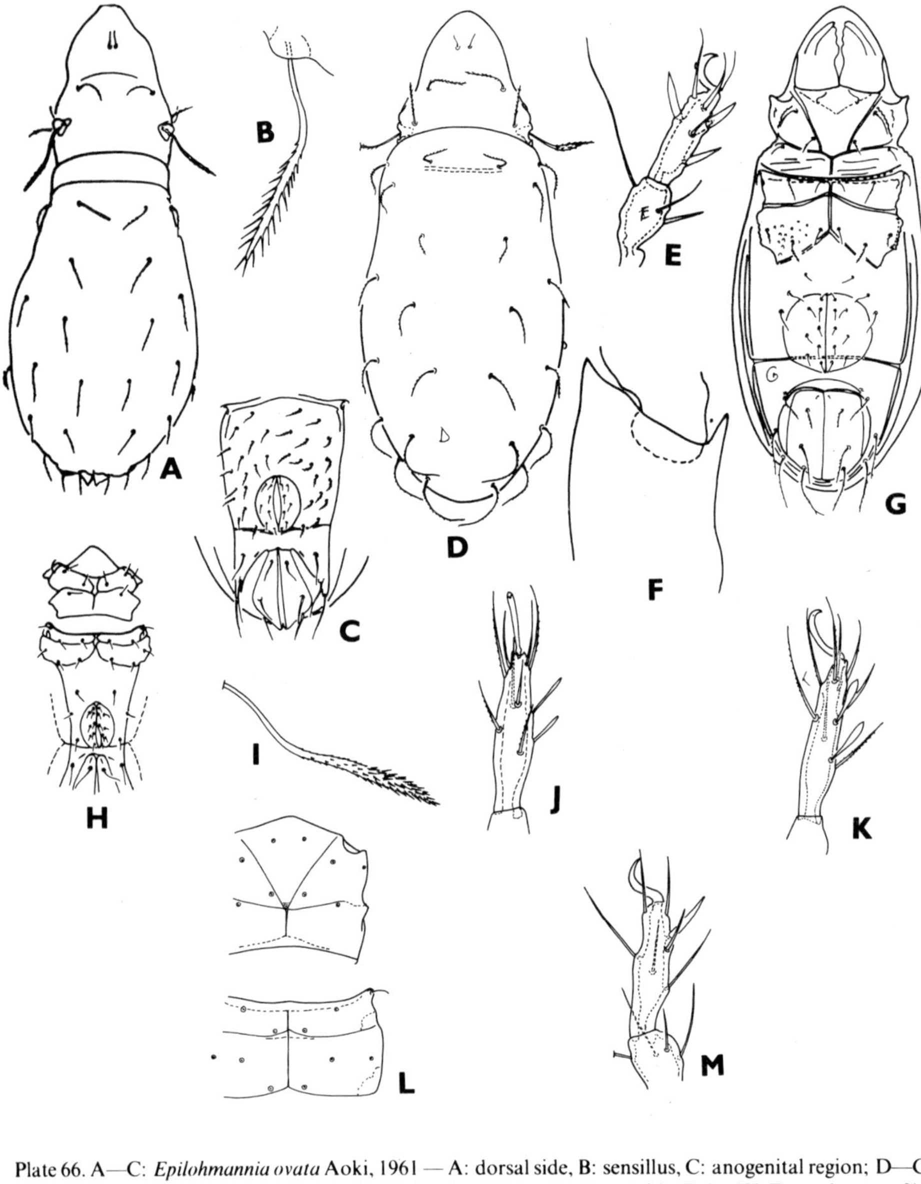

Plate 66. A—C: *Epilohmannia ovata* Aoki, 1961 — A: dorsal side, B: sensillus, C: anogenital region; D—G: *Epilohmannia pallida aegyptica* Bayoumi et Mahunka, 1976 — D: dorsal side, E: leg IV, F: trochanter of leg IV, G: ventral side; H—K: *Epilohmannia spathulata* Aoki, 1970 — H: epimeral region, I: sensillus, J—K: leg IV; L—M: *Epilohmannia inexpectata* Schuster, 1960 — L: epimeral region, M: leg IV (after Aoki: A—C, H—K and Schuster: L—M; D—G original)

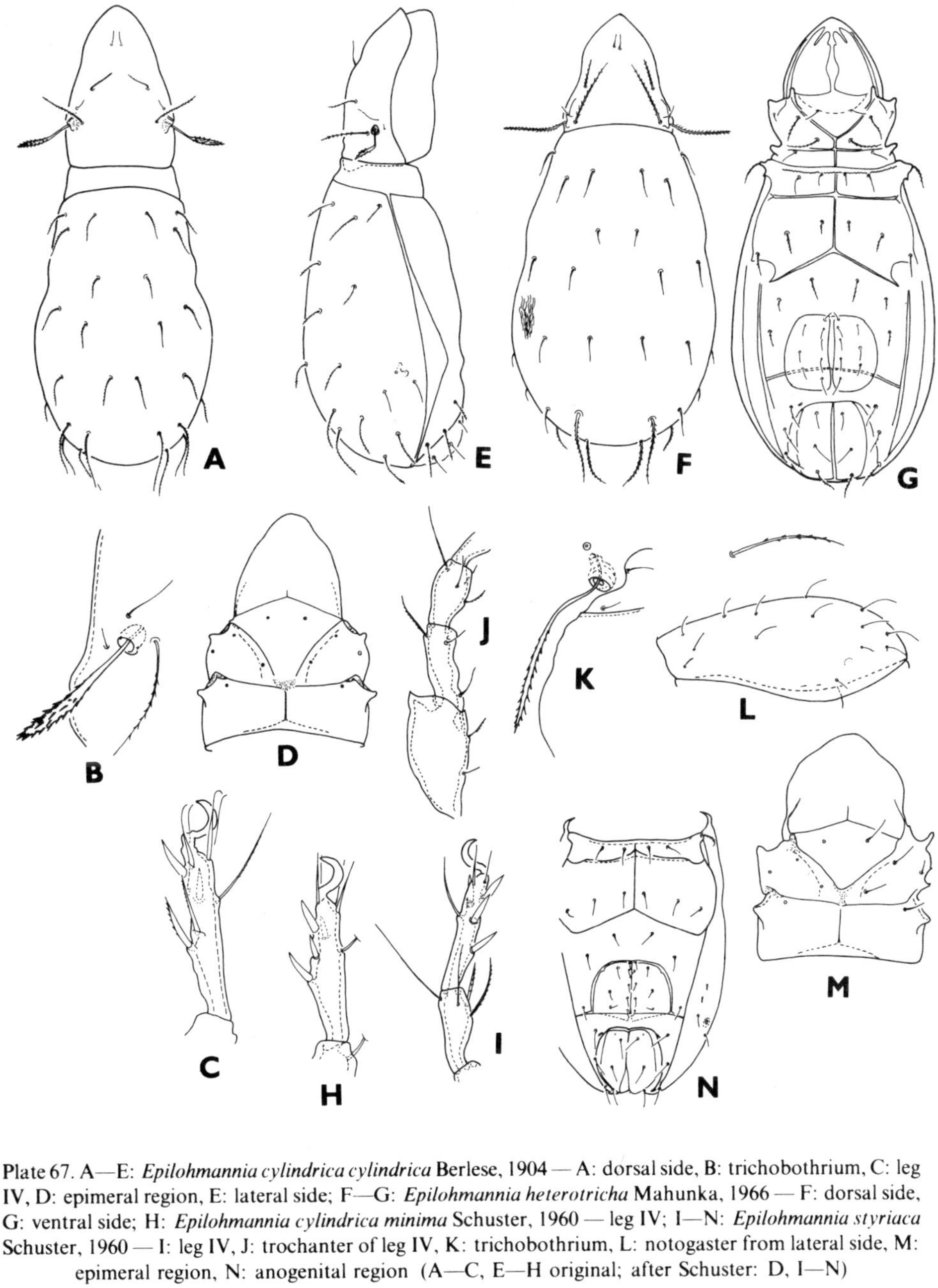

Plate 67. A—E: *Epilohmannia cylindrica cylindrica* Berlese, 1904 — A: dorsal side, B: trichobothrium, C: leg IV, D: epimeral region, E: lateral side; F—G: *Epilohmannia heterotricha* Mahunka, 1966 — F: dorsal side, G: ventral side; H: *Epilohmannia cylindrica minima* Schuster, 1960 — leg IV; I—N: *Epilohmannia styriaca* Schuster, 1960 — I: leg IV, J: trochanter of leg IV, K: trichobothrium, L: notogaster from lateral side, M: epimeral region, N: anogenital region (A—C, E—H original; after Schuster: D, I—N)

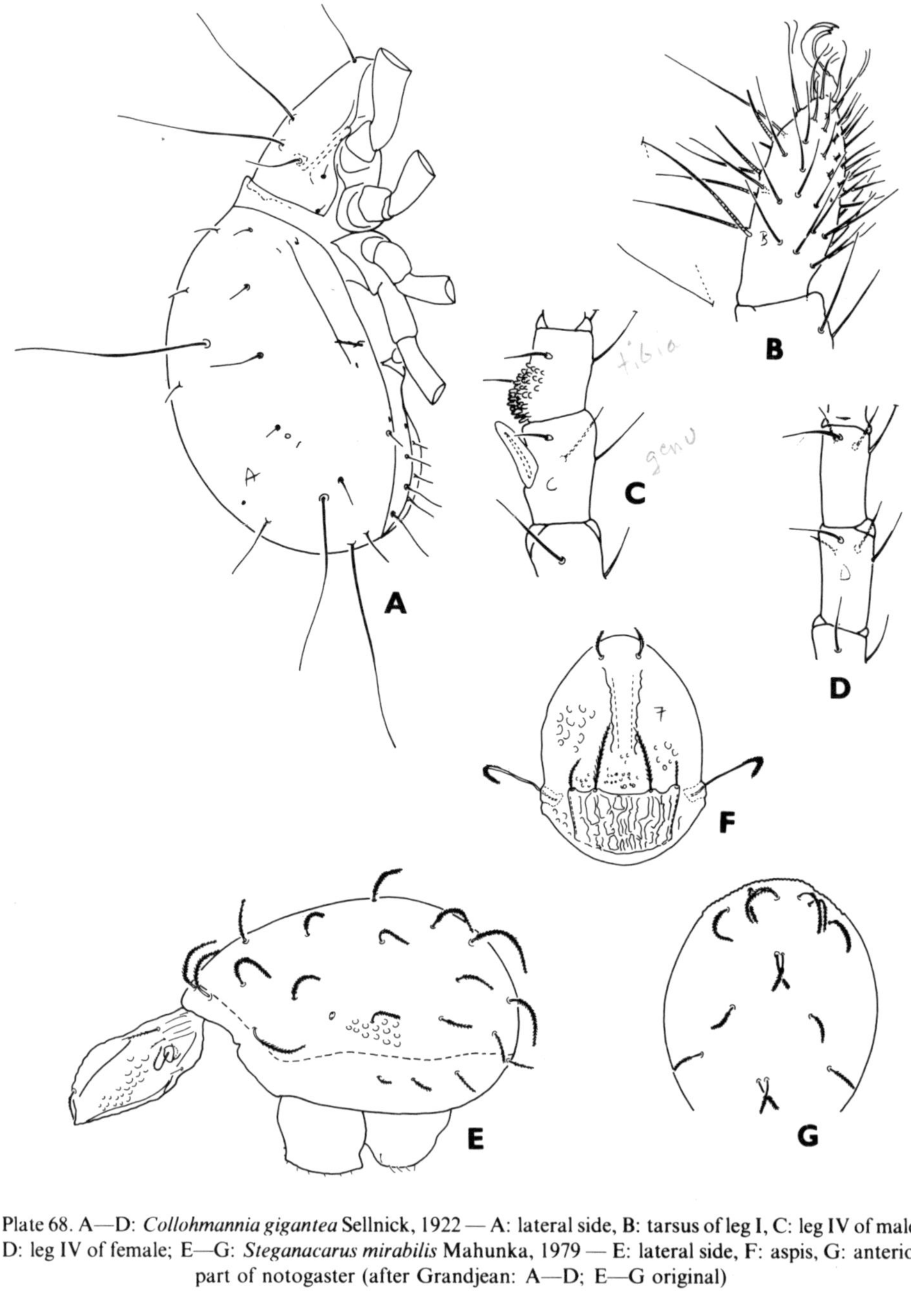

Plate 68. A—D: *Collohmannia gigantea* Sellnick, 1922 — A: lateral side, B: tarsus of leg I, C: leg IV of male, D: leg IV of female; E—G: *Steganacarus mirabilis* Mahunka, 1979 — E: lateral side, F: aspis, G: anterior part of notogaster (after Grandjean: A—D; E—G original)

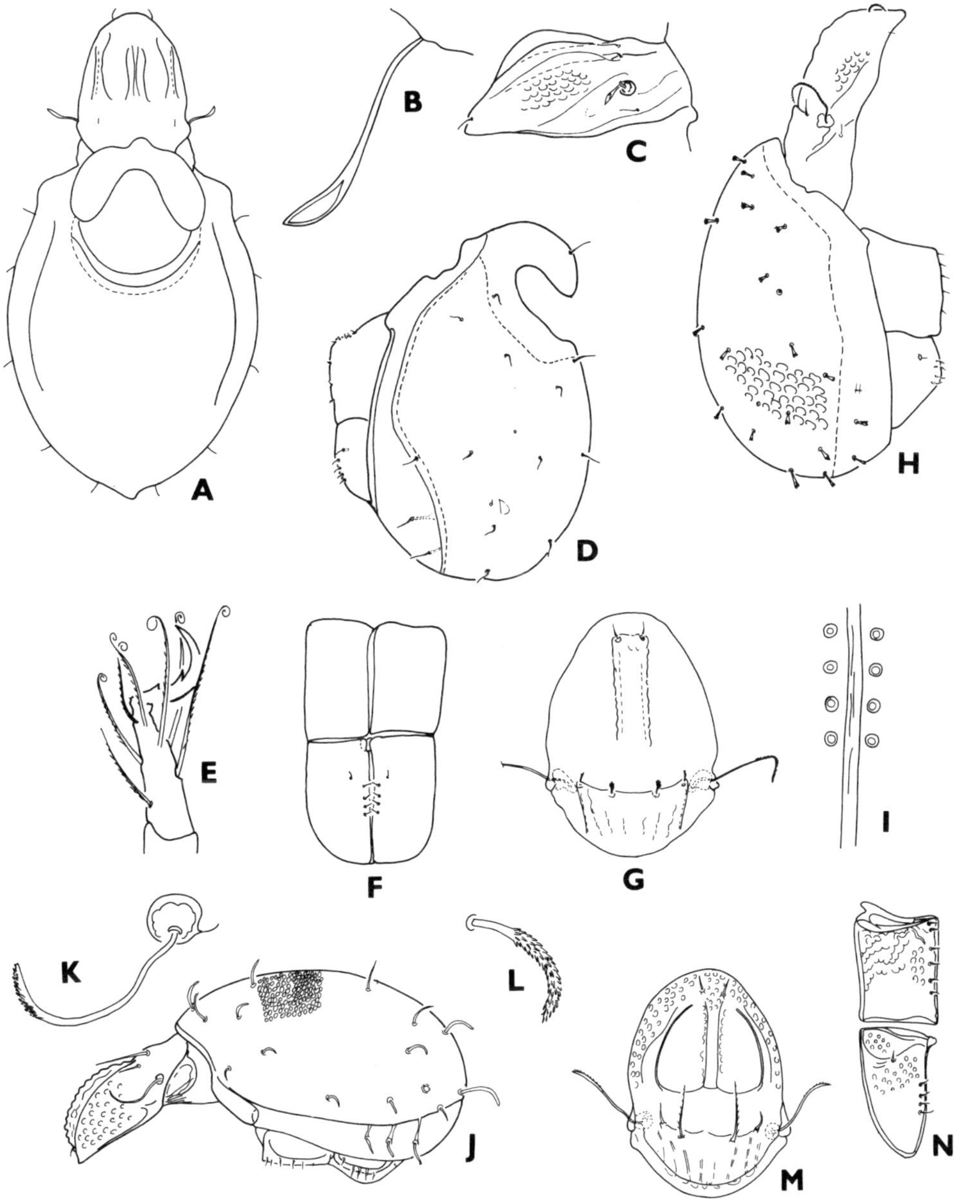

Plate 69. A—F: *Steganacarus grandjeani* Mahunka, 1977 — A: dorsal side, B: sensillus, C: aspis, D: notogaster from lateral side, E: leg IV, F: anogenital region; G—I: *Steganacarus clavigerus* (Berlese, 1904) — G: aspis, H: lateral side, I: position of anal hairs; J—N: *Steganacarus serratus* Feider et Suciu, 1957 — J: lateral side, K: sensillus, L: seta h_1, M: aspis, N: anogenital region (A—I original; after Feider and Suciu: J—N)

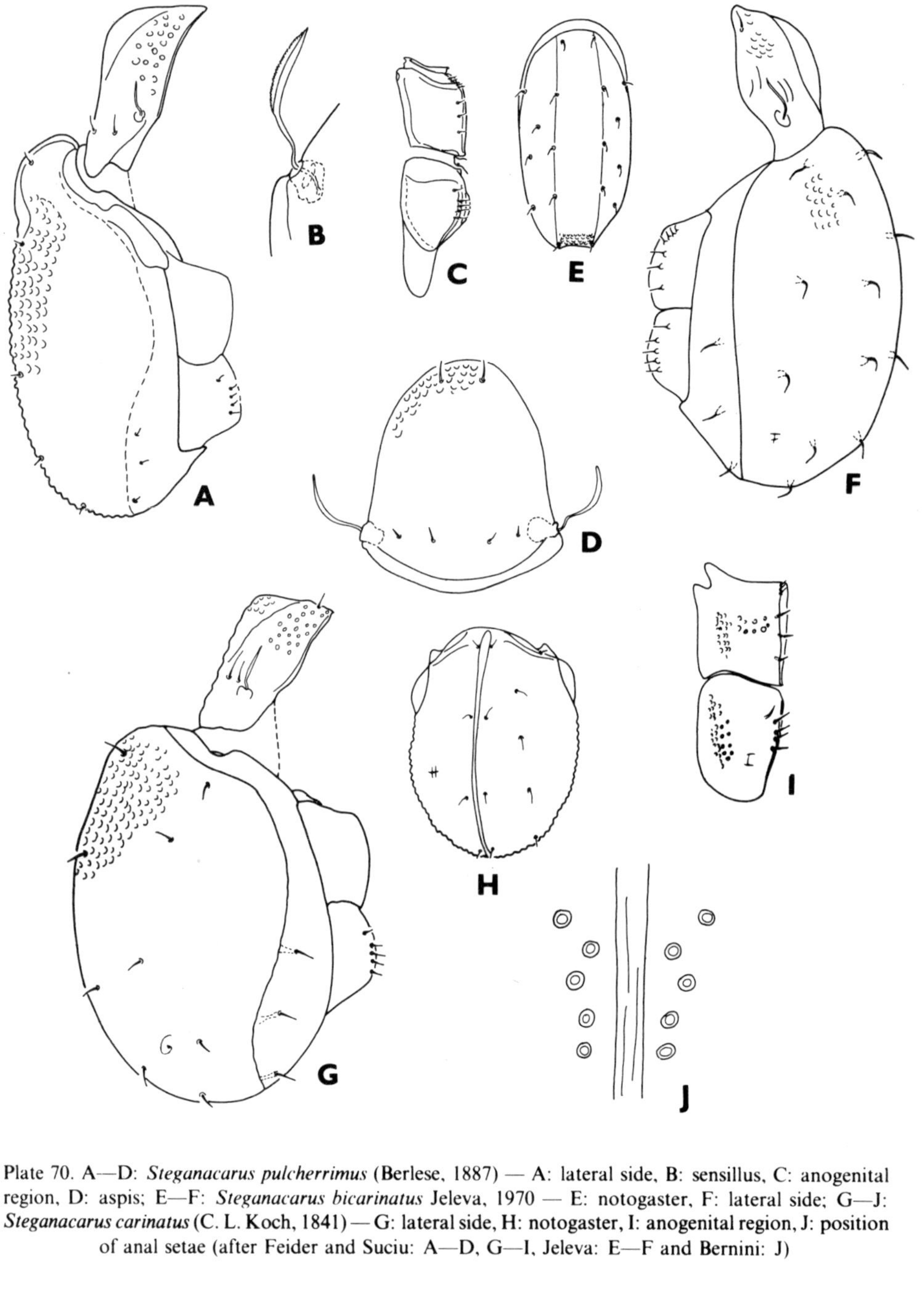

Plate 70. A—D: *Steganacarus pulcherrimus* (Berlese, 1887) — A: lateral side, B: sensillus, C: anogenital region, D: aspis; E—F: *Steganacarus bicarinatus* Jeleva, 1970 — E: notogaster, F: lateral side; G—J: *Steganacarus carinatus* (C. L. Koch, 1841) — G: lateral side, H: notogaster, I: anogenital region, J: position of anal setae (after Feider and Suciu: A—D, G—I, Jeleva: E—F and Bernini: J)

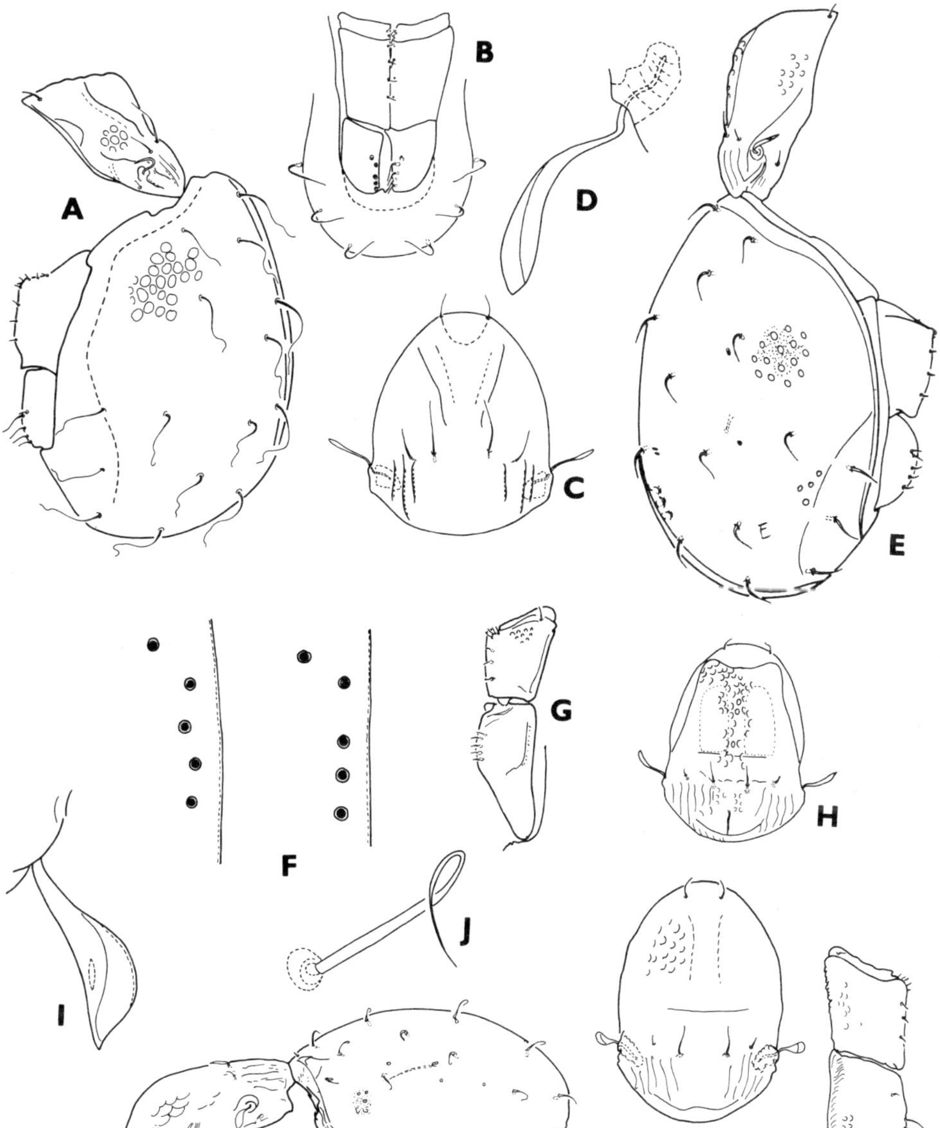

Plate 71. A—C: *Steganacarus lasithiensis* Mahunka, 1979 — A: lateral side, B: anogenital region, C: aspis; D—H: *Steganacarus brevipilus* (Berlese, 1923) — D: sensillus, E: lateral side, F: positions of anal setae, G: anogenital region, H: aspis; I—M: *Steganacarus travei* Lions, 1968 — I: sensillus, J: notogastral seta, K: lateral side, L: aspis, M: anogenital region (A—C original; after Bernini: D—H and Lions: I—M)

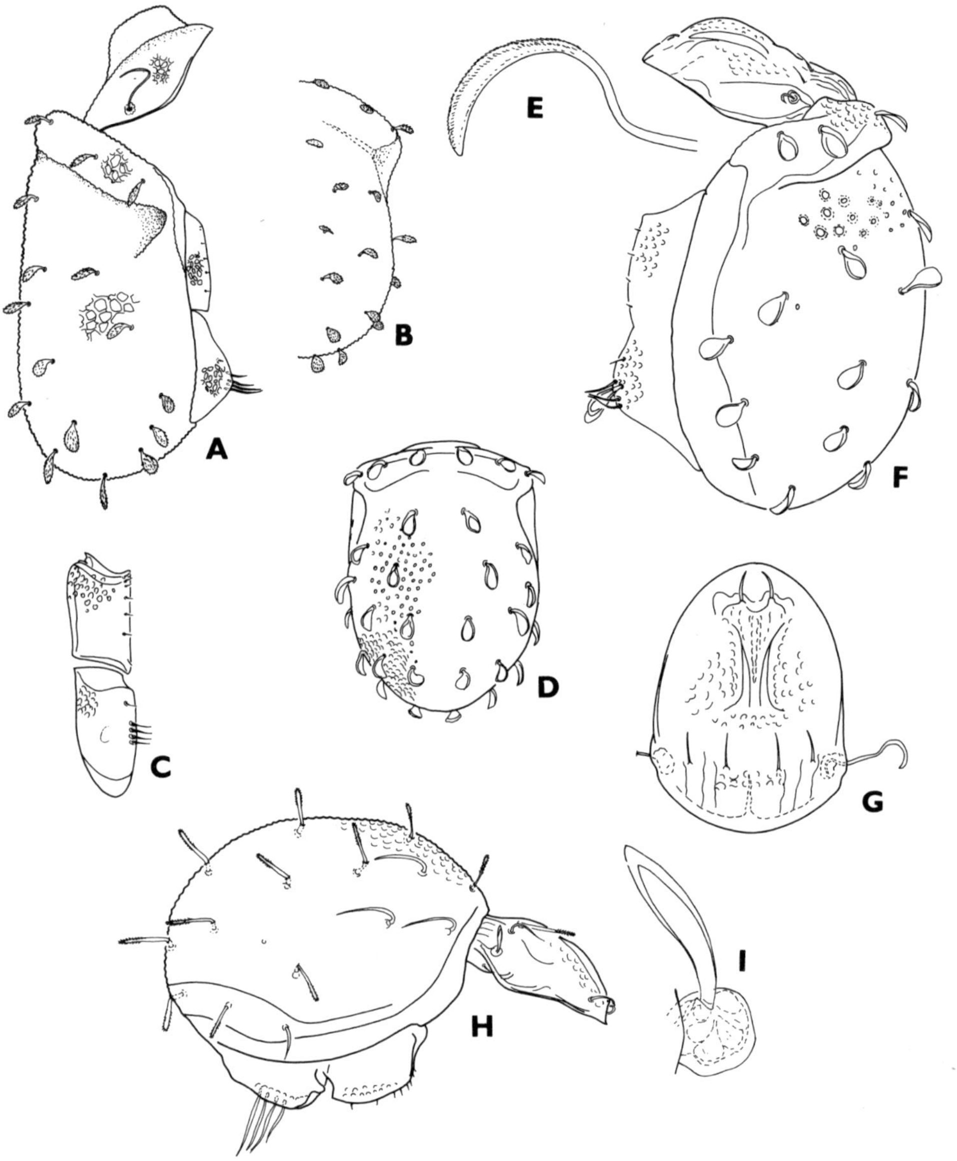

Plate 72. A—C: *Steganacarus phyllophorus* (Berlese, 1904) — A: lateral side, B: notogaster, C: anogenital region; D—G: *Steganacarus cucullatus* (Ewing, 1909) — D: notogaster, E: sensillus, F: lateral side, G: aspis; H—I: *Steganacarus teneriffensis* Pérez-Iñigo, 1972 — H: lateral side, I: sensillus (after Schuster: A—C, Aoki: D—G and Pérez-Iñigo: H—I)

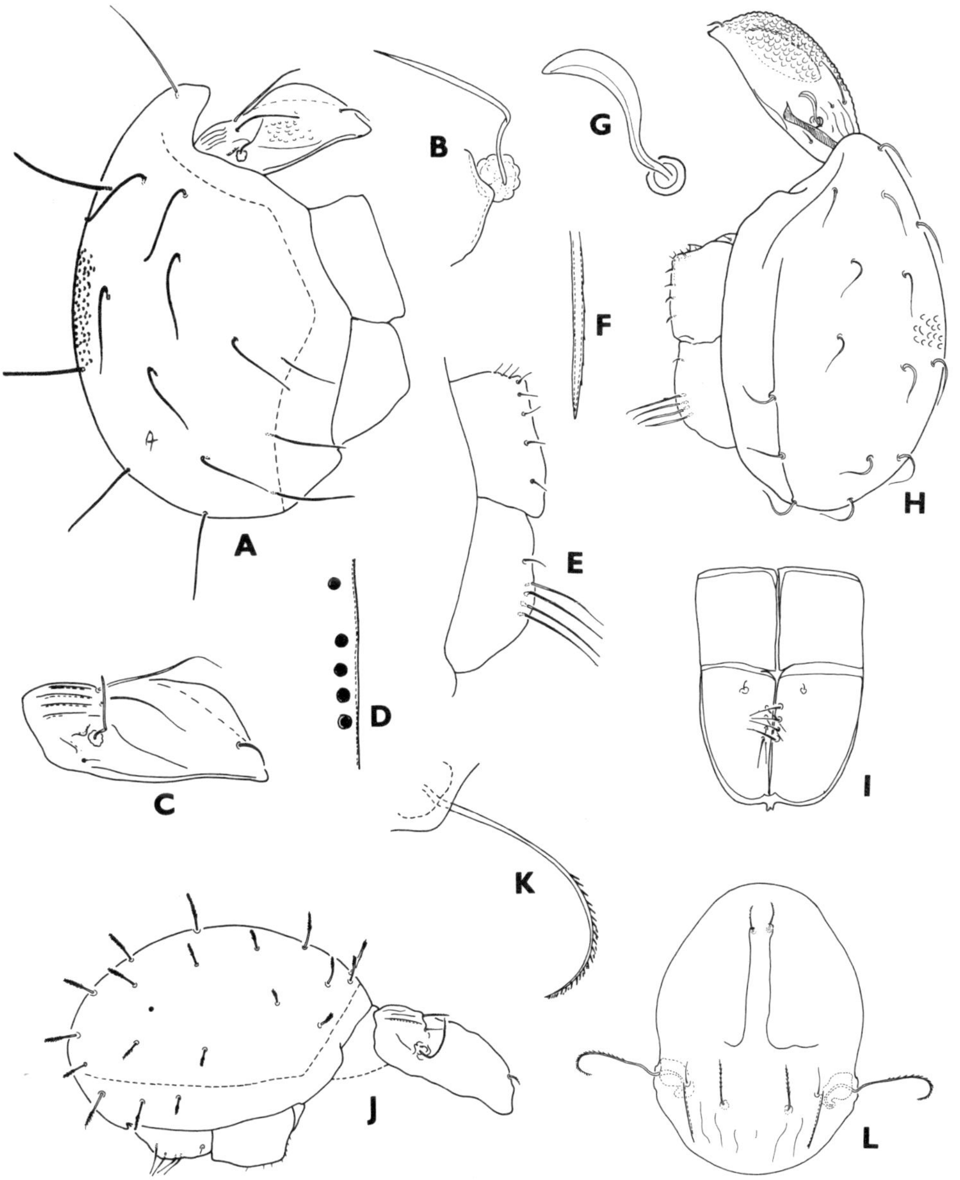

Plate 73. A—F: *Steganacarus anomalus* (Berlese, 1883) — A: lateral side, B: sensillus, C: apsis, D: position of anal setae, E: anogenital region, F: distal end of sensillus; G—H: *Steganacarus balearicus* Pérez-Iñigo, 1969 — G: sensillus, H: lateral side; I—L: *Steganacarus csiszarae* Balogh et Mahunka, 1979 — I: anogenital region, J: lateral side, K: sensillus, L: aspis (A—F, I—L original; after Pérez-Iñigo: G—H)

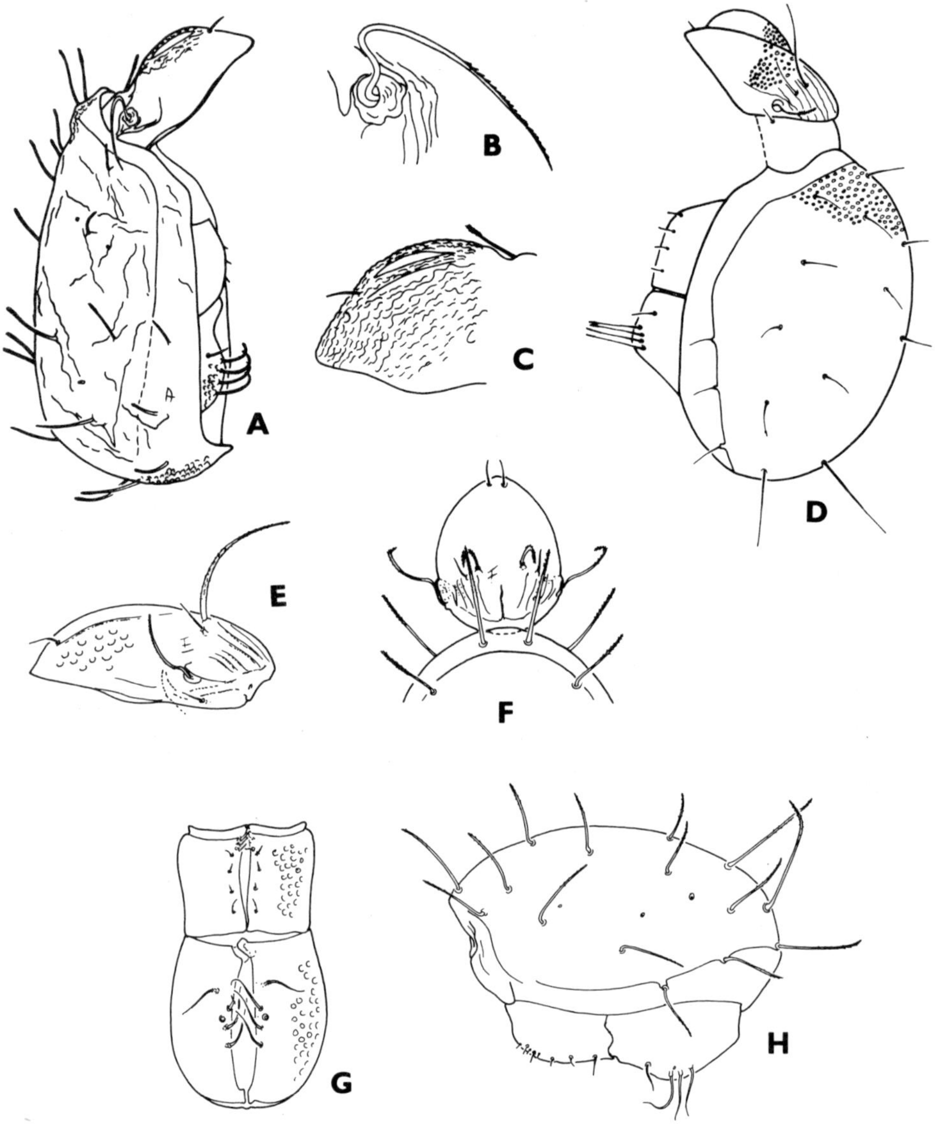

Plate 74. A—C: *Steganacarus senex* Aoki, 1958 — A: lateral side, B: sensillus, C: aspis; D: *Steganacarus herculeanus* Willmann, 1953 — lateral side; E—H: *Steganacarus ortizi* Pérez-Iñigo, 1969 — E: aspis, F: anterior part of idiosoma, G: anogenital region, H: lateral side of notogaster (after Aoki: A—C, Willmann: D and Pérez-Iñigo: E—H)

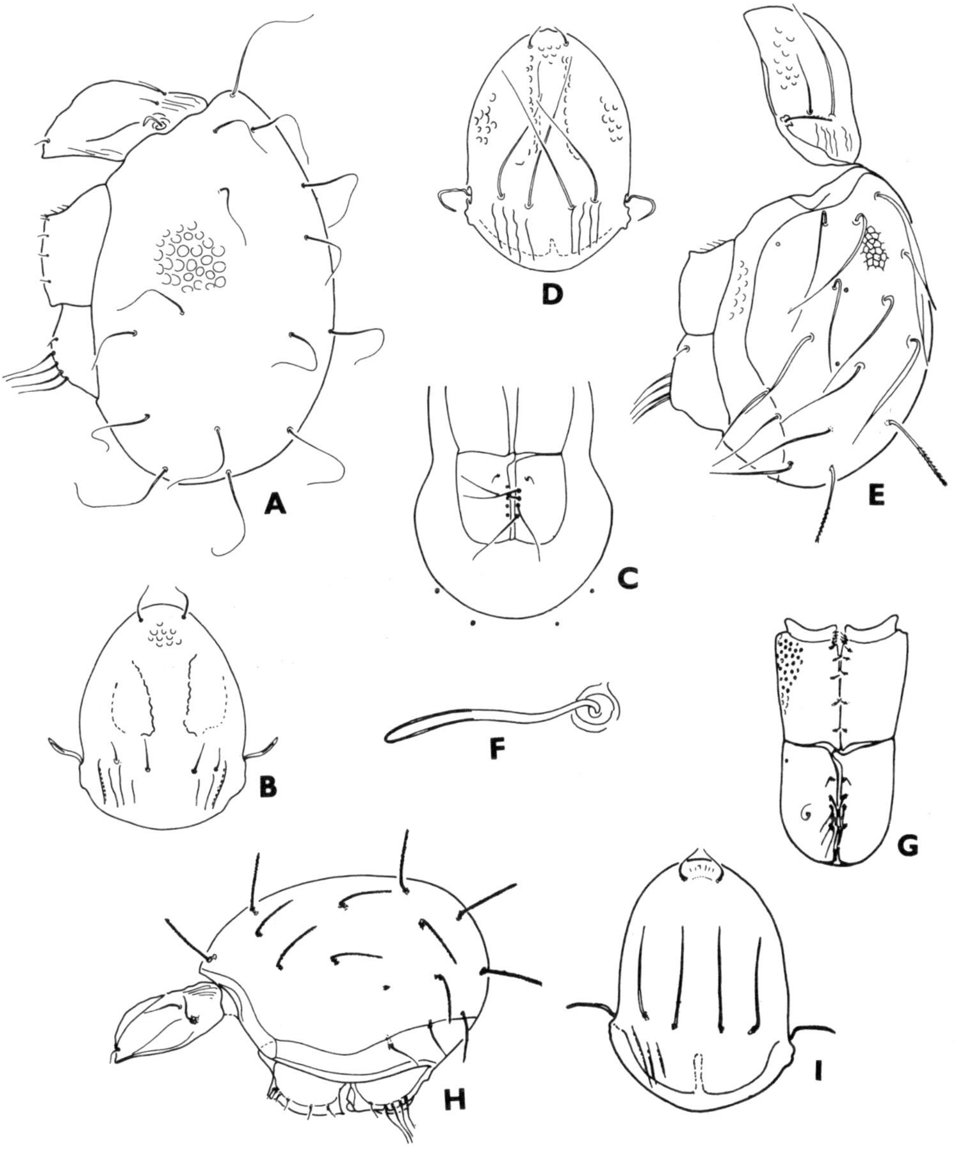

Plate 75. A—C: *Steganacarus flagellatissimus* Mahunka, 1979 — A: lateral side, B: aspis, C: anogenital region; D—E: *Steganacarus applicatus* (Sellnick, 1920) — D: apsis, E: lateral side, F—I: *Steganacarus magnus* (Nicolet, 1855) — F: sensillus, G: anogenital region, H: lateral side, I: aspis (A—E original; after Pérez-Iñigo: F—I)

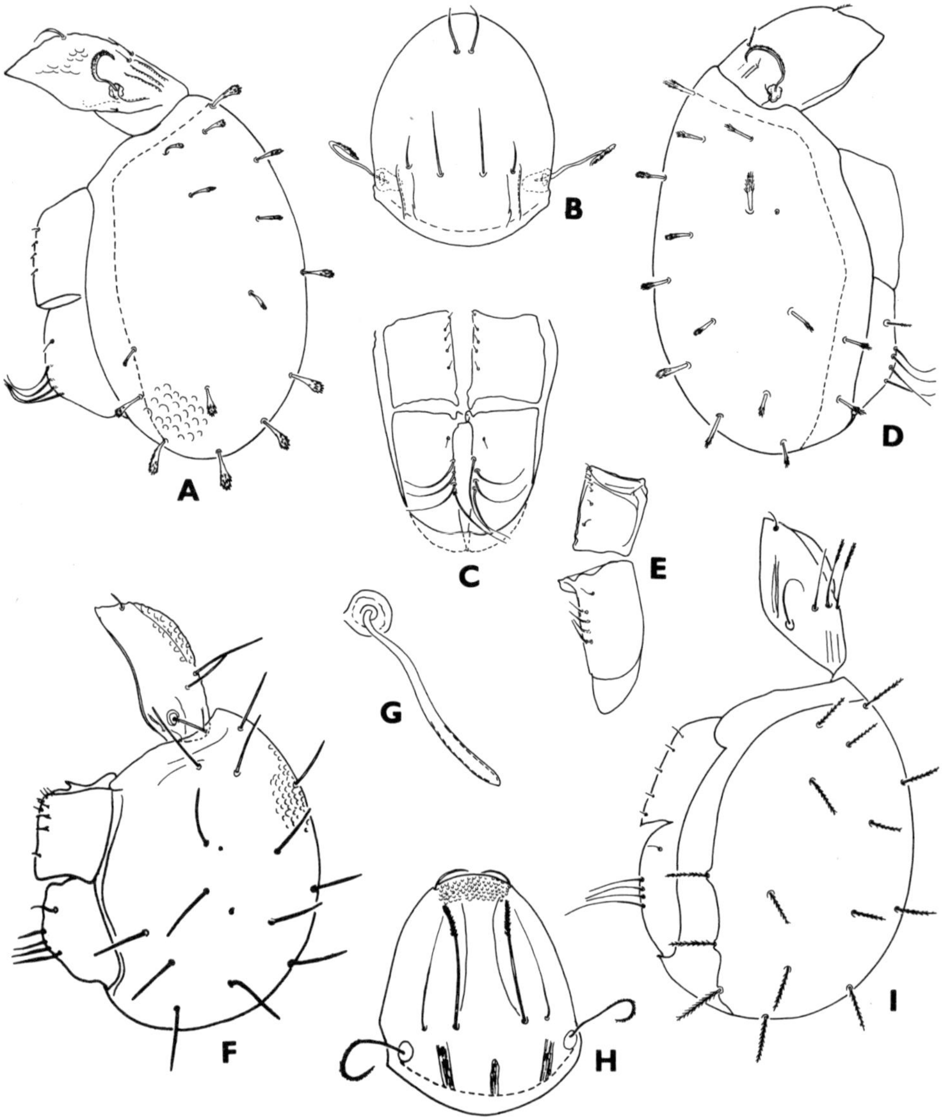

Plate 76. A—C: *Steganacarus platakisi* Mahunka, 1979 — A: lateral side, B: apsis, C: anogenital region; D—E: *Steganacarus striculus* (C. L. Koch, 1836) — D: lateral side, E: anogenital region; F—G: *Steganacarus hirsutus* Pérez-Iñigo, 1974 — F: lateral side, G: sensillus; H—I: *Steganacarus similis* Willmann, 1939 — H: aspis, I: lateral side (A—E original; after Pérez-Iñigo: F—G and Willmann: H—I)

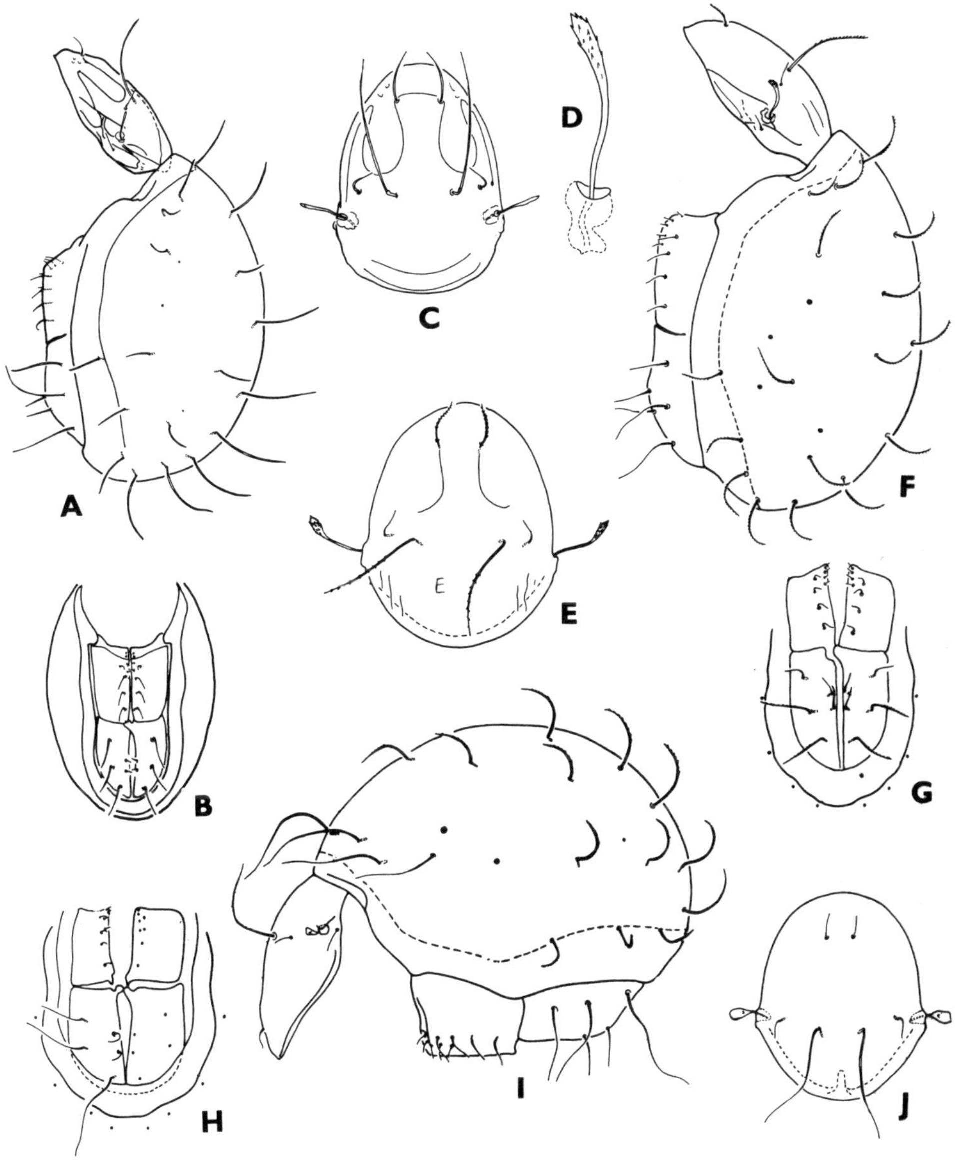

Plate 77. A—C *Hoplophthiracarus costai* Macfarlane et Sheals, 1965 — A: lateral side, B: anogenital region, C: aspis; D—G: *Hoplophthiracarus cretensis* Mahunka, 1979 — D: sensillus, E: aspis, F: lateral side, G: anogenital regio; H—J: *Holophthiracarus heterotricha* Mahunka, 1979— H: anogenital region, I: lateral side, J: aspis (after Macfarlane and Sheals: A—C; D—J original)

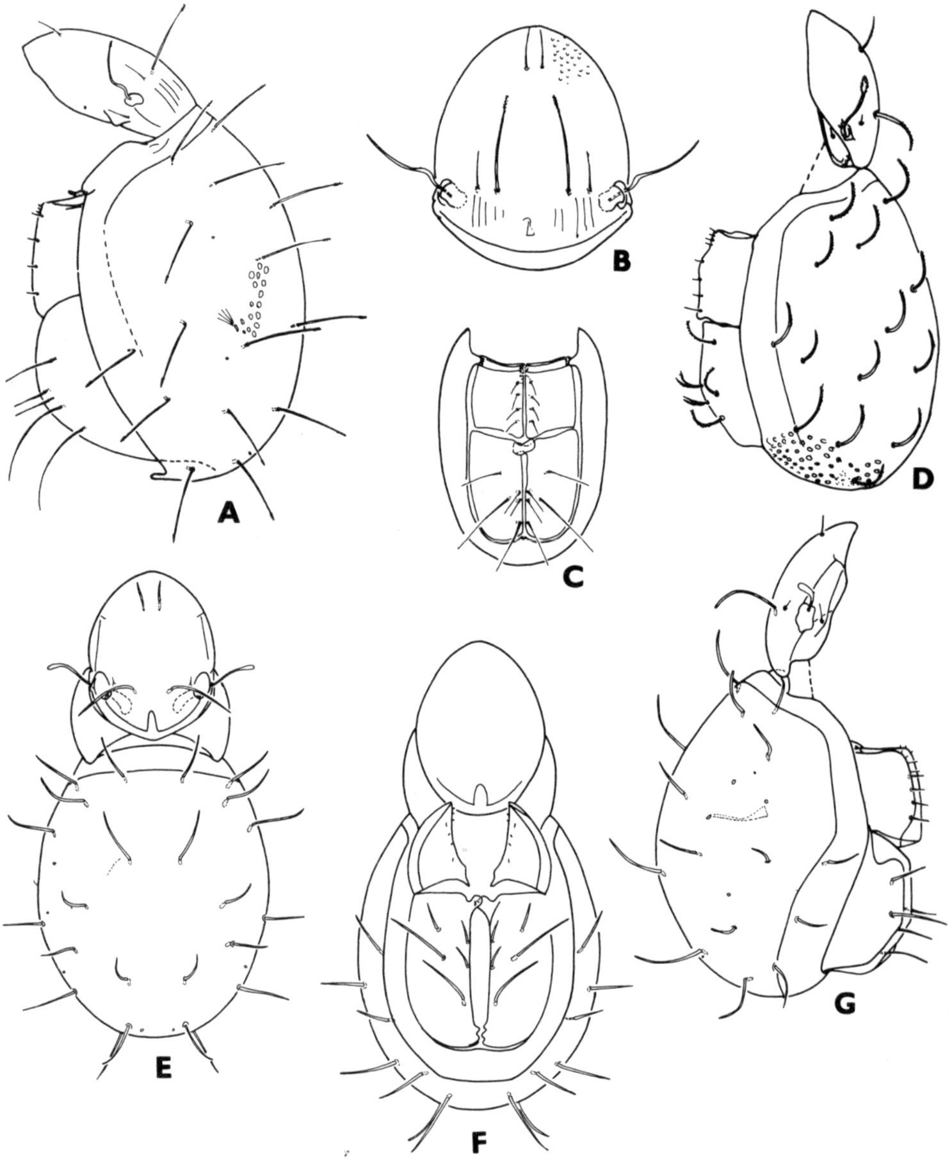

Plate 78. A—C: *Hoplophthiracarus nepalensis* Sheals, 1965 — A: lateral side, B: aspis, C: anogenital region; D: *Hoplophthiracarus minus* (D. Krivolutsky, 1965) — lateral side; E—G: *Hoplophthiracarus pavidus* (Berlese, 1913) — E: dorsal side, F: ventral side, G: lateral side (after Sheals: A—C, D. Krivolutsly: D and Van der Hammen: E—G)

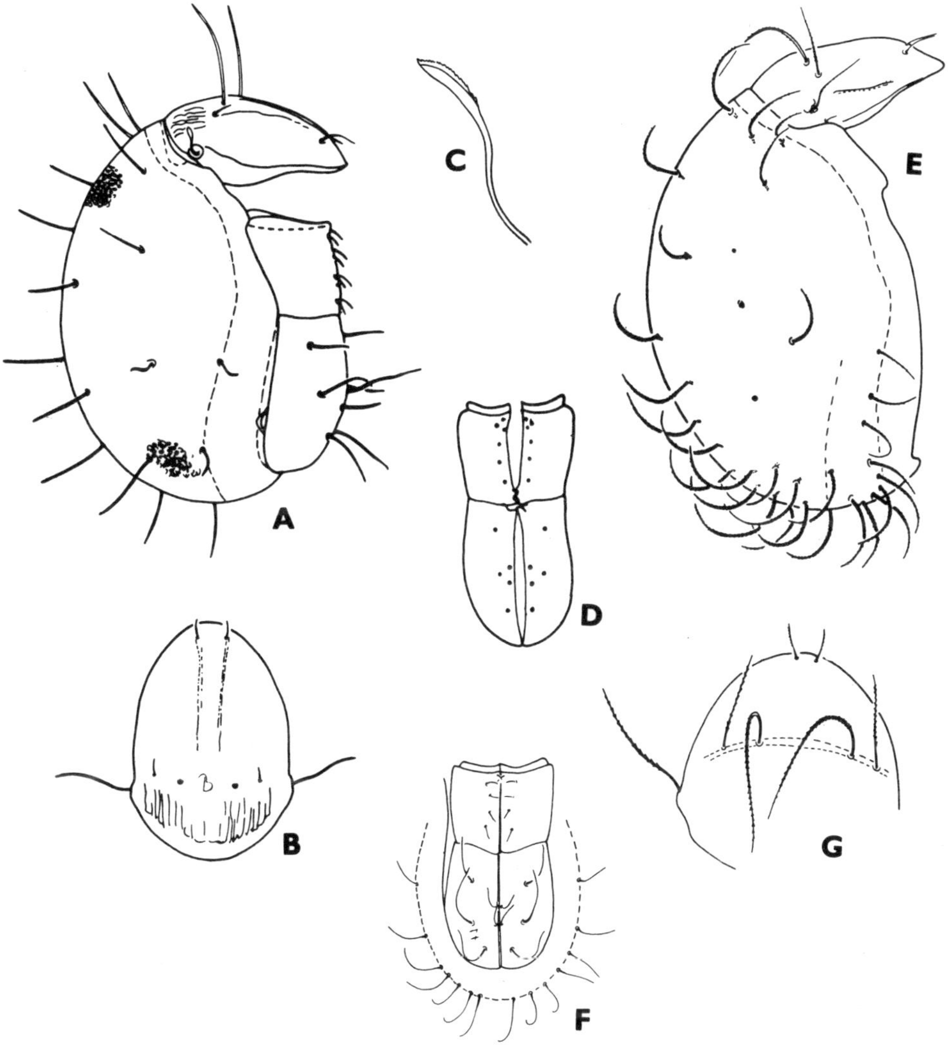

Plate 79. A—D: *Hoplophthiracarus kugohi* Aoki, 1959 — A: lateral side, B: aspis, C: sensillus, D: anogenital region; E—G: *Hauserophthiracarus oenipontanus* Mahunka, 1982—E: lateral side, F: anogenital region, G: aspis (after Aoki: A—D; E—G original)

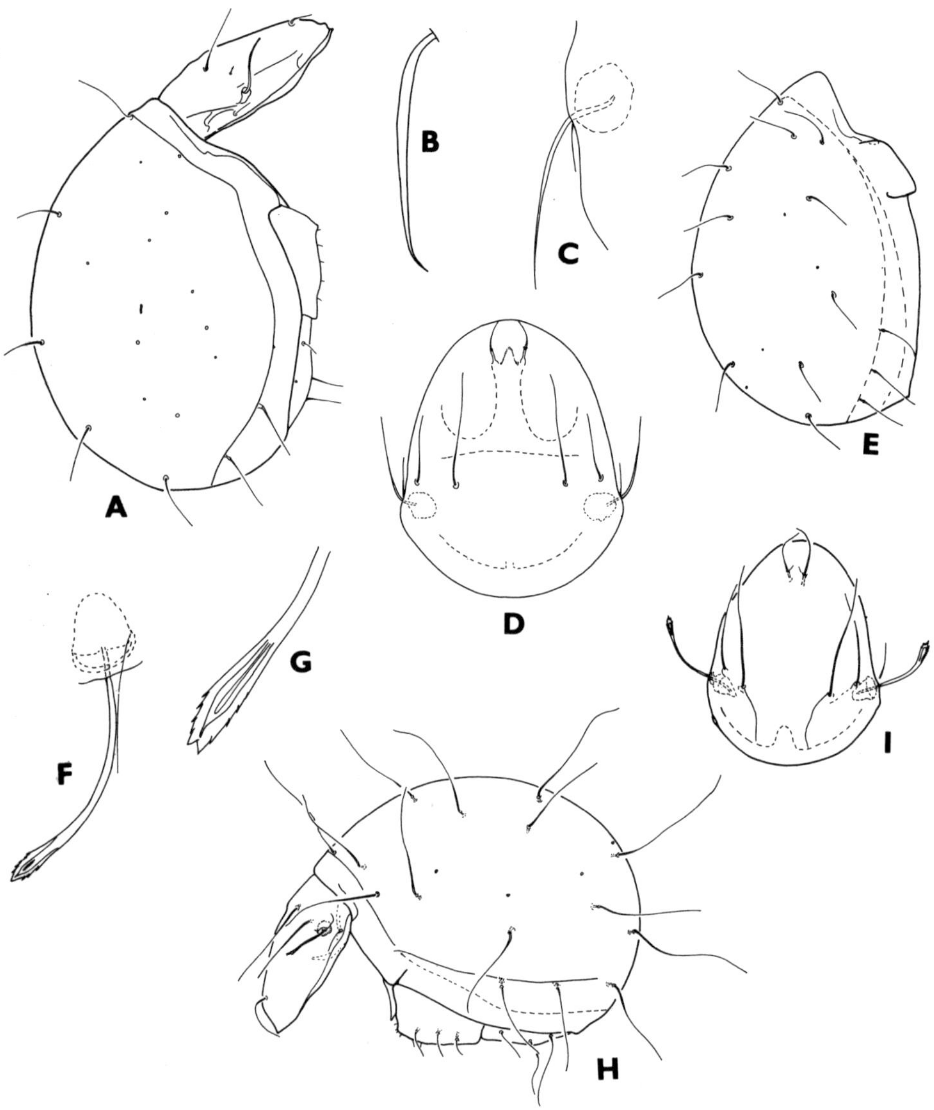

Plate 80. A—B: *Phthiracarus testudienus* (C. L. Koch, 1841) — A: lateral side B: sensillus; C—E: *Phthiracarus juvenalis* Parry, 1979 — C: sensillus, D: aspis, E: notogaster; F—I: *Phthiracarus kaszabi* Mahunka, 1981 — E: sensillus, G: distal end of sensillus, H: lateral side, I: aspis (after Jacot: A—B and Parry: C—E; F—I original)

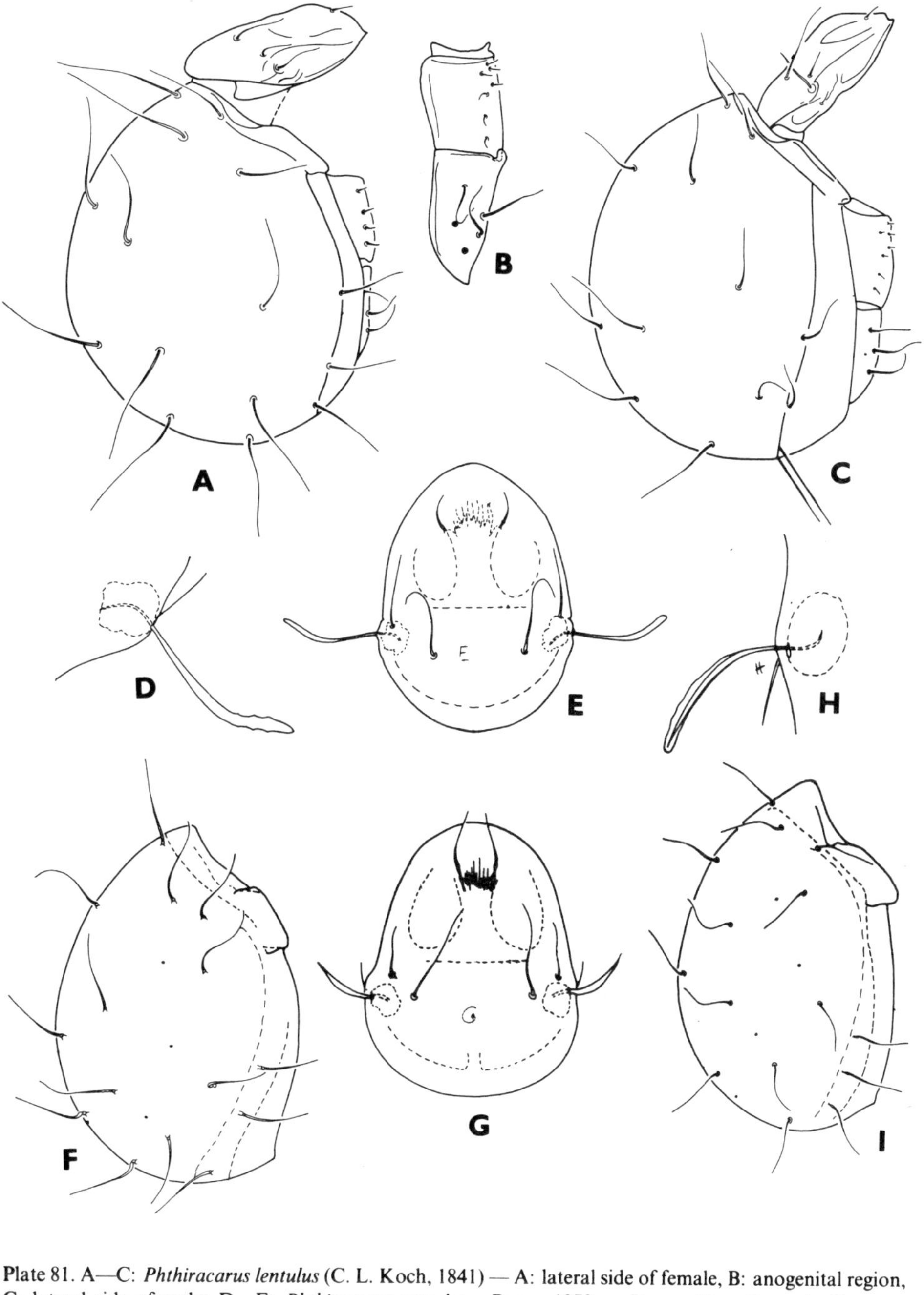

Plate 81. A—C: *Phthiracarus lentulus* (C. L. Koch, 1841) — A: lateral side of female, B: anogenital region, C: lateral side of male; D—F: *Phthiracarus serrulatus* Parry, 1979 — D: sensillus, E: aspis, F: notogaster, G—I: *Phthiracarus membranifer* Parry, 1979 — G: aspis, H: sensillus, I: notogaster (after Jacot: A—C and Parry: D—I)

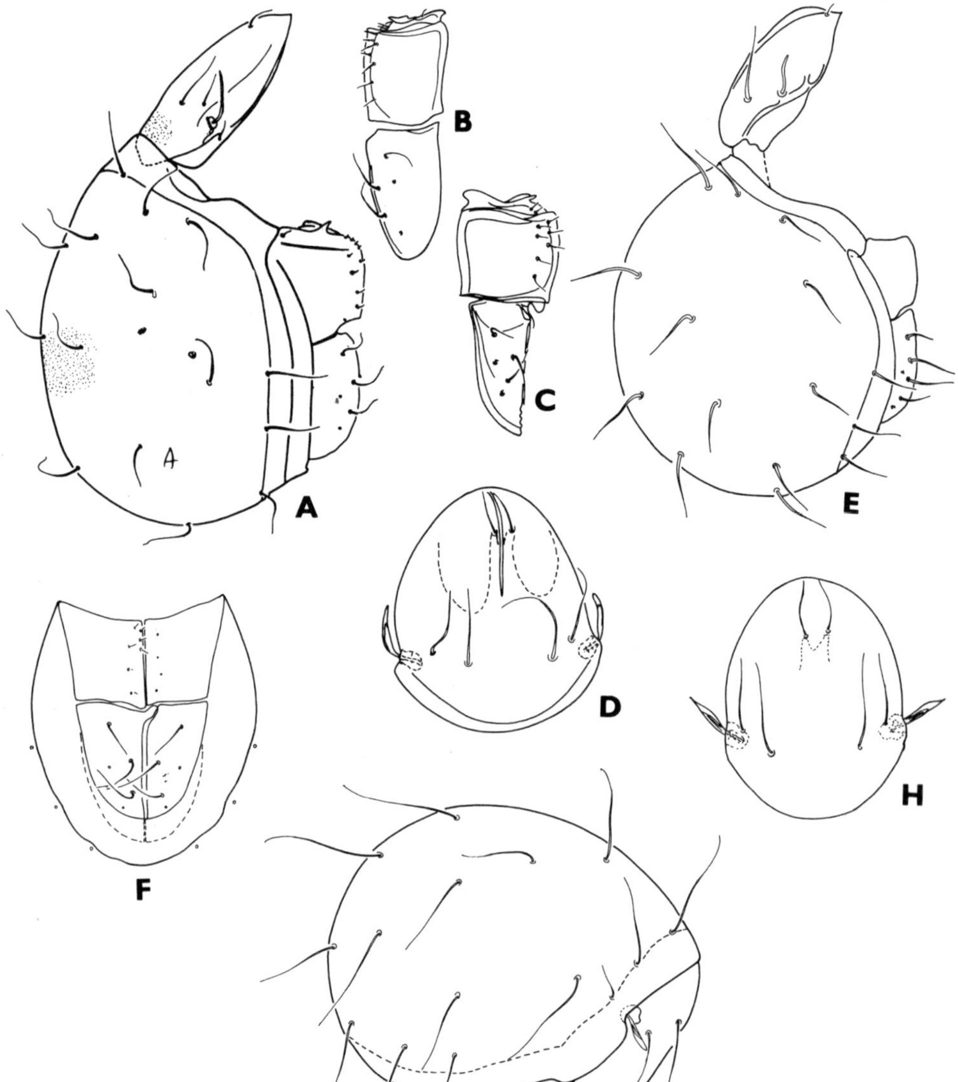

Plate 82. A—B: ***Phthiracarus*** *italicus* (Oudemans, 1906) — A: lateral side B: anogenital region; C—E: ***Phthiracarus*** *globosus* (C. L. Koch, 1841) — C: anogenital region, D: aspis, E: lateral side; F—H: ***Phthiracarus*** *zicmani* nom. nov. — F: anogenital region, G: lateral side, H: aspis (after Feider and Suciu: A—E; F—H original)

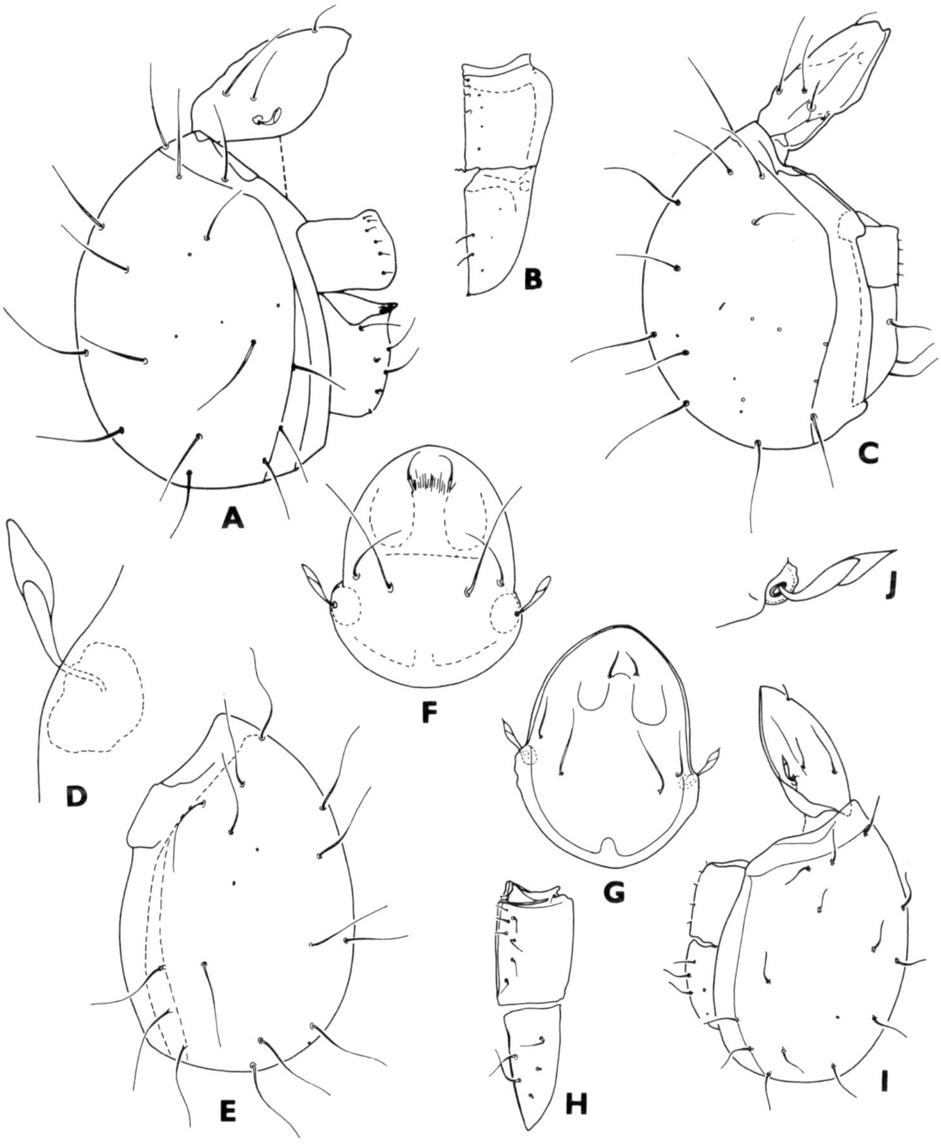

Plate 83. A: *Phthiracarus crenophilus* Willmann, 1951 — lateral side; B—C: *Phthiracarus crinitus* (C. L. Koch, 1841) — B: anogenital region, C: lateral side; D—F: *Phthiracarus flexisetosus* Parry, 1979 — D: sensillus, E: notogastsre, F: aspis; G—J: *Phthiracarus pallidus* Feider et Suciu, 1958 — G: aspis, H: anogenital region, I: lateral side, J: sensillus (after Willmann: A and Feider and Suciu: B—J)

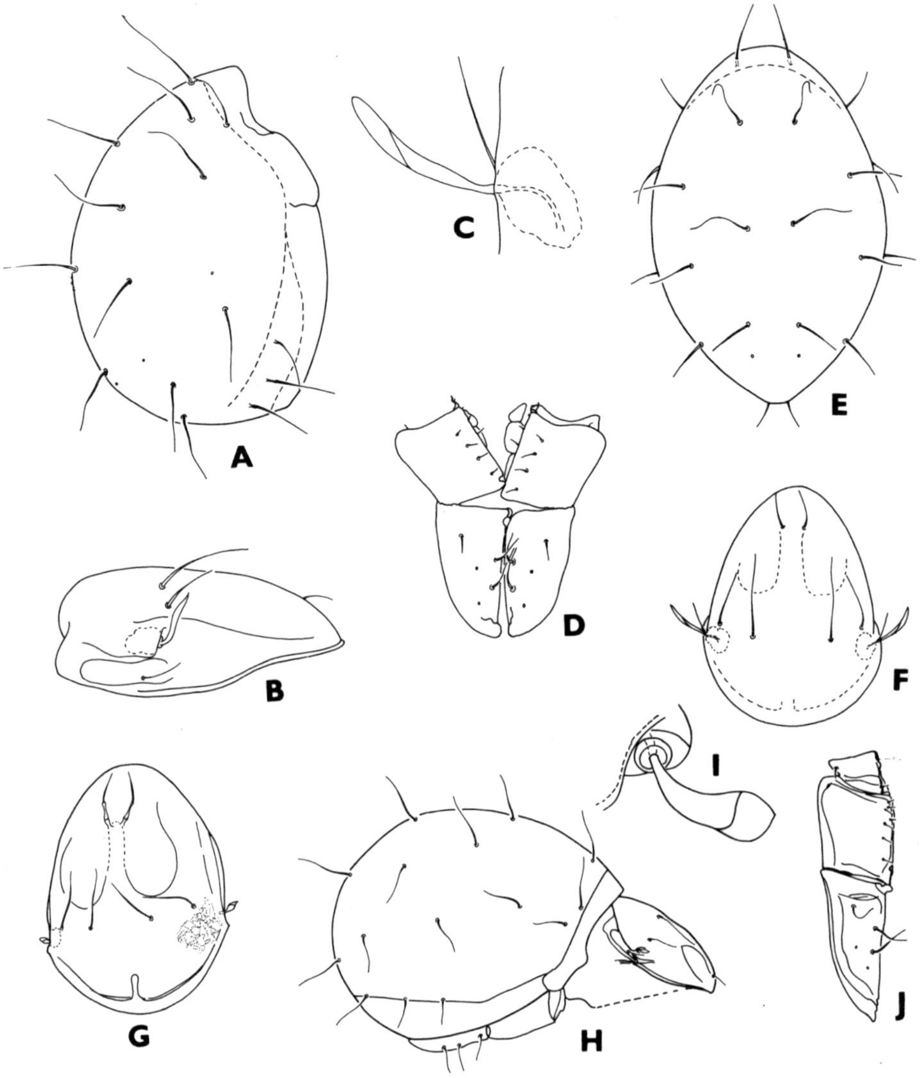

Plate 84. A—F: *Phthiracarus affinis* (Hull, 1914) — A: notogaster, B: aspis, C: sensillus, D: anogenital region, E: notogaster from dorsal side, F: aspis; G—I: *Phthiracarus jacoti* Feider et Suciu, 1958 — G: aspis, H: lateral side, I: sensillus, J: anogenital region (after Parry: A—F and Feider and Suciu: G—J)

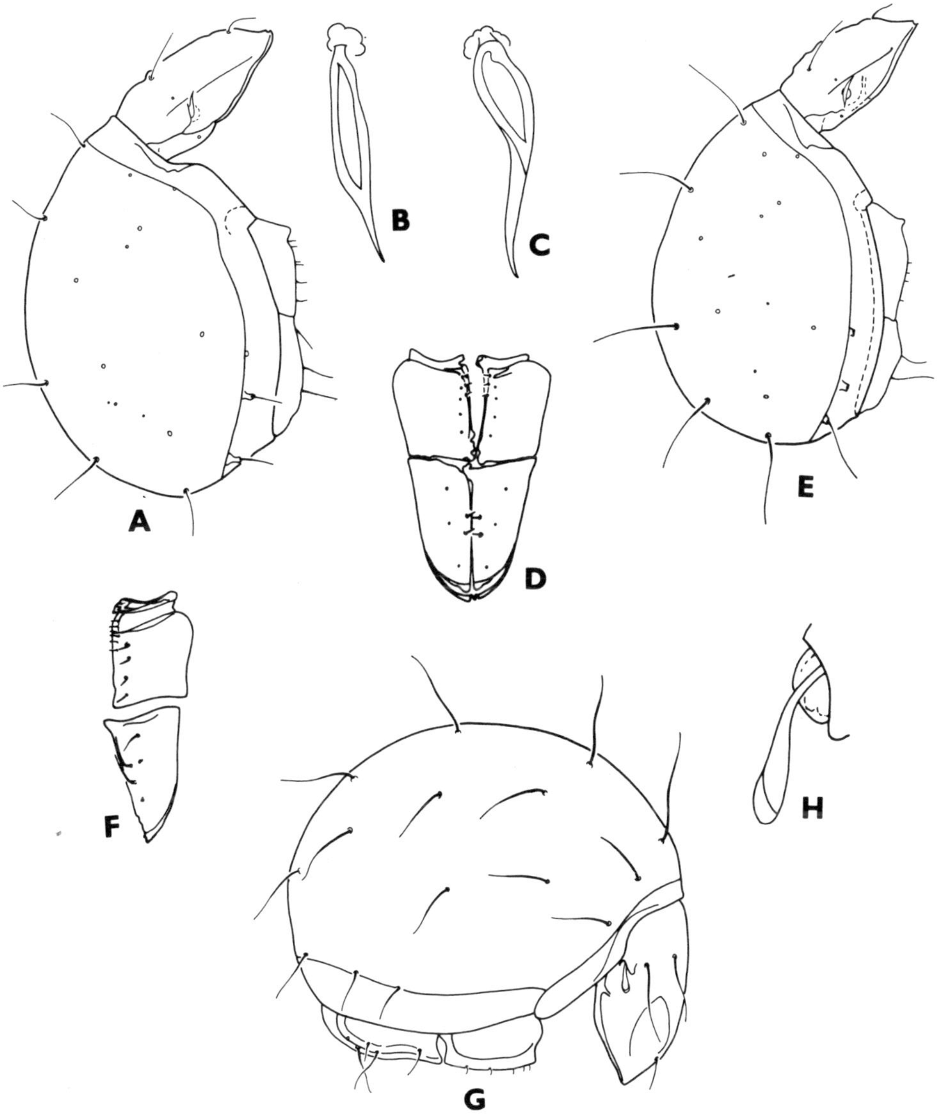

Plate 85. A—E: A: *Phthiracarus ferrugineus* (C. L. Koch, 1841) — lateral side of male, B—C: variations of sensillus, D: anogenital region, E: lateral side of female; F—H: *Phthiracarus dubinini* Feider et Suciu, 1958 — F: anogenital region, G: lateral side, H: sensillus (after Jacot: A—E and Feider and Suciu: F—H) Plate

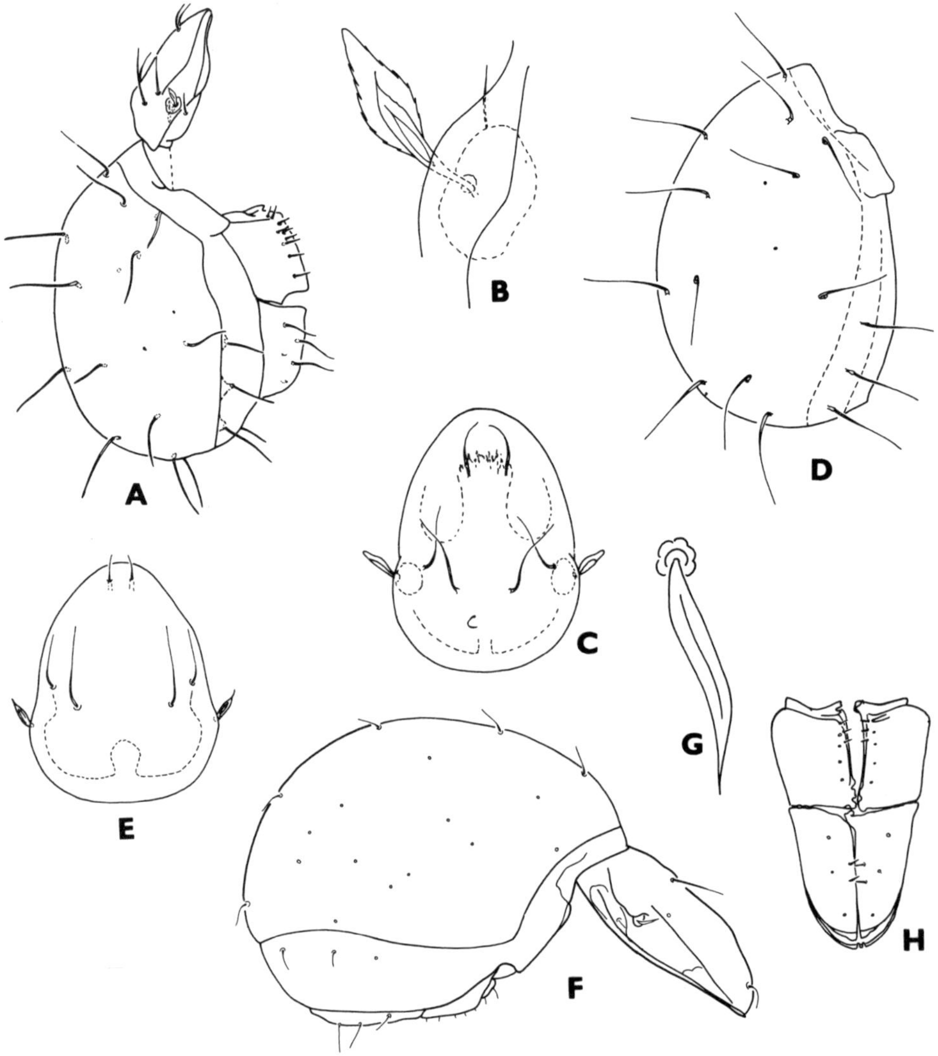

Plate 86. A—D: *Phthiracarus tardus* Forsslund, 1956 — A: lateral side, B: sensillus, C: aspis, D: notogaster; E—H: *Phthiracarus laevigatus* (C. L. Koch, 1841) — E: aspis, F: lateral side, G: sensillus, H: anogenital region (after Forsslund: A, Parry: B—D and Jacot: E—H)

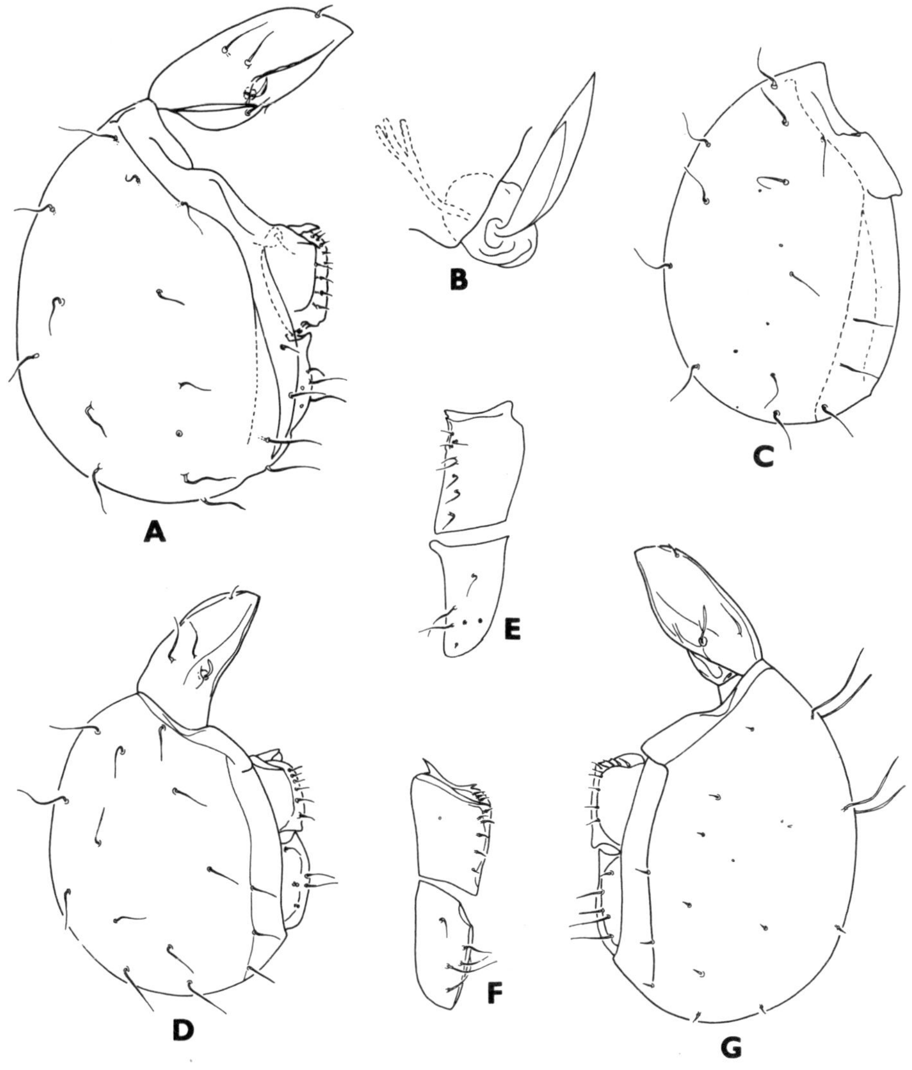

Plate 87. A—C: *Phthiracarus nitens* (Nicolet, 1855) — A: lateral side, B: sensillus, C: notogaster; D—E: *Phthiracarus piger* (Scopoli, 1763) — D: lateral side, E: anogenital region; F—G *Archiphthiracarus baloghi* Feider et Suciu, 1957) — F: anogenital region, G: lateral side (after Feider and Suciu: A, D—G and Parry: B—C)

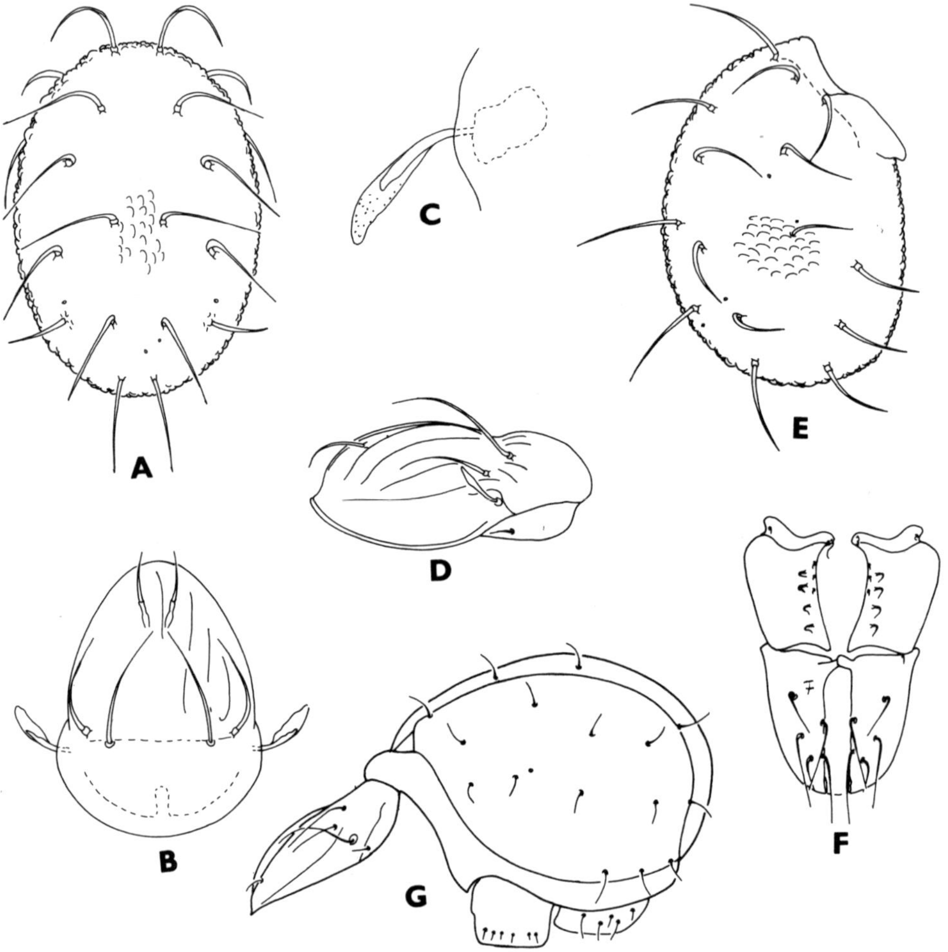

Plate 88. A—F: *Archiphthiracarus papillosus* Parry, 1979 — A: dorsal side of notogaster, B: aspis, C: sensillus, D: aspis from lateral side, E: notogaster from lateral side, F: anogenital region; G: *Archiphthiracarus torosus* (Willmann, 1939) from lateral side (after Parry: A—F and Willmann: G)

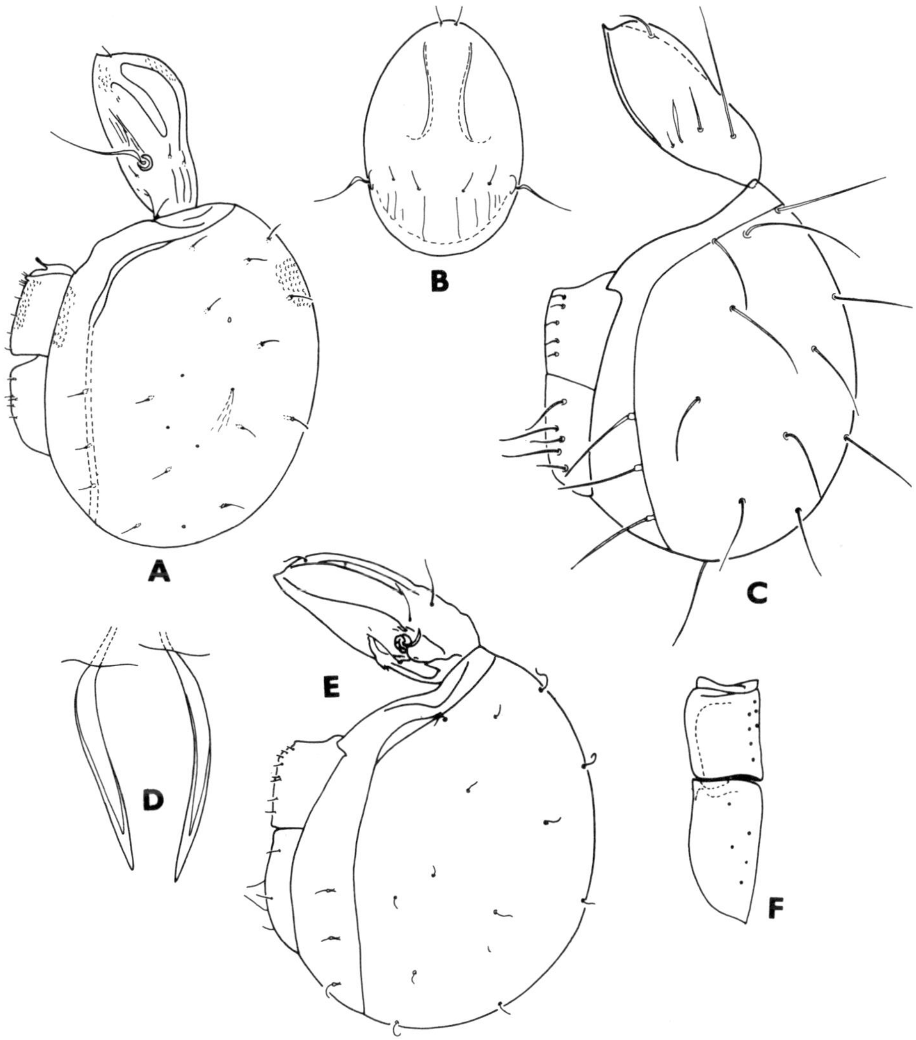

Plate 89. A—B: *Archiphtiracarus robertsi* (Sheals, 1965) — A: lateral side, B: aspis; C: *Archipthiracarus peristomaticus* (Willmann, 1954) — lateral side; D—F: *Archiphthiracarus clemens* (Aoki, 1963) — D: sensilli, E: lateral side, F: anogenital region (after Sheals: A—B, Willmann: C and Aoki: D—F)

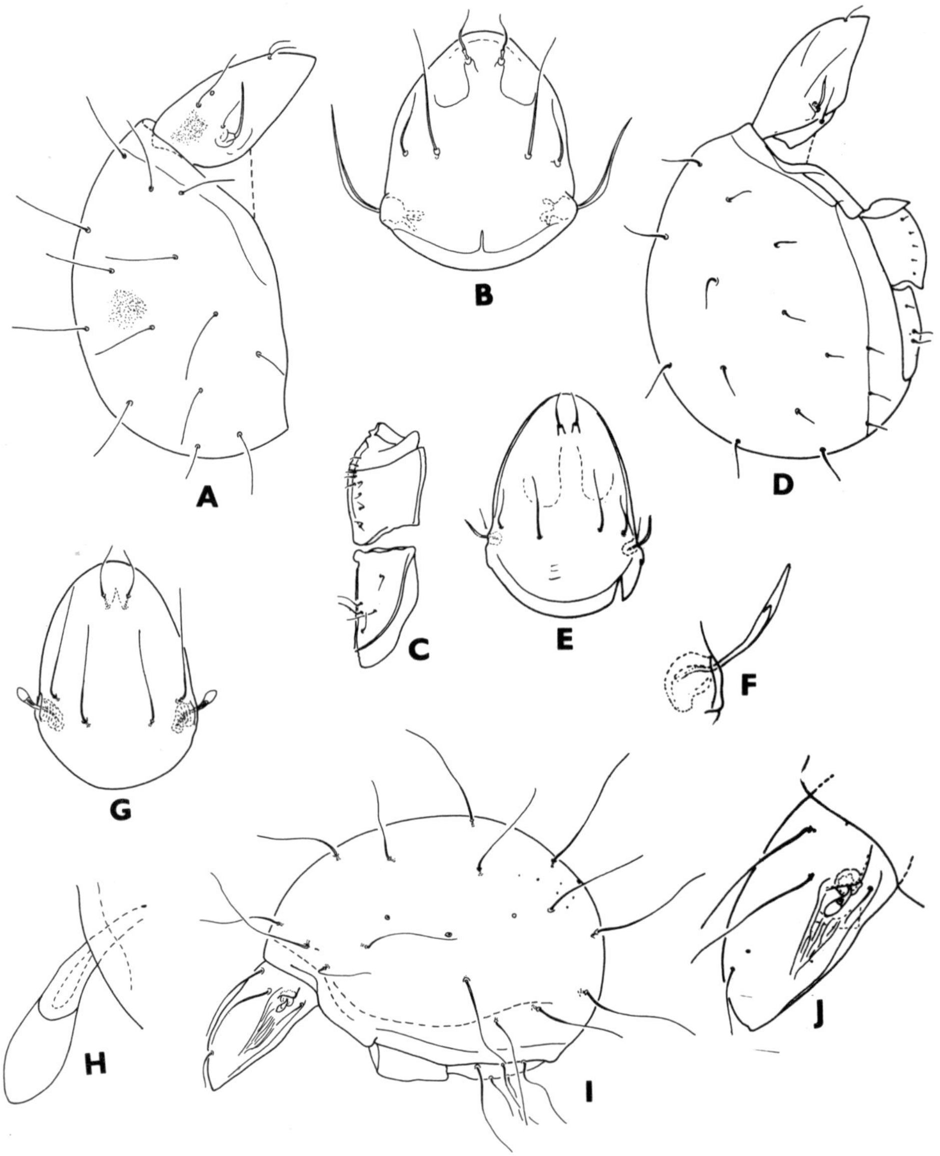

Plate 90. A—B: *Archiphthiracarus ligneus* (Willmann, 1931) — A: lateral side, B: aspis; C—F: *Archiphthiracarus parabotrichus* (Feider et Suciu, 1957) — C: anogenital region, D: lateral side, E: aspis, F: sensillus; G—J: *Archiphthiracarus gobiensis* Mahunka, 1981 — G: aspis, H: sensillus, I: lateral side, J: aspis from lateral side (after Feider and Suciu: A—D; E—J original)

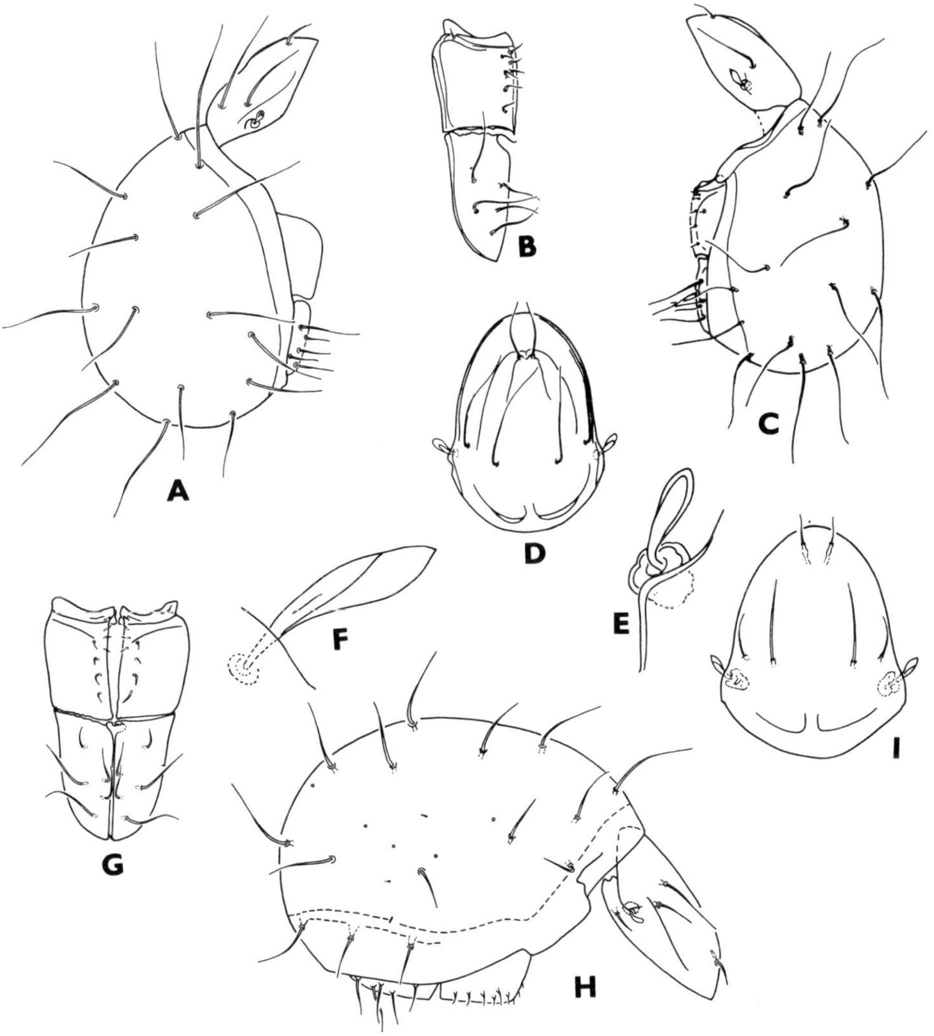

Plate 91. A: *Archiphthiracarus crinitosimilis* (Willmann, 1939) — lateral side B—E: *Archiphthiracarus lanatus* (Feider et Suciu, 1957) — B: anogenital region, B: lateral side, D: aspis, E: sensillus; F—I: *Archiphthiracarus murphyi* (Harding, 1976) — F: sensillus, G: anogenital region, H: lateral side, I: aspis (after Willmann: A and Feider and Suciu: B—I)

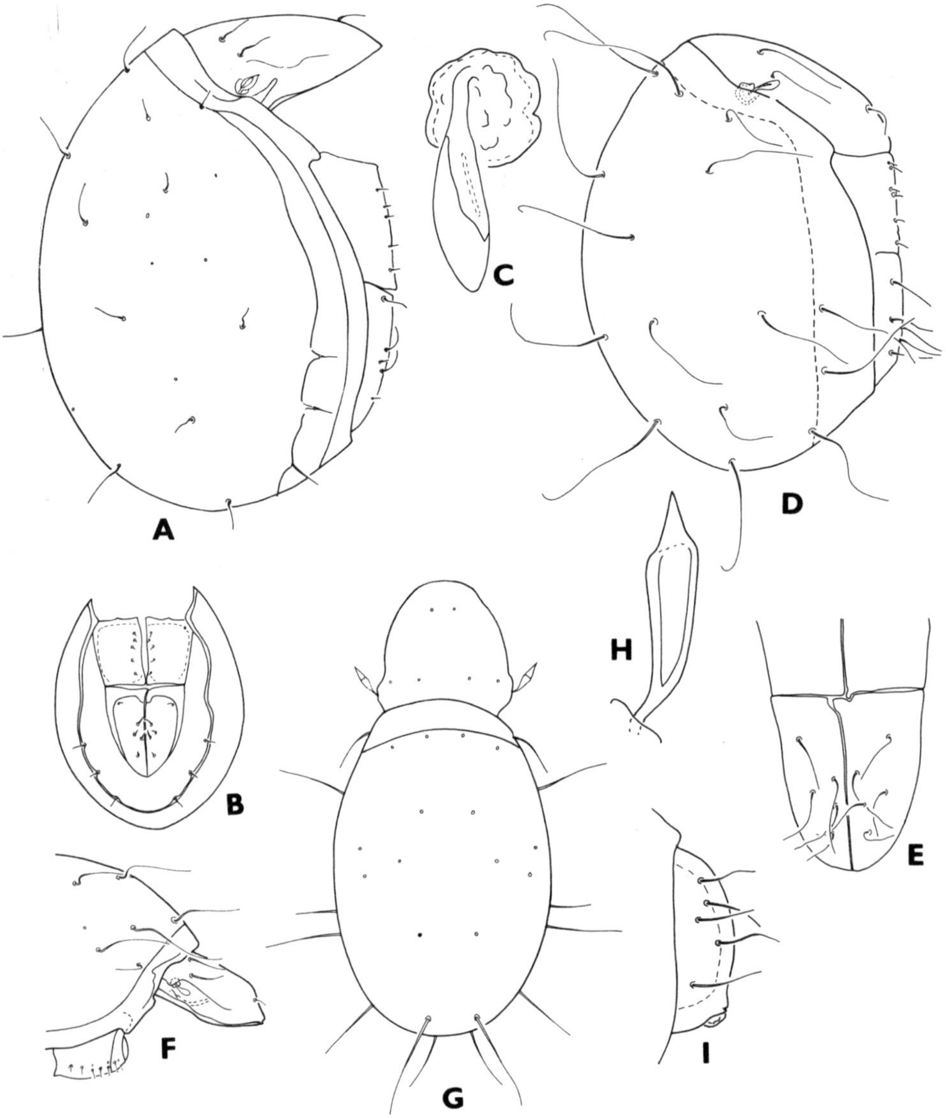

Plate 92. A—B: *Archiphthiracarus japonicus* (Aoki, 1958) — A: lateral side, B: anogenital region; C—E: *Archiphthiracarus hungaricus* Balogh et Mahunka, 1979 — C: sensillus, D: lateral side, E: anogenital region; F—I: *Archiphthiracarus stramineus* (C. L. Koch, 1841) — F: anterior part of idiosoma, G: dorsal side, H: sensillus, I: anal plate (after Aoki: A—B and Jacot: C—E; F—I original)

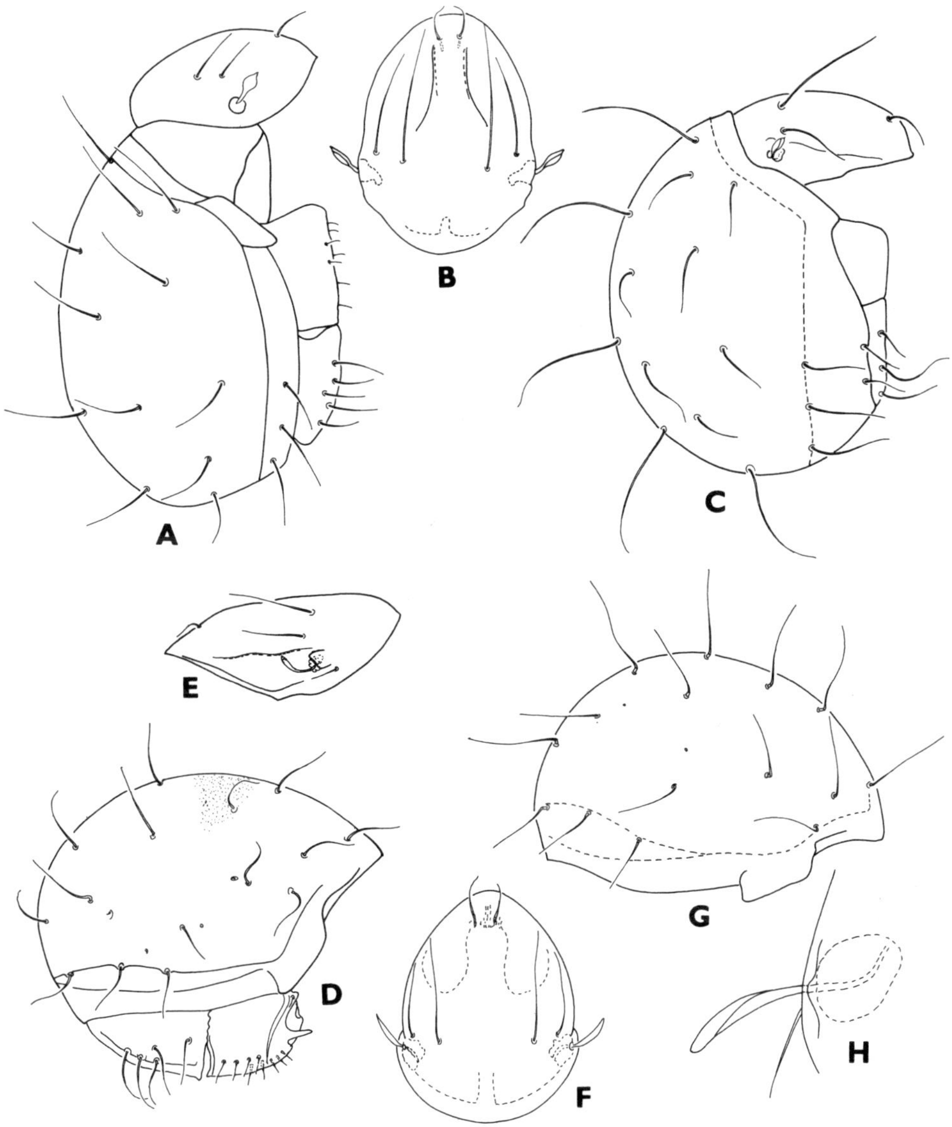

Plate 93. A: *Archiphthiracarus lucidus* (C. L. Koch, 1841) — lateral side, B—C: *Archiphthiracarus hammeni* Balogh et Mahunka, 1979 — B: aspis, C: lateral side; D—E: *Archiphthiracarus montanus* (Pérez-Iñigo, 1969) — D: notogaster, E: aspis, F—H: *Archiphthiracarus globus* (Parry, 1979) — F: aspis, G: notogaster, H: sensillus (after Willmann: A, Pérez-Iñigo: D—E and Parry: F—H; B—C original)

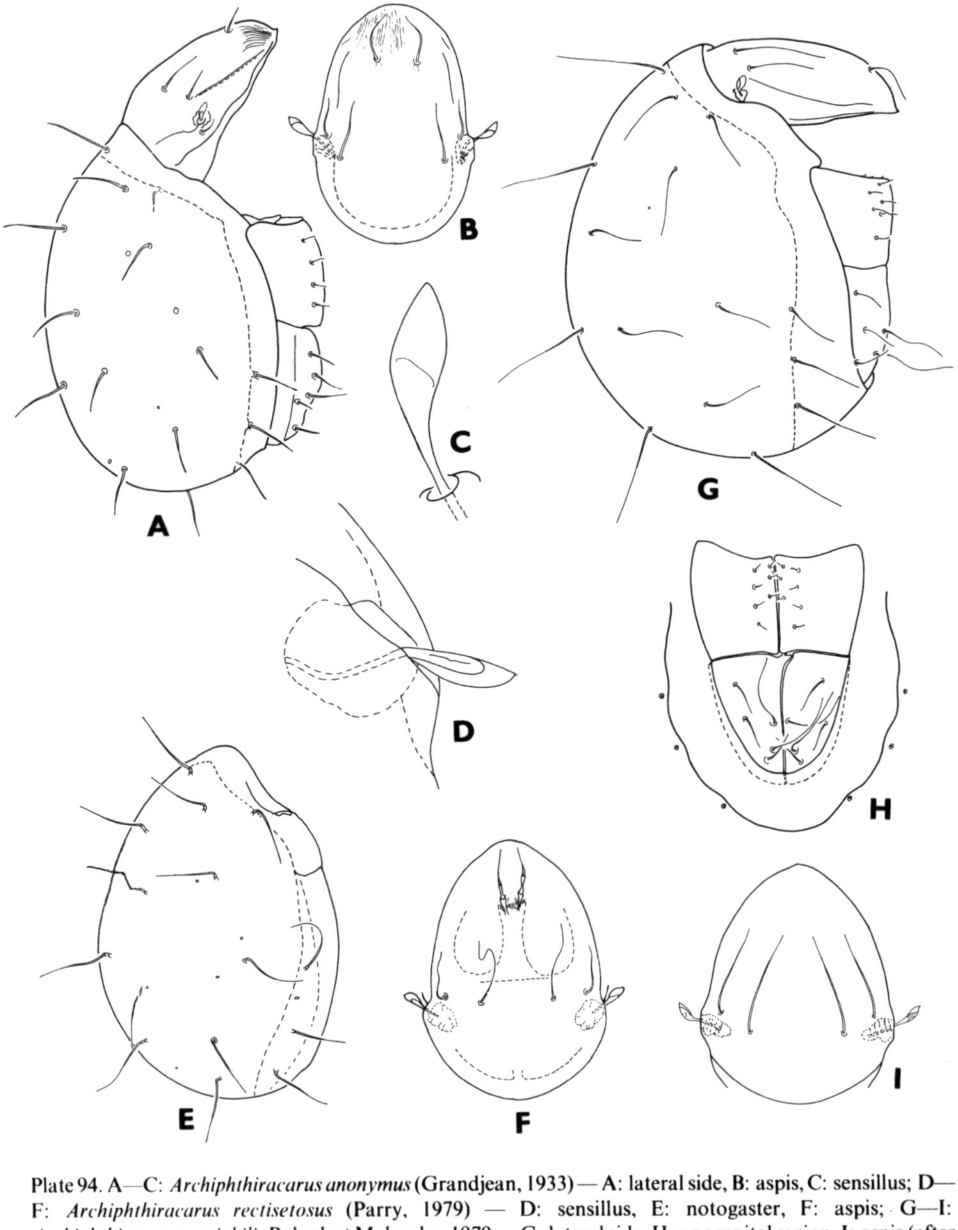

Plate 94. A—C: *Archiphthiracarus anonymus* (Grandjean, 1933) — A: lateral side, B: aspis, C: sensillus; D—F: *Archiphthiracarus rectisetosus* (Parry, 1979) — D: sensillus, E: notogaster, F: aspis; G—I: *Archiphthiracarus variabilis* Balogh et Mahunka, 1979 — G: lateral side, H: anogenital region, I: aspis (after Grandjean: A—C and Parry: D—E; F—H original)

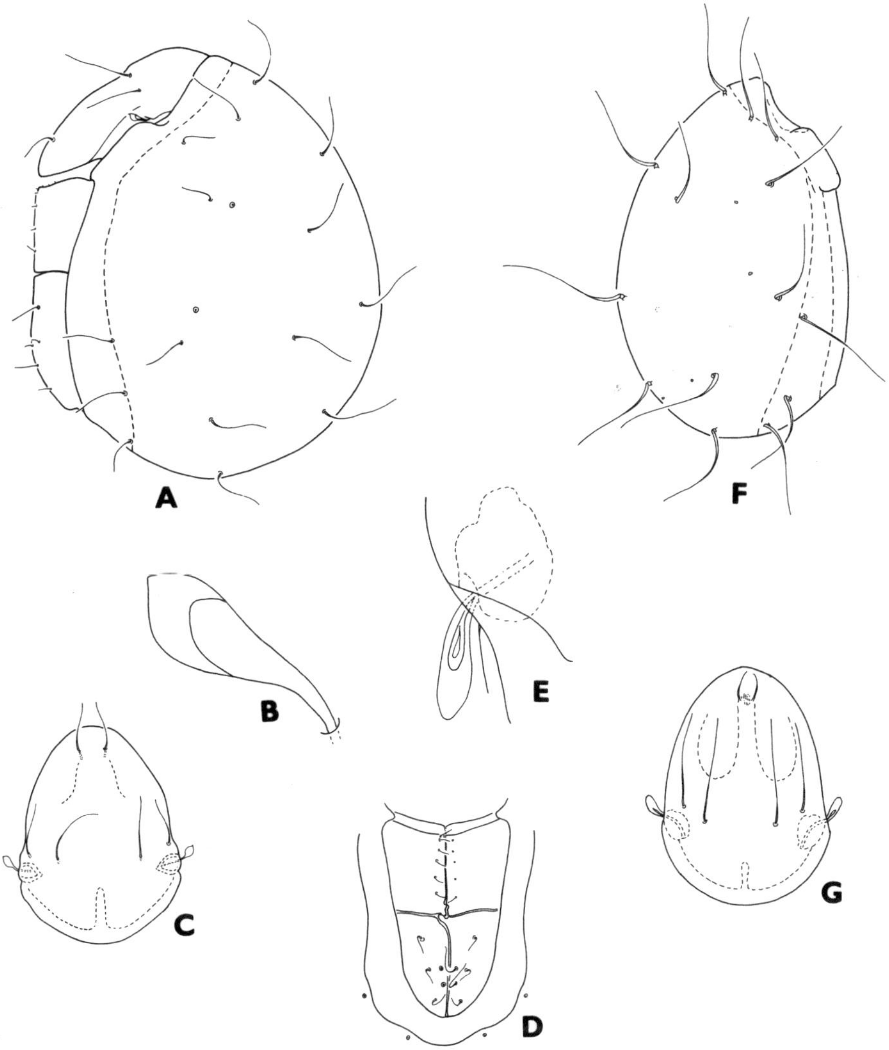

Plate 95. A—D: *Archiphthiracarus tzanoudakisi* Mahunka, 1979 — A: lateral side, B: sensillus, C: aspis, D: anogenital region; E—G: *Archiphthiracarus clavatus* (Parry, 1979): — E: sensillus, F: notogaster, G: aspis (A—D original; after Parry: E—G)

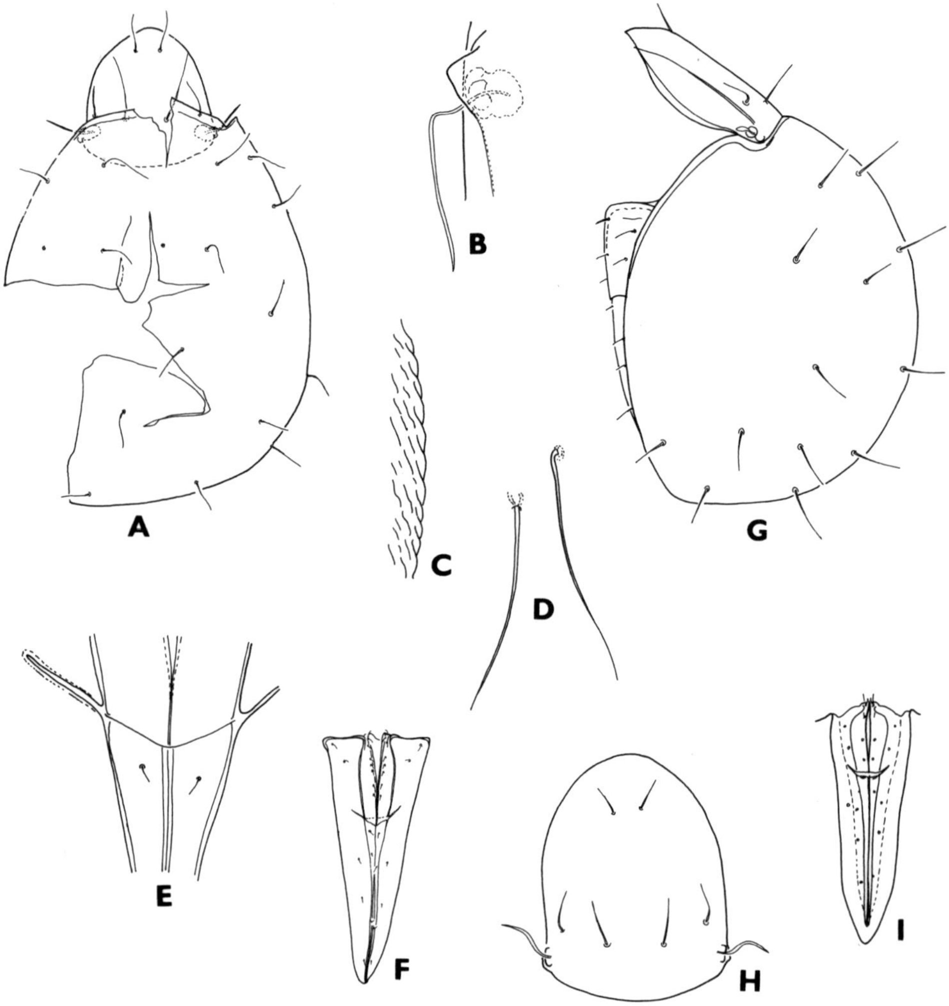

Plate 96. A—F: *Oribotritita tokukoae* Aoki, 1973 — A: dovsel side, B: sensillus, C: sculpture of notogaster D: rostral and notogastral setae E—F: anogenital region and detail, G—I: *Oribotritia mollis* Aoki, 1959 — G: lateral side, H: aspis, I: anogenital region (after Aoki)

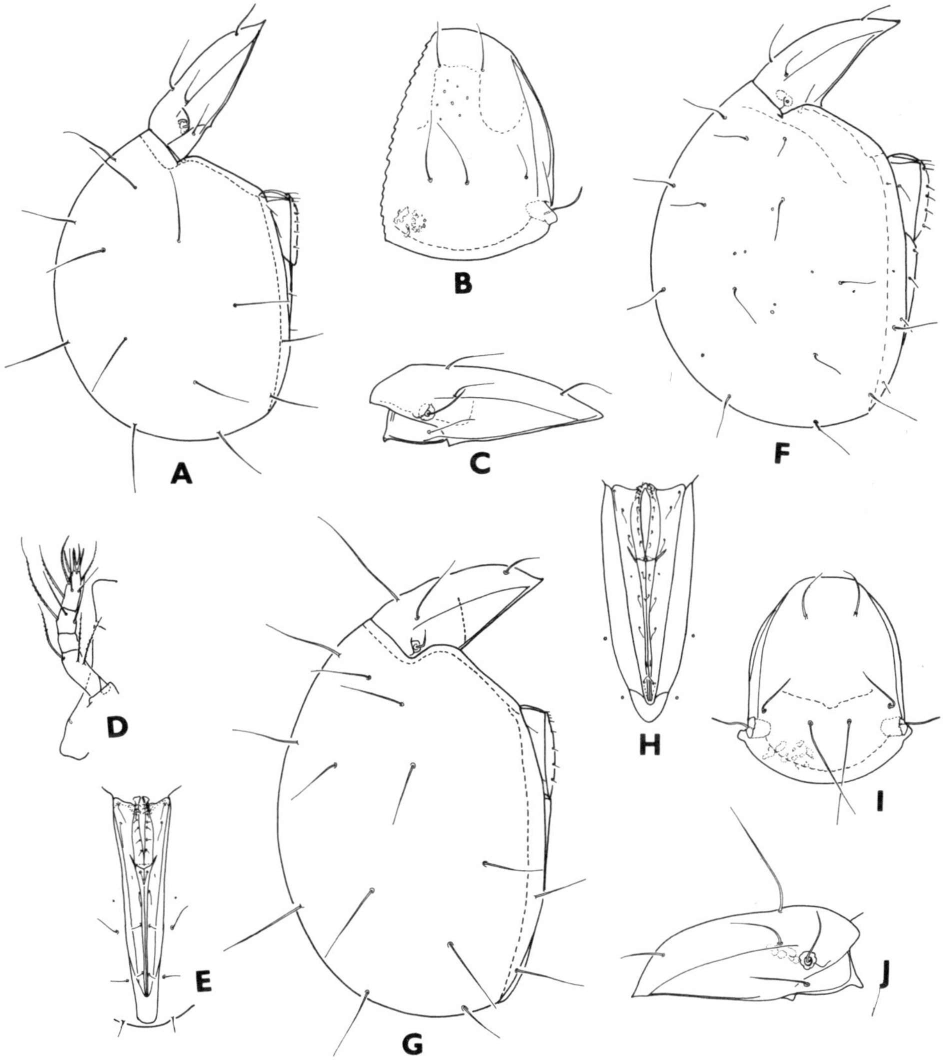

Plate 97. A—E: *Oribotritia berlesei* (Michael, 1898) — A: lateral side, B: aspis, C: aspis from lateral side, D: palp, E: anogenital region, F: *Oribotritia hermanni* Grandjean, 1967 — lateral side, G—J: *Oribotritia fennica* Forsslund et Märkel, 1963 — G: lateral side, H: anogenital region, I: aspis, J: aspis from lateral side (after Märkel: A—E, Grandjean: F and Forsslund and Märkel: G—J)

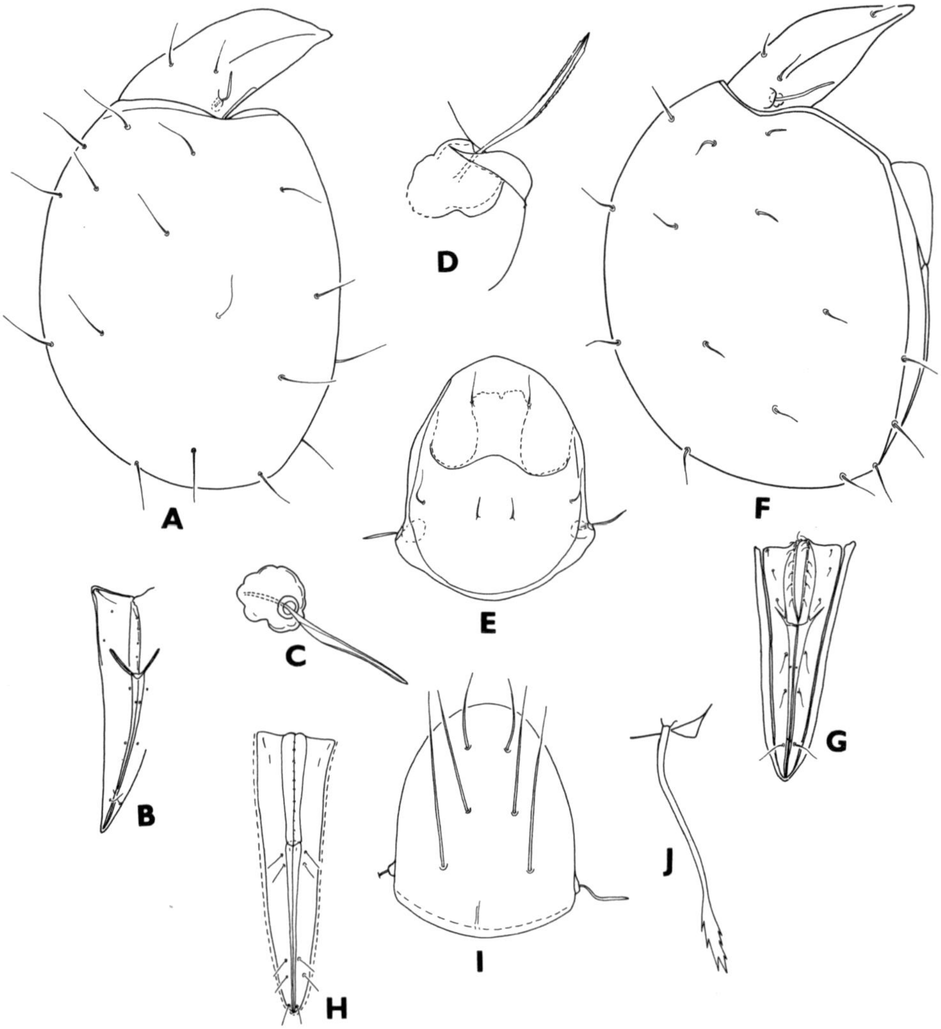

Plate 98. A—C: *Oribotritia storkani* Feider et Suciu, 1957 — A: lateral side, B: anogenital region C: sensillus; D—G: *Oribotritia serrata* Feider et Suciu, 1958 — D: lateral side, E: aspis, F: lateral side, G: anogenital region; H—J: *Mesotritia faroensis* (Sellnick, 1923) — H: anogenital region, I: aspis, J: sensillus (after Feider and Suciu: A—G and Sellnick: H—J)

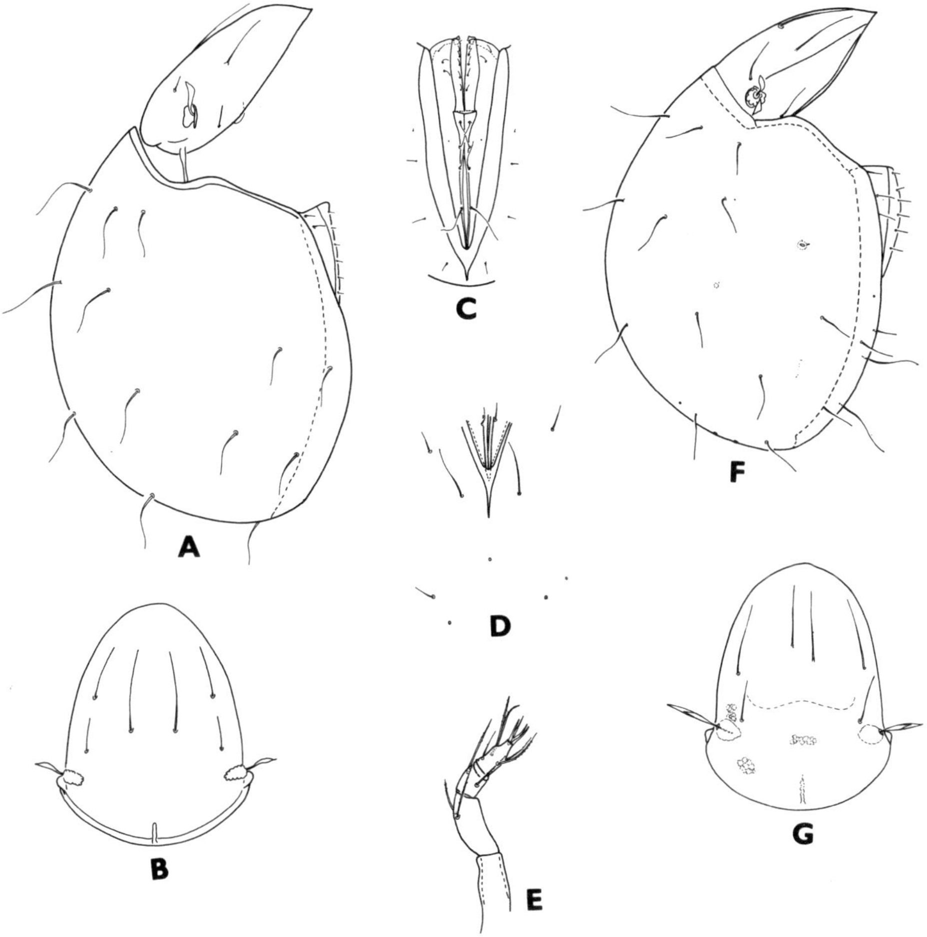

Plate 99. A—B: *Mesotritia maerkeli* Sheals, 1965 — A: lateral side, B: aspis; C—G: *Mesotritia testacea* Forsslund, 1963 — C: anogenital region, D: pygidium, E: palp, F: lateral side, G: aspis (after Sheals: A—B and Märkel: C—G)

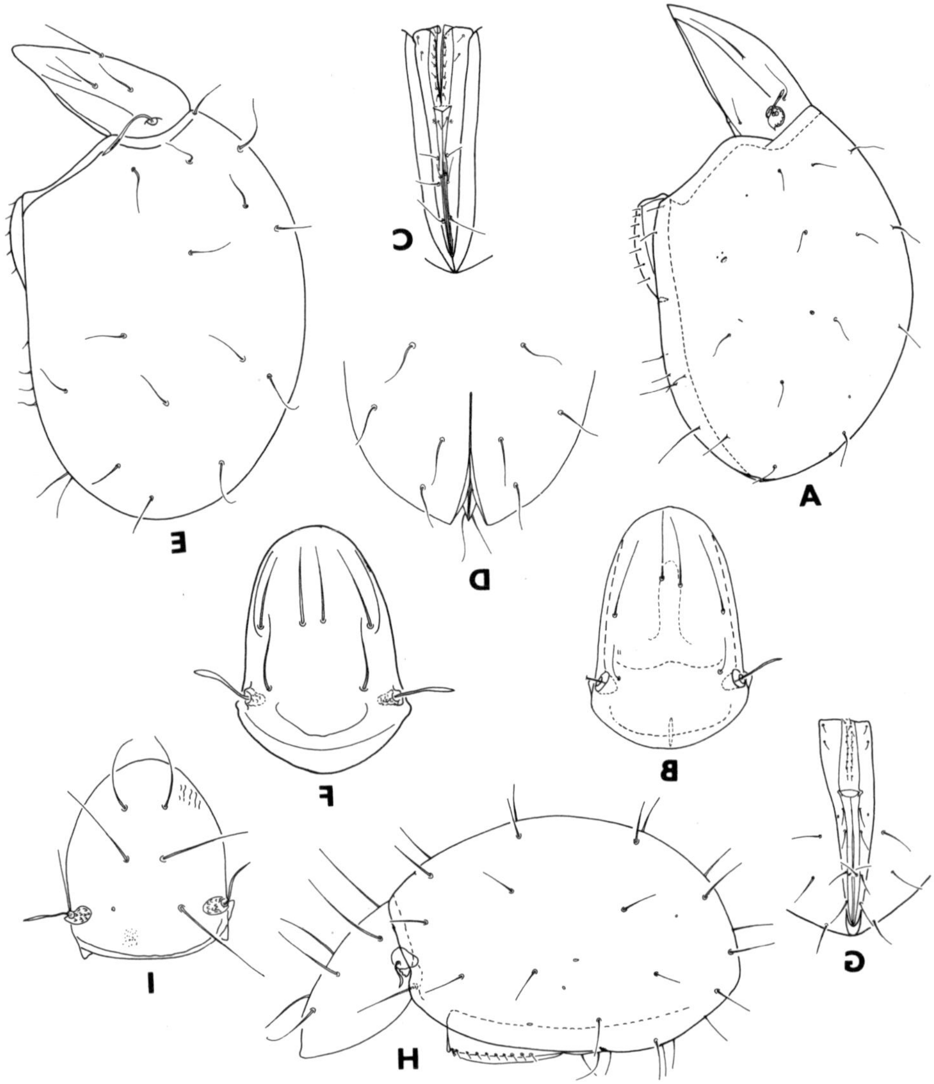

Plate 100. A—D: *Mesotritia piffli* Märkel, 1964 — A: lateral side, B: aspis, C: anogenital region, D: pygidium; E—F: *Mesotritia grandjeani* (Feider et Suciu, 1957) — E: lateral side, F: aspis; G—I: *Maerkelotrtitia alaskensis* Hammer, 1967 — G: anogenital region, H: lateral side, I: aspis (after Märkel: A—D, Feider and Suciu: E—F and Hammer: G—I)

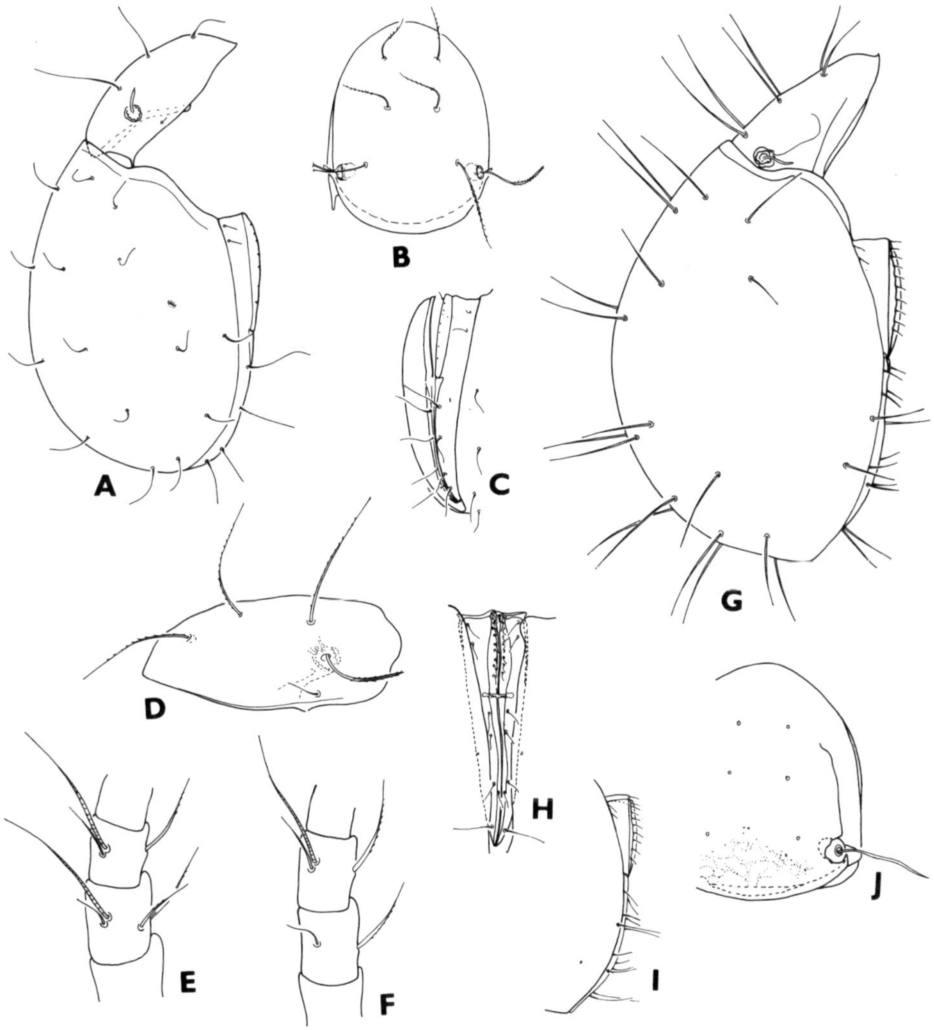

Plate 101. A—C: *Maerkelotritia krivolutzkii krivolutzkii* Märkel, 1968 — A: lateral side, B: aspis, C: anogenital region; D—F: *Maerkelotritia krivolutzkii mongolica* Mahunka, 1981 — D: aspis, E: genu of leg III, F: genu of leg IV; G—J: *Maerkelotritia kishidai* (Aoki, 1958) — G: lateral side, H: anogenital region, I: anogenital region from lateral side, J: aspis (after Märkel: A—C and Aoki: G—J; D—F original)

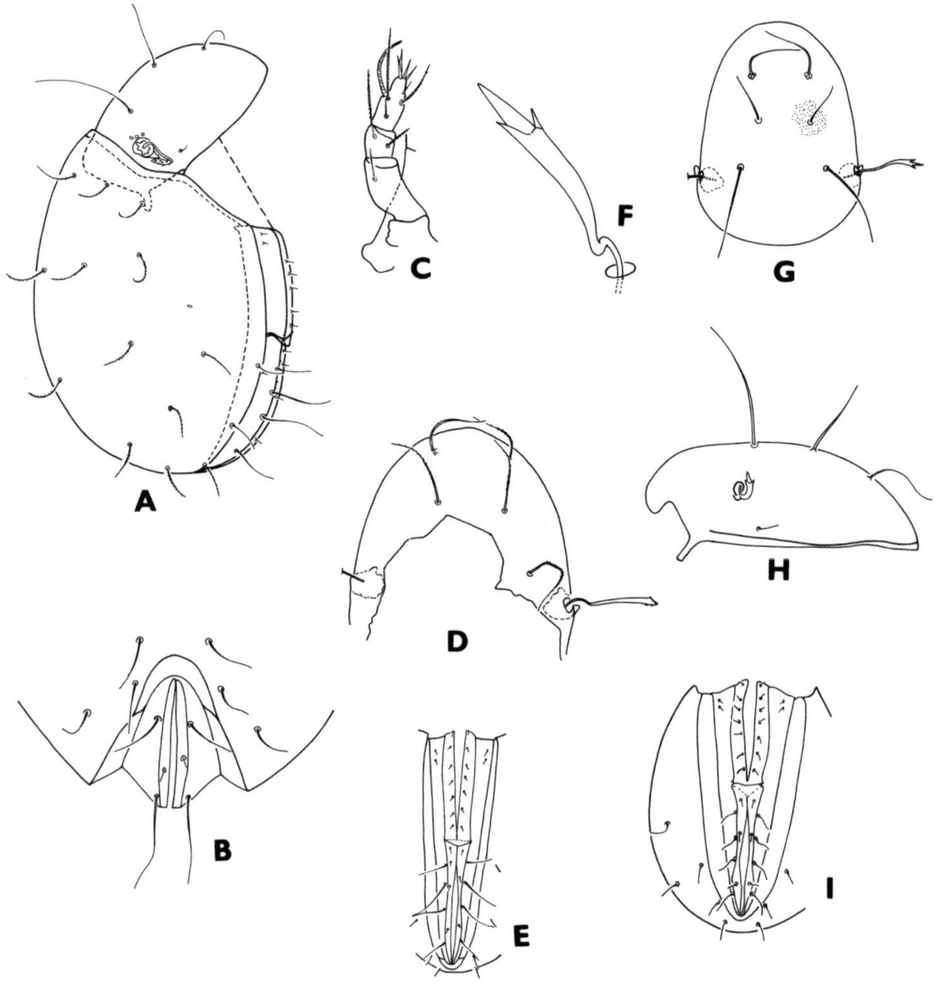

Plate 102. A—E: *Protoribotritia aberrans* (Märkel et Meyer, 1959) — A: lateral side, B: pygidium, C: palp, D: aspis, E: anogenital region; F—I: *Protoribotritia ensifer* Aoki, 1969 — F: sensillus, G: aspis, H: aspis from lateral side, I: anogenital region (after Märkel: A—E and Aoki: F—I)

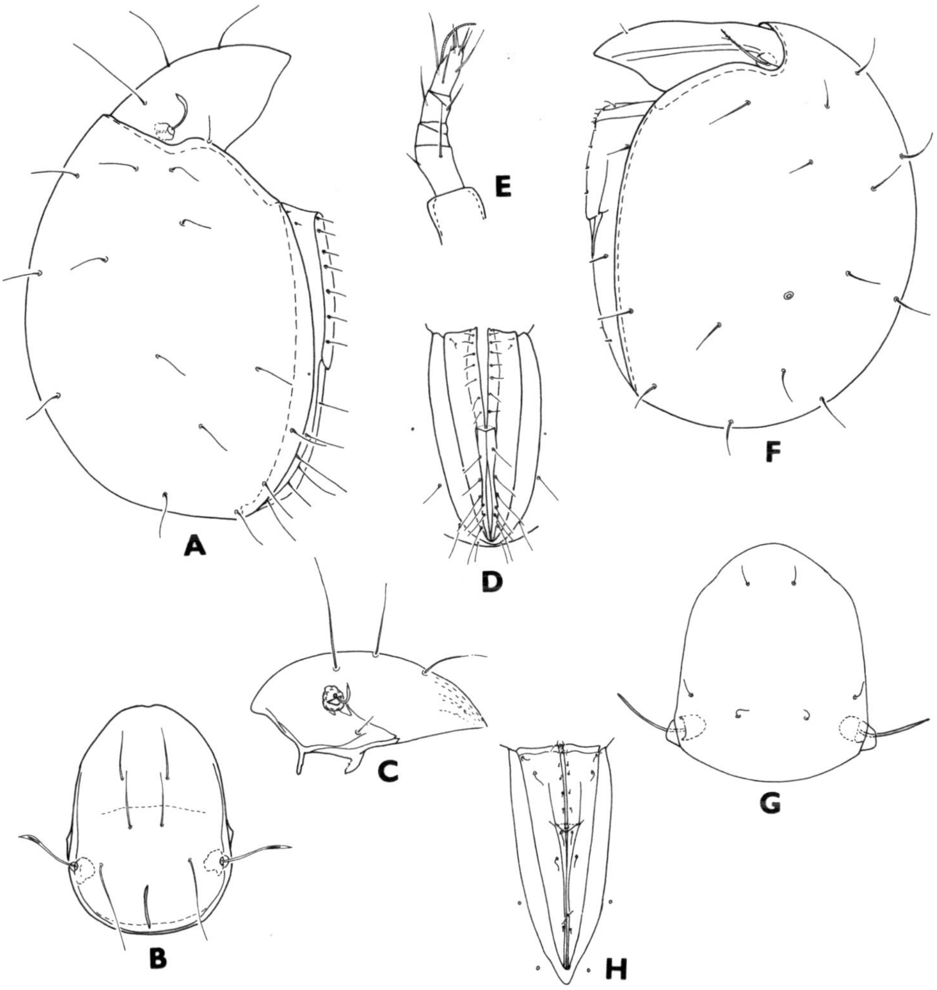

Plate 103. A—E: *Protoribotritia oligotricha* Märkel, 1963 — A: lateral side, B: aspis, C: aspis from lateral side, D: anogenital region, E: palp; F—H: *Indotritia consimilis* Märkel, 1964 — F: lateral side, G: aspis, H: anogenital region (after Märkel)

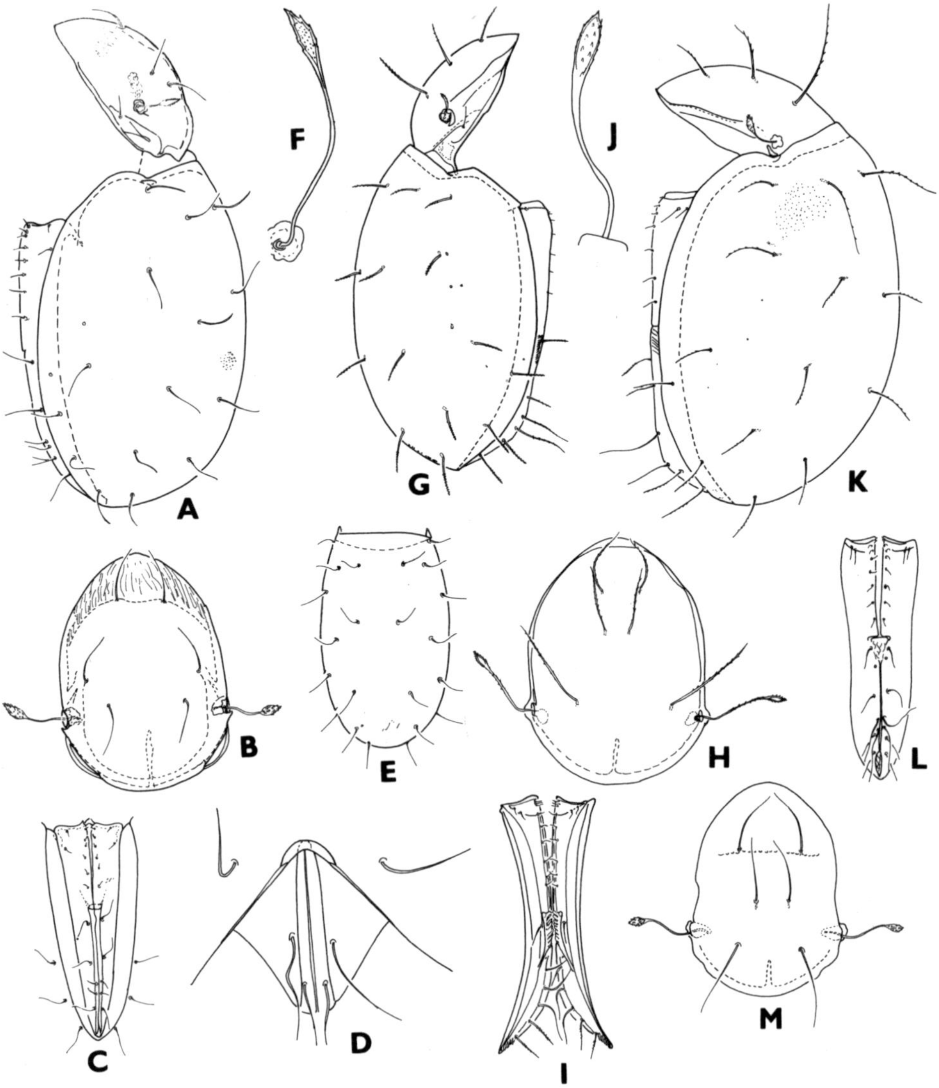

Plate 104. A—E: *Paratritia baloghi* Moritz, 1966 — A: lateral side, B: aspis, C: anogenital region, D: pygidium, E: notogaster; F—I: *Euphthiracarus monodactylus* (Willmann, 1919) — F: sensillus, G: lateral side, H: aspis I: anogenital region; J—M: *Euphthiracarus mixtus* Mahunka, 1979 — J: sensillus, K: lateral side, L: anogenital region, M: aspis (after Moritz: A—E, Feider and Suciu: F—I and Mahunka: J—M)

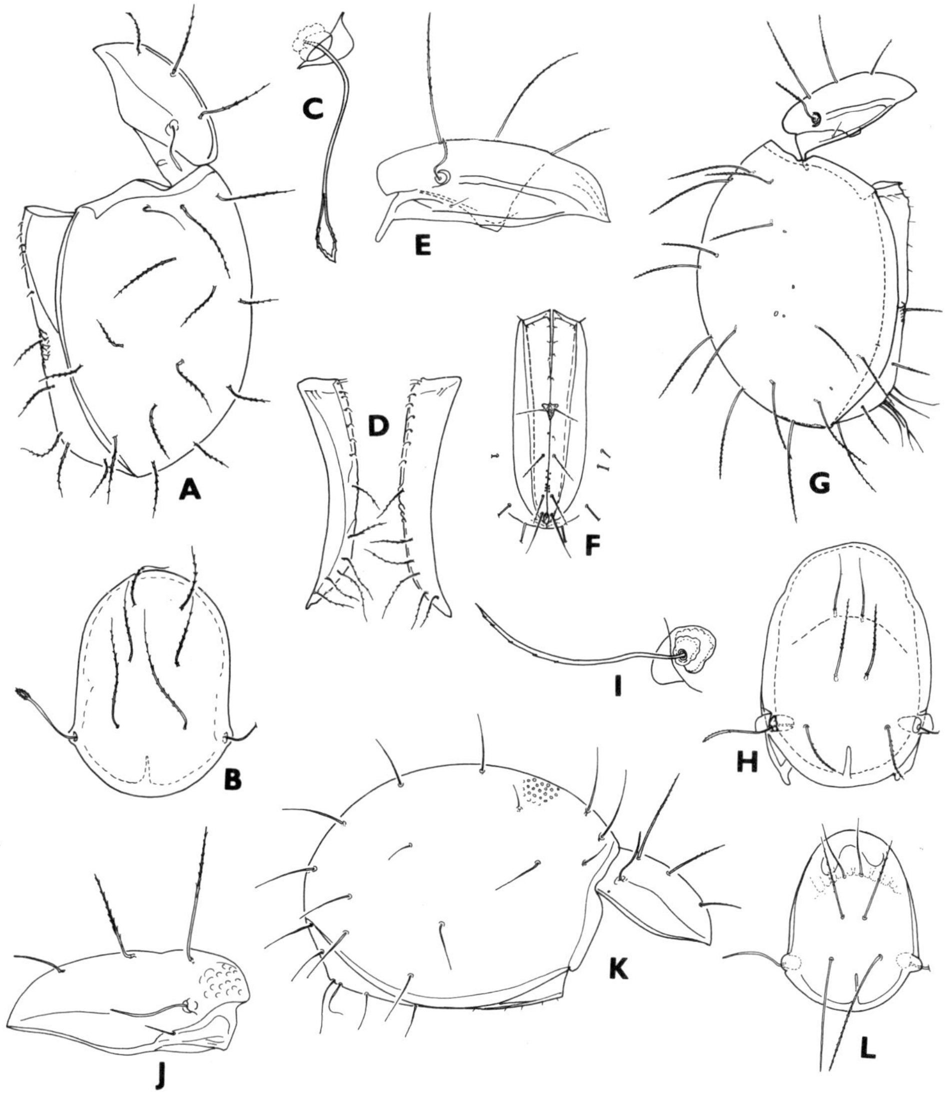

Plate 105. A—D: *Euphthiracarus antipai* Feider et Calugar, 1969 — A: lateral side, B: aspis, C: sensillus, D: anogenital region; E—H: *Euphthiracarus reticulatus alpinus* Märkel, 1964 — E: aspis, F: anogenital region, G: lateral side, H: aspis from drosal side; I—L: *Euphthiracarus intermedius* Feider et Suciu, 1958 — I: sensillus, J: aspis from lateral side, K: lateral side, L: aspis from dorsal side (after Feider and Calugar: A—D, Märkel: E—H and Feider and Suciu: I—L)

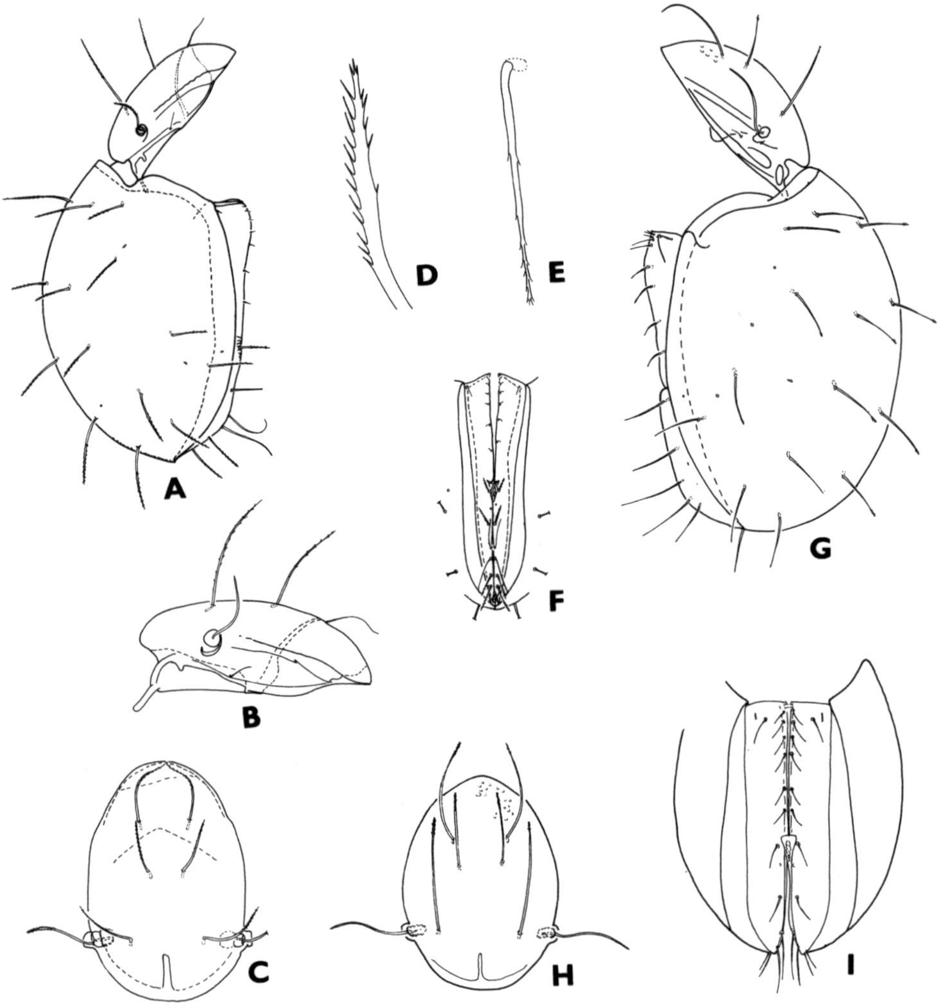

Plate 106. A—F: *Euphthiracarus cribrarius* (Berlese, 1904) — A: lateral side, B: aspis from lateral side, C: aspis from dorsal side, D: distal part of sensillus, E: seta h_2, F: anogenital region; G—I: *Euphthiracarus inglisi* Sheals, 1965 — G: lateral side, H: aspis, I: anogenital region (after Märkel: A—C, F, Bernini: D, E and Sheals: G—I)

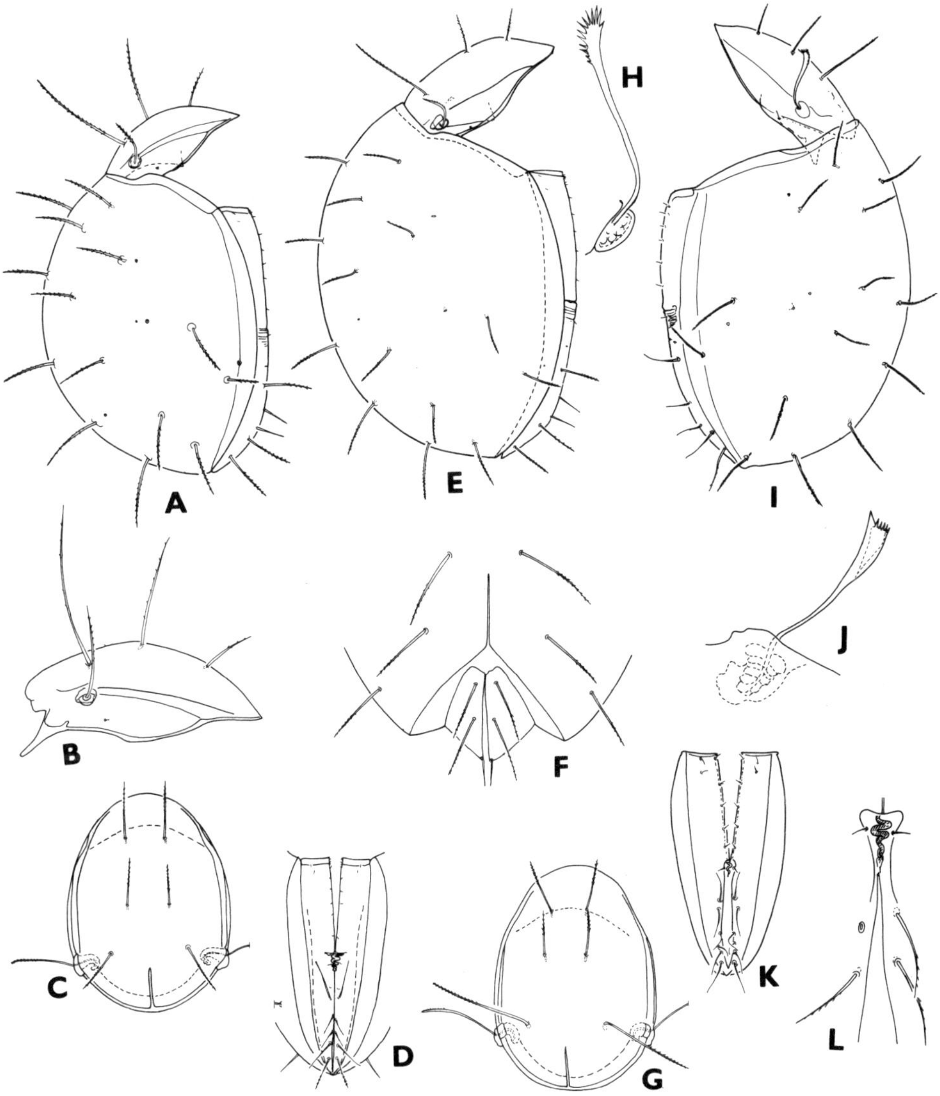

Plate 107. A—D: *Rhysotritia duplicata* (Grandjean, 1953) — A: lateral side, B: aspis from lateral side, C: aspis from dorsal side, D: anogenital region; E—G: *Rhysotritia ardua ardua* (C. L. Koch, 1841) — E: lateral side, F: pygidium, G: aspis; H—I: *Rhysotritia ardua penicillata* Pérez-Iñigo, 1969 — H: sensillus, I: lateral side; J—L: *Rhysotritia clavata sextiana* Lions, 1966 — J: sensillus, K: anogenital region, L: triangle of anogenital region (after Märkel: A—G, Pérez-Iñigo: H—I and Lions J—L)

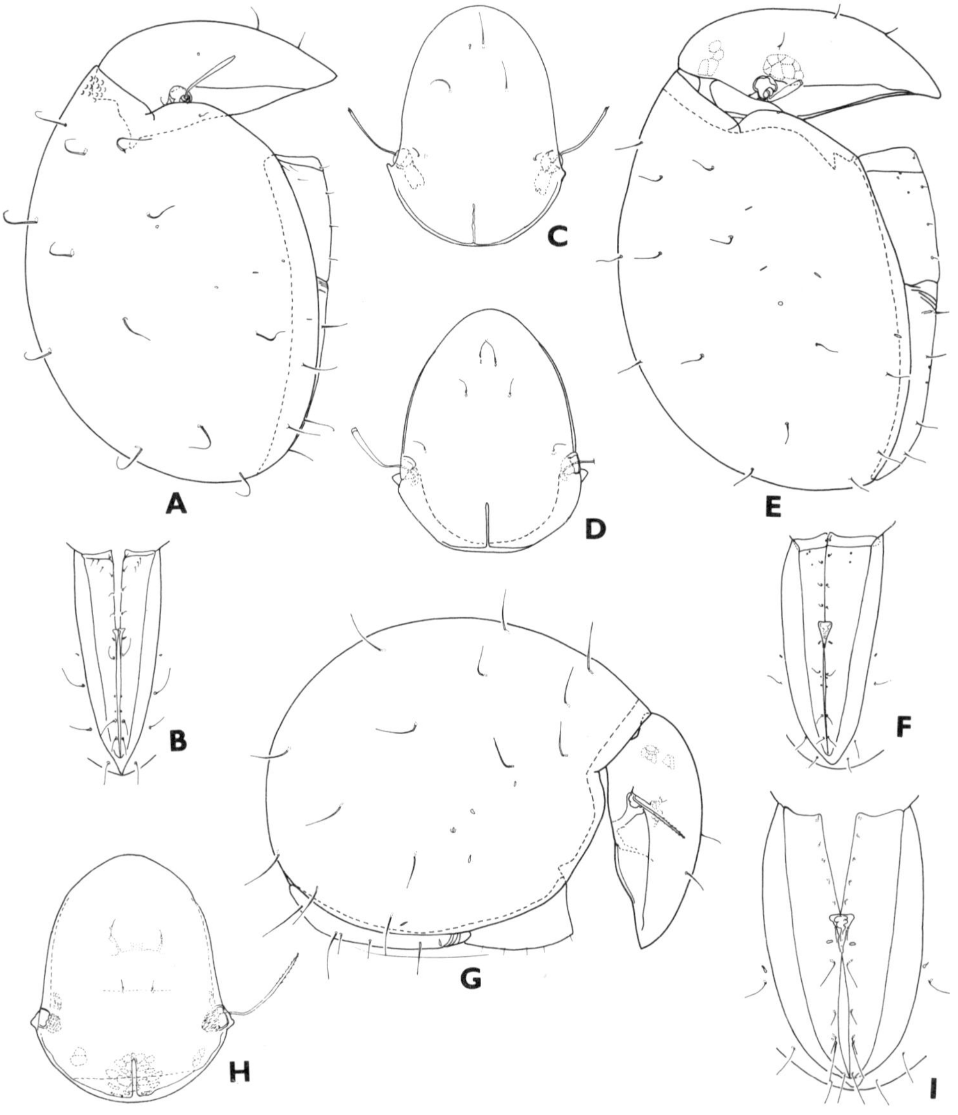

Plate 108. A—C: *Microtritia fissurata* Märkel, 1968 — A: lateral side, B: anogenital region, C: aspis; D—F: *Microtritia minima* (Berlese, 1904) — D: aspis, E: lateral side, F: anogenital region; G—I: *Microtritia tropica* Märkel, 1964 — G: lateral side, H: aspis, I: anogenital region (after Märkel)

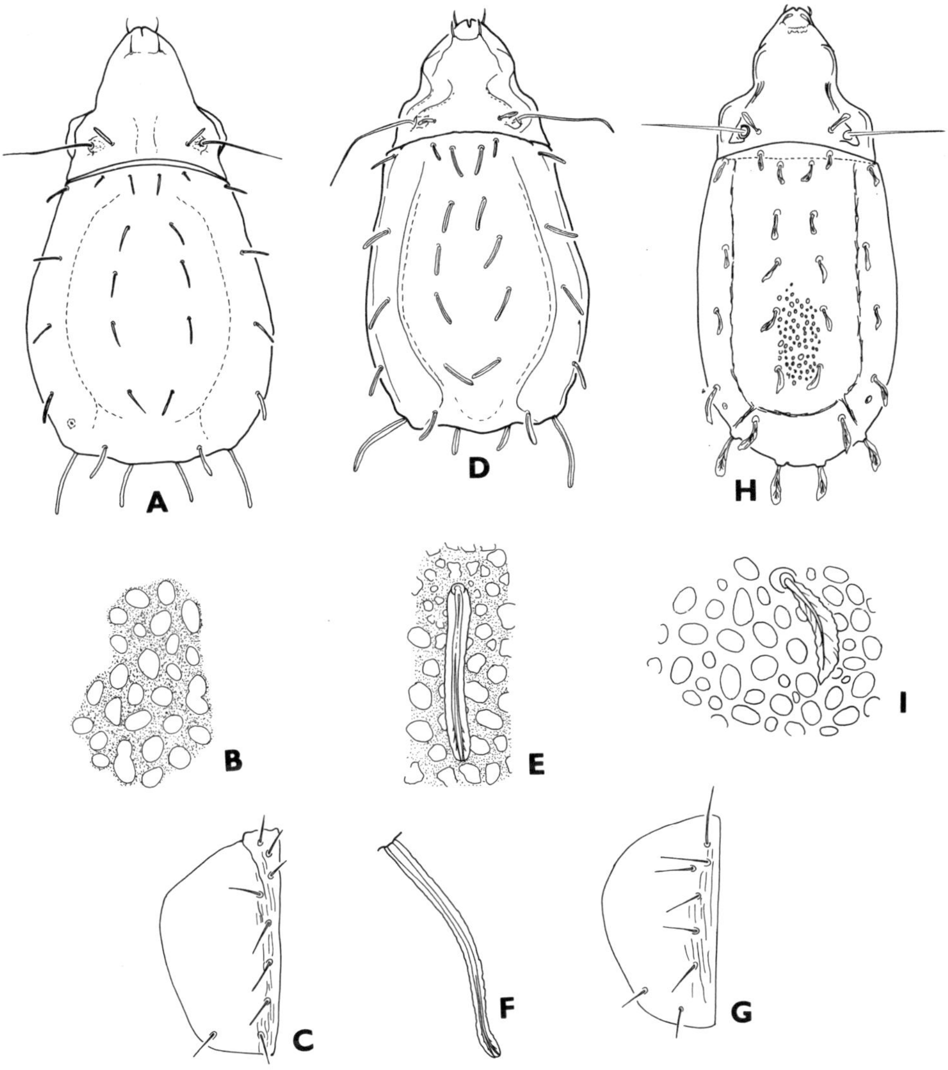

Plate 109. A—C: *Nothrus pratensis* Sellnick, 1929 — A: dorsal side, B: sculpture notogaster, C: genital plate; D—G: *Nothrus silvestris* Nicolet, 1855 — D: dorsal side, E: sculpture of notogaster, F: seta k_1, G: genital plate; H—I: *Nothrus perezinigoi* Mahunka, 1980 — H: dorsal side, I: sculpture of notogaster (after Sellnick and Forsslund: A—G and Pérez-Iñigo: H—I)

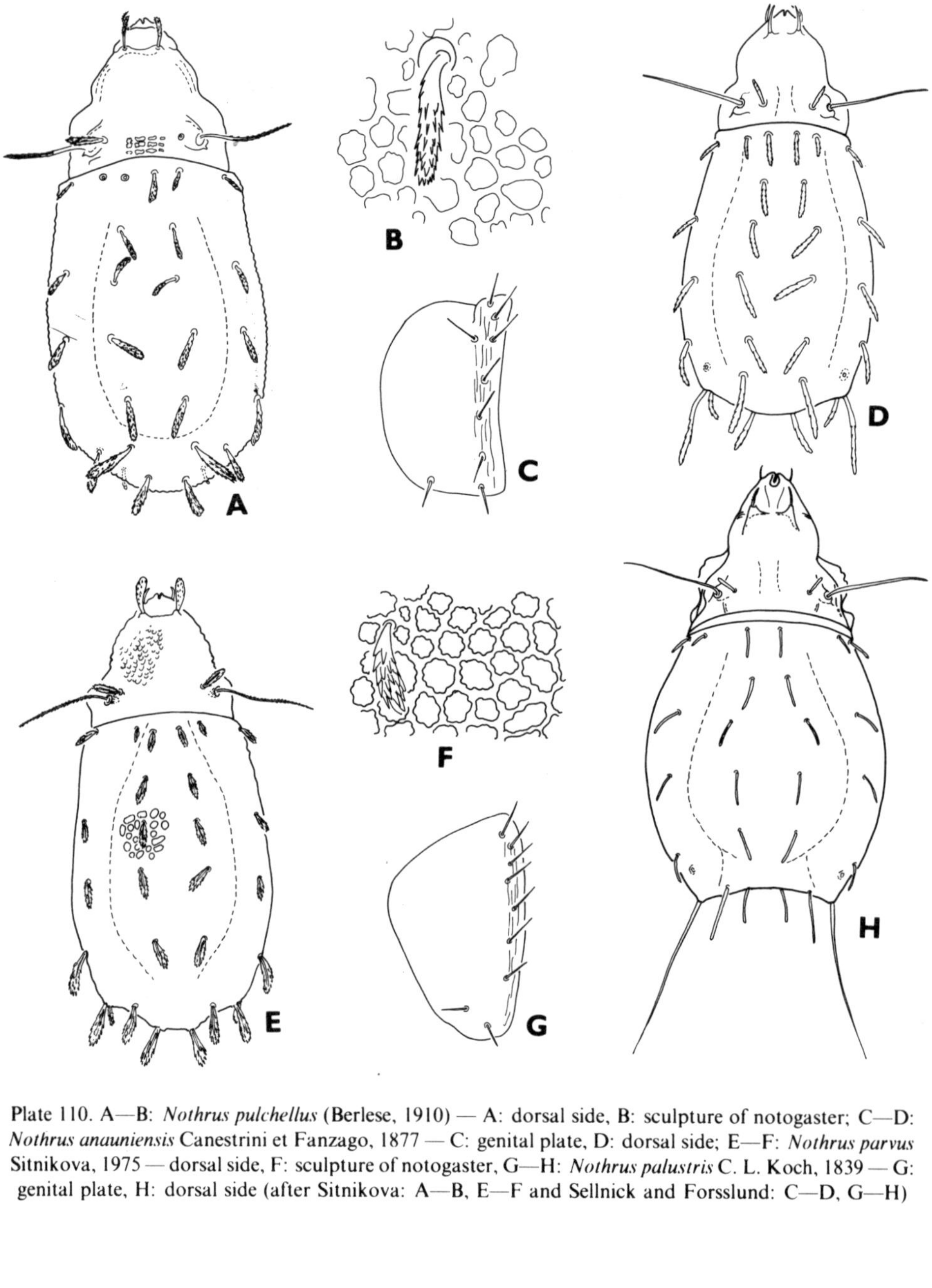

Plate 110. A—B: *Nothrus pulchellus* (Berlese, 1910) — A: dorsal side, B: sculpture of notogaster; C—D: *Nothrus anauniensis* Canestrini et Fanzago, 1877 — C: genital plate, D: dorsal side; E—F: *Nothrus parvus* Sitnikova, 1975 — dorsal side, F: sculpture of notogaster, G—H: *Nothrus palustris* C. L. Koch, 1839 — G: genital plate, H: dorsal side (after Sitnikova: A—B, E—F and Sellnick and Forsslund: C—D, G—H)

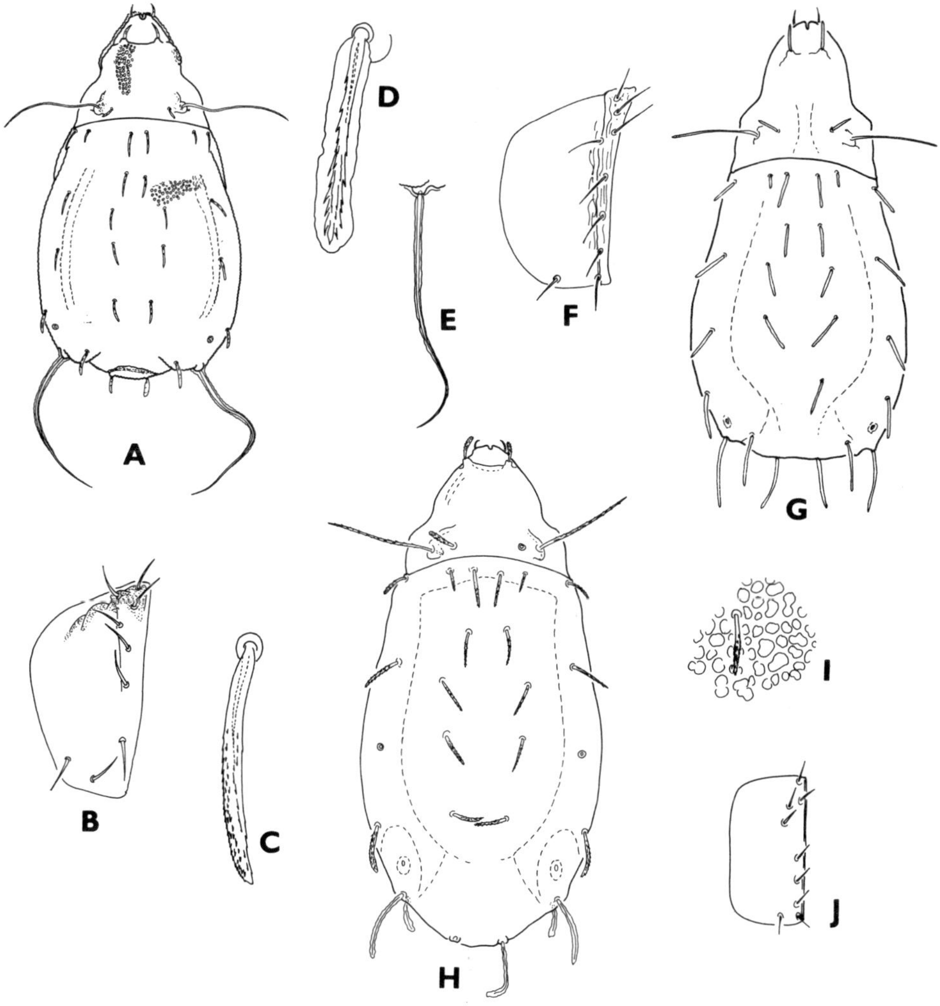

Plate 111. A—E: *Nothrus asiaticus* Aoki et Ohnishi, 1974 — A: dorsal side, B: genital plate, C: seta c_1, D: seta pn_2, E: seta k_1, F—G: *Nothrus borussicus* Sellnick, 1929 — F: genital plate, G: dorsal side; H—J: *Nothrus longipilus* (Berlese, 1910). — H: dorsal side, I: sculpture of notogaster, J: genital plate (after Aoki: A—E and Sellnick and Forsslund: F—G; H—J original)

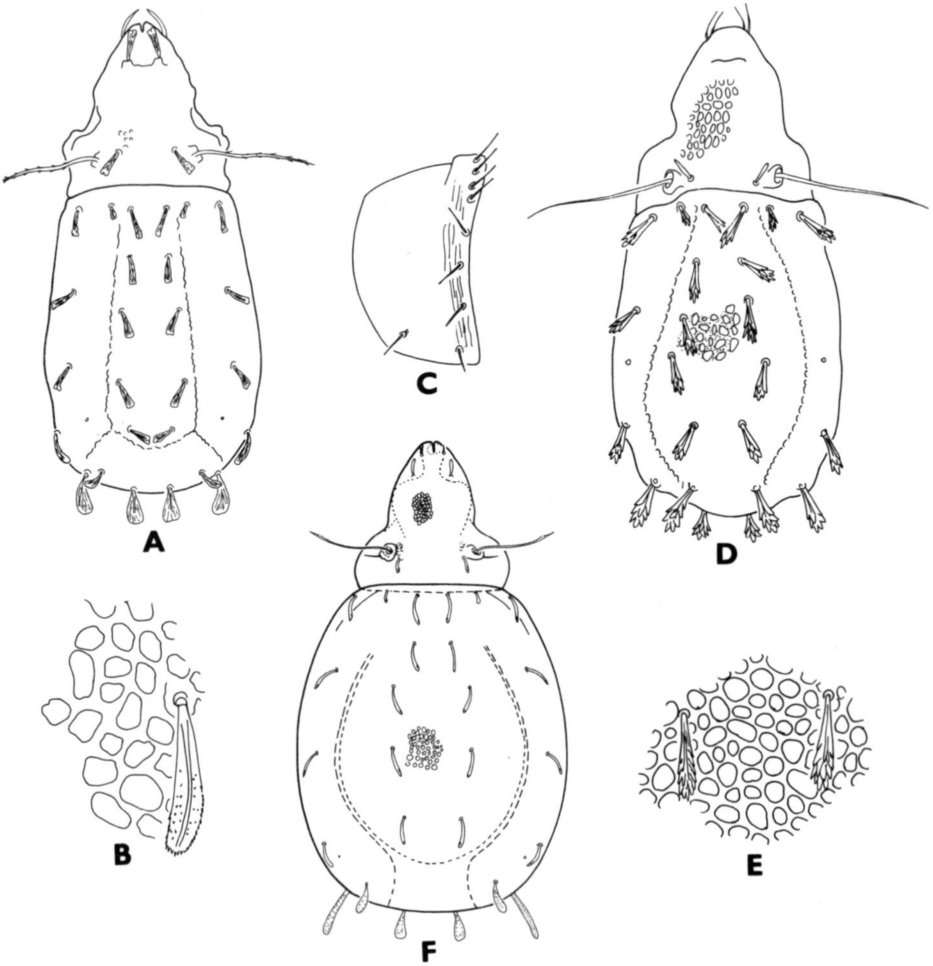

Plate 112. A—C: *Nothrus biciliatus* C. L. Koch, 1841 — A: dorsal side, B: sculpture of notogaster, C: genital plate; D—E: *Nothrus reticulatus* Sitnikova, 1975 — D: dorsal side, E: sculpture of notogaster; F: *Nothrus springsmythi* Sheals, 1965 — dorsal side (after Sellnick and Forsslund: A—C, Sitnikova: D—E and Sheals: F)

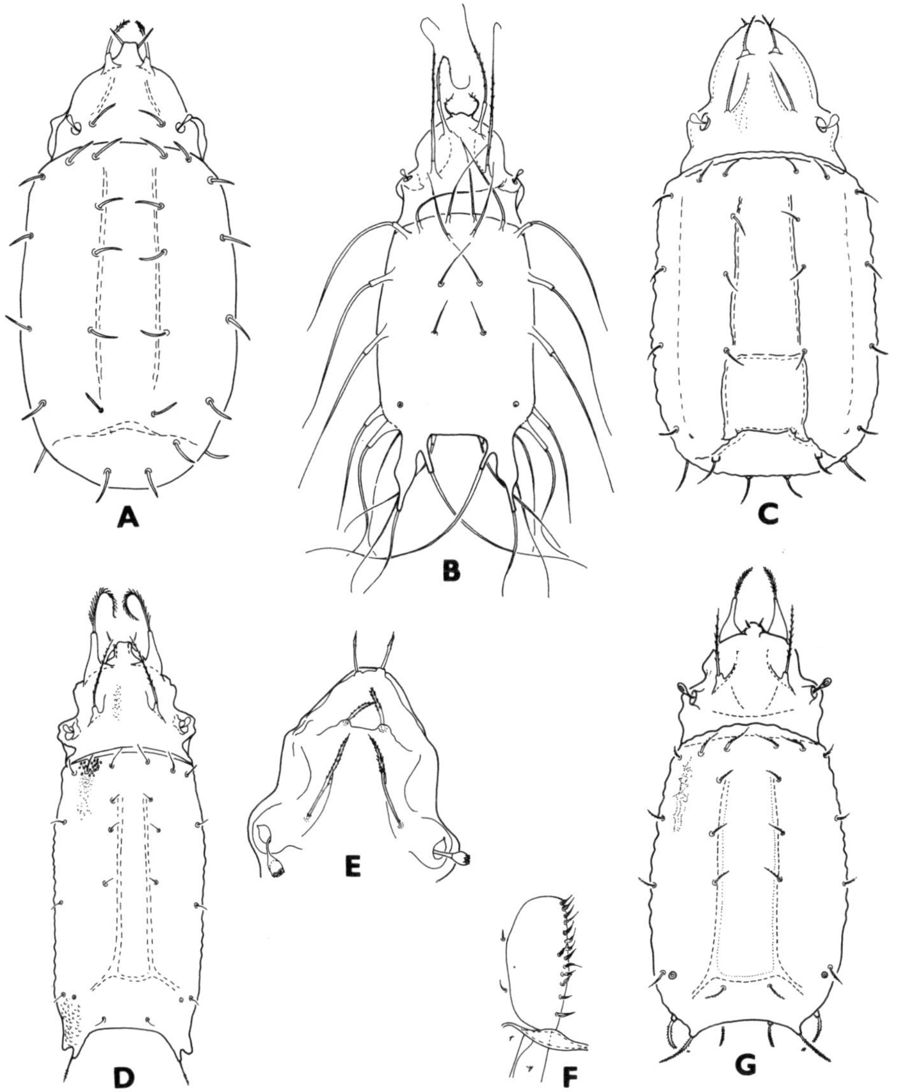

Plate 113. A: *Camisia lapponica* (Trägårdh, 1910) — dorsal side; B: *Camisia spinifer* (C. L. Koch, 1836) — dorsal side; C: *Camisia invenusta* (Michael, 1888) — dorsal side; D: *Camisia biurus* (C. L. Koch, 1839) — dorsal side; E—F: *Camisia polytricha* Piffl, 1971 — E: prodorsum, F: genital plate; G: *Camisia segnis* (Hermann, 1804) — dorsal side (after Sellnick and Forsslund: A—D, G and Piffl: E—F)

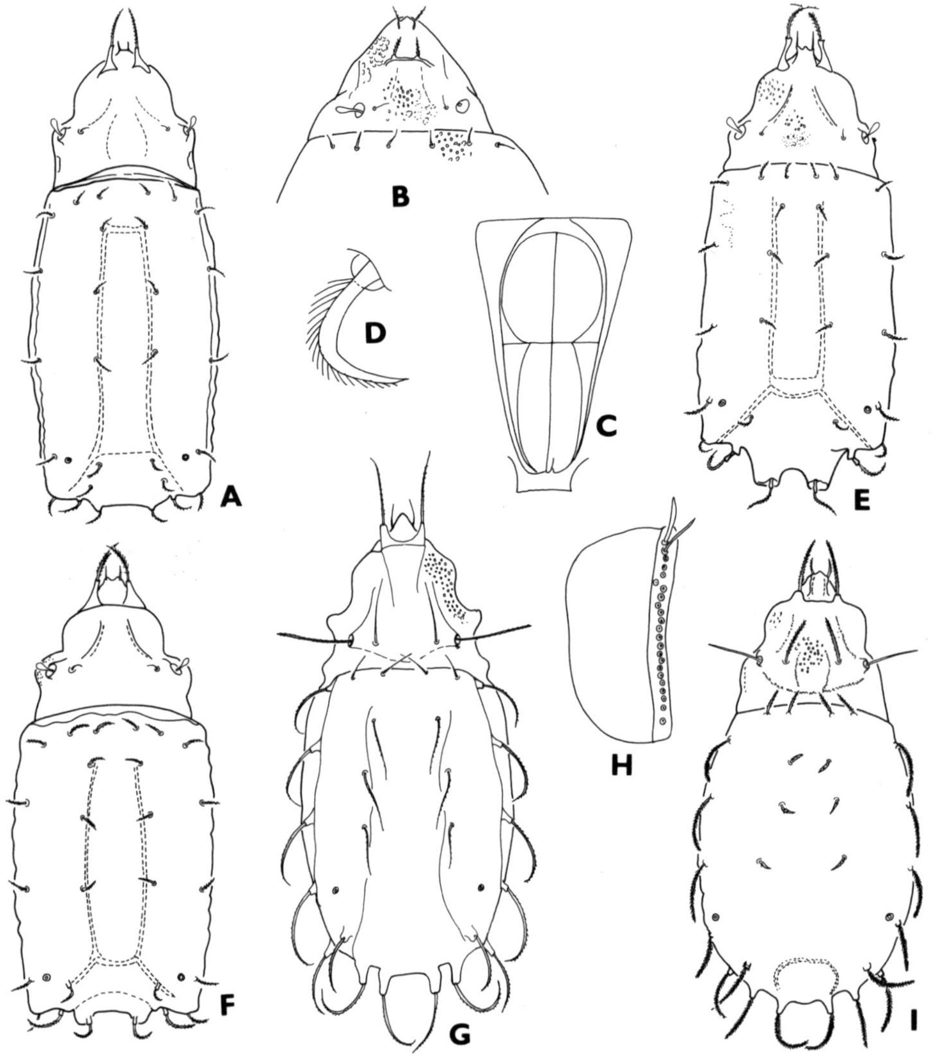

Plate 114. A: *Camisia borealis* (Thorell, 1872) — dorsal side; B—D: *Camisia foveolata* Hammer, 1955 — B: anterior part of idiosoma, C: anogenital region, D: seta from posterior part of notogaster; E: *Camisia biverrucata* (C. L. Koch, 1839) — dorsal side; F: *Camisia horrida* (Hermann, 1804) — dorsal side; G: *Heminothrus paolianus* Berlese, 1913 — dorsal side, H—I: *Heminothrus targionii* (Berlese, 1885) — H: genital plate, I: dorsal side (after Sellnick and Forsslund: A, E—F, H—I, Hammer: B—D and Berlese: G)

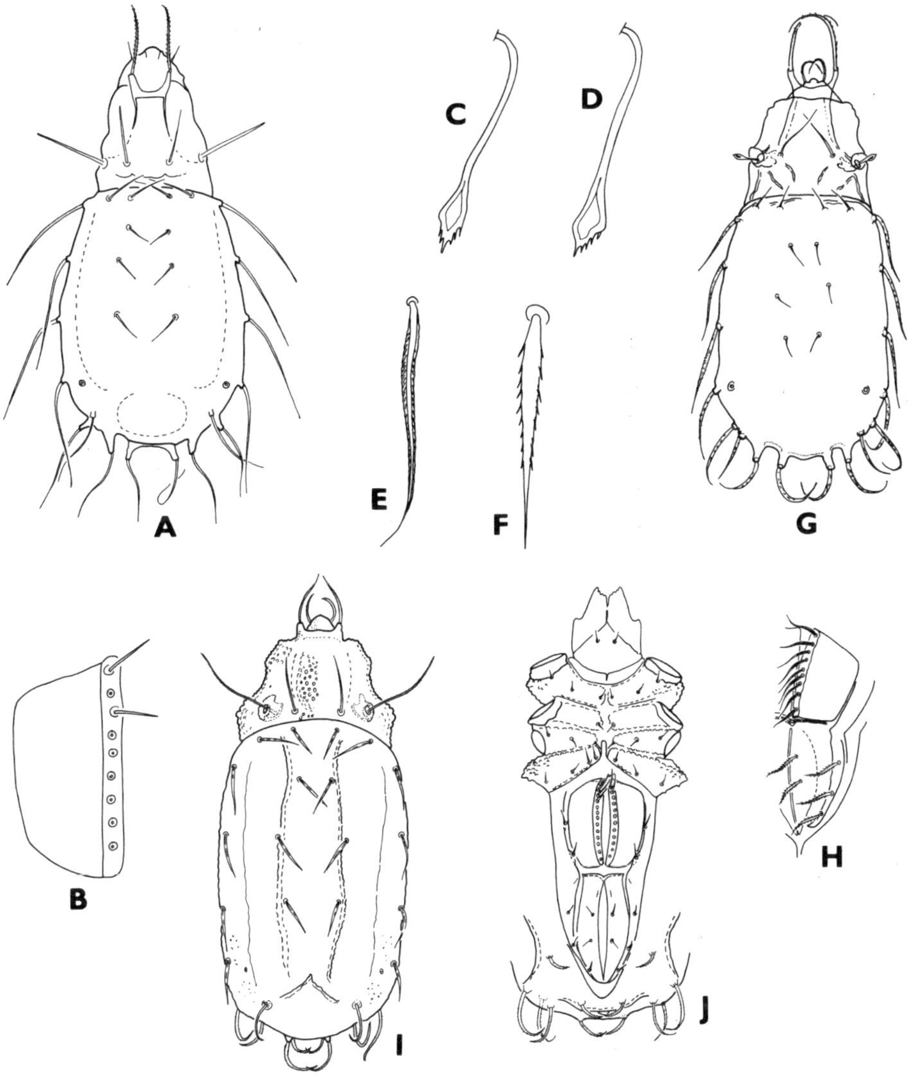

Plate 115. A—B: *Heminothrus longisetosus* Willmann, 1925 — A: dorsal side, B: genital plate; C—H: *Heminothrus minor* Aoki, 1969 — C—D: sensilli, E: lateral seta of notogaster; G: dorsal side, H: anogenital region; I—J: *Heminothrus humicola* (Forsslund, 1955): I: dorsal side, J: ventral side (after Sellnick and Forsslund: A—B, I—J and Aoki: C—H)

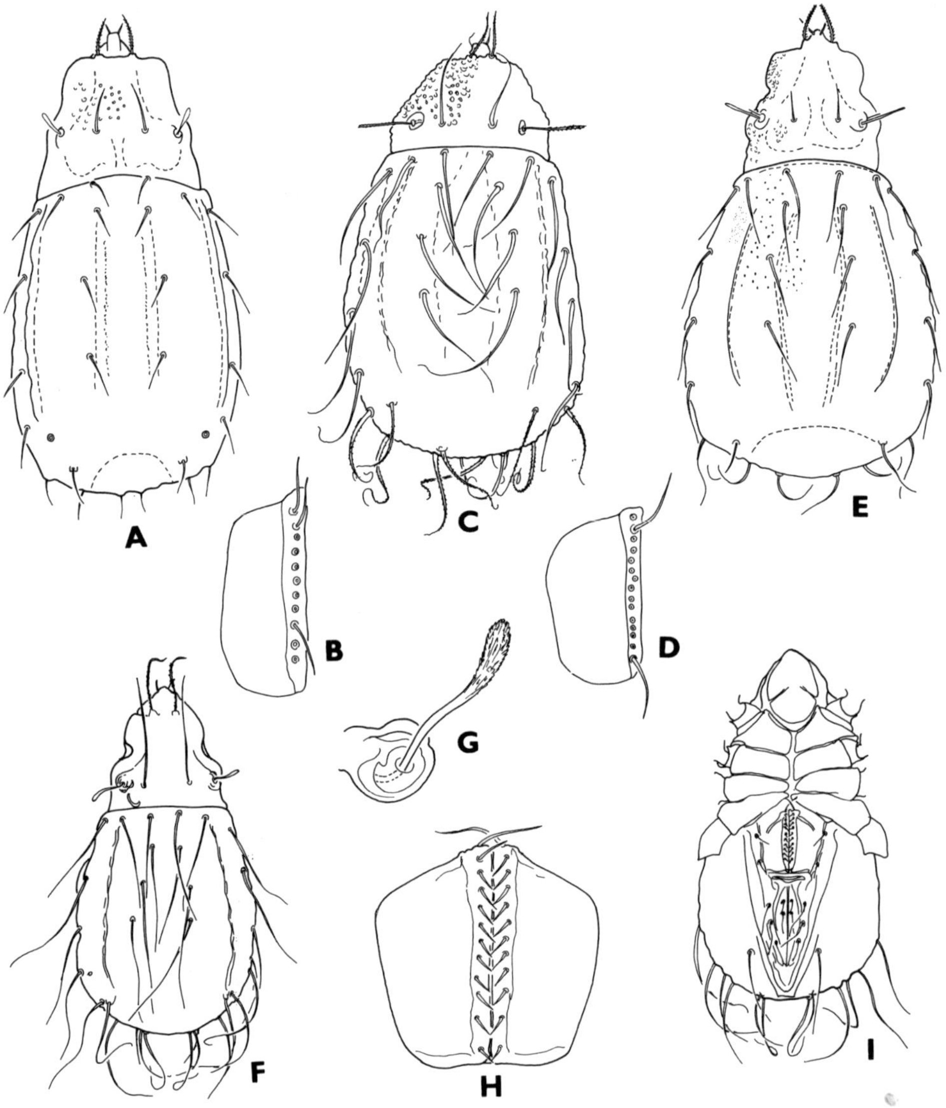

Plate 116. A—B: *Heminothrus punctatus* (L. Koch, 1879) — A: dorsal side, B: genital plate; C: *Heminothrus sibiricus* Sitnikova, 1975 — dorsal side; D—E: *Heminothrus peltifer* (C. L. Koch, 1839) — D: genital plate, E: dorsal side; F—I: *Heminothrus yamasakii* Aoki, 1958 — F: dorsal side, G: sensillus, H: genital plate, I: ventral side (after Sellnick and Forsslund: A—B, D—E, Sitnikova: C and Aoki: F—I)

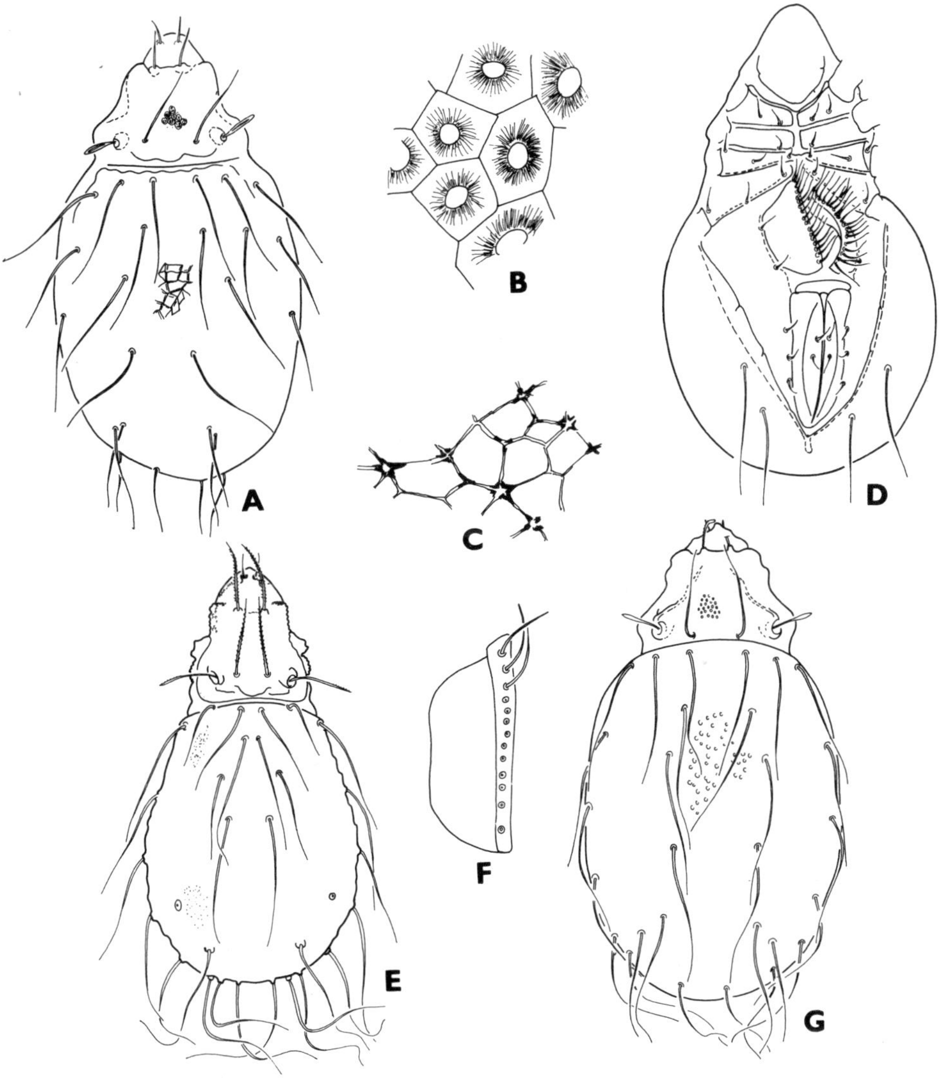

Plate 117. A—D: *Heminothrus nevadensis* (Pérez-Iñigo, 1969) — A: dorsal side, B: sculpture of prodorsum, C: sculpture of notogaster, D: ventral side; E—F: *Heminothrus thori* (Berlese, 1904) — E: dorsal side, F: genital plate; G: *Heminothrus capillatus* (Berlese, 1914) — dorsal side (after Pérez-Iñigo: A—D, Sellnick and Forsslund: E—F; G original)

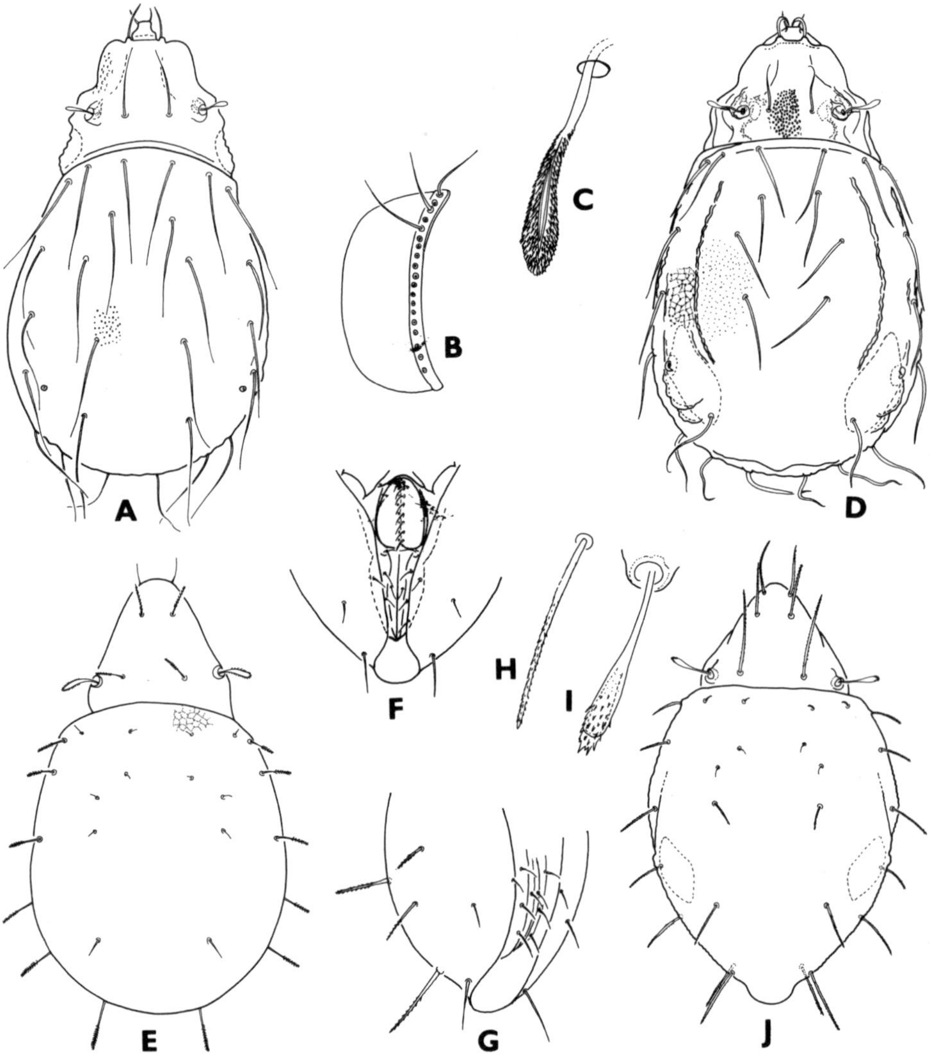

Plate 118. A—B: *Heminothrus septentrion s* (Sellnick, 1944) — A: dorsal side, B: genital plate; C—D: *Heminothrus numatai* (Aoki, 1965) — C: sensillus, D: dorsal side; E: *Trhypochthonius tectorum* (Berlese, 1896) — dorsal side; F—J: *Trhypochthonius japonicus* Aoki, 1970 — F: anogenital region, G: pygidium from lateral side, H: seta *in* I: sensillus, J: dorsal side (after Sellnick and Forsslund: A—B, Aoki: C—D, F—J; E original)

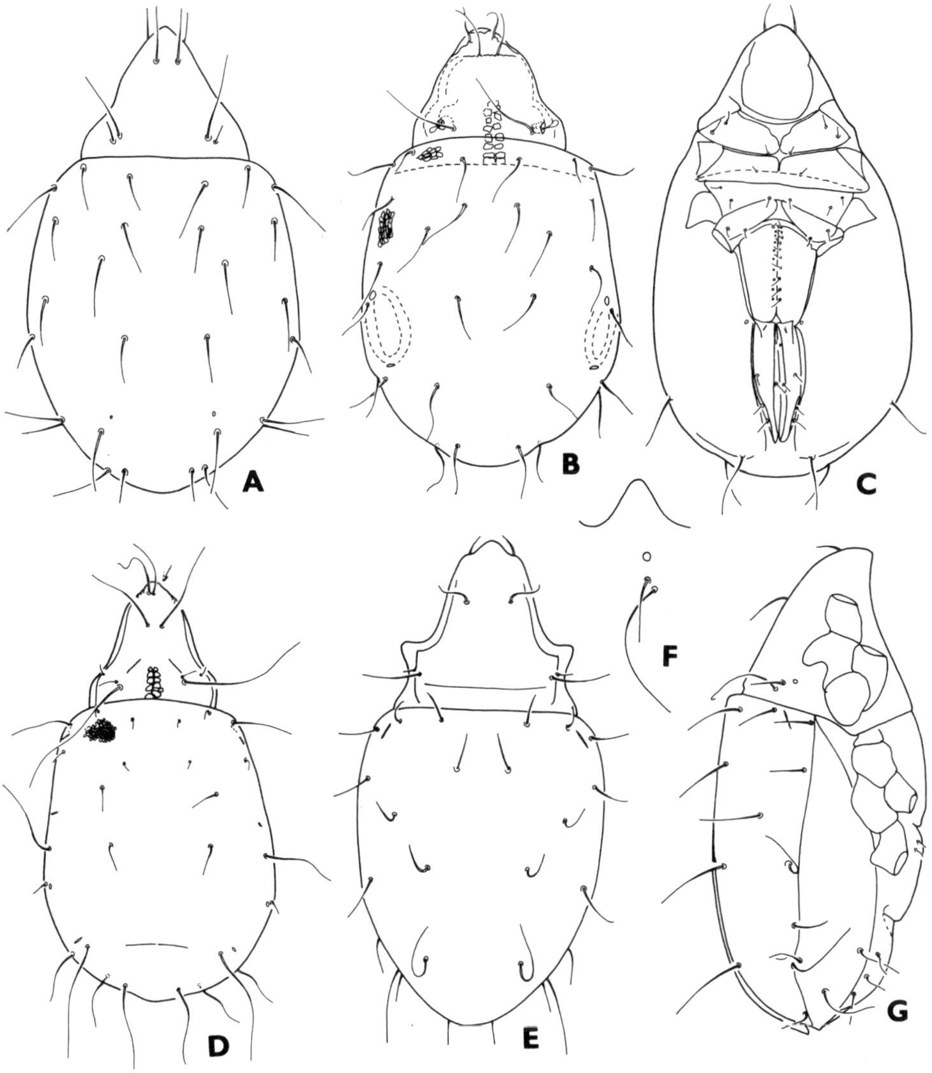

Plate 119. A: *Trhypochthoniellus setosus* Willmann, 1928 — dorsal side; B—C: *Trhypochthoniellus excavatus* (Willmann, 1919) — B: dorsal side, C: ventral side; D: *Mucronothrus nasalis* (Willmann, 1929) — dorsal side; E—G: *Malaconothrus processus* Van der Hammen, 1952 — E: dorsal side, F: prodorsum from lateral side, G: lateral side (after Willmann: A, Hammer: B—D and Van der Hammen: E—G)

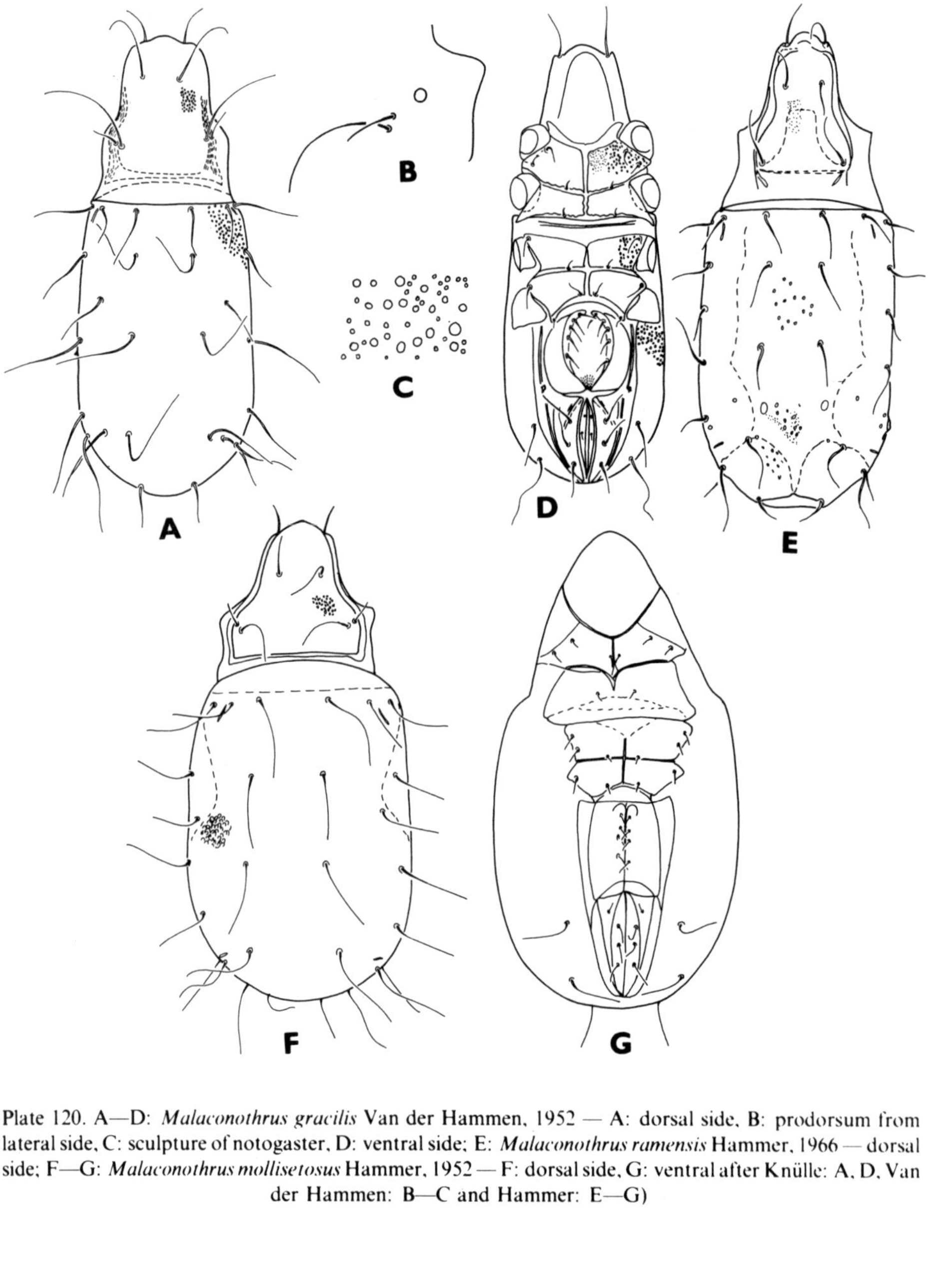

Plate 120. A—D: *Malaconothrus gracilis* Van der Hammen, 1952 — A: dorsal side, B: prodorsum from lateral side, C: sculpture of notogaster, D: ventral side; E: *Malaconothrus ramensis* Hammer, 1966 — dorsal side; F—G: *Malaconothrus mollisetosus* Hammer, 1952 — F: dorsal side, G: ventral after Knülle: A, D, Van der Hammen: B—C and Hammer: E—G)

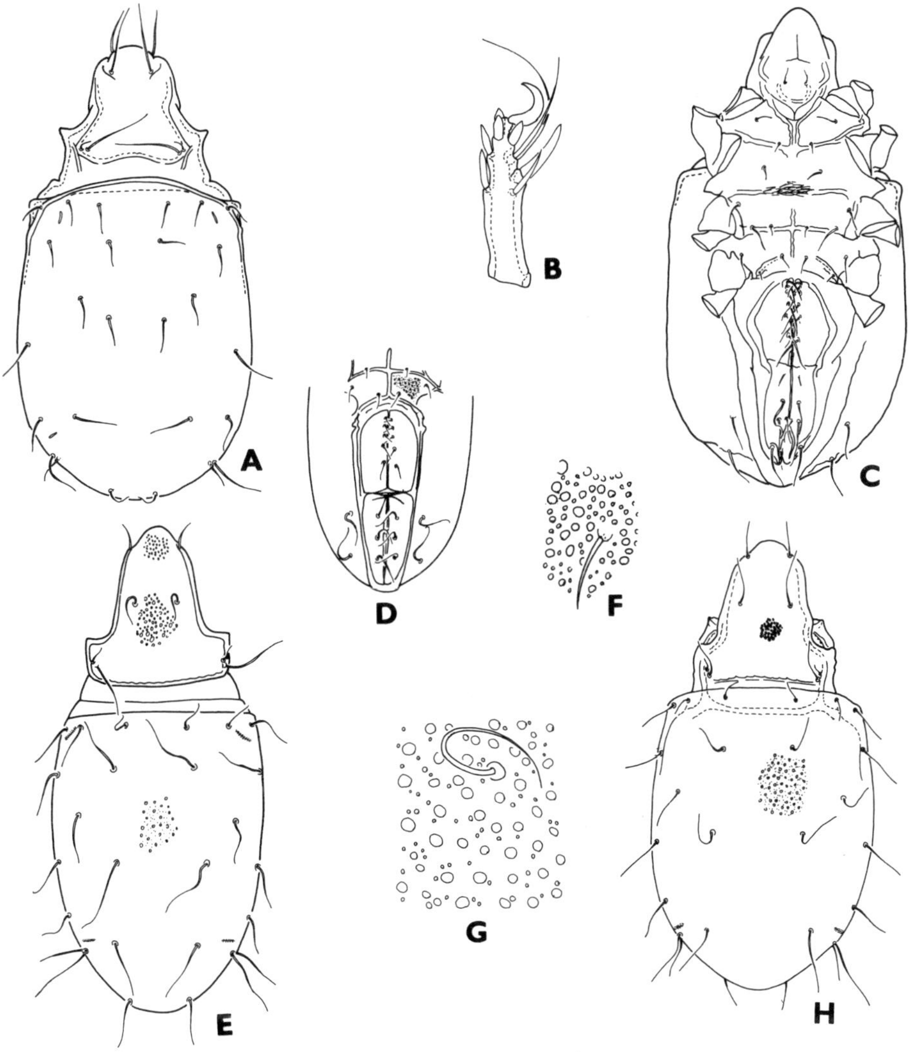

Plate 121. A—C: *Malaconothrus japonicus* Aoki, 1966 — A: dorsal side, B: leg I, C: ventral side; D—E: *Malaconothrus egregius* Berlese, 1904 — D: anogenital region, E: dorsal side; F—H: *Malaconothrus pygmaeus* Aoki, 1969 — F: sculpture of prodorsum, G: sculpture of notogaster, H: dorsal side (after Aoki: A—C, F—H and Van der Hammen: D—E)

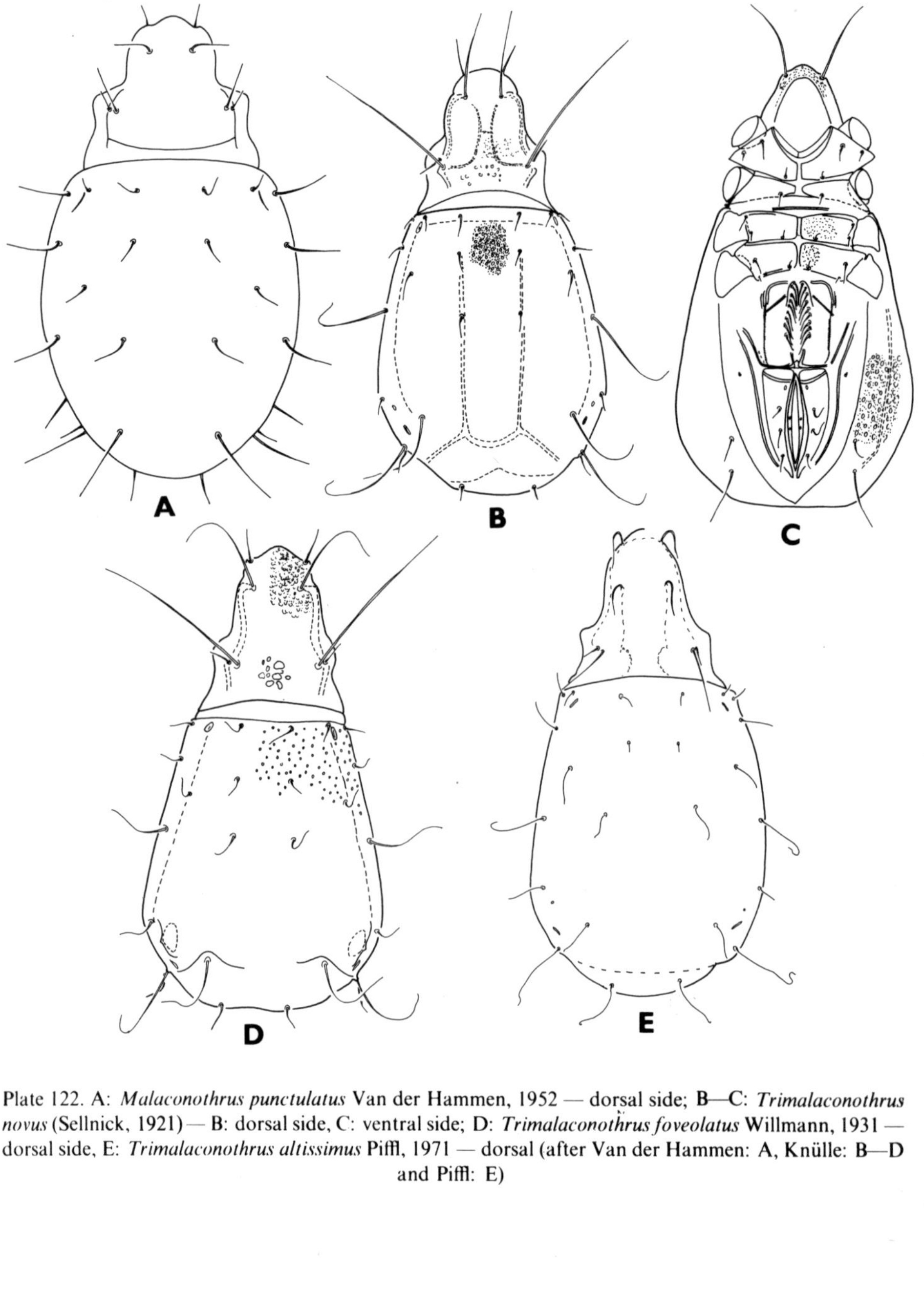

Plate 122. A: *Malaconothrus punctulatus* Van der Hammen, 1952 — dorsal side; B—C: *Trimalaconothrus novus* (Sellnick, 1921) — B: dorsal side, C: ventral side; D: *Trimalaconothrus foveolatus* Willmann, 1931 — dorsal side, E: *Trimalaconothrus altissimus* Piffl, 1971 — dorsal (after Van der Hammen: A, Knülle: B—D and Piffl: E)

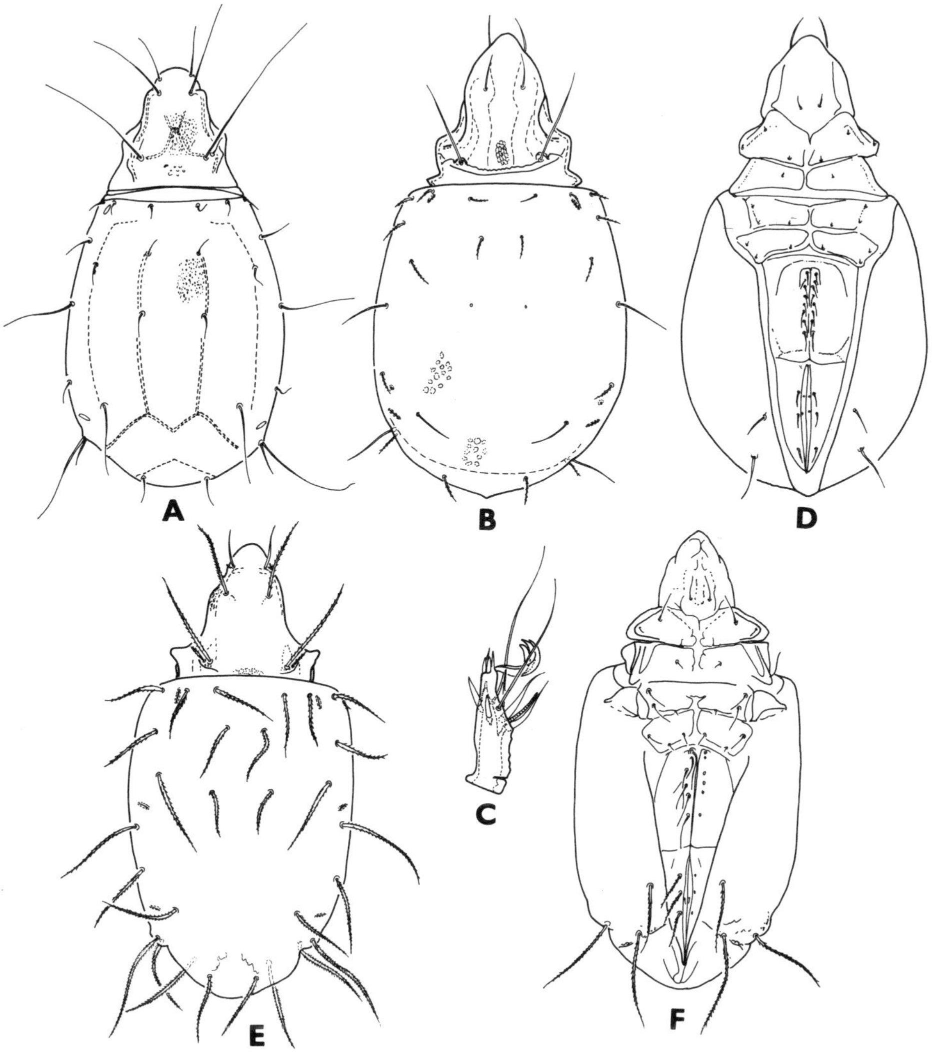

Plate 123. A: *Trimalaconothrus glaber* (Michael, 1888) — dorsal side; B—D: *Trimalaconothrus nipponicus* Yamamoto et Aoki, 1971 — B: dorsal side, C: tarsus of leg I, D: ventral side; E—F: *Trimalaconothrus barbatus* Yamamoto, 1977 — E: dorsal side, F: ventral side (after Knülle: A, Yamamoto and Aoki: B—D and Yamamoto: E—F)

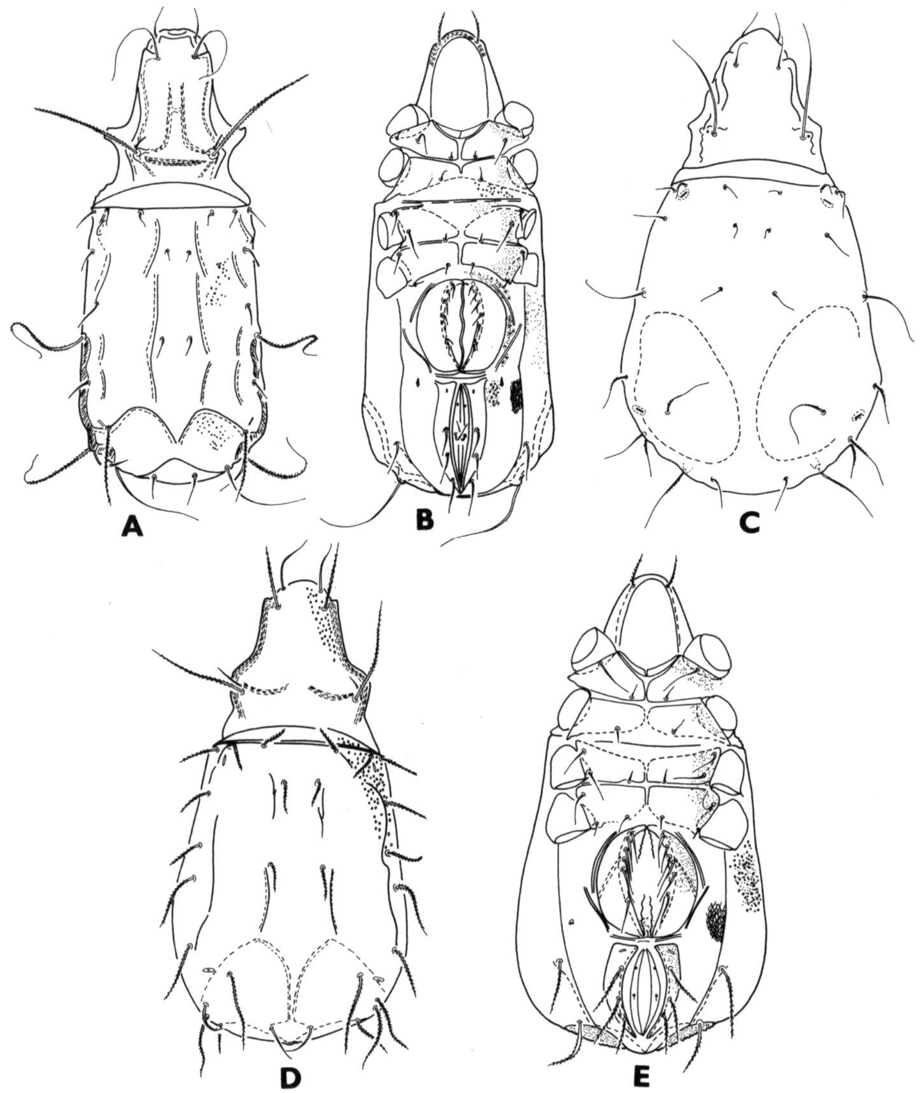

Plate 124. A—B: *Trimalaconothrus sculptus* Knülle, 1957 — A: dorsal side, B: ventral side; C: *Trimalaconothrus indusiatus* Berlese, 1916 — dorsal side; D—E: *Trimalaconothrus saxosus* Knülle, 1957 — D: dorsal side, E: ventral side (after Knülle: A—B, D—E; and Pérez-Iñigo: C)

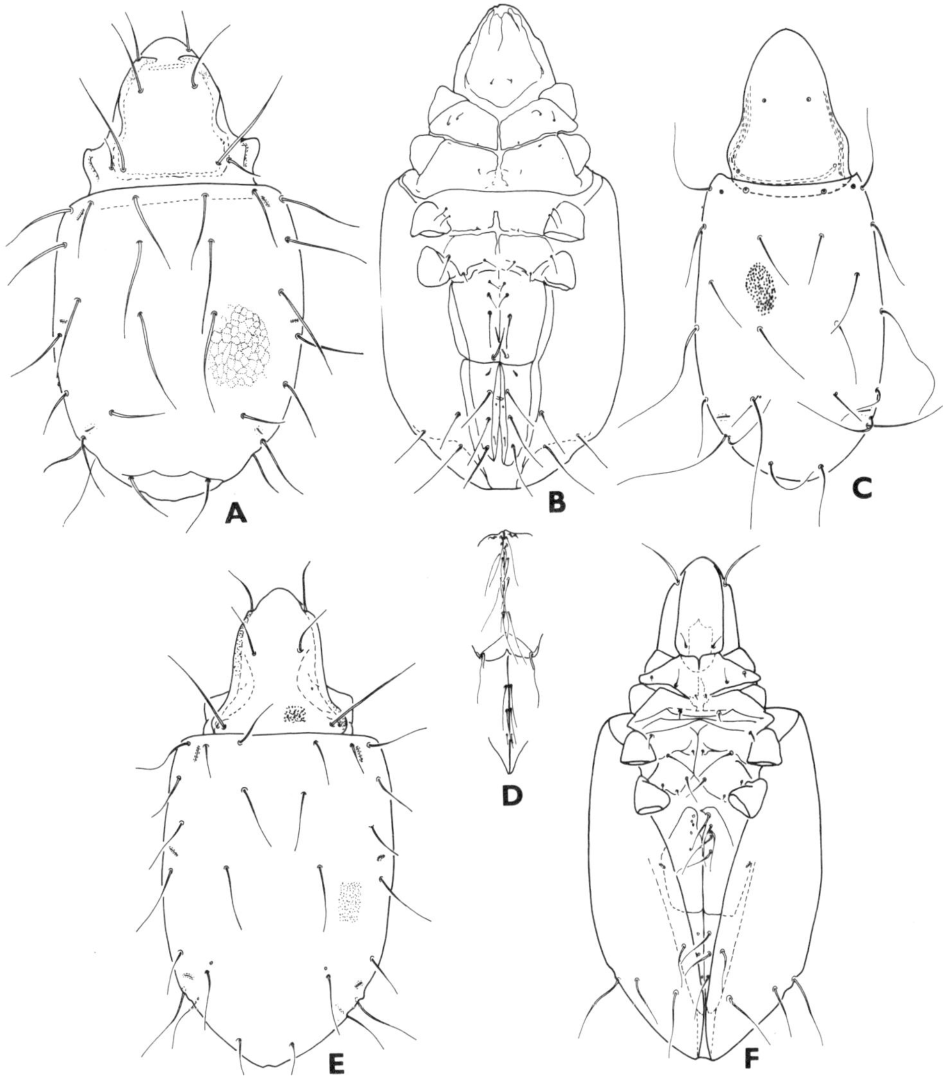

Plate 125. A—B: *Trimalaconothrus reticulatus* Yamamoto, 1977 — A: dorsal side, B: ventral side; C—D: *Trimalaconothrus tardus* (Michael, 1888) — C: dorsal side, D: anogenital region; E—F: *Trimalaconothrus hakonensis* Yamamoto, 1977: E: dorsal side, F: ventral side (after Yamamoto: A—B, E—F and Knülle: C—D)

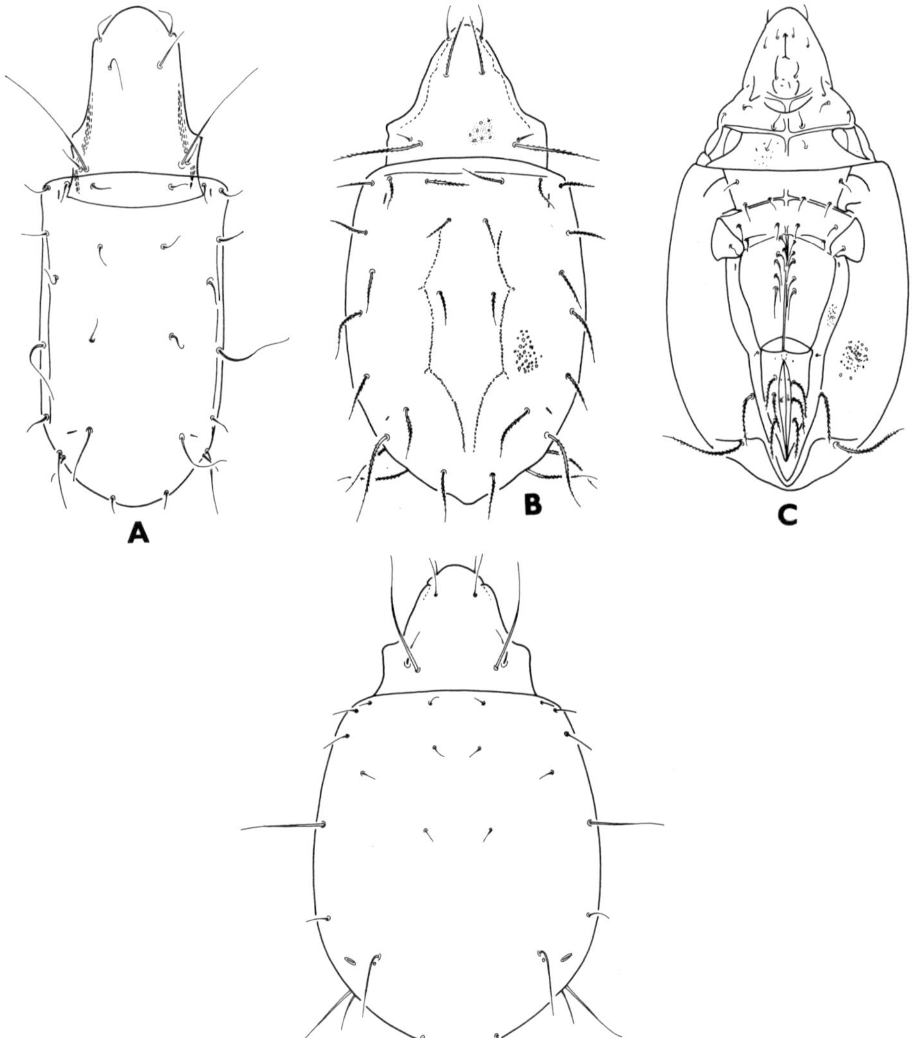

Plate 126. A: *Trimalaconothrus vietsi* Willmann, 1925 — dorsal side; B—C: *Trimalaconothrus buresi* Kunst, 1959 — B: dorsal side, C: ventral side; D: *Trimalaconothrus grandis* Van der Hammen, 1952 (after Knülle: A, Kunst: B—C and Van der Hammen: D)

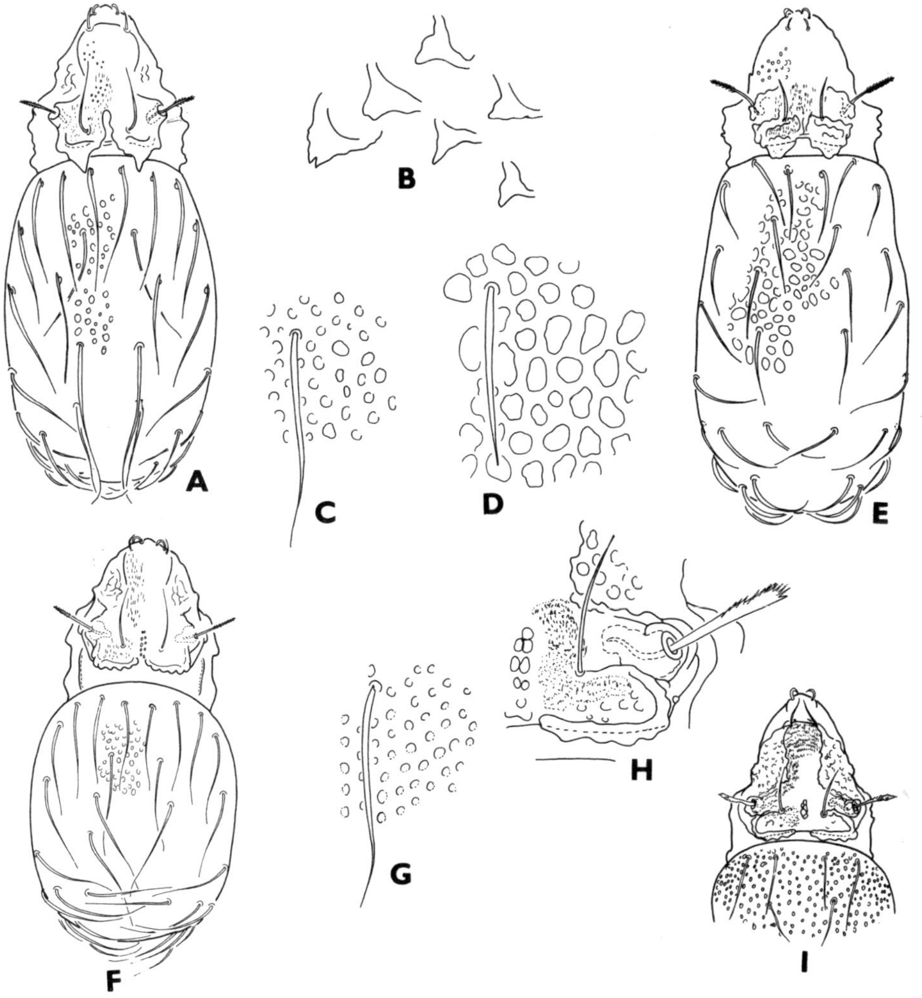

Plate 127. A—C: *Nanhermannia nana* (Nicolet, 1855) — A dorsal side, B: variation of prodorsal condyles, C: sculpture of notogaster; D—E: *Nanhermannia elegantula* Berlese, 1913 — D: sculpture of notogaster, E: dorsal side; F—I: *Nanhermannia sellnicki* Forsslund, 1958 — F: dorsal side, G: sculpture of notogaster, H: lateral part of prodorsum I: anterior part of idiosoma (A—H: original; after Sitnikova: I)

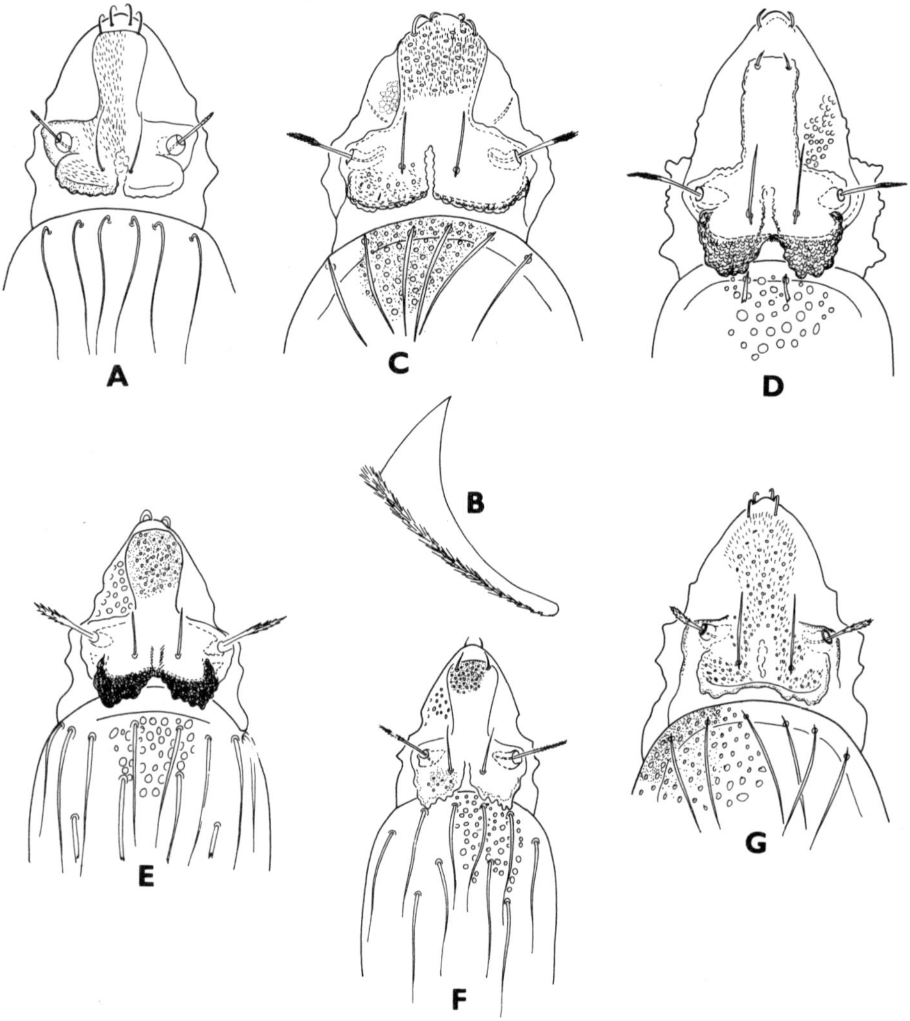

Plate 128. A: *Nanhermannia laevis* Sitnikova, 1975 — anterior part of idiosoma; B—C: *Nanhermannia comitalis* Berlese, 1916 — B: seta of genu of leg I; C: anterior part of idiosoma; D: *Nanhermannia pectinata* Strenzke, 1953 — anterior part of idiosoma; E: *Nanhermannia forsslundi* Karppinen, 1958 — anterior part of idiosoma; F: *Nanhermannia komareki* Kunst, 1956 — anterior part of idiosoma; G: *Nanhermannia gorodkovi* Sitnikova, 1975 — anterior part of idiosoma (after Sitnikova: A—C, E, G, Strenzke: D and Kunst: F)

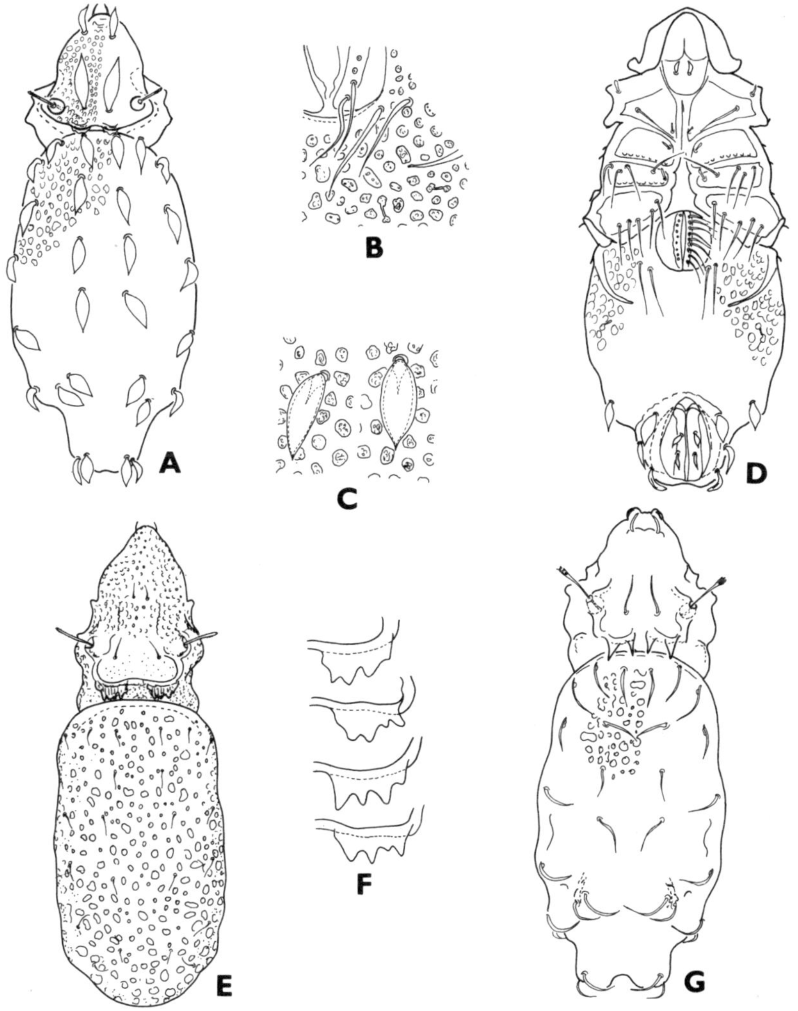

Plate 129. A—D: *Cosmohermannia frondosa* Aoki et Yoshida, 1970 — A: dorsal side, B: sculpture of ventral plate, C: sculpture of notogaster, D: ventral side; E—F: *Cyrthermannia parallela* (Aoki, 1961) — E: dorsal side, F: variation of prodorsal condyles; G: *Cyrthermannia ezzati* Bayoumi et Mahunka, 1977 — dorsal side (after Aoki and Yoshida: A—D and Aoki: E—F; G original)

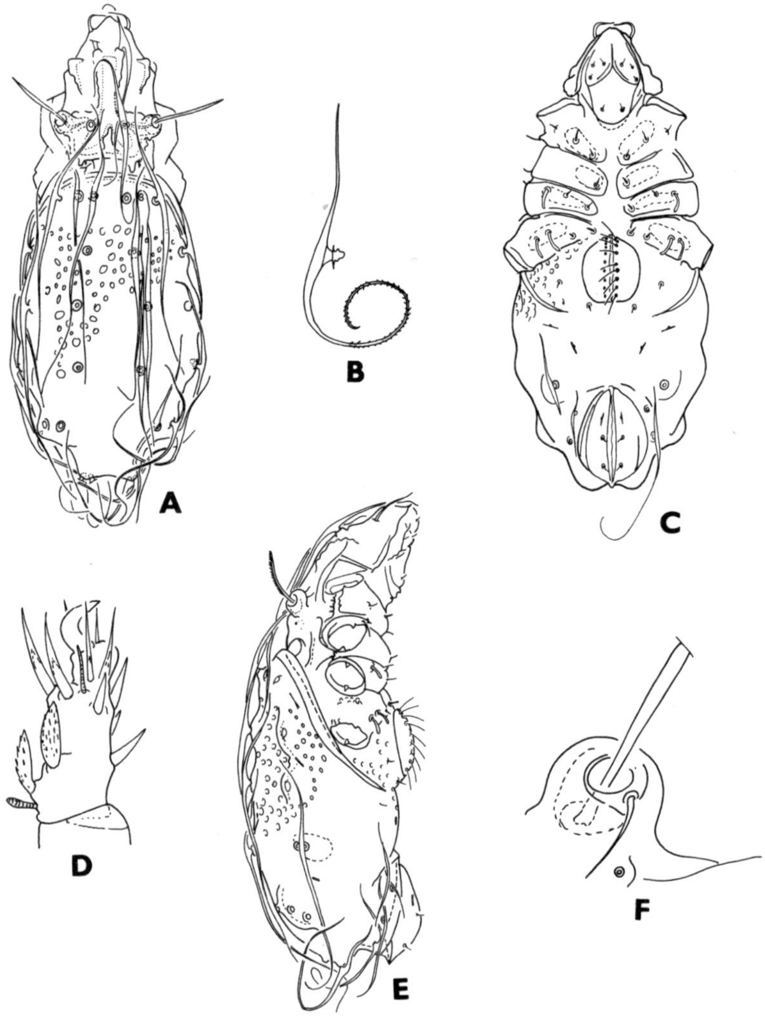

Plate 130. A—F: *Masthermannia mammillaris* (Berlese, 1904) — A: dorsal side, B: seta of notogaster, C: ventral side, D: leg II, E: lateral side, F: bothridium (after Grandjean)

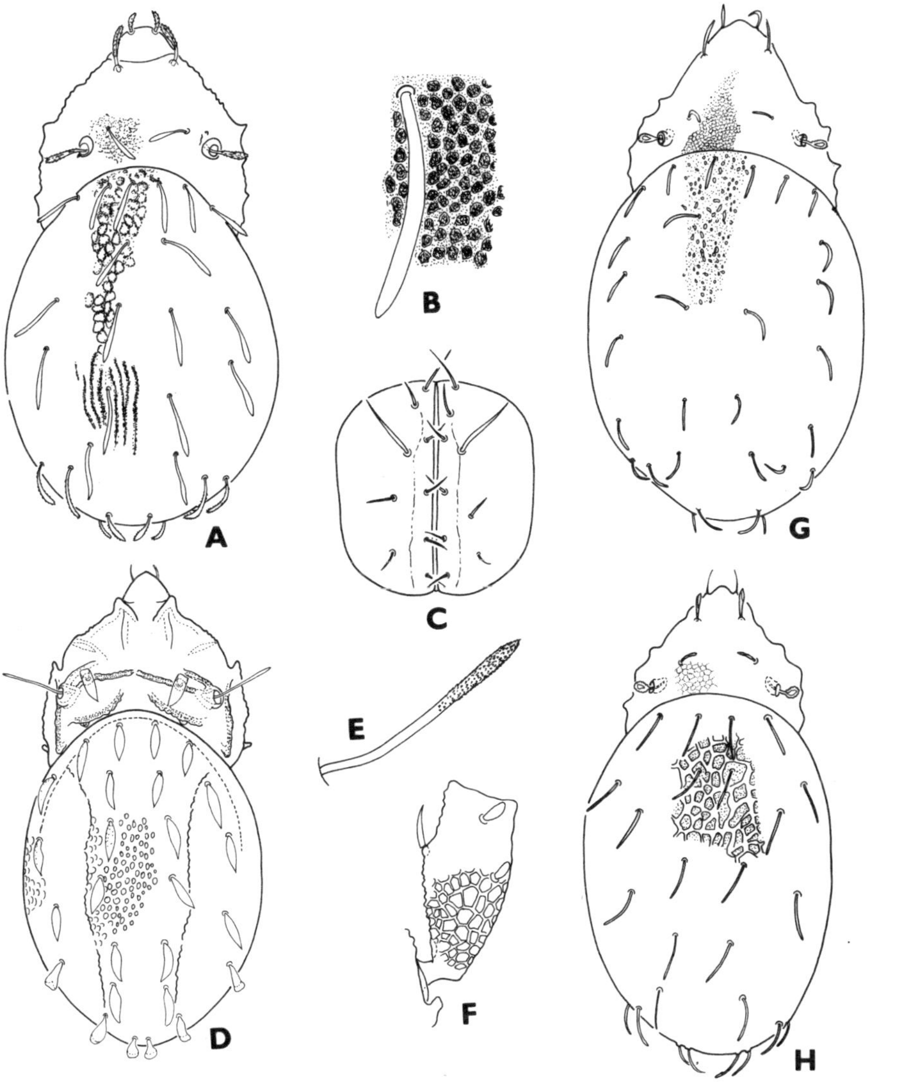

Plate 131. A: *Hermannia clara* Sitnikova, 1975 — dorsal side; B: *Hermannia convexa* (C. L. Koch, 1840) — sculpture of notogaster; C—F: *Hermannia pulchra* (Aoki, 1973) — C: genital plates, D: dorsal side, E: sensillus, F: femur of leg I; G: *Hermannia gigantea* Sitnikova, 1975 — dorsal side; H: *Hermannia reticulata* Thorell, 1871 — dorsal side (after Sitnikova: A—B, G—H and Aoki, C—F)

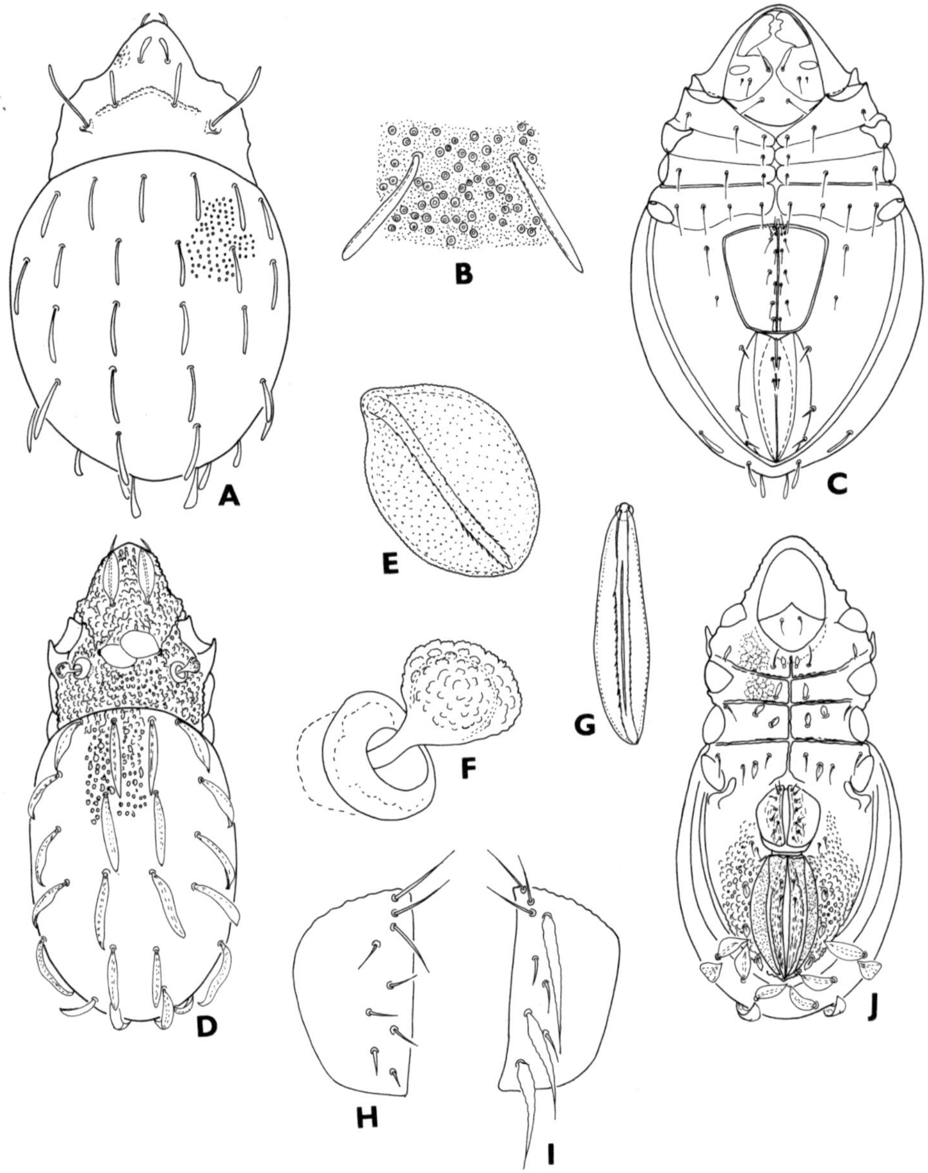

Plate 132. A—C: *Hermannia gibba* (C. L. Koch, 1840) — A: dorsal side, B: sculpture of notogaster, C: ventral side; D—J: *Hermannia areolata* (Aoki, 1970) — D: dorsal side, E: seta *in*, F: sensillus, G: seta of notogaster, H—I: variation of genital plates, J: ventral side (after Bäumler: A, C, Sitnikova: B and Aoki: D—J)

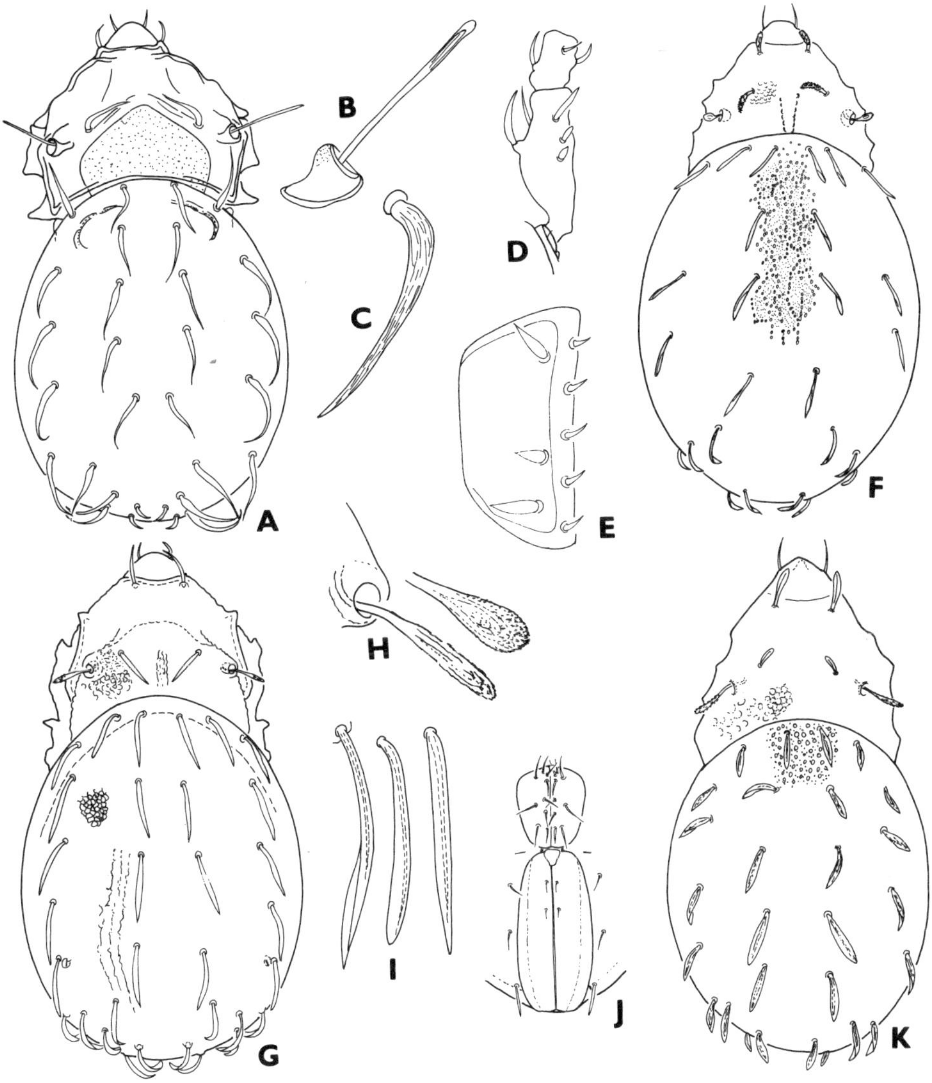

Plate 133. A—E: *Hermannia kanoi* (Aoki, 1959): A: dorsal side, B: sensillus, C: seta *in*, D: femur of leg I, E: genital plate; F: *Hermannia scabra* (L. Koch, 1879) — dorsal side; G—J: *Hermannia hokkaidensis* Aoki et Ohnishi, 1974 — G: dorsal side, H: sensillus, I: setae of notogaster, J: anogenital region; K: *Hermannia subglabra* Berlese, 1910 — dorsal side (after Aoki: A—E, Sitnikova: F, K and Aoki and Ohnishi: G—J)

INDEX

The italicized names are either synonyms or species inquirendae. The bold face numbers refer to names figured in identification keys.

Higher categories

Species and subspecific categories